Frontiers in Statistical Quality Control 10

Hans-Joachim Lenz • Peter-Theodor Wilrich
Wolfgang Schmid
Editors

Frontiers in Statistical Quality Control 10

Physica-Verlag
A Springer Company

Editors

Hans-Joachim Lenz
Peter-Theodor Wilrich
Freie Universität Berlin
Institut für Statistik und Ökonometrie
FU Berlin
Berlin
Germany

Wolfgang Schmid
Europa-Universität Viadrina
LS für Quantitative Methoden
insbesondere Statistik
Frankfurt/Oder
Germany

ISBN 978-3-7908-2845-0 ISBN 978-3-7908-2846-7 (eBook)
DOI 10.1007/978-3-7908-2846-7
Springer Heidelberg New York Dordrecht London

Library of Congress Control Number: 2012944564

Mathematics Subject Classification: 62-06, 62P30, 62L99, 62N05, 62K99

Printed on acid-free paper

Physica-Verlag is a brand of Springer-Verlag Berlin Heidelberg
Springer-Verlag is part of Springer Science+Business Media (www.springer.com)

Editorial

The Xth International Workshop on *Intelligent Statistical Quality Control* took place in Seattle, USA, Aug 18–20, 2010. It was hosted by Professor C. M. Mastrangelo, Department of Industrial and Systems Engineering, University of Washington, Seattle. The workshop was jointly organized by Professors H.-J. Lenz, C. M. Mastrangelo, W. Schmid and P.-T. Wilrich. The 27 papers in this volume were carefully selected by the scientific program committee, reviewed by its members, revised by the authors and, finally, adapted by the editors for this volume.

The book is divided into two parts: Part "On-line Control" which covers fields like *control charting, monitoring* and *surveillance* as well as *acceptance sampling*. Part "Off-line Control" is devoted to *experimental design, process capability analysis* and *data quality*.

Part I: On-line Control

Reynolds, Jr., and ***Lou*** start with a paper entitled "A GLR Control Chart for Monitoring the Process Variance". They present a performance evaluation of a generalized likelihood ratio (GLR) control charting under a Gaussian regime. The likelihood ratio is based on a moving window of past observations. It is shown that the overall performance of the GLR chart is better than or equal to other options like Shewhart or CUSUM charts.

An old problem of control charting is the robustness with respect to different types of dependencies. ***Hryniewicz*** uses the concept of copulas in "On the Robustness of the Shewhart Control Chart to Different Types of Dependencies in Data" to model this new type of dependency. He investigates the impact of type and strength of dependence on the ARL of Shewhart charts using Monte Carlo simulation.

Ramos, Morais, Pacheco and ***Schmid*** are concerned with misleading signals in their paper on "Assessing the Impact of Autocorrelation in Misleading Signals in Simultaneous Residual Schemes for the Process Mean and Variance: a Stochastic

Ordering Approach". The authors analyze the probability of misleading signals when residuals are autocorrelated stationary processes. For this type of process Shewhart and EWMA residual charts for the mean and variance are simultaneously applied.

It is well-known that trend dependency of the moments is a trouble maker of control charting besides robustness and auto-correlation. ***Knoth*** picks up this frequently existing quality process type in "More on Control Charting under Drift". Mean level control charts are suited for detecting drifts. His analysis provides numerical results, includes a competitor to an algorithm of Gan, and presents various charts to allow for drift.

Morais, Okhrin, and ***Schmid*** show in their paper on "Limit Properties of EWMA Charts for Stationary Processes" that that for stationary processes the distribution of the run length of the EWMA chart based on the exact variance converges to the distribution of the run length of the repeated significance test if the smoothing parameter converges to zero. However, for the EWMA chart based on the limit variance the asymptotic distribution degenerates.

Saniga, Lucas, Davis, and ***Williams*** in "Economic Control Chart Policies for Monitoring Variables when there are Two Components of Variance" determine regions of cost advantages of the CUSUM chart versus the Shewhart $\bar{X}$ chart. Dominance regions are found with a large cost advantage of CUSUM charts and regions where the Shewhart $\bar{X}$ chart is not too bad.

Khediri and ***Weihs*** are concerned with "Process Monitoring Using an Online Nonlinear Data Reduction Based Control Chart". The study advocates for an on-line Kernel PCA chart transforming a nonlinear system to a nearly linear one. Simulations show that the control chart is robust and provides a reduced rate of false alarms with high fault detection strength.

Nishina, Higashide, Kawamura, and ***Ishii*** reconsider the relationship between statistical (SPC) and automatic process control (APC) in their contribution entitled "On the Integration of SPC and APC: APC can be a Convenient Support for SPC". From experiences with case studies the authors propose using the process rate to control the between subgroup variation and principal component analysis to control the within-subgroup variation.

A different perspective on APC and SPC is taken by ***Kawamura, Nishina*** and ***Suzuki*** in: "Process Adjustment Control Chart for Simultaneous Monitoring of Process Capability and State of Statistical Control". The authors propose a method for switching from SPC to APC by combining the EWMA chart and process capability analysis. The usefulness of such an approach is demonstrated using real data.

Gan, Woodall, and ***Szarka*** in "Adaptive Threshold Methods for Monitoring Rates in Public Health Surveillance" critically look at the W2r method for disease control developed by the U.S. Centers for Disease Control and Prevention. They develop a specially tailored adaptive threshold monitoring method which is based on a negative binomial distribution of counts, and its conversion to Z-scores through p-values. The results give evidence that this method dominates the W2r method.

The contribution of ***Han, Jian***, and ***Tsui*** on "Spatiotemporal Bio Sur-veillance Under Non-homogeneous Population" discusses the spatiotemporal surveillance problem of detecting a change in the mean of Poisson count data in a non-homogeneous population environment. The authors investigate several likelihood ratio-based approaches and compare them by Monte-Carlo simulations.

"Monitoring Hospital-Associated Infections with Control Charts" authored by ***Mastrangelo*** and ***Gillan*** compares g-type and negative binomial control charts. They find out that the g-type chart performs best if the average time between signal events is chosen as a reasonable metric for comparison.

Yashchin thoroughly investigates the problem of monitoring warranty data streams of computer components in his paper on "Design and Implementation of Systems for Monitoring Lifetime Data". In this domain the underlying lifetime distribution undergoes abrupt changes. The emphasis is on a Cusum – based approach and its implementation at a computer manufacturer's site.

Abrupt change point detection is the topic of "A Robust Detection Procedure for Multiple Change Points of Linear Trends" authored by ***Yasui, Noguchi*** and ***Ojima***, too. They propose a robust procedure for detecting multiple change points based on a Epanechnikov kernel. In a simulation study the procedure performs well in terms of the power of jump detection even when drift and outliers exist. The bandwidth is optimized using a trial and error approach.

Gan, Lin, and ***Loke*** in "Risk-adjusted Cumulative Sum Charting Procedures" generalize the risk adjusted cumulative sum (RA-CUSUM) chart based on the odds ratio developed by Steiner et al. A sensitivity analysis shows that both types of RA-CUSUM charts are sensitive to changes in the underlying distribution.

In "Bayesian Sampling Plans for Inspection by Variables" ***Wilrich*** designs a truly Bayesian sampling plan for inspection by variables. The lot acceptance decision is directly based on the a posteriori distribution of the fraction of nonconforming units in the lot. A Monte-Carlo study shows that the performance of the ISO 3951 procedure (with switching rules) and the new Bayesian method do not show essential differences, but the Bayesian competitor delivers detailed information about the estimated probability of the fraction defectives in the lot being larger than p_{AQL} and the process curve.

Expensive cost of sampling in photovoltaics motivated ***Meisen, Pepelyshev***, and ***Steland*** to construct asymptotically optimal sampling plans with small sample sizes for non-normal measurements. In their paper on "Quality Assessment in the Presence of Additional Data in Photovoltaics" additional data is assumed to be available, and the new plans are investigated by MC-simulation.

We close Part I *On-line control* with ***Iwersen's*** paper "On practical Uses of ISO standards – Two Case Studies". It presents two case studies from the pharmaceutical domain. The first case presents an algorithm which reduces the sample size for In-Process Control. The second one considers an automatic vision system running in 100%-control.

Part II: Off-line Control

This part is started with experimental design and a paper on "Hybrid Space-Filling Designs for Computer Experiments" written by ***Johnson, Montgomery***, and ***Kennedy***. The authors give a survey and insight into how hybrid space-filling designs perform. The designs are surveyed and compared to both solely space-filling and solely optimal designs. A new hybrid design is proposed.

Englert, Rigdon, Borror, Montgomery, and ***Pan*** consider life testing designs in their paper "Optimal Design for Multifactor Life Testing Experiments for Exponentially Distributed Lifetime". They apply genetic algorithms to find a near optimal design for life testing experiments when there are two predictor variables, data are as usually censored, and the response is exponentially distributed.

Göb, ***Lurz***, and ***Heinemann*** review in their contribution entitled "Accelerated Lifetime Testing of Thermal Insulation Elements" the physical models for thermal insulation elements (TIEs) degradation over time. A new mixed nonlinear regression model of degradation as a function of time and ambient temperature is proposed. Göb et al. investigate inferential techniques for parameter estimation and lifetime prediction, and study the design of accelerated experiments on TIEs.

Suzuki, Kawamura, Yasui, and ***Ojima's*** paper on "Proposal of Advanced Taguchi's Linear Graphs for Split-Plot Experiments" is devoted to Taguchi's linear graphs. The authors investigate how to list all the possible linear graphs that can be applied when using L16 orthogonal arrays. A proposal is made and many new linear graphs are presented. They may help the practitioner for controlling the source of variation so that the standard errors of the estimated effects can hit fixed targets.

Noguchi, Ojima, and ***Yasui*** are concerned with model selection for experimental designs. In their paper "A Practical Variable Selection for Linear Models" they extend the LASSO method to identify significant interaction terms mainly focusing on the heredity principle. The proposed method is compared with ordinary LASSO and a traditional variable selection approach.

Tsutsumi, Kawamura, and ***Suzuki*** are concerned with the detection of toxic substances. In their paper "Capability of Detection for Poisson Distributed Measurements by Normal Approximations" the capability of detection for Poisson distributed measurements is evaluated. A best approximation method is proposed and compared with the exact method.

Lenz and ***Borowski*** investigate data quality in business. Their paper "Business Data Quality Control – a Step by Step Procedure" presents a step by step procedure for data quality assurance. A methodology and a workflow for data quality control is developed, a corresponding metadata model discussed, and ***DaRT*** – a business data quality reporting tool on top of Oracle's Warehouse Builder – presented.

Sparks and ***OkuGami*** investigate data quality in science. Their paper on "Data Quality: Algorithms for Automatic Detection of Unusual Measurements" offers simple robust, but computationally efficient algorithms for checking the consistency of large volumes of measured data. Estimated expected values and variances are used to judge measurement consistency. Three-sigma control limits are applied to flag inconsistent measurements. CUSUM and EWMA charts are advocated for flagging consistently small biased measurements.

The paper of ***von Collani*** reconsiders uncertainty as an important concept for quality control. In his paper "Uncertainty and Quality Control" he carefully reviews Probability Theory, Belief, Credibility or Uncertainty Theory, and Bernouilli's Probability Theory. He sceptically closes by "But the various uncertainty theories represent . . . the attempt to establish belief as a scientific category".

The quality level of a workshop on *Intelligent Statistical Quality Control* is determined by the quality of its papers. We believe that this volume truly represents the frontiers of statistical on-line and off-line control. The editors would like to express their deep gratitude to the members of the scientific program committee, who carefully invited researchers from around the world and refereed all papers submitted:

David H. Baillie, U.K., Elart von Collani, Germany, Olgierd Hryniewicz, Poland, Hans-J. Lenz, Germany, Christina M. Mastrangelo, U.S.A, Yoshikazu Ojima, Japan, W. Schmid, Germany, Peter-Th. Wilrich, Germany, William H. Woodall, U.S.A.

We would like to cordially thank our host at Seattle, Christina M. Mastrangelo, who efficiently organized the tenth workshop. Moreover, we again thank Springer/Physica-Verlag, Heidelberg, for the continuing collaboration.

Berlin, Germany — Hans-J. Lenz
Frankfurt, Germany — Wolfgang Schmid
Berlin, Germany — Peter-Th. Wilrich

Contents

Contributors

Borowski, Esther Institute of Information Science,Freie Universität Berlin, Berlin, Germany

Borror, Connie M. School of Computing, Informatics and Decision Systems Engineering, Arizona State University, Tempe, AZ, USA

Collani, Elart von University Würzburg, Würzburg, Germany

Davis, Darwin Department of Business Administration, University of Delaware, Newark, DE, USA

Englert, Brandon R. Department of Mathematics and Statistics, Southern Illinois University Edwardsville, Edwardsville, IL, USA

Gan, Fah F. Department of Statistics and Applied Probability, National University of Singapore, Singapore, Singapore

Gan, Linmin Virginia Polytechnic Institute and State University, Blacksburg, VA, USA

Gillan, Anna M. Industrial Systems Engineering, University of Washington, Seattle, WA, USA

Göb, Rainer Institute for Applied Mathematics and Statistics, University of Würzburg, Würzburg, Germany

Han, Sung Won Department of Biostatistics and Epidemiology, University of Pennsylvania, Philadelphia, PA, USA

Heinemann, Ulrich Bavarian Center for Applied Energy Research, Am Hubland, Würzburg, Germany

Higashide, Masanobu Renesas Electronics Co., Nakahara-ku, Kawasaki, Japan

Hryniewicz, Olgierd Systems Research Institute, Warsaw, Poland

Ishii, Naru Nagoya Institute of Technology Gokiso-cho, Showa-ku, Nagoya, Japan

Iwersen, Jørgen Novo Nordisk A/S, Novo Alle, Bagsvaerd, Denmark

Jiang, Wei Department of Operations Management, Shanghai Jiaotong University, Shanghai, P.R. China

Johnson, Rachel T. Naval Postgraduate School, Operations Research Department, Monterey, CA, USA

Kawamura, Hironobu Tokyo University of Science, Noda, Chiba, Japan

Kennedy, Kathryn S. Arizona State University, School of Computing, Informatics and Decision Systems Engineering, Tempe, AZ, USA

Khediri, Issam Ben Department of Statistics, TU Dortmund University, Dortmund, Germany

Knoth, Sven Institute of Mathematics and Statistics, Department of Economics and Social Sciences, Helmut Schmidt University Hamburg, Germany

Lenz, Hans-J. Institute for Statistics and Econometrics, Freie Universität Berlin, Berlin, Germany

Lin, Lin Department of Statistics and Applied Probability, National University of Singapore, Singapore, Singapore

Loke, Chok K. Department of Statistics and Applied Probability, National University of Singapore, Singapore, Singapore

Lou, Jianying Medpace, Cincinnati, OH, USA

Lucas, James J. M. Lucas and Associates, Wilmington, DE, USA

Lurz, Kristina Institute for Applied Mathematics and Statistics, University of Würzburg, Würzburg, Germany

Mastrangelo, Christina M. Industrial Systems Engineering, University of Washington, Seattle, WA, USA

McWilliams, Thomas Department of Decision Sciences, Bennett S. LeBow College of Business, Drexel University, Philadelphia, PA, USA

Meisen, Sabine Institute for Medical Statistics, RWTH Aachen University, Aachen, Germany

Montgomery, Douglas C. Arizona State University, School of Computing, Informatics and Decision Systems Engineering, Tempe, AZ, USA

Morais, Manuel Cabral Department of Mathematics and CEMAT Instituto Superior Tcnico, Technical University of Lisbon, Lisboa, Portugal

Nishina, Ken Nagoya Institute of Technology, Gokiso-cho, Showa-ku, Nagoya, Aichi, Japan

Noguchi, Hidehisa Department of Industrial Administration, Tokyo University of Science, Yamazaki, Noda, Chiba, Japan

Ojima, Yoshikazu Department of Industrial Administration, Tokyo University of Science, 2641 Yamazaki, Noda, Chiba, Japan

Okhrin, Yarema Department of Statistics, University of Augsburg, Augsburg, Germany

OkuGami, Chris CSIRO Mathematics, Informatics and Statistics, North Ryde, NSW, Australia

Pacheco, António Department of Mathematics and CEMAT Instituto Superior Tcnico, Technical University of Lisbon, Lisboa, Portugal

Pan, Rong School of Computing, Informatics and Decision Systems Engineering, Arizona State University, Tempe, AZ, USA

Pepelyshev, Andrey Institute of Statistics, RWTH Aachen University, Aachen, Germany

Ramos, Patrícia Ferreira CEMAT Instituto Superior Tcnico, Technical University of Lisbon, Lisboa, Portugal

Reynolds, Marion R. Virginia Polytechnic Institute and State University, Department of Statistics and Department of Forest Resources and Environmental Conservation, Blacksburg, VA, USA

Rigdon, Steven E. Department of Mathematics and Statistics, 1333 Science Building, Southern Illinois University Edwardsville, Edwardsville, IL, USA

Saniga, Erwin Department of Business Administration, University of Delaware, Newark, DE, USA

Schmid, Wolfgang Department of Statistics, European University Viadrina, Frankfurt (Oder), Germany

Sparks, Ross CSIRO Mathematics, Informatics and Statistics, North Ryde, NSW, Australia

Steland, Ansgar Institute of Statistics, RWTH Aachen University, Aachen, Germany

Suzuki, Tomomichi Department of Industrial Administration, Tokyo University of Science, Noda, Chiba, Japan

Szarka, John L. W. L. Gore & Associates Inc., Elkton, MD, USA

Tsui, Kwok-Leung Department of Manufacturing Engineering and Engineering Management, City University of Hong Kong, Hong Kong, P.R. China

School of Industrial and Systems Engineering, Georgia Institute of Technology, Atlanta, GA, USA

Tsutsumi, Yusuke Mitsubishi Tanabe Pharma corporation, Chuoku Tokyo, Japan

Weihs, Claus Department of Statistics, TU Dortmund University, Dortmund, Germany

Wilrich, Peter-Th. Institute for Statistics and Econometrics, Freie Universität Berlin, Berlin, Germany

Woodall, William H. Virginia Polytechnic Institute and State University, Blacksburg, VA, USA

Yashchin, Emmanuel IBM, Thomas J. Watson Research Ctr., Yorktown Heights, NY, USA

Yasui, Seiichi Department of Industrial Administration, Tokyo University of Science, Noda, Chiba, Japan

Part I
On-line Control

A GLR Control Chart for Monitoring the Process Variance

Marion R. Reynolds Jr. and Jianying Lou

Abstract This paper considers the problem of monitoring the variance of a normally distributed process variable when the objective is to effectively detect both small and large increases in the variance. The performance of a generalized likelihood ratio (GLR) control chart is evaluated, where the likelihood ratio is based on a moving window of past observations. The performance of the GLR chart is compared to the performance of other options such as Shewhart charts, CUSUM charts, and combinations of two CUSUM charts. It is shown that the overall performance of the GLR chart is as good as or better than these other options. A CUSUM chart has a tuning parameter which allows for the chart to be tuned to be sensitive to a certain shift of interest. However, the GLR chart does not require users to specify the values of any tuning parameters other than the size of the window and the control limit. We recommend a specific window size, and provide a table of control limits corresponding to specified values of the in-control average number of samples to signal, so the GLR chart has the advantage that it can be easily designed for use in applications. Simulating the performance of the GLR chart is time consuming, but the GLR chart can be very well approximated with a set of CUSUM charts, and this provides a fast method for evaluating the performance of the GLR chart.

Keywords Average time to signal • CUSUM chart • Generalized likelihood ratio • Shewhart chart • Statistical process control • Steady state • Surveillance

---

M.R. Reynolds Jr. (✉)
Department of Statistics and Forest Resources and Environmental Conservation, Virginia Polytechnic Institute and State University, Blacksburg, VA 24061-0439, USA
e-mail: mrr@vt.edu

J. Lou
Medpace, 5375 Medpace Way, Cincinnati, OH 45212, USA
e-mail: j.lou@medpace.com

H.-J. Lenz et al. (eds.), *Frontiers in Statistical Quality Control 10*,
Frontiers in Statistical Quality Control 10, DOI 10.1007/978-3-7908-2846-7_1,

1 Introduction

In many process monitoring applications it is assumed that the distribution of observations from the process is normal with mean μ and variance σ^2. Here we consider the problem of detecting increases in σ^2. The traditional Shewhart charts used for this problem will not be very effective if the size of the increase in σ^2 is small. CUSUM and EWMA charts can be tuned to be very effective for detecting small shifts in σ^2, but then these charts will not be very effective for detecting large shifts. In applications the size of the shift in σ^2 that occurs will be unknown, so it is desirable to be able to effectively detect a wide range of shift sizes.

One option for obtaining good performance over a wide range of shift sizes is to use two or more control charts together in combination. For example, Lorden (1971) originally investigated the use of multiple CUSUM charts, and more recent work includes Sparks (2000) and Han et al. (2007). A disadvantage of using multiple charts is that designing such a scheme requires that control chart parameters be determined for multiple charts.

Another approach to obtaining a control chart that will detect different sizes of process changes is to base the control chart on a likelihood ratio test. Such charts are usually called generalized likelihood ratio (GLR) charts, and have been shown to be very effective in a wide variety of settings (see, for example, Willsky and Jones 1976; Siegmund and Venkatraman 1995; Lai 1995, 2001; Apley and Shi 1999; Hawkins and Zamba 2005; Han et al. 2007; Reynolds and Lou 2010). GLR charts are usually perceived to be computationally intensive, and have not received as much attention in statistical process control (SPC) applications as Shewhart, CUSUM, and EWMA charts.

The objective of this paper is to evaluate a GLR control chart for monitoring σ^2, and show that this GLR chart provides a very attractive option for detecting a wide range of shifts in σ^2. The GLR chart is very easy to design for use in applications because it does not require that users specify control chart parameters other than the control limit which can be found from a table provided in this paper. The fact that multiple control chart parameters do not need to be specified means that there is no flexibility in the design of the GLR chart, but this lack of flexibility can actually be an advantage because practitioners can obtain very effective detection of shifts in σ^2 without having to specify multiple control chart design parameters.

2 The GLR Control Chart for Monitoring the Process Variance

Suppose that the process variable X being measured has a $N(\mu, \sigma^2)$ distribution, and when the process is in control the distribution is a $N(\mu_0, \sigma_0^2)$ distribution. We assume that the in-control values μ_0 and σ_0^2 are known or have been estimated accurately enough during a Phase I period that any error in estimation can be neglected. Consider the problem of real-time monitoring in Phase II, where the objective is to detect any special cause that produces a small or large shift in σ^2 to a value above σ_0^2. We are assuming here that there is no change in μ; a GLR chart for detecting changes in μ was recently evaluated by Reynolds and Lou (2010).

Suppose that independent samples each consisting of $n \geq 1$ observation are taken from the process using a sampling interval of d between samples. Let $\mathbf{X}_k = (X_{k1}, X_{k2}, \ldots, X_{kn})$ represent the sample at sampling point k. Numerical results are presented here for two values of n; $n = 1$ corresponding to the common practice of taking individual observations from the process, and $n = 4$ which is representative of the traditional practice of taking small samples of around 4 or 5 from the process. When $n = 1$ we use $d = 1.0$, and when $n = 4$ we use $d = 4.0$, so that in both cases the sampling rate is 1.0 observations per unit time.

After sample k is obtained the available data consists of $\mathbf{X}_1, \mathbf{X}_2, \ldots, \mathbf{X}_k$. Consider the hypothesis that a shift in σ^2 to some value $\sigma_1^2 > \sigma_0^2$ has occurred at some time between samples τ and $\tau + 1$, where $\tau < k$ (with no change in μ from μ_0). Then the observations in samples $\mathbf{X}_1, \mathbf{X}_2, \ldots, \mathbf{X}_\tau$ have a $N(\mu_0, \sigma_0^2)$ distribution, and the observations in samples $\mathbf{X}_{\tau+1}, \mathbf{X}_{\tau+2}, \ldots, \mathbf{X}_k$ have a $N(\mu_0, \sigma_1^2)$ distribution. Under the hypothesis of no shift in σ^2, the observations in all k samples have a $N(\mu_0, \sigma_0^2)$ distribution, and a log likelihood ratio statistic for testing for a shift in σ^2 is

$$R_k = \ln \max_{0 \leq \tau < k, \sigma_0^2 < \sigma_1^2 < \infty} \frac{(2\pi)^{-n(k-\tau)/2} \left(\sigma_1^2\right)^{-n(k-\tau)/2} \exp\left(-\frac{1}{2\sigma_1^2} \sum_{i=\tau+1}^{k} \sum_{j=1}^{n} (X_{ij} - \mu_0)^2\right)}{(2\pi)^{-n(k-\tau)/2} \left(\sigma_0^2\right)^{-n(k-\tau)/2} \exp\left(-\frac{1}{2\sigma_0^2} \sum_{i=\tau+1}^{k} \sum_{j=1}^{n} (X_{ij} - \mu_0)^2\right)}$$

$$= \max_{0 \leq \tau < k, \sigma_0^2 < \sigma_1^2 < \infty} \sum_{i=\tau+1}^{k} \frac{n}{2} \left[\left(\sigma_0^{-2} - \sigma_1^{-2}\right) \left(\frac{1}{n} \sum_{j=1}^{n} (X_{ij} - \mu_0)^2 \right) - \ln\left(\frac{\sigma_1^2}{\sigma_0^2}\right) \right]. \quad (1)$$

If there has been a shift to some unknown σ_1^2 between samples τ and $\tau + 1$, then the maximum likelihood estimator of σ_1^2 under the restriction $\sigma_1^2 \geq \sigma_0^2$ is

$$\hat{\sigma}^2_{1,\tau,k} = \max\left\{ \sigma_0^2, \frac{1}{n(k-\tau)} \sum_{i=\tau+1}^{k} \sum_{j=1}^{n} (X_{ij} - \mu_0)^2 \right\}. \quad (2)$$

Using Eq. 2, R_k in Eq. 1 reduces to

$$R_k = \max_{0 \leq \tau < k} \frac{n(k-\tau)}{2} \left[\frac{\hat{\sigma}^2_{1,\tau,k}}{\sigma_0^2} - 1 - \ln\left(\frac{\hat{\sigma}^2_{1,\tau,k}}{\sigma_0^2}\right) \right]. \quad (3)$$

Using R_k in this form requires taking the maximum over $0 \leq \tau < k$, and, to ease the computational burden when k is large, we use a window of the past m samples (see Willsky and Jones 1976). In particular, let $R_{m,k}$ be the value of likelihood ratio when the maximum is taken over $0 \le \tau < k$ when $k \le m$ and over $k - m \le \tau < k$ when $k > m$. This means that a maximum of m past sample values need to be stored when applying the GLR chart. A signal is given at sample k if $R_{m,k} > h_{GLR}$, where the control limit h_{GLR} can be chosen to give specified in-control performance.

3 The Average Time to Signal

When the process is in control we use the average time to signal (ATS), which is the expected time from the start of monitoring until a signal, as the measure of the rate of false alarms. When there is a shift in σ^2 we use the steady state ATS (SSATS) as the measure of the time required to detect the shift. The SSATS is based on the assumption that control chart statistics have reached their steady state distributions by the time that the shift in σ^2 occurs. The SSATS is also based on the assumption that when the shift in sigma occurs in the interval between samples τ and $\tau + 1$, the time at which the shift occurs in this interval is uniformly distributed on this interval.

Most of the evaluations of ATS and SSATS values in this paper were done using simulation with 1,000,000 runs. In the evaluations and comparisons done here we choose the control limits of the charts so that the in-control ATS is 1481.6 time units (the value 1481.6 has been recently used, for example, by Reynolds and Stoumbos (2004) and Reynolds and Lou (2010)). We assumed that $\tau = 400/n$ in simulating the SSATS values given here.

4 Choosing the Window Size of the GLR Chart

For the case of $n = 4$ and $d = 4.0$, Table 1 gives SSATS values for shifts in σ^2 for values of m ranging from 1 to 10,000. The size of the shift is expressed in terms of ψ, where

$$\psi = \frac{\sigma}{\sigma_0}.$$

For each value of m the control limit h_{GLR} was adjusted to give an in-control ATS of 1481.6 (the actual values given in the row of Table 1 corresponding to $\psi = 1.0$ vary slightly from 1481.6 due to simulation error). The column labeled [1] in Table 1 is for the case of $m = 1$ and in this case the GLR chart is equivalent to a Shewhart chart based on $\sum_{j=1}^{n} (X_{kj} - \mu_0)^2/n$. Column [9] corresponds to the very large value $m = 10,000$, and the performance of the GLR chart in this case is presumably close to what would be obtained without using a window (using $m = \infty$).

Table 1 The effect of m on the SSATS of the GLR chart when $n = 4$ and $d = 4.0$

$m =$	1	5	10	25	50	100	200	300	10,000
ψ	[1]	[2]	[3]	[4]	[5]	[6]	[7]	[8]	[9]
1.00	1481.71	1481.58	1481.58	1481.60	1481.59	1481.68	1481.61	1481.55	1481.53
1.10	427.13	316.24	279.48	243.71	226.90	217.61	214.50	214.26	214.26
1.20	168.60	109.04	93.48	82.25	79.02	78.37	78.35	78.37	78.37
1.40	47.23	30.52	27.96	27.18	27.20	27.22	27.23	27.24	27.24
1.60	20.94	15.13	14.74	14.80	14.84	14.86	14.87	14.87	14.87
1.80	12.01	9.69	9.69	9.78	9.81	9.83	9.83	9.83	9.83
2.00	8.07	7.07	7.14	7.21	7.23	7.24	7.25	7.25	7.25
2.40	4.80	4.63	4.68	4.72	4.74	4.74	4.74	4.75	4.75
3.00	3.19	3.22	3.25	3.27	3.28	3.28	3.28	3.28	3.28
5.00	2.18	2.20	2.21	2.21	2.21	2.21	2.21	2.21	2.21
7.00	2.05	2.06	2.06	2.06	2.06	2.06	2.06	2.06	2.06
10.00	2.01	2.01	2.01	2.02	2.02	2.02	2.02	2.02	2.02
15.00	2.00	2.00	2.00	2.00	2.00	2.00	2.00	2.00	2.00
$h_{GLR} =$	3.3250	3.9722	4.1105	4.2023	4.2303	4.2410	4.2443	4.2448	4.2451

Table 2 The effect of m on the SSATS of the GLR chart when $n = 1$ and $d = 1.0$

$m =$	1	20	40	100	200	400	800	1,200	40,000
ψ	[1]	[2]	[3]	[4]	[5]	[6]	[7]	[8]	[9]
1.00	1481.61	1481.61	1481.51	1481.54	1481.68	1481.58	1481.68	1481.54	1481.58
1.10	500.43	339.13	307.17	272.87	254.24	242.61	238.04	237.54	237.49
1.20	216.53	118.25	103.02	90.34	86.25	85.28	85.21	85.22	85.22
1.40	65.45	32.56	29.77	28.69	28.65	28.65	28.65	28.65	28.66
1.60	29.26	15.88	15.36	15.32	15.34	15.35	15.36	15.36	15.36
1.80	16.47	10.05	9.98	10.01	10.03	10.03	10.04	10.04	10.04
2.00	10.71	7.24	7.25	7.29	7.31	7.31	7.31	7.31	7.31
2.40	5.88	4.59	4.62	4.65	4.66	4.66	4.66	4.66	4.66
3.00	3.39	2.97	2.99	3.01	3.01	3.01	3.01	3.01	3.01
5.00	1.51	1.49	1.50	1.50	1.50	1.50	1.50	1.50	1.50
7.00	1.09	1.10	1.11	1.11	1.11	1.11	1.11	1.11	1.11
10.00	0.86	0.87	0.88	0.88	0.88	0.88	0.88	0.88	0.88
15.00	0.72	0.73	0.73	0.73	0.73	0.73	0.73	0.73	0.73
$h_{GLR} =$	4.0557	4.9931	5.0686	5.1187	5.1341	5.1402	5.1422	5.1425	5.1427

Table 2 gives SSATS values for the case of $n = 1$ and $d = 1.0$ for m ranging from 1 up to 40,000. Except for column [1] in Tables 1 and 2 where $m = 1$, the value of m in columns [2]–[9] in Table 2 is four times the corresponding value in Table 1, so that the total number of observations in a window (nm) is the same in both tables.

From Tables 1 and 2 we see that using $m = 1$ (which is a Shewhart chart) gives very bad performance for detecting small increases in σ^2, and performance for detecting small increases in σ^2 improves as m increases. Detecting small increases in σ^2 requires a relatively large number of observations, so a relatively large value of m is required. It appears that $m = 200$ in Table 1 and $m = 800$ in Table 2 (each corresponding to a total of 800 observations) is large enough that detection of shifts as small as $\psi = 1.10$ is essentially as fast as when m is much larger.

The values of h_{GLR} in the last row of Tables 1 and 2 increase as m increases. Detecting very large shifts in σ^2 requires only a few observations, so a small value of m with the corresponding small value of h_{GLR} will be better in this case. However, a large value of m is only slightly worse than small values of m for detecting very large shifts. We conclude that the best overall performance for detecting a wide range of shifts in σ^2 is achieved by using a relatively large value of m, such as $m = 800/n$. Once the appropriate software is available, the GLR statistic with this value of m could be calculated and plotted very quickly as each new sample is obtained, so computational issues should not be a problem when actually applying the GLR chart.

5 Choosing the Control Limit of the GLR Chart

For the case in which $m = 800/n$ and $n = 1$–5, Table 3 gives values of h_{GLR} corresponding to some in-control values of the average number of samples to signal (ANSS) ranging from 50 up to 2,000 (the in-control ATS is then given by ATS $= d \times$ ANSS). The value of h_{GLR} is the only parameter that needs to be determined in order to use the GLR chart (assuming that $m = 800/n$ will be used). It turns out that h_{GLR} is approximately a linear function of the log of the in-control ANSS, so linear interpolation using the log of the ANSS can be used when necessary.

Table 3 Values of h_{GLR} for the GLR chart for an increase in σ^2 with $m = 800/n$ corresponding to specified values of the in-control ANSS

	$n = 1$	$n = 2$	$n = 3$	$n = 4$	$n = 5$
In-control ANSS	$m = 800$	$m = 400$	$m = 266$	$m = 200$	$m = 160$
50	1.6168	1.8796	2.0075	2.0885	2.1453
75	1.9961	2.2810	2.4203	2.5070	2.5678
100	2.2760	2.5750	2.7209	2.8118	2.8755
200	2.9791	3.3079	3.4657	3.5639	3.6322
300	3.4053	3.7459	3.9092	4.0105	4.0808
500	3.9531	4.3072	4.4740	4.5774	4.6483
750	4.3944	4.7540	4.9244	5.0285	5.1007
1,000	4.7073	5.0718	5.2430	5.3470	5.4207
1,500	5.1559	5.5194	5.6913	5.7962	5.8673
2,000	5.4727	5.8364	6.0085	6.1125	6.1843
3,000	5.9162	6.2806	6.4531	6.5565	6.6275
5,000	6.4716	6.8390	7.0098	7.1140	7.1869
7,500	6.9108	7.2818	7.4536	7.5572	7.6254
10,000	7.2232	7.5935	7.7637	7.8696	7.9359

6 CUSUM Control Charts for Monitoring the Process Variance

We now consider the performance of the GLR chart relative to traditional Shewhart and CUSUM charts. CUSUM charts (see, for example, Hawkins and Olwell 1998 or Montgomery 2009) for monitoring σ^2 are equivalent to applying a sequence of sequential probability ratio tests for testing $H_0 : \sigma^2 = \sigma_0^2$ against $H_1 : \sigma^2 = \sigma_1^2$, where σ_1^2 corresponds to a shifted value of σ^2 that should be detected quickly. Note that in the CUSUM chart some value for σ_1^2 must be specified, even though the actual value of the shift is unknown, so σ_1^2 serves as a tuning parameter for the CUSUM chart. In contrast, σ_1^2 in the GLR chart is estimated using $\hat{\sigma}^2_{1,\hat{\tau},k}$ as samples are obtained from the process.

The increment accumulated in the CUSUM statistic at sample k is

$$\ln \frac{f(\mathbf{X}_k|\sigma_1^2)}{f(\mathbf{X}_k|\sigma_0^2)} = \frac{n}{2}\left[\left(\sigma_0^{-2} - \sigma_1^{-2}\right)\left(\frac{1}{n}\sum_{j=1}^{n}(X_{kj} - \mu_0)^2\right) - \ln\left(\frac{\sigma_1^2}{\sigma_0^2}\right)\right] \quad (4)$$

The CUSUM chart signals if the sum of these increments from the previous minimum to the current sample exceeds a control limit h_C. Thus, the chart signals if $C_k > h_C$, where

$$C_k = \max_{0\le\tau<k} \sum_{i=\tau+1}^{k} \frac{n}{2}\left[\left(\sigma_0^{-2} - \sigma_1^{-2}\right)\left(\frac{1}{n}\sum_{j=1}^{n}(X_{ij} - \mu_0)^2\right) - \ln\left(\frac{\sigma_1^2}{\sigma_0^2}\right)\right]. \quad (5)$$

Define C_0 as $C_0 = 0$ so that this CUSUM statistic can conveniently be written as

$$C_k = \max\{0, C_{k-1}\} + \frac{n}{2}\left[\left(\sigma_0^{-2} - \sigma_1^{-2}\right)\left(\frac{1}{n}\sum_{j=1}^{n}(X_{kj} - \mu_0)^2\right) - \ln\left(\frac{\sigma_1^2}{\sigma_0^2}\right)\right]. \quad (6)$$

In practice the CUSUM statistic for monitoring σ^2 is usually expressed in a different form by dividing all terms in Eq. 6 by the constant $n(\sigma_0^{-2} - \sigma_1^{-2})/2$. This results in a CUSUM statistic, say C_k', given by

$$C_k' = \frac{2C_k}{n(\sigma_0^{-2} - \sigma_1^{-2})} = \max\{0, C_{k-1}'\} + \left[\left(\frac{1}{n}\sum_{j=1}^{n}(X_{kj} - \mu_0)^2\right) - \frac{\ln(\sigma_1^2/\sigma_0^2)}{(\sigma_0^{-2} - \sigma_1^{-2})}\right]. \quad (7)$$

A signal is given if $C_k' > h_C'$, where $h_C' = 2h_C/(n(\sigma_0^{-2} - \sigma_1^{-2}))$. Here we use the CUSUM statistic in the form Eqs. 5 and 6 with control limit h_C to show the relationship to the GLR chart statistic R_k in Eq. 1. We will call this control chart the CUSUM(X^2) chart because it is based on the squares of the deviations of the observations from the target μ_0.

When $n > 1$, control charts for monitoring σ^2 are frequently based on the sample variances $S_1^2, S_2^2, \ldots, S_k^2$. The CUSUM chart based on these statistics is of the same form as in Eqs. 6 or 7 except that $\sum_{j=1}^{n}(X_{kj} - \mu_0)^2/n$ is replaced with $S_k^2 = \sum_{j=1}^{n}(X_{kj} - \bar{X}_k)^2/(n-1)$. Call this chart the CUSUM(S^2) chart. Shewhart charts can be based on S_k^2 or on $\sum_{j=1}^{n}(X_{kj} - \mu_0)^2/n$, and here we represent these Shewhart charts as special cases of CUSUM charts with $h_C = 0$ and the appropriate value of σ_1^2.

7 Comparisons of Charts

Table 4 gives SSATS values for the GLR chart (in column [1]), five CUSUM(S^2) charts (columns [2]–[6]), and a Shewhart chart based on S_k^2 (column [7]) for the case of $n = 4$ and $d = 4.0$. The tuning parameter σ_1^2 in the CUSUM charts is expressed in terms of

$$\psi_1 = \frac{\sigma_1}{\sigma_0}.$$

We see that the GLR chart is uniformly better than all of the other charts in Table 4.

Table 5 gives SSATS values for the GLR chart, five CUSUM(X^2) charts, and a Shewhart chart based on $\sum_{j=1}^{n}(X_{kj} - \mu_0)^2/n$ for the case of $n = 4$ and $d = 4.0$. We see that if ψ_1 is very small then the CUSUM(X^2) will perform a little better than the GLR chart for small shifts, but will be worse than the GLR chart for intermediate

Table 4 SSATS values for the GLR chart, CUSUM(S^2) charts, and the Shewhart(S^2) chart for the case of $n = 4$ and $d = 4.0$

	GLR	CUSUM(S^2)					Shewhart(S^2)
$\psi_1 =$	–	1.20	1.50	2.00	3.00	4.00	10.35
ψ	[1]	[2]	[3]	[4]	[5]	[6]	[7]
1.00	1481.61	1481.65	1481.47	1481.69	1481.61	1481.55	1481.60
1.10	214.50	224.36	298.30	372.56	432.97	454.18	469.33
1.20	78.35	86.10	105.97	138.02	171.81	185.60	197.40
1.40	27.23	33.53	33.96	39.99	49.44	54.33	59.42
1.60	14.87	19.84	18.44	19.71	22.89	24.84	27.23
1.80	9.83	13.80	12.33	12.48	13.72	14.60	15.84
2.00	7.25	10.48	9.20	9.03	9.54	9.98	10.67
2.40	4.74	7.00	6.08	5.81	5.90	6.03	6.28
3.00	3.28	4.71	4.15	3.95	3.91	3.94	4.01
5.00	2.21	2.68	2.51	2.45	2.42	2.42	2.42
7.00	2.06	2.26	2.20	2.17	2.16	2.16	2.16
10.00	2.02	2.10	2.07	2.06	2.06	2.06	2.06
15.00	2.00	2.03	2.02	2.02	2.02	2.02	2.02
$h_{GLR} =$	4.2443	–	–	–	–	–	–
$h_C =$	–	3.0505	3.7292	3.7213	3.1399	2.5300	0

Table 5 SSATS values for the GLR chart, CUSUM(X^2) charts, and the Shewhart(X^2) chart for the case of $n = 4$ and $d = 4.0$

	GLR	CUSUM(X^2)					Shewhart(X^2)
$\psi_1 =$	–	1.20	1.50	2.00	3.00	4.00	7.34
ψ	[1]	[2]	[3]	[4]	[5]	[6]	[7]
1.00	1481.61	1481.66	1481.51	1481.51	1481.48	1481.63	1481.60
1.10	214.50	192.79	266.57	342.91	401.17	418.34	425.74
1.20	78.35	70.85	89.04	119.93	151.24	162.39	167.96
1.40	27.23	27.09	27.27	32.54	40.82	44.64	47.10
1.60	14.87	15.95	14.61	15.65	18.29	19.79	20.91
1.80	9.83	11.06	9.73	9.81	10.78	11.43	12.00
2.00	7.25	8.37	7.23	7.06	7.45	7.76	8.06
2.40	4.74	5.58	4.79	4.57	4.62	4.70	4.80
3.00	3.28	3.79	3.33	3.18	3.15	3.17	3.19
5.00	2.21	2.33	2.23	2.19	2.18	2.18	2.18
7.00	2.06	2.10	2.06	2.05	2.05	2.05	2.05
10.00	2.02	2.03	2.02	2.01	2.01	2.01	2.01
15.00	2.00	2.01	2.00	2.00	2.00	2.00	2.00
$h_{GLR} =$	4.2443	–	–	–	–	–	–
$h_C =$	–	3.2395	3.8466	3.7247	2.9185	2.0979	0

Table 6 SSATS values for the GLR chart, CUSUM(X^2) charts, and the Shewhart(X^2) chart for the case of $n = 1$ and $d = 1.0$

	GLR	CUSUM(X^2)					Shewhart(X^2)
$\psi_1 =$	–	1.20	1.50	2.00	3.00	4.00	323.26
ψ	[1]	[2]	[3]	[4]	[5]	[6]	[7]
1.00	1481.68	1481.59	1481.37	1481.39	1481.61	1481.59	1481.60
1.10	238.04	192.92	265.98	342.15	407.50	437.07	599.96
1.20	85.21	71.10	89.09	119.75	154.14	171.58	216.33
1.40	28.65	27.33	27.52	32.79	41.80	47.35	65.41
1.60	15.36	16.15	14.83	15.91	18.83	20.94	29.25
1.80	10.04	11.22	9.91	10.02	11.14	12.09	16.47
2.00	7.31	8.49	7.36	7.21	7.68	8.16	10.71
2.40	4.66	5.62	4.81	4.57	4.65	4.81	5.88
3.00	3.01	3.69	3.17	2.97	2.94	2.98	3.39
5.00	1.50	1.80	1.59	1.50	1.46	1.46	1.51
7.00	1.11	1.28	1.17	1.11	1.09	1.08	1.09
10.00	0.88	0.98	0.92	0.88	0.87	0.86	0.86
15.00	0.73	0.79	0.75	0.73	0.72	0.72	0.72
$h_{GLR} =$	5.1422	–	–	–	–	–	–
$h_C =$	–	3.5009	4.4225	6.2921	4.6077	4.3942	0

and large shifts. On the other hand, a CUSUM(X^2) chart with a large value of ψ_1 will perform well for large shifts but not for small shifts.

Table 6 gives SSATS values of the GLR chart, five CUSUM(X^2) charts, and a Shewhart chart based on $(X_k - \mu_0)^2$ for the case of $n = 1$ and $d = 1.0$. When

$n = 1$ charts based on S^2 cannot, of course, be used. The conclusion from Table 6 is similar to the conclusion from Table 5; the CUSUM(X^2) chart can be tuned to be better than the GLR chart in a relatively narrow range of values of ψ, but then the GLR chart will be better for other values of ψ.

Comparing the SSATS values for the GLR and CUSUM charts in Tables 5 and 6 shows that using $n = 4$ and $d = 4.0$ is a little better than using $n = 1$ and $d = 1.0$ for small shifts, but is much worse for large shifts. It appears that it would be better overall to use $n = 1$ and $d = 1.0$ unless large shifts are considered to be unlikely to occur.

8 A Combination of Two CUSUM(X^2) Charts

The CUSUM(X^2) does not perform well over a wide range of shifts, so next consider the use of a combination of two CUSUM(X^2) charts, where one chart is tuned to detect small shifts and the other to detect larger shifts. Table 7 gives some SSATS values for this combination for the case of $n = 4$ and $d = 4.0$, where the two values of ψ_1 are labeled ψ_{11} and ψ_{12}. The values of h_C for the two CUSUM(X^2) charts are taken to be equal.

From Table 7 we see that using the combination of two CUSUM(X^2) charts gives performance close to the GLR chart, with the combination a little better for some values of ψ and the GLR chart a little better for other value of ψ. Table 8 has

Table 7 SSATS values for the GLR chart and a combination of two CUSUM(X^2) charts for the case of $n = 4$ and $d = 4.0$

	GLR	Two CUSUM(X^2) Charts			
$\psi_{11} =$	–	1.2	1.2	1.5	1.5
$\psi_{12} =$	–	2.0	3.0	2.0	3.0
ψ	[1]	[2]	[3]	[4]	[5]
1.00	1481.61	1481.51	1481.59	1481.60	1481.67
1.10	214.50	210.59	225.20	217.20	227.98
1.20	78.35	78.23	85.97	75.97	79.24
1.40	27.23	26.77	28.93	26.57	27.43
1.60	14.87	14.78	15.00	14.86	14.92
1.80	9.83	9.94	9.72	10.01	9.81
2.00	7.25	7.43	7.12	7.49	7.21
2.40	4.74	4.94	4.66	4.97	4.73
3.00	3.28	3.41	3.24	3.43	3.27
5.00	2.21	2.24	2.21	2.25	2.21
7.00	2.06	2.07	2.06	2.07	2.06
10.00	2.02	2.02	2.01	2.02	2.02
15.00	2.00	2.00	2.00	2.00	2.00
$h_{GLR} =$	4.2443	–	–	–	–
$h_C =$	–	3.8225	3.9187	3.8751	4.0483

Table 8 SSATS values for the GLR chart and a combination of two CUSUM(X^2) charts for the case of $n = 1$ and $d = 1.0$

	GLR	Two CUSUM(X^2) Charts			
$\psi_{11} =$	–	1.2	1.2	1.5	1.5
$\psi_{12} =$	–	2.0	3.0	2.0	3.0
ψ	[1]	[2]	[3]	[4]	[5]
1.00	1481.68	1481.65	1481.60	1481.54	1481.67
1.10	238.04	217.53	242.54	221.74	241.81
1.20	85.21	79.62	90.80	77.19	83.09
1.40	28.65	26.99	29.45	26.85	28.16
1.60	15.36	14.96	15.21	15.02	15.14
1.80	10.04	10.09	9.87	10.15	9.92
2.00	7.31	7.54	7.22	7.58	7.28
2.40	4.66	4.95	4.66	4.94	4.69
3.00	3.01	3.27	3.05	3.28	3.07
5.00	1.50	1.63	1.54	1.63	1.55
7.00	1.11	1.19	1.13	1.19	1.14
10.00	0.88	0.93	0.90	0.93	0.90
15.00	0.73	0.76	0.74	0.76	0.74
$h_{GLR} =$	5.1422	–	–	–	–
$h_C =$	–	4.2970	4.6604	4.3348	4.7310

SSATS values for the same situation as Table 7, except that $n = 1$ and $d = 1.0$. The conclusion from Table 8 is similar to the conclusion from Table 7.

9 The GLR Chart as an Infinite Combination of CUSUM Charts

We see from Eq. 1 that the GLR statistic is the maximum over σ_1^2 of the CUSUM statistics in Eq. 5, so the GLR chart without the window is equivalent to applying an infinite number of CUSUM charts and signaling if any CUSUM chart signals. Lorden (1971) has shown that the GLR chart is optimal in the sense that, asymptotically as $h_{GLR} \rightarrow \infty$, the expected time for the GLR chart to detect a shift to a specific σ_1^2 is the same as the expected detection time for the CUSUM chart which has been tuned to detect this specific shift.

The control limit h_{GLR} used for the GLR chart is the one control limit for each of the infinite number of CUSUM charts that make up the GLR chart. Thus, the GLR chart is equivalent to an infinite number of CUSUM charts, with the CUSUM charts based on different values of σ_1^2, but having the same control limit h_{GLR}. Note, however, that the CUSUM statistic is traditionally expressed in the form C_k' in Eq. 7 with control limit $h_C' = 2h_C/(n(\sigma_0^{-2} - \sigma_1^{-2}))$. If h_C, which corresponds to the form for h_{GLR} in the GLR chart, is constant for all σ_1^2, then h_C' for the form C_k' changes as σ_1^2 changes.

The fact that the GLR chart is equivalent to the use of an infinite number of CUSUM charts suggests that the GLR chart could be approximated by a finite number of CUSUM charts. Here we consider using many CUSUM charts to obtain a very close approximation to the GLR chart.

We first show that the GLR chart is equivalent to a countable set of CUSUM charts with σ_1^2 values that fall within a specific range. From Eq. 1 we see that the GLR chart without the window signals at sample k if, for some $0 \leq \tau < k$ and some $\sigma_1^2 > \sigma_0^2$, we have

$$\frac{n(k-\tau)}{2}\left(\sigma_0^{-2} - \sigma_1^{-2}\right)\left(\frac{1}{n(k-\tau)}\sum_{i=\tau+1}^{k}\sum_{j=1}^{n}(X_{ij}-\mu_0)^2\right)$$
$$-\frac{n(k-\tau)}{2}\ln\left(\frac{\sigma_1^2}{\sigma_0^2}\right) > h_{GLR},$$

which is equivalent to

$$\frac{2h_{GLR} + n(k-\tau)\ln(\sigma_1^2/\sigma_0^2)}{n(k-\tau)(\sigma_0^{-2} - \sigma_1^{-2})} < \frac{1}{n(k-\tau)}\sum_{i=\tau+1}^{k}\sum_{j=1}^{n}(X_{ij}-\mu_0)^2. \tag{8}$$

For a given value of $k-\tau$, the left hand side of Eq. 8 is minimized with respect to σ_1^2 when

$$\frac{n(k-\tau)}{2}\left(\frac{\sigma_1^2}{\sigma_0^2} - 1 - \ln\left(\frac{\sigma_1^2}{\sigma_0^2}\right)\right) = h_{GLR}. \tag{9}$$

If the GLR chart signals at sample k because some σ_1^2 satisfies Eq. 8 for some value of $k-\tau$, then it follows that there must be a σ_1^2 satisfying Eq. 9 that also satisfies Eq. 8 for the given value of $k-\tau$ (because the solution of Eq. 9 gives the minimum of the left hand side of Eq. 8). The possible values of $k-\tau$ are $1, 2, 3, \ldots$, so this implies that the GLR chart without a window will be equivalent to a countable set of CUSUM charts based on values of σ_1^2 satisfying

$$\frac{ni}{2}\left(\frac{\sigma_1^2}{\sigma_0^2} - 1 - \ln\left(\frac{\sigma_1^2}{\sigma_0^2}\right)\right) = h_{GLR},\ i = 1, 2, 3, \ldots. \tag{10}$$

A signal by the GLR chart may correspond to a signal by many of the CUSUM charts that make up the GLR chart, but among these CUSUM charts that signal, there will always be at least one that has σ_1^2 of the form in Eq. 10 for some $i = 1, 2, 3, \ldots$.

The maximum value of the σ_1^2 values from Eq. 10, say $\sigma_{1,\max}^2$, occurs when $i = 1$, so the values of σ_1^2 that need to be considered when approximating the GLR chart with a set of CUSUM charts are values in the interval $(\sigma_0^2, \sigma_{1,\max}^2]$. In approximating the GLR chart we cannot use all of the σ_1^2 values in a countable set of values, so instead we use a finite set of σ_1^2 values spaced over the interval $(\sigma_0^2, \sigma_{1,\max}^2]$. It appears that the σ_1^2 values need to be spaced more closely together when σ_1^2 is close to σ_0^2, but can be more widely spaced for larger values of σ_1^2. To investigate this approximation we considered a spacing of σ_1^2 values determined by the function $\sigma_1^2(i) = \sigma_0^2 + (\sigma_{1,\max}^2 - \sigma_0^2)i^2/r^2$, $i = 1, 2, \ldots, r$, where r is the number of values of σ_1^2 in $(\sigma_0^2, \sigma_{1,\max}^2]$.

We simulated this approximation to the GLR chart using various values of r, and the results are given in Table 9 for the case of $n = 4$, and in Table 10 for the case of $n = 1$. Column [1] in Table 9 has ATS and SSATS values for the GLR chart with $m = 200$ that was used in previous tables. The GLR chart that is being represented by an infinite number of CUSUM charts is the GLR chart without a window ($m = \infty$), so column [2] has ATS and SSATS values for the GLR chart with m = 10,000, which is presumably essentially the same as the GLR chart with $m = \infty$. Columns [3]–[8] in Table 9 have ATS and SSATS values for sets of CUSUM charts with values of r ranging from 5 up to 200. The control limit used for these sets of CUSUM charts is the same control limit used for the GLR chart in column [2]. In Table 10 column [2] has the GLR chart with m = 40,000, and columns [3]–[8] have sets of CUSUM charts with values of r ranging from 20 up to 800.

We see from Tables 9 and 10 that an excellent approximation to the GLR chart can be obtained as long as r is not too small. For practical applications, a relatively small value of r would be quite adequate, but a larger value would be needed

Table 9 The SSATS values for the GLR chart and approximating CUSUM charts for the case of $n = 4$ and $d = 4.0$

	GLR	GLR	A set of r CUSUM(X^2) charts					
$m =$	200	10,000	–	–	–	–	–	–
$r =$	–	–	5	10	25	50	100	200
ψ	[1]	[2]	[3]	[4]	[5]	[6]	[7]	[8]
1.00	1481.61	1481.53	1586.90	1505.54	1486.82	1482.39	1481.76	1481.60
1.10	214.50	214.26	229.05	218.49	214.94	214.37	214.22	214.12
1.20	78.35	78.37	81.77	79.07	78.47	78.37	78.30	78.33
1.40	27.23	27.24	27.68	27.37	27.23	27.21	27.21	27.21
1.60	14.87	14.87	15.03	14.89	14.88	14.88	14.88	14.88
1.80	9.83	9.83	9.91	9.86	9.84	9.84	9.84	9.84
2.00	7.25	7.25	7.28	7.25	7.25	7.25	7.25	7.24
2.40	4.74	4.75	4.76	4.75	4.74	4.75	4.75	4.75
3.00	3.28	3.28	3.28	3.28	3.28	3.28	3.28	3.28
5.00	2.21	2.21	2.21	2.21	2.21	2.21	2.21	2.21
7.00	2.06	2.06	2.06	2.06	2.06	2.06	2.06	2.06
10.00	2.02	2.02	2.02	2.02	2.02	2.02	2.02	2.02
15.00	2.00	2.00	2.00	2.00	2.00	2.00	2.00	2.00
$h_{GLR} =$	4.2443	4.2451	–	–	–	–	–	–
$h_C =$		–	4.2451	4.2451	4.2451	4.2451	4.2451	4.2451

Table 10 The SSATS values for the GLR chart and approximating CUSUM charts for the case of $n = 1$ and $d = 1.0$

	GLR	GLR	A Set of r CUSUM(X^2) Charts					
$m =$	800	40,000	–	–	–	–	–	–
$r =$	–	–	20	40	100	200	400	800
ψ	[1]	[2]	[3]	[4]	[5]	[6]	[7]	[8]
1.00	1481.68	1481.58	1495.04	1485.07	1482.04	1481.70	1481.60	1481.58
1.10	238.04	237.49	241.42	238.49	237.66	237.55	237.33	237.32
1.20	85.21	85.22	85.76	85.24	85.05	85.05	85.17	85.17
1.40	28.65	28.66	28.76	28.68	28.66	28.64	28.63	28.63
1.60	15.36	15.36	15.39	15.37	15.36	15.36	15.38	15.38
1.80	10.04	10.04	10.04	10.05	10.04	10.04	10.03	10.03
2.00	7.31	7.31	7.32	7.30	7.32	7.32	7.32	7.32
2.40	4.66	4.66	4.66	4.66	4.65	4.65	4.66	4.66
3.00	3.01	3.01	3.02	3.01	3.02	3.01	3.02	3.02
5.00	1.50	1.50	1.50	1.51	1.51	1.51	1.51	1.51
7.00	1.11	1.11	1.11	1.11	1.11	1.11	1.11	1.11
10.00	0.88	0.88	0.88	0.88	0.88	0.88	0.88	0.88
15.00	0.73	0.73	0.73	0.73	0.73	0.73	0.73	0.73
$h_{GLR} =$	5.1422	5.1427	–	–	–	–	–	–
$h_C =$	–	–	5.1427	5.1427	5.1427	5.1427	5.1427	5.1427

for a research study doing precise comparisons of control charts. The advantage to representing the GLR chart as an infinite set of CUSUM charts is that it is much faster to simulate the set of CUSUM charts. Note however, that practitioners would not need to do any simulation in order to use the GLR chart, so the issue of simulation pertains to research studies.

For practical applications of the GLR chart, software could be designed to store the past m sample values and use these to compute the GLR statistic as each new sample is obtained. Alternately, software could be designed to use r CUSUM charts in the form Eq. 6, with r large enough to give a good approximation to the GLR chart. The form Eq. 6 allows the r CUSUM charts to have the same control limit ($h_C = h_{GLR}$). It would not be feasible or desirable to plot all of the r CUSUM statistics, so the maximum of these statistics could be plotted on the control chart.

10 Conclusions

We have shown that the GLR chart is effective for detecting a wide range of shifts in the process variance. Its overall performance is at least as good as the performance of other options for detecting a wide range of shifts, such as a combination of two charts.

The GLR chart has the advantage that it does not require the user to specify the values of multiple control chart parameters, and this greatly simplifies the design of the GLR chart. To design the GLR chart for an application all that the user has to do is specify the in-control ATS, and then look up the control limit in Table 3 of this paper. Once software is in place to actually compute and plot the GLR statistic, using the GLR chart requires no particular sophistication on the part of the user.

The approach used here for developing a GLR chart for monitoring the process variance can be used for other process parameters. For example, Reynolds and Lou (2010) recently evaluated a GLR chart for monitoring the process mean. Work is in progress on developing a GLR chart for simultaneously monitoring the mean and variance.

References

Apley, D. W., & Shi, J. (1999). The GLRT for statistical process control of autocorrelated processes. *IIE Transactions, 31*, 1123–1134.

Han, D., Tsung, F., Hu, X., & Wang, K. (2007). CUSUM and EWMA multi-charts for detecting a range of mean shifts. *Statistica Sinica, 17*, 1139–1164.

Hawkins, D. M., & Olwell, D. H. (1998). *Cumulative sum control charts and charting for quality improvement*. New York: Springer.

Hawkins, D. M., & Zamba, K. D. (2005). Statistical process control for shifts in mean or variance using a changepoint formulation. *Technometrics, 47*, 164–173.

Lai, T. L. (1995). Sequential changepoint detection in quality control and dynamical systems (with discussion). *Journal of the Royal Statistical Society, Series B, 57*, 613–658.

Lai, T. L. (2001). Sequential analysis: Some classical problems and new challenges (with discussion). *Statistica Sinica, 11*, 303–408.

Lorden, G. (1971). Procedures for reacting to a change in distribution. *The Annals of Mathematical Statistics, 42*, 1897–1908.

Montgomery, D. C. (2009). *Introduction to statistical quality control* (6th ed.). New York: Wiley.

Reynolds, M. R., Jr., & Lou, J. (2010). An evaluation of a GLR control chart for monitoring the process mean. *Journal of Quality Technology, 42*, 287–310.

Reynolds, M. R., Jr., & Stoumbos, Z. G. (2004). Should observations be grouped for effective process monitoring? *Journal of Quality Technology, 36*, 343–366.

Siegmund, D., & Venkatraman, E. S. (1995). Using the Generalized likelihood ratio statistic for sequential detection of a change-point. *The Annals of Statistics, 23*, 255–271.

Sparks, R. S. (2000). CUSUM charts for signaling varying locations shifts. *Journal of Quality Technology, 32*, 157–171.

Willsky, A. S., & Jones, H. L. (1976). A generalized likelihood ratio approach to the detection and estimation of jumps in linear systems. *IEEE Transactions on Automatic Control, 21*, 108–112.

On the Robustness of the Shewhart Control Chart to Different Types of Dependencies in Data

Olgierd Hryniewicz

Abstract Shewhart control charts were originally designed under the assumption of independence of consecutive observations. In the presence of dependence the authors usually assume dependencies in the form of autocorrelated and normally distributed data. However, there exist many other types of dependencies which are described by other mathematical models. The question arises then, how classical control charts are robust to different types of dependencies. This problem has been sufficiently well discussed for the case of autocorrelated and normal data. In the paper we use the concept of copulas to model dependencies of other types. We use Monte Carlo simulation experiments to investigate the impact of type and strength of dependence in data on the value of the ARL of Shewhart control charts.

Keywords Shewhart control charts • Correlated data • Copulas • ARL

1 Introduction

Statistical process control (SPC) is a collection of statistical methods used by thousands of practitioners who are striving to achieve continuous improvement in quality. This objective is accomplished by continuous monitoring of the process under study in order to quickly detect the occurrence of assignable causes. The Shewhart $\bar{X}$ control chart, known for more than 80 years, is the most popular SPC method used to detect whether observed process is under control. Its classical and internationally standardized version is designed under the assumption that process measurements are described by independent and identically distributed random variables. In the majority of practical cases these assumptions are fulfilled at

O. Hryniewicz (✉)
Systems Research Institute, Newelska 6, 01-447 Warsaw, Poland
e-mail: hryniewi@ibspan.waw.pl

H.-J. Lenz et al. (eds.), *Frontiers in Statistical Quality Control 10*,
Frontiers in Statistical Quality Control 10, DOI 10.1007/978-3-7908-2846-7_2,

least approximately. However, there exist production processes where consecutive observations are obviously correlated, e.g. in case of certain continuous production processes. The presence of correlations between consecutive measurements should be taken into account during the design of control charts. This need was already noticed in the 1970s, see e.g. the papers by Johnson and Bagshaw (1974) and by Vasilopoulos and Stamboulis (1978), but the problem was widely discussed in papers published in the late 1980s and in the 1990s.

One of the visible effects of autocorrelation in observed process data is the significant difference between statistical properties of control charts designed for independent and dependent data. There exist several approaches for dealing with this problem. First approach, historically the oldest one, consists in dealing with original data and adjusting control limits of classical control charts. This approach was used, for example, in papers by Reynolds Jr. and co-authors (Lu and Reynolds, 1999a,b, 2001; VanBrackle and Reynolds, 1997; Schmid, 1995, 1997; Vasilopoulos and Stamboulis, 1978; Zhang, 1998). Other approaches are based on the concept of residuals (see the papers by Alwan and Roberts (1988) or by Montgomery and Mastrangelo (1991)) or on monitoring statistics related to autocorrelations (see the papers by Yourstone and Montgomery (1991) or by Jiang et al. (2000)). There also exist more sophisticated methods for dealing with SPC autocorrelated data. An overview of SPC methods used for autocorrelated data can be found in papers by Wardell et al. (1994), Lu and Reynolds (1999a) and Knoth et al. (2001).

While dealing with correlated data we cannot rely, even in the case of classical control charts, on the methods used for the estimation of their parameters in case independent observations. Some corrections are necessary, as it was mentioned e.g. in the paper by Maragah and Woodall (1992). Another problem with the application of the procedures designed to control autocorrelated data is the knowledge of the structure of correlation. In the majority of papers it is assumed that the type of a stochastic process that describes the process data is known. Moreover, it is also assumed that the parameters of this stochastic process are also known. However, Lu and Reynolds (1999a, 2001) have shown that precise estimation of such parameters requires at least hundreds of observations.

All these problems, noticed by many authors, make the SPC with dependent data very difficult, especially for not well-trained in statistics practitioners who need efficient tools to discriminate between complicated problems with dependent data and relatively simple problems when observed data are independent. This problem was considered in the paper by Hryniewicz and Szediw (2010) who proposed a relatively simple and efficient nonparametric tool, named by them the Kendall control chart, for testing hypotheses about independence of SPC data. While discussing the properties of this tool they noticed that the type of existing dependence plays a crucial role. In this paper we continue the work along that line by analyzing the properties of Shewhart control charts when data are generated by different variants of a simple autoregression model. The mathematical model that describes serial dependence between consecutive observations of a process in terms of copulas is described in the second section of the paper. In the third section we present the results of Monte Carlo simulation experiments which show very strong dependence of statistical properties of control charts upon the type and the strngth of dependence. Conclusions derived from these results are presented in the last section of the paper.

2 Mathematical Models of Dependence Between Consecutive Observations on a Control Chart

Mathematical models used for the description of dependent random variables are well known for many years. In the simplest two-dimensional case we are interested in the description of dependence between two random variables X and Y having marginal distributions described by cumulative probability functions $F(x)$ and $G(y)$, respectively. In the context of the considered in this paper time-dependent observations we can, in the simplest case, set $X = X_t$ and $Y = X_{t+1}$, where $X_t, t = 1, 2, \ldots$ is the time series representing consecutive observations of the process under consideration. In his fundamental work Sklar (1959) showed that for a two-dimensional probability distribution function $H(X, Y)$ with marginal distribution functions $F(X)$ and $G(Y)$ there exists a copula C such that $H(x, y) = C(F(x), G(y))$. This result has been later extended to the case of multivariate probability distributions. For more information about copulas the reader should refer e.g. to the book by Nelsen (2006).

All well known multivariate probability distributions, the multivariate normal distribution included, can be generated by parametric families C_α of copulas, where real- or vector-valued parameter α describes the strength of dependence between the components of the random vector. Thus, copulas have found many interesting practical applications. The number of papers devoted to the theory and applications of copulas is still growing rapidly, thanks to the increasing interest coming from e.g. the analysis of financial risks and the survival analysis. For more recent results the reader should consult already mentioned book by Nelsen (2006).

In this paper we focus our attention on three types of copulas. First is the normal copula, which in the two-dimensional case is defined as follows:

$$C(u_1, u_2; \rho) = \Phi_N(\Phi^{-1}(u_1), \Phi^{-1}(u_2); \rho) \tag{1}$$

where $\Phi_N(u_1, u_2)$ is the cumulative probability distribution function of the bivariate normal distribution, $\Phi^{-1}(u)$ is the inverse of the cumulative probability function of the univariate normal distribution (the quantile function). Parameter ρ in case of marginals described by the normal distribution is equal to the well known coefficient of linear correlation introduced by Pearson. It is worth noticing that the values of the linear correlation coefficient depend upon the type of marginals. Therefore, for the same value of the parameter ρ of the normal copula, the values of the Pearson's correlation may be different for different distributions of X and Y.

Second copula considered in this paper is the Farlie-Gumbel-Morgenstern (FGM) copula who is frequently used for modelling weak dependencies. This copula is defined by the following formula:

$$C(u_1, u_2; \theta) = u_1 u_2 + \theta u_1 u_2 (1 - u_1)(1 - u_2), |\theta| \leq 1 \tag{2}$$

The remaining three copulas considered in this paper belong to a general class of symmetric copulas, named the Archimedean copulas. They are generated using a class Φ of functions $\phi : [0, 1] \rightarrow [0, \infty]$, named generators, that have two continuous derivatives on $(0, 1)$ and fulfill the following conditions: $\phi(1) = 1$, $\phi'(t) < 0$, and $\phi''(t) > 0$ for all $0 < t < 1$ (these conditions guarantee that ϕ has an inverse ϕ^{-1} that also has two derivatives). Every member of this class generates a multivariate distribution function. In this paper we consider three two-dimensional Archimedean copulas defined by the following formulae (copulas and their respective generators):

- Clayton's

$$C(u, v) = \max\left(\left[u^{-\alpha} + v^{-\alpha} - 1\right]^{-1/\alpha}, 0\right), \alpha \in [-1, \infty) \setminus 0 \tag{3}$$

$$\phi(t) = (t^{-\alpha} - 1)/\alpha, \alpha \in [-1, \infty) \setminus 0 \tag{4}$$

- Frank's

$$C(u, v) = -\frac{1}{\alpha} \ln\left(1 + \frac{(e^{-\alpha u} - 1)(e^{-\alpha v} - 1)}{e^{-\alpha} - 1}\right), \alpha \in (-\infty, \infty) \setminus 0 \tag{5}$$

$$\phi(t) = \ln\left(\frac{1 - e^{-\alpha}}{1 - e^{-\alpha t}}\right), \alpha \in (-\infty, \infty) \setminus 0 \tag{6}$$

- Gumbel's

$$C(u, v) = \exp\left(-\left[(-\ln u)^{1+\alpha} + (-\ln v)^{1+\alpha}\right]^{\frac{1}{1+\alpha}}\right), \alpha \in (0, \infty) \tag{7}$$

$$\phi(t) = (-\ln(t))^{\alpha+1}, \alpha \in (0, \infty) \tag{8}$$

In case of independence the dependence parameter α_{ind} adopts the value of 0 (in Clayton's and Frank's copulas as an appropriate limit). The copulas mentioned above are sometimes presented using different parametrization, and in such cases independence is equivalent to other values of α.

As it has been already mentioned above, a well known coefficient of linear correlation cannot be used for measuring the strength of dependence between random variables whose dependence is described by a given copula. Nonparametric measures of dependence, such as Spearman's ρ or Kendall's τ can be used for this purpose. For the copulas considered in this paper the values of Kendall's τ are easier to calculate, and for this reason we use this measure of dependence in further analyses.

Genest and McKay (1986) considered the population version of the Kendall's coefficient od dependence (association) τ. This characteristic can be used for the description of the strength of dependence in copulas, and its importance in characterizations of copulas has been shown recently in papers by Nelsen et al. (2009). Let $K(t)$ be the cumulative probability function of the random variable $T = C(U_1, U_2)$, where U_1 and U_2 are random variables uniformly distributed on $[0, 1]$. The following relation links a copula with Kendall's τ:

$$\tau = 3 - 4 \int_0^1 K(t)dt \tag{9}$$

Estimation of $K(t)$ for the case of two-dimensional copulas, and thus the estimation of τ, was considered by Genest and Rivest (1993).

Closed formulae for Kendall's τ are available only for some copulas. In the case of the normal copula we have the following expression

$$\tau_{Norm} = arcsin(\rho)/(\pi/2). \tag{10}$$

For the FGM copula we can compute Kendall's τ from a very simple formula

$$\tau_{FGM} = 2\theta/9. \tag{11}$$

For the family of Archimedean copulas there exists the following general formula that links Kendall's τ with the generator function ϕ:

$$\tau_{Arch} = 1 + 4 \int_0^1 \frac{\phi(v)}{\phi'(v)} dv. \tag{12}$$

For specific cases of the considered in this paper Archimedean copulas we have:

- Clayton's copula

$$\tau = \frac{\alpha}{\alpha + 2}, \tag{13}$$

- Frank's copula

$$\tau = 1 + 4 \left(\frac{1}{\alpha} \int_0^\alpha \frac{t}{e^t - 1} dt - 1 \right) / \alpha, \tag{14}$$

- Gumbel's copula

$$\tau = \frac{\alpha}{\alpha + 1}. \tag{15}$$

Each copula can be looked upon as a multivariate probability distribution whose all marginal distributions are uniform. However, by using an inverse probability distribution function (a quantile function) we can transform each uniformly distributed random variable to a variable with any continuous probability distribution. In this paper we will consider the case when such transformation will lead to marginals described by the standard normal distribution $N(0, 1)$. This assumption definitely restricts generality of inferred conclusions, but – on the other hand – allows to compare our new results with those presented by other authors who usually made this assumption.

3 Basic Properties of the Shewhart Control Chart in Case of Dependencies of Different Types

The most frequently used statistical characteristic of a control chart is its Average Run Length ARL. This characteristic describes the expected number of observations (points plotted on a chart) until the occurrence of an alarm (e.g. when the first point beyond 3-σ control limits has been observed). When consecutive observations are independent, and their probability distribution is known, the random variable which describes the waiting time till the moment of the first observation beyond the control limits is distributed according to the geometric distribution, and the value of the ARL can be calculated analytically. However, when observations are dependent (serially correlated) and/or their probability distributions are only partially known (e.g. the class of the distribution is known, but its parameters are estimated) this characteristic usually cannot be calculated from a closed formula. Therefore, we need to use statistical Monte Carlo simulation in order to evaluate the value of the ARL.

In our simulation experiments we have generated consecutive observations using conditional probability distributions derived from two-dimensional copulas. In order to arrive at comparable results we have generated serially correlated processes described by a fixed in advance value of Kendall's τ. By having the same normal marginal distributions, and the same values of the measure of the strength of dependence we can detect a possible influence of the type of dependence related to the type of the underlying copula. In the following two subsections we will present the results of experiments for two cases:

- Parameters of the normal distribution (design parameters) are known,
- Parameters of the normal distribution (design parameters) are estimated from an initial sample.

In both cases we consider only one type of the process deterioration: the shift of the process level by $k\sigma$. When $k = 0$ (i.e. when there is no shift) the value of the ARL represents the average time to a false alarm. When $k = 1$ we have the case of a small deterioration. Significant deterioration of the process is in our experiment modelled by setting $k = 3$.

Table 1 ARLs–Shewhart control chart (design parameters known), TEST 1: "3-σ" rule, no shift

Kendall's τ	Normal	FGM	Clayton	Frank	Gumbel
0.8	1391.0	x	759.3	385.4	1301.8
0.5	466.8	x	622.4	373.1	633.57
0.3	389.0	x	496.0	373.4	528.6
0.1	371.1	369.6	384.4	370.5	456.4
0.05	370.5	369.3	374.7	372.1	443.3
0.01	370.6	369.7	371.6	372.2	430.7
0	**370.5**	**370.5**	**370.5**	**370.5**	**370.5**
−0.01	372.31	369.28	370.19	372.97	x
−0.05	371.3	370.8	370.0	371.7	x
−0.1	371.3	368.9	370.6	371.2	x
−0.3	390.0	x	384.3	374.1	x
−0.5	468.4	x	433.9	375.0	x
−0.8	1379.4	x	898.6	384.0	x

3.1 Known Design Parameters

The results of the simulation experiment for known design parameters are presented in Tables 1–3 for different values of the shift of the process level (mean value). Each number in these tables has been obtained after averaging the results of 200,000 simulation runs. The maximal length of each simulation run varied from 10,000 to 100,000 observations (for strongly dependent observations).

In Table 1 we present the average times to a false alarm. In all considered cases the expected time to a false alarm in presence of dependent data is always larger than in the case of independence, and this difference increases with increasing strength of dependence. However, the way how the ARL depends on the value of τ strongly depends on the type of the copula that describes the data. It is interesting to see that for all considered copulas, with a noticeable exception of Gumbel's copula, the values of the ARL change insignificantly for weakly dependent data. However, for moderate and strong dependencies these values are changing in a completely different way depending on the type of a copula. For the normal copula (i.e. for an ordinary Gaussian autoregression $AR(1)$ process) the value of the ARL increases for the increasing absolute value of the strength of dependence measured by Kendall's τ. The dependence of the ARL on the value of τ is symmetrical and these values become very large for large values of τ. In case of the FGM copula, which is used for modelling weak dependencies, the influence of the value of τ on the ARL is practically non-existing. The similar situation, but extended to larger values of τ, has been observed for Frank's copula.

In the case of Clayton's copula the dependence of the ARL upon the value of τ is not symmetric. In case of positive dependence ($\tau > 0$), and small and moderate strength of dependence, the ARL in this case is larger than in the case of the normal copula. However, in case of very strong positive dependence this value of the ARL is significantly smaller than in the normal case. In case of negative dependence

Table 2 ARLs–Shewhart control chart (design parameters known), TEST 1: "3-σ" rule, shift of 1σ

Kendall's τ	Normal	FGM	Clayton	Frank	Gumbel
0.8	273.21	x	90.28	86.74	653.04
0.5	71.95	x	48.55	54.18	126.65
0.3	52.25	x	45.18	47.7	73.07
0.1	45.11	43.97	43.74	45.42	50.39
0	**43.78**	**43.78**	**43.78**	**43.78**	**43.78**
−0.1	43.62	43.73	43.74	44.75	x
−0.3	44.36	x	44.63	44.55	x
−0.5	50.96	x	50.15	45.53	x
−0.8	135.18	x	101.04	52.44	x

Table 3 ARLs–Shewhart control chart (design parameters known), TEST 1: "3-σ" rule, shift of 3σ

Kendall's τ	Normal	FGM	Clayton	Frank	Gumbel
0.8	10.58	x	24.94	9.75	9.1
0.5	3.29	x	4.05	4.51	3.16
0.3	2.49	x	2.59	3.55	2.44
0.1	2.1	1.99	2.12	3.13	2.11
0	**1.99**	**1.99**	**1.99**	**1.99**	**1.99**
−0.1	1.92	2.0	1.92	2.91	x
−0.3	1.79	x	1.83	2.76	x
−0.5	1.69	x	1.77	2.65	x
−0.8	1.57	x	1.71	2.57	x

($\tau < 0$) the ARL for Clayton's copula is always smaller than the ARL in the normal case. The case of Gumbel's copula requires special comments. This copula describes only positive dependence, and even for very weak dependencies the corresponding values of the ARL are significantly greater than in the case of independence. Only in case of very strong dependence the behaviour of the Shewhart control chart seems to be similar to that described by the normal copula. It means that the Shewhart control chart is very sensitive to this type of dependence, even if this dependence is very weak, and thus difficult to be confirmed.

In case of small shifts (equal to 1σ) of the process level the dependence of the ARL upon τ looks different. The values of the ARL in presence of dependent data are nearly always greater than in the case of independence. It means that the dependence in data has negative impact on discrimination abilities of the Shewhart control chart, and this unpleasant feature does not depend upon the type of dependence. In case of strongly dependent data the values of the ARL may be so large (especially for normal and Gumbel's copulas) that the chart becomes practically insensitive to relatively small deterioration of the process. However, in the case of Frank's copula the value of the ARL remains reasonable even for strongly dependent data (especially in case of negative dependence).

When the shift of the process is large (e.g. equal to 3σ) the situation is different. First of all, in case of negative dependence described by the normal and Clayton's copulas the chart reacts faster than in the case of independence. Positive dependence in all considered cases has negative influence on the ability of the chart to detect shifts. The worse situation is in the case of Frank's copula, and this is somewhat unexpected because for small shifts this copula seems to be the most favourable. A similar situation is with Clayton's copula which usually behaves quite well except for the case of large shifts and strong positive dependence.

3.2 Estimated Design Parameters

Let us consider the case when parameters of the probability distribution (mean value and standard deviation) that are used for the design of a control chart are estimated from a process (its Phase I, as the sampling period is sometimes called) with possibly dependent consecutive observations. This assumption leads to significant consequences. First of all, random character of control lines which are estimated from a sample adds some variability resulting in wider (on average) in-control area on a control chart. This problem has been considered by many authors, and some conclusions from that research may be found in the paper by Woodall and Montgomery (1999) or in the paper by Albers and Kallenberg (2004). Second, the autocorrelation between sample observations influences the properties of estimators, as it was noticed already in the paper by Vasilopoulos and Stamboulis (1978). Variability related to both these two sources is difficult to be assessed analytically. Thus, simulation experiments are needed in order to evaluate the properties of control charts designed in such a way.

Our simulation experiment has two phases as in actual applications. First we simulate a sample of n elements, and the results of this simulation are used for the design of a control chart. The minimal number of observations which is suggested for designing a chart should be, according to many authors, such as e.g. Quesenberry (1993), not smaller than 300. However, in the majority of popular textbooks on quality control this minimal value is proposed to be equal to 100. Having in mind our main purpose, i.e. to investigate the influence of different types of dependence on the performance of control charts actually used in practice, in our experiments we set the sample size (the number of consecutive observations that are used for the design of a chart) as equal to 100. In the experiment we have simulated 500 different control charts, and for each of them we have simulated 500 production runs. Thus, for each experiment described by the chosen copula and the given value of Kendall's τ we have had altogether 250,000 simulation runs. These runs have been used for the estimation of the ARL, and other statistical properties of the chart.

Table 4 contains the results of the simulation experiment similar to those presented in Table 1, i.e. in presence of no shift in the level of a process. Somewhat unexpectedly these results are different not only with respect to the simulated values of the ARL. The dependence of the ARL upon the value of τ is also somewhat

Table 4 ARLs–Shewhart control chart (design parameters estimated), TEST 1: "3-σ" rule, no shift

Kendall's τ	Normal	FGM	Clayton	Frank	Gumbel
0.8	2111.72	x	346.65	568.74	1192.92
0.5	967.84	x	974.28	437.75	1142.76
0.3	540.38	x	730.8	487.09	799.15
0.1	473.11	490.15	560.42	508.98	586.61
0	**486.14**	**486.14**	**486.14**	**486.14**	**486.14**
−0.1	503.28	458.77	447.48	458.49	x
−0.3	589.39	x	537.13	512.63	x
−0.5	1242.9	x	861.8	593.1	x
−0.8	9966.6	x	6255.6	1253.5	x

different than that represented in Table 1. The results of the experiment displayed in Table 4 show that the existence of dependence of any type results, in general, in increasing value of the ARL. Only in few cases, in presence of weak dependence, the values of ARL are slightly smaller than in the case of independence. When the strength of dependence is low, the values of the ARL are similar. Only for Gumbel's copula this value is visibly larger than in the case of independence. For moderate values of Kendall's τ practically acceptable worsening of the value of the ARL can be noticed only in the case of Frank's copula. In the case of strong dependence, both positive and negative, the values of the ARL are large enough to make the chart insensible to the process deterioration of that magnitude. An interesting, and difficult to explain, exception is the case of Clayton's copula where the large value of ARL for $\tau = 0.5$ decreases to a low value for $\tau = 0.8$. A phenomenon of a similar type is also seen in the case of Gumbel's copula.

When the magnitude of the process deterioration is large (i.e. when the shift in the process level is equal to 3σ) the picture is anew different. First of all, it can be noticed that in the case of small and moderate negative dependencies the value of the ARL may be smaller than in the case of independence. It means that negative dependence, unless it is not too strong, has a positive impact on the ability of the chart to detect deteriorations of large magnitude. In case of strong negative dependence the situation is different, and the value of the ARL usually becomes too large. In the case of the normal copula this value becomes completely unacceptable. In case of positive dependence good properties of the chart are observed for Frank's and Gumbel's copulas.

The results presented in Tables 4–6 show a very complicated situation. Only in the case of Frank's copula the performance of the Shewhart control chart is more or less robust to the existence of dependence between consecutive observations. In all remaining cases one can observe situations which are difficult to describe and explain. Only in the case of the normal copula the dependence of the ARL on the strength of dependence can be described in a relatively simple way: the chart is completely insensible to process shifts only in the case of strong, both positive and negative, dependence.

Table 5 ARLs–Shewhart control chart (design parameters estimated), TEST 1: "3-σ" rule, shift of 1σ

Kendall's τ	Normal	FGM	Clayton	Frank	Gumbel
0.8	352.88	x	55.77	80.64	206.43
0.5	84.7	x	102.46	55.72	205.43
0.3	59.61	x	57.34	51.55	92.76
0.1	54.88	51.4	50.51	49.12	58.54
0	**48.09**	**48.09**	**48.09**	**48.09**	**48.09**
−0.1	48.9	50.72	50.15	47.39	x
−0.3	56.1	x	56.07	53.08	x
−0.5	78.25	x	80.61	59.4	x
−0.8	769.18	x	433.52	89.26	x

Table 6 ARLs–Shewhart control chart (design parameters estimated), TEST 1: "3-σ" rule, shift of 3σ

Kendall's τ	Normal	FGM	Clayton	Frank	Gumbel
0.8	7.35	x	11.95	4.80	2.00
0.5	3.27	x	3.94	2.68	2.54
0.3	2.54	x	2.70	2.26	2.55
0.1	2.15	2.05	2.18	2.15	2.19
0	**2.06**	**2.06**	**2.06**	**2.06**	**2.06**
−0.1	1.92	2.07	1.96	1.87	x
−0.3	1.85	x	1.95	2.00	x
−0.5	1.92	x	2.09	2.02	x
−0.8	21.47	x	4.22	2.45	x

Table 7 Skewness of the run length – Shewhart control chart (design parameters estimated), shift of 3σ

Kendall's τ	Normal	FGM	Clayton	Frank	Gumbel
0.8	9.53	x	7.18	5.89	2.1
0.5	3.7	x	4.05	4.11	7.16
0.3	3.02	x	7.47	3.63	3.98
0.1	2.7	2.54	3.3	2.7	2.94
0	**2.49**	**2.49**	**2.49**	**2.49**	**2.49**
−0.1	2.12	2.54	2.6	2.8	x
−0.3	3.11	x	3.75	2.32	x
−0.5	6.87	x	6.47	2.72	x
−0.8	27.96	x	10.59	6.08	x

ARL is the most frequently used statistical characteristic of control charts. Another characteristic which is often calculated is the variance of the run length. Specialist are fully aware of the fact that the run length is a highly skewed random variable, and these two characteristics are not sufficient for the comprehensive description of the statistical properties of control charts. The coefficient of skewness whose value equal to 2 is well known for the chart with known design parameters is rarely calculated for other cases. In Table 7 we present the values of the coefficient of skewness of the run length for the case of estimated design parameters and shift equal to 3σ.

The values given in Table 7 show that the times to alarm are highly skewed, especially in case of strong (both positive and negative) dependence. In practice it means that despite reasonable values of the ARL there is quite substantial possibility that even significant process deterioration may not be detected sufficiently quickly.

4 Tests Based on Runs in Case of Dependent Data

Classical Shewhart control chart has been supported by additional decisions rules based on runs. Different rules have been proposed by many authors, but the most popular ones were proposed in the Western Electric handbook in 1956. They are also described in the international standard ISO 8258 and in the paper by Nelsen (1984). These rules are designed with the aim to detect deteriorations of different type. Statistical properties of control charts with supporting run rules can be computed using the Markov chain approach. A general solution of this problem has been proposed in the paper by Champ and Woodall (1987). This methodology has been successfully implemented for the calculations made under the assumption of independence of observations, and full knowledge of the values of design parameters. However, in case of dependent observations, and for estimated values of design parameters such computations are very difficult or even hardly possible. Therefore, in our analysis we used the results of the Monte Carlo simulation experiments. The settings of these experiments are the same as in the cases described in the previous sections of this paper.

One of the most popular rule, known as Test 3 or "6 increasing (decreasing) in a row", is used for the detection of harmful trends. The properties of this test do not depend upon the design parameters, and may be evaluated using recently published results of Ferguson et al. (2000). In Table 8 we present the values of the ARL for this particular test when the process is in the in-control state.

These results show that dependencies have detrimental impact on the properties of this test. In case of positive dependence the average time to a false alarm becomes unacceptably small. On the other hand, the negative dependence (especially the strong one) may decrease the ability of the test to detect trends in data. Similar results, which are not presented in this paper because of its limited volume, have been observed in preliminary experiments for the case of deteriorated processes.

Another popular additional decision rule, known as Test 5 or "2 out of 3 in a row observation in an outer zone", is used to improve the ability to detect small shifts of the process level. In Table 9 we present the values of the ARL for this test when the process is in the in-control state, and the design parameters are estimated from (possibly dependent) observations.

The dependence of the value of the ARL and the value of τ cannot be easily explained at the current stage of our research. For example, in case of the normal

Table 8 ARLs–Shewhart control chart (design parameters known), TEST 3: "6 in a row" rule, no shift

Kendall's τ	Normal	FGM	Clayton	Frank	Gumbel
0.8	24.8	x	27.3	26.1	24.3
0.5	30.1	x	32.4	30.9	30.26
0.3	48.2	x	47.4	48.4	46.1
0.1	97.1	147.3	95.3	96.1	92.4
0.05	119.2	146.4	117.1	118.5	115.0
0.01	140.4	147.1	140.1	141.1	138.0
0	**147.1**	**147.1**	**147.1**	**147.1**	**147.1**
−0.01	153.66	148.50	153.85	155.59	x
−0.05	183.2	147.1	182.6	186.2	x
−0.1	225.7	146.6	225.8	236.7	x
−0.3	539.14	x	507.63	665.26	x
−0.5	1228.96	x	1237.65	1831.29	x
−0.8	4723.89	x	13773.52		x

Table 9 ARLs–Shewhart control chart (design parameters estimated), TEST 5, no shift

Kendall's τ	Normal	FGM	Clayton	Frank	Gumbel
0.8	165.63	x	206.31	102.07	169.82
0.5	95.92	x	184.54	159.38	111.92
0.3	143.0	x	155.32	249.16	130.95
0.1	323.26	510.36	245.75	413.61	222.92
0	**510.14**	**510.14**	**510.14**	**510.14**	**510.14**
−0.1	321.9	509.96	620.63	598.52	x
−0.3	401.61	x	335.36	567.58	x
−0.5	193.24	x	165.45	379.85	x
−0.8	209.73	x	134.81	201.11	x

copula this dependence is highly non-monotonic. On the other hand, in case of Frank's copula the largest value of the ARL is observed for small negative dependence, and then the value of the ARL decreases with increasing (decreasing) values of τ. Interesting is the case of the FGM copula where in contrast to other considered cases the existing weak dependence does not influence the value of the ARL.

In our experiments we have also calculated the properties of chart with combined decision rules. When Test 1 is combined with Test 5 the ARL in case of estimated design parameters and independence has been evaluated as equal to 293.11. The exact calculations performed for this case, but for known values of design parameters, by Champ and Woodall (1987) gave the value of the ARL equal to 225.44.

The results presented in Table 10 confirm this value, and additionally show how the ARL in the case of this combination of tests depends on the type and the strength of dependence.

Table 10 ARLs–Shewhart control chart (design parameters known), TEST 1+TEST 5, no shift

Kendall's τ	Normal	FGM	Clayton	Frank	Gumbel
0.8	165.44	x	180.76	91.68	168.74
0.5	93.46	x	151.39	122.71	105.92
0.3	120.51	x	130.32	161.28	115.87
0.1	187.27	226.65	169.16	207.24	165.62
0	**225.45**	**225.45**	**225.45**	**225.45**	**225.45**
−0.1	250.41	226.58	244.68	240.57	x
−0.3	222.3	x	201.0	235.94	x
−0.5	168.26	x	145.03	202.07	x
−0.8	209.72	x	131.67	149.55	x

5 Conclusions

The results presented in this paper confirm without any doubts the findings of many authors who considered the bahaviour of Shewhart control charts in case of dependent data described by autoregressive stochastic processes. What seems to be new is the demonstration that the type of dependence, encapsulated in the type of respective copula, plays important role. Moreover, it becomes very clear that the knowledge of the strength of dependence, measured using popular statistical measures of dependence such as Kendall's τ is not sufficient for the evaluation of the properties of the Shewhart control chart.

From the results presented in this paper one can derive the following recommendations. First, it is necessary to detect the existence of dependence in data. This can be done using the Kendall control chart proposed by Hryniewicz and Szediw (2010). Then, it is necessary to indicate the copula which fits to the observed data. Unfortunately, the appropriate tests, such as presented e.g. in the paper by Fermanian (2005), seem to be not simple enough to be used by quality control practitioners. Therefore, a lot has to be done in order to propose even approximate but simple methods for the identification of an actual copula. Then, the future investigations should be concentrated on finding appropriate corrections to classical procedures, similar in spirit to those that has been proposed in case of dependencies described by the normal copula.

References

Albers, W., & Kallenberg, W. C. M. (2004). Estimation in shewhart control charts: Effects and corrections. *Metrika, 59*, 207–234.

Alwan, L. C., & Roberts, H. V. (1988). Time-series modeling for statistical process control. *Journal of Business and Economic Statistics, 6*, 87–95.

Champ, C. W., & Woodall, W. H. (1987). Exact results for shewhart control charts with supplementary runs rules. *Technometrics, 29*, 393–399.

Ferguson, T. S., Genest, C., & Hallin, M. (2000). Kendall's tau for serial dependence. *The Canadian Journal of Statistics, 28*, 587–604.

Fermanian, J.-D. (2005). Goodness-of-fit tests for copulas. *Journal of Multivariate Analysis, 95*, 119–152.
Genest, C., & McKay, R. J. (1986). The joy of copulas: Bivariate distributions with uniform marginals. *American Statistician, 88*, 1034–1043.
Genest, C., Rivest, & L.-P. (1993). Statistical inference procedures for bivariate archimedean copulas. *Journal of the American Statistical Association, 88*, 1034–1043.
Hryniewicz, O., & Szediw, A. (2010). Sequential signals on a control chart based on nonparametric statistical tests. In H.-J. Lenz, P.-T. Wilrich, & W. Schmid (Eds.), *Frontiers in statistical quality control 9* (pp. 99–118). Heidelberg: Physica-Verlag.
Jiang, W., Tsui, K.-L., & Woodall, H. W. (2000). A new SPC monitoring method: The ARMA chart. *Technometrics, 42*, 399–410.
Johnson, R. A., & Bagshaw, M. (1974). The effect of serial correlation on the performance of CUSUM tests. *Technometrics, 16*, 103–112.
Knoth, S., Schmid, W., & Schone, A. (2001). Simultaneous Shewhart-type charts for the mean and the variance of a time series. In H. J. Lenz & P. T. Wilrich (Eds.), *Frontiers in statistical quality control 6* (pp. 61–79). Heidelberg: Physica-Verlag.
Lu, C. W., & Reynolds, M. R., Jr. (1999a). EWMA control charts for monitoring the mean of autocorrelated processes. *Journal of Quality Technology, 31*, 166–188.
Lu, C.W., & Reynolds, M. R., Jr. (1999b). Control charts for monitoring the mean and variance of autocorrelated processes. *Journal of Quality Technology, 31*, 259–274.
Lu, C. W., & Reynolds, M. R., Jr. (2001). CUSUM charts for monitoring an autocorrelated process. *Journal of Quality Technology, 33*, 316–334.
Maragah, H. D., & Woodall, W. H. (1992). The effect of autocorrelation on the retrospective X-chart. *Journal of Statistical Computation and Simulation, 40*, 29–42.
Montgomery, D. C., & Mastrangelo, C. M. (1991). Some statistical process control methods for autocorrelated data. *Journal of Quality Technology, 23*, 179–193.
Nelson, L. S. (1984). The Shewhart control chart – tests for special causes. *Journal of Quality Technology, 16*, 237–239.
Nelsen, R. B. (2006). *An introduction to copulas*. New York: Springer.
Nelsen, R. B., Quesada-Molina, J. J., Rodriguez-Lallena, J. A., & Úbeda-Flores, M. (2009). Kendall distribution functions and associative copulas. *Fuzzy Sets and Systems, 160*, 52–57.
Quesenberry, C. P. (1993). The effect of sample size on estimated limits of $\bar{X}$ and X control charts. *Journal of Quality Technology, 25*, 237–247
Schmid, W. (1995). On the run length of a Shewhart control chart for correlated data. *Statistical Papers, 36*, 111–130.
Schmid, W. (1997). On EWMA charts for time series, In H. J. Lenz & P. T. Wilrich (Eds.), *Frontiers in statistical quality control 5* (pp. 115–137). Heidelberg: Physica-Verlag.
Sklar, A. (1959). Fonctions de répartition á n dimensions et leurs marges. *Publications de l'Intitut de Statistique de l'Universit de Paris, 8*, 229–231
VanBrackle, L. N., III. & Reynolds, M. R., Jr. (1997). EWMA and CUSUM control charts in presence of correlations. *Communications in Statistics Simulation and Computation, 26*, 979–1008.
Vasilopoulos, A. V., & Stamboulis, A. P. (1978). Modification of control limits in the presence of correlation. *Journal of Quality Technology, 10*, 20–30.
Wardell, D. G., Moskowitz, H., & Plante, R. D. (1994). Run-length distributions of special-cause control charts for correlated processes. *Technometrics, 36*, 3–17.
Woodall, W. H., & Montgomery, D. C. (1999). Research issues and ideas in statistical process control. *Journal of Quality Technology, 25*, 188–198.
Yourstone, S. A., & Montgomery, D. C. (1991). Detection of process upsets sample autocorrelation control chart and group autocorrelation control chart applications. *Quality and Reliability Engineering International, 7*, 133–140.
Zhang, N. F. (1998). Statistical control chart for stationary process data. *Technometrics, 40*, 24–38.

Assessing the Impact of Autocorrelation in Misleading Signals in Simultaneous Residual Schemes for the Process Mean and Variance: A Stochastic Ordering Approach

Patrícia Ferreira Ramos, Manuel Cabral Morais, António Pacheco, and Wolfgang Schmid

Abstract Misleading signals (MS) correspond to the misinterpretation of a shift in the process mean (variance) as a shift in the process variance (mean). MS occur when:

- The individual chart for the mean triggers a signal before the one for the variance, even though the process mean is on-target and the variance is off-target;
- The individual chart for the variance triggers a signal before the one for the mean, although the variance is in-control and the process mean is out-of-control.

MS can lead to a misdiagnosis of assignable causes and to incorrect actions to bring the process back to target. Unsurprisingly, the performance assessment of simultaneous schemes for the process mean and variance requires not only the use of run length (RL) related performance measures, but also the probability of misleading signals (PMS). We assess the impact of autocorrelation on the PMS of simultaneous Shewhart and EWMA residual schemes for the mean and variance of stationary AR(1), AR(2) and ARMA(1,1) processes. This assessment is done by means of some stochastic ordering results and some illustrations.

P.F. Ramos (✉)
CEMAT Instituto Superior Técnico, Technical University of Lisbon, Av. Rovisco Pais, 1049-001 Lisboa, Portugal
e-mail: patriciaferreira@ist.utl.pt

M.C. Morais · A. Pacheco
Department of Mathematics and CEMAT Instituto Superior Técnico, Technical University of Lisbon, Av. Rovisco Pais, 1049-001 Lisboa, Portugal
e-mail: maj@math.ist.utl.pt; apacheco@math.ist.utl.pt

W. Schmid
Department of Statistics, European University Viadrina, D-15207 Frankfurt(Oder), Germany
e-mail: schmid@euv-frankfurt-o.de

H.-J. Lenz et al. (eds.), *Frontiers in Statistical Quality Control 10*,
Frontiers in Statistical Quality Control 10, DOI 10.1007/978-3-7908-2846-7_3,

Keywords Statistical process control • Misleading signals • Time series • Simultaneous residual schemes • Stochastic ordering

1 Introduction

Control charts are primarily designed to detect changes like shifts in the process mean or variance. In practice, one individual chart for the process mean and another for the variance are run simultaneously because it is not realistic to believe that only one of the parameters is subject to shifts.

The resulting simultaneous scheme triggers a signal whenever one (or both) of the individual charts triggers a signal. Therefore, it is possible that a shift in one parameter results in a signal triggered by the individual chart designed to monitor the other parameter. This is what John and Bragg (1991) called a misleading signal. These authors identified three types of misleading signals that may occur when we are using a simultaneous scheme with a two-sided chart for the mean (μ) and an upper one-sided chart for the variance (σ^2):

I. The process mean increases but the signal is given by the chart for σ^2 or it is observed in the negative part of the chart for μ;
II. The process mean decreases but the signal is given by the chart for σ^2 or it is observed in the positive part of the chart for μ;
III. The process variance increases but the signal is given by the chart for μ.

Later, Morais and Pacheco (2000) defined a fourth type of error which summarizes types I and II previously described:

IV. The process mean changes but the signal is given by the chart for σ^2.

According to Morais and Pacheco (2000), types III and IV correspond to pure misleading signals since they correspond to a misinterpretation of a shift in the process mean (resp. variance) as a shift in the process variance. As mentioned by Morais and Pacheco (2006) and Knoth et al. (2009), the diagnostic and correction procedures that follow a signal can differ depending on which chart triggers the alarm. Therefore, the occurrence of a misleading signal can lead to inappropriate diagnose and correction measures and consequently to an increase in production (and inspection) costs.

The main question regarding misleading signals should not be whether they occur or not, but rather how frequently they take place. Unsurprisingly, the probability of a misleading signal (PMS) should be considered as an additional performance measure for simultaneous schemes. The PMS has been addressed for i.i.d. and Gaussian output by a few authors (Morais and Pacheco, 2000; Reynolds and Stoumbos, 2001, 2004; Morais, 2002; Morais and Pacheco, 2006). More recently, Antunes (2009) and Knoth et al. (2009) numerically assessed the impact of autocorrelation in misleading signals in simultaneous residual schemes for the process mean and variance of AR(1) output. Ramos et al. (2010) used stochastic ordering to prove that the PMS of Type IV of simultaneous Shewhart (resp. EWMA) residual schemes for the process mean and variance of stationary AR(1) output increases with the autoregressive parameter $\phi \in (-1, 1)$ (resp. $\phi \in (0, 1)$). This paper is a natural extension of Ramos et al. (2010) to general stationary Gaussian processes.

The remainder of this paper is structured as follows. In Sect. 2 we make a brief review of simultaneous residual schemes for the process mean and variance of stationary Gaussian processes. In Sect. 3 we establish stochastic monotonicity properties concerning the run lengths of the constituent individual residual charts. In Sect. 4 monotonicity properties of the probabilities of misleading signals are derived. We end the paper with some concluding remarks in Sect. 5.

2 Simultaneous Residual Schemes for the Process Mean and Variance of Stationary Gaussian Processes

In what follows we consider residual Shewhart and EWMA individual upper one-sided charts to monitor the mean and the variance of a process; these charts are only suited to detect inflations on those two parameters. We assume that the output is governed by a general stationary Gaussian process and later on we focus on AR(1), AR(2) and ARMA(1,1) stationary processes.

2.1 *Output Process and Residuals*

Let us denote by $\{X_{i,j}\}$ the observed process, where i is the sample number and j is the index of the observation within the sample, with the sample size being fixed and equal to n. We shall assume that different samples are independent, however, there is an autocorrelation structure within the sample.

Let $\{Y_{i,j}\}$ represent the target process, that is, the in-control process. $\{Y_{i,j}\}$ is assumed to be a stationary Gaussian process with mean μ_0 and autocovariance function $\{\gamma_h\}$; both μ_0 and $\gamma_0, \gamma_1, \ldots, \gamma_{n-1}$ are known nominal values.

We can write the observed process in terms of the target process as follows:

$$X_{i,j} = \begin{cases} Y_{i,j}, & i < 1 \\ \theta(Y_{i,j} - \mu_0) + \mu_0 + \delta\sqrt{\gamma_0}, & i \geq 1, \end{cases} \tag{1}$$

where $\delta = \frac{E(X_{i,j})-\mu_0}{\sqrt{\gamma_0}}$, $\delta \geq 0$, and $\theta = \sqrt{\frac{Var(X_{i,j})}{\gamma_0}}$, $\theta \geq 1$.

If $\delta > 0$ (resp. $\theta > 1$) then a shift from μ_0 to $\mu_0 + \delta\sqrt{\gamma_0}$ (resp. from γ_0 to $\theta^2\gamma_0$) has occurred just before the first sample has been collected.

According to Knoth and Schmid (2002), the residuals of this process are obtained using the best linear predictor for $X_{i,j}$ given $X_{i,j-1}, X_{i,j-2}, \ldots, X_{i,1}$, which, according to Brockwell and Davis (1991), can be written as

$$\hat{X}_{i,j} = \sum_{k=1}^{j-1} \phi_{j-1,k}(X_{j-k} - \mu_0) + \mu_0, \tag{2}$$

where: $\phi_{j-1,k}$ is the k-th entry of the vector $\phi_{j-1} = (\phi_{j-1,1}, \ldots, \phi_{j-1,j-1})' = \Gamma_{j-1}^{-1}\gamma_{j-1}$ which can be recursively calculated using the Durbin-Levinson algorithm (Brockwell and Davis 1991, p. 169); $\gamma_{j-1} = (\gamma_1, \ldots, \gamma_{j-1})'$ is a vector of covariances and Γ_{j-1} is a covariance matrix, both referring to the in-control process $\{Y_{i,j}\}$.

The corresponding standardized residuals are equal to

$$\hat{\varepsilon}_{i,j} = \frac{X_{i,j} - \hat{X}_{i,j}}{\sqrt{Var_{0,1}(X_{i,j} - \hat{X}_{i,j}))}} = \theta\hat{\xi}_{i,j} + \delta\sqrt{\gamma_0}\frac{1 - \sum_{k=1}^{j-1}\phi_{j-1,k}}{\sqrt{Var_{0,1}(X_{i,j} - \hat{X}_{i,j})}}, \tag{3}$$

where $Var_{0,1}(X_{i,j} - \hat{X}_{i,j})$ represents the in-control variance of the residuals and $\hat{\xi}_{ij} \overset{i.i.d.}{\sim} \mathcal{N}(0,1)$ are the standardized residuals of the in-control process. To simplify the notation let us write

$$\hat{\varepsilon}_{i,j} = \theta\hat{\xi}_{i,j} + \delta\sqrt{\gamma_0} \times b_j. \tag{4}$$

where $b_j = \frac{1-\sum_{k=1}^{j-1}\phi_{j-1,k}}{\sqrt{Var_{0,1}(X_{i,j}-\hat{X}_{i,j})}}$.

Since $Y_{i,j}$ is a stationary Gaussian process, the residuals are independent and normally distributed, therefore we can derive distributions of their sample mean and variance:

$$\overline{\hat{\varepsilon}_i} = \frac{1}{n}\sum_{j=1}^{n}\hat{\varepsilon}_{i,j} \overset{i.i.d.}{\sim} \mathcal{N}\left(\frac{\delta\sqrt{\gamma_0}}{n}\sum_{j=1}^{n} b_j, \frac{\theta^2}{n}\right) \tag{5}$$

$$\frac{n-1}{\theta^2}\hat{S}_i^2 = \frac{1}{\theta^2}\sum_{j=1}^{n}(\hat{\varepsilon}_{i,j} - \overline{\hat{\varepsilon}_i})^2 \overset{i.i.d.}{\sim} \chi^2_{n-1,\nu}, \tag{6}$$

where $\chi^2_{n-1,\nu}$ denotes the noncentral χ^2 distribution with $n-1$ degrees of freedom and noncentrality parameter

$$\nu = \left(\frac{\delta}{\theta}\right)^2 \gamma_0 \left(\sum_{j=1}^{n} b_j^2 - n\overline{b}^2\right), \tag{7}$$

according to Mathai and Provost (1992).

2.2 Simultaneous Shewhart and EWMA Residual Schemes

Capitalizing on the distributional properties of $\bar{\hat{\varepsilon}}_i$ and $\hat{S}_i^2$, we conclude that the run length (RL) of the individual Shewhart residual charts for μ ($\text{RL}_{S-\mu}(\delta, \theta, \mathbf{b})$) and for σ^2 ($\text{RL}_{S-\sigma}(\delta, \theta, \mathbf{b})$), where $\mathbf{b} = (b_j)$, $j = 1, \ldots n$, have geometric distributions with parameters

$$\xi_{S-\mu}(\delta, \theta, \mathbf{b}) = 1 - \Phi\left[\frac{1}{\theta}\left(\gamma_{S-\mu} - \frac{\delta\sqrt{\gamma_0}}{\sqrt{n}}\sum_{j=1}^{n} b_j\right)\right] \tag{8}$$

$$\xi_{S-\sigma}(\delta, \theta, \mathbf{b}) = 1 - F_{\chi^2_{n-1,v}}\left[\frac{n-1}{\theta^2}\left(1 + \gamma_{S-\sigma}\sqrt{\frac{2}{n-1}}\right)\right]. \tag{9}$$

In addition, since $\bar{\hat{\varepsilon}}_i$ and $\hat{S}_i^2$ are independent, the RL of the simultaneous Shewhart residual scheme also has geometric distribution with parameter

$$\xi_{S-\mu,\sigma}(\delta, \theta, \mathbf{b}) = \xi_{S-\mu}(\delta, \theta, \mathbf{b}) + \xi_{S-\sigma}(\delta, \theta, \mathbf{b}) - \xi_{S-\mu}(\delta, \theta, \mathbf{b}) \times \xi_{S-\sigma}(\delta, \theta, \mathbf{b}). \tag{10}$$

As for the EWMA individual charts and simultaneous schemes, the Markov chain approach (Brook and Evans, 1972) provides the following approximations to the survival functions of the run lengths $RL_{E-\mu}(\delta, \theta, \mathbf{b})$, $RL_{E-\sigma}(\delta, \theta, \mathbf{b})$ and $RL_{E-\mu,\sigma}(\delta, \theta, \mathbf{b})$:

$$\overline{F}_{RL_{E-\mu}(\delta,\theta,\mathbf{b})}(m) \simeq \mathbf{e}_\mu^T \times [\mathbf{Q}_\mu(\delta, \theta, \mathbf{b}; x_\mu)]^m \times \mathbf{1}_\mu \tag{11}$$

$$\overline{F}_{RL_{E-\sigma}(\delta,\theta,\mathbf{b})}(m) \simeq \mathbf{e}_\sigma^T \times [\mathbf{Q}_\sigma(\delta, \theta, \mathbf{b}; x_\sigma)]^m \times \mathbf{1}_\sigma \tag{12}$$

$$\overline{F}_{RL_{E-\mu,\sigma}(\delta,\theta,\mathbf{b})}(m) = \overline{F}_{RL_{E-\mu}(\delta,\theta,\mathbf{b})}(m) \times \overline{F}_{RL_{E-\sigma}(\delta,\theta,\mathbf{b})}(m) \tag{13}$$

$$\simeq (\mathbf{e}_\mu^T \times [\mathbf{Q}_\mu(\delta, \theta, \mathbf{b}; x_\mu)]^m \times \mathbf{1}_\mu) \times (\mathbf{e}_\sigma^T \times [\mathbf{Q}_\sigma(\delta, \theta, \mathbf{b}; x_\sigma)]^m \times \mathbf{1}_\sigma), \tag{14}$$

for $m = 0, 1, 2, \ldots$. Note that the approximations are based on $(x_\mu + 1)$ (resp. $(x_\sigma + 1)$) transient states associated with the individual EWMA residual chart for μ (resp. σ^2) (Table 1). Moreover,

- $\mathbf{e}_\mu$ (resp. $\mathbf{e}_\sigma$) denotes the first (resp. $(x_\sigma + 1)/UCL_{E-\sigma}$)th) vector of the orthonormal basis for $\mathbb{R}^{x_\mu+1}$ (resp. $\mathbb{R}^{x_\sigma+1}$), associated with the state related to the initial value of the control statistic;
- $\mathbf{1}_\mu$ (resp. $\mathbf{1}_\sigma$) is a column vector of $(x_\mu + 1)$ (resp. $(x_\sigma + 1)$) ones;
- The entries of the sub-stochastic matrix $\mathbf{Q}_\mu(\delta, \theta, \mathbf{b}; x_\mu)$ follow from an adaptation of the ones defined in Ramos et al. (2010); and
- The entries of the sub-stochastic matrix $\mathbf{Q}_\sigma(\delta, \theta, \mathbf{b}; x_\sigma))$ result from an adaptation of the ones defined in Knoth et al. (2009).

Table 1 Control statistics and upper control limits of the Shewhart ($S-\mu$ and $S-\sigma$) and EWMA ($E-\mu$ and $E-\sigma$) individual residual charts (the lower control limits are all equal to zero)

Control statistics	Control limits
$\max\{0, \bar{\hat{\varepsilon}}_i\}$	$UCL_{S-\mu} = \frac{\gamma_{S-\mu}}{\sqrt{n}}$
$\hat{S}_i^2$	$UCL_{S-\sigma} = 1 + \gamma_{S-\sigma}\sqrt{\frac{2}{n-1}}$
$W_{\bar{\hat{\varepsilon}},i} = \begin{cases} 0, & i=0 \\ \max\{0, (1-\lambda_\mu)W_{\bar{\hat{\varepsilon}},i-1} + \lambda_\mu \bar{\hat{\varepsilon}}_i\}, & i>0 \end{cases}$	$UCL_{E-\mu} = \gamma_{E-\mu}\sqrt{\frac{\lambda_\mu}{n(2-\lambda_\mu)}}$
$W_{\hat{S}_i^2,i} = \begin{cases} 1, & i=0 \\ (1-\lambda_\sigma)W_{\hat{S}_i^2,i-1} + \lambda_\sigma \hat{S}_i^2, & i>0 \end{cases}$	$UCL_{E-\sigma} = 1 + \gamma_{E-\sigma}\sqrt{\frac{2\lambda_\sigma}{(n-1)(2-\lambda_\sigma)}}$

Moreover, the corresponding left partial sums are given by:

$$a_{\mu,ij}(\delta,\theta,\mathbf{b};x_\mu) = \Phi\left\{\frac{1}{\theta}\left[\frac{\gamma_{E-\mu}[(j+1)-(1-\lambda_\mu)(i+1/2)]}{(x_\mu+1)\sqrt{\lambda_\mu(2-\lambda_\mu)}} - \frac{\delta\sqrt{\gamma_0}}{\sqrt{n}}\sum_{k=1}^{n} b_k\right]\right\}, \quad (15)$$

for $i, j = 0, \ldots, x_\mu$;

$$a_{\sigma,ij}(\delta,\theta,\mathbf{b};x_\sigma) = F_{\chi^2_{n-1,\nu}}\left[\frac{(n-1)[(j+1)-(1-\lambda_\sigma)(i+1/2)]}{\theta^2\lambda_\sigma(x_\sigma+1)}\left(1+\gamma_{E-\sigma}\sqrt{\frac{2\lambda_\sigma}{(n-1)(2-\lambda_\sigma)}}\right)\right], \quad (16)$$

for $i, j = 0, \ldots, x_\sigma$. We ought to note that results Eq. 8–16 are quite similar to the ones derived by Knoth et al. (2009) and Ramos et al. (2010): the autocorrelation parameter has been replaced by the vector $\mathbf{b} = (b_j)$, $j = 1, \ldots, n$.

2.3 *Simultaneous Shewhart and EWMA Residual Schemes for the Stationary AR(2) and ARMA(1,1) Processes*

We now address to the stationary AR(2) and ARMA(1,1) processes. The characterization of these processes and the properties of the distributions of the control statistics in Eqs. 5 and 6 are condensed in Tables 2 and 3. The expressions of: the parameters of the run lengths of the individual Shewhart residual charts and the left partial sums of the entries of the sub-stochastic matrices $\mathbf{Q}_\mu(\delta,\theta,\mathbf{b};x_\mu)$ and $\mathbf{Q}_\sigma(\delta,\theta,\mathbf{b};x_\sigma)$ for these two processes can be found in Tables 4 and 5.

Table 2 Characterization of AR(2) processes and properties of the control statistics

Equation	Stationarity conditions
$X_{i,j} = \mu_0 + \phi_1(X_{i,j-1} - \mu_0) + \phi_2(X_{i,j-2} - \mu_0) + \varepsilon_{i,j}$	$\phi_1 + \phi_2 < 1, \phi_2 - \phi_1 < 1$ and $-1 < \phi_2 < 1$
$E(\overline{\hat{\varepsilon}_i})$	$Var(\overline{\hat{\varepsilon}_i})$
$\frac{\delta}{n}\left\{1 + \frac{1-\phi_1}{\sqrt{1-\frac{\phi_1^2(1+\phi_2)}{1-\phi_2}}} + (n-2)\frac{1-\phi_1-\phi_2}{\sqrt{\frac{1+\phi_2}{1-\phi_2}[(1-\phi_2)^2-\phi_1^2]}}\right\}$	$\frac{\theta^2}{n}$
Noncentrality parameter (ν)	
$(\frac{\delta}{\theta})^2\left(1 - \frac{n}{\delta^2}E(\overline{\hat{\epsilon}_i})^2 + (1-\phi_1)^2/\left(1 - \frac{\phi_1^2(1+\phi_2)}{1-\phi_2}\right) + (n-2)\frac{(1-\phi_1-\phi_2)^2}{\frac{1+\phi_2}{1-\phi_2}[(1-\phi_2)^2-\phi_1^2]}\right)^2$	

Table 3 Characterization of ARMA(1,1) processes and properties of the control statistics

Equation	Stationarity and invertibility conditions
$X_{i,j} = \mu_0 + \phi(X_{i,j-1} - \mu_0) + \varepsilon_{i,j} - \alpha\varepsilon_{i,j-1}$	$\phi, \alpha \in (-1, 1)$
$E(\overline{\hat{\varepsilon}_i})$	$Var(\overline{\hat{\varepsilon}_i})$
$\frac{\delta}{n}\left\{1 + \frac{1}{\delta}\sum_{j=2}^{n}\frac{\delta\left[1+\left(1-\alpha^{j-1}\right)(\alpha-\phi)/(1-\alpha)\right]\sqrt{1+\alpha^2-2\alpha\phi}}{\eta}\right\}$	$\frac{\theta^2}{n}$
Noncentrality parameter (ν)	
$\frac{1}{\theta^2}\left(\left[\delta - E(\overline{\hat{\varepsilon}_i})\right]^2 + \sum_{j=2}^{n}\left\{\delta\left[1 + \left(1-\alpha^{j-1}\right)(\alpha-\phi)/(1-\alpha)\right]\sqrt{1+\alpha^2-2\alpha\phi}/\eta - E(\overline{\hat{\varepsilon}_i})\right\}^2\right)$	

where $\eta = \sqrt{1 + \alpha^{2(j-2)} + \alpha^{2(j-1)}\phi - 2\alpha^{2j-1}\phi - \phi^2 - 2\alpha^{j+1}\phi^{j-1} - 2\alpha^{j-1}\phi^{j+1} + 2(\alpha-\phi)^2(\alpha\phi)^{j-1} + 4(\alpha\phi)^j}$

Table 4 Parameters of the run lengths of the individual Shewhart residual charts and the left partial sums of the entries of the sub-stochastic matrices $\mathbf{Q}_\mu(\delta, \theta, \mathbf{b}; x_\mu)$ and $\mathbf{Q}_\sigma(\delta, \theta, \mathbf{b}; x_\sigma)$ for a stationary AR(2) process

Shewhart scheme

$$\xi_{S-\mu}(\delta,\theta,\phi_1,\phi_2) = 1 - \Phi\left[\frac{1}{\theta}\left(\gamma_{S-\mu} - \frac{\delta}{\sqrt{n}}\left\{1 + \frac{1-\phi_1}{\sqrt{1-\frac{\phi_1^2(1+\phi_2)}{1-\phi_2}}} + (n-2)\frac{1-\phi_1-\phi_2}{\sqrt{\frac{1+\phi_2}{1-\phi_2}[(1-\phi_2)^2-\phi_1^2]}}\right\}\right)\right] \tag{17}$$

$$\xi_{S-\sigma}(\delta,\theta,\phi_1,\phi_2) = 1 - F_{\chi^2_{n-1,\nu}}\left[\frac{\left(1+\gamma_{S-\sigma}\sqrt{\frac{2}{n-1}}\right)(n-1)}{\theta^2}\right] \tag{18}$$

EWMA scheme

$$a_{\mu,ij}(\delta,\theta,\phi_1,\phi_2;x_\mu) = \Phi\left\{\frac{1}{\theta}\left[\frac{\gamma_{E-\mu}[(j+1)-(1-\lambda_{E-\mu})(i+1/2)]}{(x_\mu+1)\sqrt{\lambda_{E-\mu}(2-\lambda_{E-\mu})}} - \frac{\delta}{\sqrt{n}}\left(1+\frac{1-\phi_1}{\sqrt{1-\frac{\phi_1^2(1+\phi_2)}{1-\phi_2}}} + (n-2)\frac{1-\phi_1-\phi_2}{\sqrt{\frac{1+\phi_2}{1-\phi_2}[(1-\phi_2)^2-\phi_1^2]}}\right)\right]\right\} \tag{19}$$

$$a_{\sigma,ij}(\delta,\theta,\phi_1,\phi_2;x_\sigma) = F_{\chi^2_{n-1,\nu}}\left[\frac{(n-1)[(j+1)-(1-\lambda_{E-\sigma})(i+1/2)]}{\theta^2\lambda_{E-\sigma}(x_\sigma+1)}\left(1+\gamma_{E-\sigma}\sqrt{\frac{2\lambda_{E-\sigma}}{(n-1)(2-\lambda_{E-\sigma})}}\right)\right] \tag{20}$$

Table 5 Parameters of the run lengths of the individual Shewhart residual charts and the left partial sums of the entries of the sub-stochastic matrices $\mathbf{Q}_\mu(\delta, \theta, \mathbf{b}; x_\mu)$ and $\mathbf{Q}_\sigma(\delta, \theta, \mathbf{b}; x_\sigma)$ for a stationary and invertible ARMA(1,1) process

Shewhart scheme

$$\xi_{S-\mu}(\delta,\theta,\phi,\alpha) = 1 - \Phi\left\{\frac{1}{\theta}\left[\gamma_{S-\mu} - \sqrt{n}E(\overline{\hat{\varepsilon}}_i)\right]\right\} \tag{21}$$

$$\xi_{S-\sigma}(\delta,\theta,\phi,\alpha) = 1 - F_{\chi^2_{n-1,\nu}}\left[\frac{\left(1+\gamma_{S-\sigma}\sqrt{\frac{2}{n-1}}\right)(n-1)}{\theta^2}\right] \tag{22}$$

EWMA scheme

$$a_{\mu,ij}(\delta,\theta,\phi,\alpha;x_\mu) = \Phi\left(\frac{1}{\theta}\left\{\frac{\gamma_{E-\mu}[(j+1)-(1-\lambda_{E-\mu})(i+1/2)]}{(x_\mu+1)\sqrt{\lambda_{E-\mu}(2-\lambda_{E-\mu})}} - \sqrt{n}E(\overline{\hat{\varepsilon}}_i)\right\}\right) \tag{23}$$

$$a_{\sigma,ij}(\delta,\theta,\phi,\alpha;x_\sigma) = F_{\chi^2_{n-1,\nu}}\left[\left\{\frac{(n-1)[(j+1)-(1-\lambda_{E-\sigma})(i+1/2)]}{\theta^2\lambda_{E-\sigma}(x_\sigma+1)}\left[1+\gamma_{E-\sigma}\sqrt{\frac{2\lambda_{E-\sigma}}{(n-1)(2-\lambda_{E-\sigma})}}\right]\right\}\right] \tag{24}$$

2.4 Probability of a Misleading Signal

When a simultaneous scheme is used to monitor the mean and variance of a process, an alarm is classified as a misleading signal of Type III if μ is on-target ($\delta = 0$) and σ^2 is off-target ($\theta > 1$) but the chart for μ is the first to signal. The PMS of Type III is defined as follows

$$PMS_{III}(\theta,\mathbf{b}) = P[RL_\mu(0,\theta,\mathbf{b}) < RL_\sigma(0,\theta,\mathbf{b})] \tag{25}$$

$$= \sum_{i=1}^{+\infty}\left[\overline{F}_{RL_\mu(0,\theta,\mathbf{b})}(i-1) - \overline{F}_{RL_\mu(0,\theta,\mathbf{b})}(i)\right] \times \overline{F}_{RL_\sigma(0,\theta,\mathbf{b})}(i), \tag{26}$$

for $\theta > 1$.

Similarly, a MS of Type IV is said to have happened if μ is off-target ($\delta > 0$) and σ^2 is on-target ($\theta = 1$) but the chart for σ^2 is the first to signal. Thus, the PMS of Type IV equals

$$PMS_{IV}(\delta,\mathbf{b}) = P[RL_\sigma(\delta,1,\mathbf{b}) < RL_\mu(\delta,1,\mathbf{b})] \tag{27}$$

$$= \sum_{i=1}^{+\infty}\left[\overline{F}_{RL_\sigma(\delta,1,\mathbf{b})}(i-1) - \overline{F}_{RL_\sigma(\delta,1,\mathbf{b})}(i)\right] \times \overline{F}_{RL_\mu(\delta,1,\mathbf{b})}(i), \tag{28}$$

for $\delta > 0$.

We should note that $\text{PMS}_{III}(\theta, \mathbf{b})$ does not depend on $\mathbf{b}$. In fact, as the noncentrality parameter is null when $\delta = 0$, then the survival functions in Eq. 26 depend only on θ for both Shewhart and EWMA schemes. For this reason this probability is simply denoted by $\text{PMS}_{III}(\theta)$ from now on.

3 Stochastic Monotonicity Properties

The three following stochastic ordering results play a major role in the assessment of the stochastic monotonicity properties of the RL of the individual residual charts, and therefore in the monotonicity properties of the PMS. These results have been previously proved by Morais (2002, p. 16); Morais and Pacheco (1998) and Ramos et al. (2010), respectively, and are stated without proof in the next subsection.

3.1 Some Auxiliary Results

Lemma 1 (Morais 2002, p. 16). – *Let $RL(\Delta) \sim geometric(\xi(\Delta))$ be the RL of a Shewhart type control chart where $\xi(\Delta)$ represents the probability of detecting a shift of magnitude Δ or of triggering a signal when a model parameter is equal to Δ. Then, if $\xi(\Delta)$ increases with Δ, $RL(\Delta) \downarrow_{st}$ with Δ, i.e., $\overline{F}_{RL(\Delta)}(m) \downarrow$ with Δ, for all m.*

Lemma 2 (Morais and Pacheco 1998). – *Let $\{S_N(\Delta),\ N \in \mathbb{N}_0\}$ be an absorbing Markov chain with state space $\{0, 1, \ldots, x, x+1\}$, initial value $S_0(\Delta) = u$ and governed by the transition matrix $P(\Delta) = [p_{ij}(\Delta)]$. If all left partial sums of $P(\Delta)$, $\sum_{j=1}^{k} p_{ij}(\Delta)$, decrease with i and decrease with Δ then $RL(\Delta) \downarrow_{st}$ with Δ.*

Lemma 3 (Ramos et al. 2010). – *Let $X_\nu \sim \chi^2_{n-1,\nu}$ be a continuous random variable with noncentral chi-squared distribution. Then, $X_\nu \uparrow_{st}$ with ν, that is, $F_{\chi^2_{n-1,\nu}}(x)$ is a decreasing function of ν for any x.*

Besides these three lemmas we need to state and prove that the sum $\sum_{k=1}^{n} b_k$ is positive.

Lemma 4. *For any stationary Gaussian process, $\sum_{k=1}^{n} b_k > 0$.*

Proof. First we note that

$$b_1 = \left[E(X_1 - \hat{X}_1)^2\right]^{-1} = \gamma_0^{-1} > 0, \tag{29}$$

according to Brockwell and Davis (1991, p. 170). Then note that, applying the Durbin-Levinson algorithm (Brockwell and Davis, 1991), b_k can be written as follows:

$$b_k = \frac{1 - \sum_{v=1}^{k-1} \phi_{k-1,v}}{\sqrt{Var(X_k - \hat{X}_k)}} \tag{30}$$

$$= \frac{1 - \sum_{v=1}^{k-2} (\phi_{k-2,v} - \phi_{k-1,k-1}\phi_{k-2,k-1-v}) - \phi_{k-1,k-1}}{\sqrt{Var(X_k - \hat{X}_k)}} \tag{31}$$

$$= b_{k-1}\frac{\sqrt{Var(X_{k-1} - \hat{X}_{k-1})}}{\sqrt{Var(X_k - \hat{X}_k)}} - \phi_{k-1,k-1}b_{k-1}\frac{\sqrt{Var(X_{k-1} - \hat{X}_{k-1})}}{\sqrt{Var(X_k - \hat{X}_k)}} \tag{32}$$

$$= b_{k-1}\frac{\sqrt{Var(X_{k-2} - \hat{X}_{k-2})}}{\sqrt{Var(X_{k-1} - \hat{X}_{k-1})}}(1 - \phi_{k-1,k-1}) \geq 0. \tag{33}$$

by induction, using the fact that $b_1 > 0$. Thus, we get $\sum_{k=1}^{n} b_k > 0$.

3.2 *The Influence of Autocorrelation on the RL of the Individual Residual Charts – AR(2) Process*

Now we state some stochastic monotonicity properties of the run length of the individual charts in terms of the parameters ϕ_1 and ϕ_2 of a stationary AR(2) process. Recall that this process is stationary if $(\phi_1, \phi_2) \in S_{AR(2)} = \{(\phi_1, \phi_2) : \phi_1 + \phi_2 < 1, \phi_2 - \phi_2 < 1, -1 < \phi_2 < 1\}$ and that Lemma 1 plays an important role in the proof of those stochastic monotonicity properties.

Theorem 1. *The following results are valid for the RL of upper one-sided individual Shewhart and EWMA residual charts for the mean and variance of an AR(2) process.*

	μ	σ
Shewhart	*(2.1)* $RL_{S-\mu}(\delta, \theta, \phi_1, \phi_2) \uparrow_{st}$ *with* ϕ_1	*(2.4)* $RL_{S-\sigma}(\delta, \theta, \phi_1, \phi_2) \uparrow_{st}$ *with* ϕ_1, $(\phi_1, \phi_2) \in C$
	(2.2) $RL_{S-\mu}(\delta, \theta, \phi_1, \phi_2) \downarrow_{st}$ *with* ϕ_2, $(\phi_1, \phi_2) \in A$	*(2.5)* $RL_{S-\sigma}(\delta, \theta, \phi_1, \phi_2) \downarrow_{st}$ *with* ϕ_1, $(\phi_1, \phi_2) \in D$
	(2.3) $RL_{S-\mu}(\delta, \theta, \phi_1, \phi_2) \uparrow_{st}$ *with* ϕ_2, $(\phi_1, \phi_2) \in B$	*(2.6)* $RL_{S-\sigma}(\delta, \theta, \phi_1, \phi_2) \uparrow_{st}$ *with* ϕ_2, $(\phi_1, \phi_2) \in E$
		(2.7) $RL_{S-\sigma}(\delta, \theta, \phi_1, \phi_2) \downarrow_{st}$ *with* ϕ_2, $(\phi_1, \phi_2) \in F$
EWMA	*(2.8)* $RL_{E-\mu}(\delta, \theta, \phi_1, \phi_2) \uparrow_{st}$ *with* ϕ_1	*(2.11)* $RL_{E-\sigma}(\delta, \theta, \phi_1, \phi_2) \uparrow_{st}$ *with* ϕ_1, $(\phi_1, \phi_2) \in C$
	(2.9) $RL_{E-\mu}(\delta, \theta, \phi_1, \phi_2) \downarrow_{st}$ *with* ϕ_2, $(\phi_1, \phi_2) \in A$	*(2.12)* $RL_{E-\sigma}(\delta, \theta, \phi_1, \phi_2) \downarrow_{st}$ *with* ϕ_1, $(\phi_1, \phi_2) \in D$
	(2.10) $RL_{E-\mu}(\delta, \theta, \phi_1, \phi_2) \uparrow_{st}$ *with* ϕ_2, $(\phi_1, \phi_2) \in B$	*(2.13)* $RL_{E-\sigma}(\delta, \theta, \phi_1, \phi_2) \uparrow_{st}$ *with* ϕ_2, $(\phi_1, \phi_2) \in E$
		(2.14) $RL_{E-\sigma}(\delta, \theta, \phi_1, \phi_2) \downarrow_{st}$ *with* ϕ_2, $(\phi_1, \phi_2) \in F$

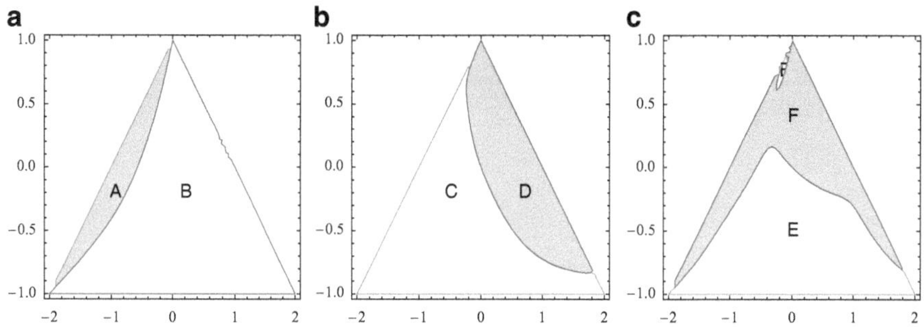

Fig. 1 Subregions of the stationary region of the $AR(2)$ process (the *shaded areas* correspond to sign=1). (**a**) sign $\left(\frac{\partial \xi_{S-\mu}(\delta,\theta,\phi_1,\phi_2)}{\partial \phi_2}\right)$; (**b**) sign $\left(\frac{\partial \xi_{RL-\sigma}(\delta,\theta,\phi_1,\phi_2)}{\partial \phi_1}\right)$; (**c**) sign $\left(\frac{\partial \xi_{RL-\sigma}(\delta,\theta,\phi_1,\phi_2)}{\partial \phi_2}\right)$

where the subregions A through F are defined as follows:

Subregion
$A = \{(\phi_1, \phi_2) \in S_{AR(2)} : \frac{\partial \xi_{S-\mu}(\delta,\theta,\phi_1,\phi_2)}{\partial \phi_2} > 0\}$
$B = \{(\phi_1, \phi_2) \in S_{AR(2)} : \frac{\partial \xi_{S-\mu}(\delta,\theta,\phi_1,\phi_2)}{\partial \phi_2} < 0\}$
$C = \{(\phi_1, \phi_2) \in S_{AR(2)} : \frac{\partial \xi_{S-\sigma}(\delta,\theta,\phi_1,\phi_2)}{\partial \phi_1} < 0\}$
$D = \{(\phi_1, \phi_2) \in S_{AR(2)} : \frac{\partial \xi_{S-\sigma}(\delta,\theta,\phi_1,\phi_2)}{\partial \phi_1} > 0\}$
$E = \{(\phi_1, \phi_2) \in S_{AR(2)} : \frac{\partial \xi_{S-\sigma}(\delta,\theta,\phi_1,\phi_2)}{\partial \phi_2} < 0\}$
$F = \{(\phi_1, \phi_2) \in S_{AR(2)} : \frac{\partial \xi_{S-\sigma}(\delta,\theta,\phi_1,\phi_2)}{\partial \phi_2} > 0\}$

These subregions were obtained by using the Mathematica function RegionPlot.

where the shaded areas in Fig. 1 represent the subregions where the derivatives of $\xi_{S-\mu}(\delta,\theta,\phi_1,\phi_2)$ and $\xi_{S-\sigma}(\delta,\theta,\phi_1,\phi_2)$ of the individual Shewhart residual charts and of the left partial sums defined in Eqs. 4–4 are positive and the remaining areas correspond to the set of values (ϕ_1,ϕ_2) for which those derivatives are negative. Please note that the sign of the derivatives of $\xi_{S-\mu}(\delta,\theta,\phi_1,\phi_2)$ and $\xi_{S-\sigma}(\delta,\theta,\phi_1,\phi_2)$ is symmetrical to the sign of the derivatives of the corresponding left partial sums.

Proof. To prove result (2.1) we must analyse the sign of $\frac{\partial \xi_{S-\mu}(\delta,\theta,\phi_1,\phi_2)}{\partial \phi_1}$. A close inspection to the expression of this probability defined in Eq. 4 leads us to the conclusion that the sign of $\frac{\partial \xi_{S-\mu}(\delta,\theta,\phi_1,\phi_2)}{\partial \phi_1}$ is equal to the sign of the derivative

$$\frac{\partial}{\partial \phi_1}\left[1 + \frac{1-\phi_1}{\sqrt{1 - \frac{\phi_1^2(1+\phi_2)}{1-\phi_2}}} + (n-2)\frac{1-\phi_1-\phi_2}{\sqrt{\frac{1+\phi_2}{1-\phi_2}[(1-\phi_2)^2 - \phi_1^2]}}\right] \tag{34}$$

which is equal to

$$\frac{(n-2)\phi_1(1-\phi_1-\phi_2)(1+\phi_2)}{(1-\phi_2)\sqrt{\left(\frac{[(1-\phi_2)^2-\phi_1^2](1+\phi_2)}{1-\phi_2}\right)^3}} - \frac{n-2}{\sqrt{\frac{[(1-\phi_2)^2-\phi_1^2](1+\phi_2)}{1-\phi_2}}} + \frac{(1-\phi_1)\phi_1(1+\phi_2)}{(1-\phi_2)\sqrt{\left(1-\frac{\phi_1^2(1+\phi_2)}{1-\phi_2}\right)^3}} - \frac{1}{\sqrt{1-\frac{\phi_1^2(1+\phi_2)}{1-\phi_2}}}. \quad (35)$$

On one hand, simplifying the first two summands of Eq. 35, we get

$$\frac{(n-2)(1+\phi_2)(\phi_1+\phi_2-1)}{\sqrt{\left(\frac{[(1-\phi_2)^2-\phi_1^2](1+\phi_2)}{1-\phi_2}\right)^3}} \quad (36)$$

which is negative since one of the stationarity conditions is $\phi_1 + \phi_2 - 1 < 0$ and all other factors are positive. On the other hand, the simplification the last two summands of Eq. 35 leads to

$$\frac{\phi_1+\phi_2+\phi_1\phi_2-1}{(1-\phi_2)\sqrt{\left(1-\frac{\phi_1^2(1+\phi_2)}{1-\phi_2}\right)^3}} \quad (37)$$

whose numerator is also negative when the stationarity conditions are valid. Thus proving result (2.1).

The proof of result (2.8) is quite similar. The sign of the derivative of the left partial sums in Eq. 4 is symmetrical to the sign of Eq. 36. Therefore, $RL_{E-\mu}(\delta,\theta,\phi_1,\phi_2) \uparrow_{st}$ with ϕ_1.

The remaining results follow immediately from Lemma 1. However, we were unable to fully simplify conditions such as $\frac{\partial \xi_{S-\mu}(\delta,\theta,\phi_1,\phi_2)}{\partial \phi_2} > 0$ and had to rely on the Mathematica function *RegionPlot* to identify the subregions A to F.

We ought to point out that results for the AR(1) process can be derived by considering $\phi_2 = 0$ in the previous theorem.

3.3 The Impact of Autocorrelation on the RL of the Individual Residual Charts – ARMA(1,1) Process

The derivation of the stochastic monotonicity properties of the RL of the individual charts for the ARMA(1,1) process is similar to the one for the AR(2) process. These stochastic monotonicity properties refer to the parameters ϕ and α in the stationarity and invertibility region $S_{ARMA(1,1)} = \{(\phi,\alpha) : -1 < \phi,\alpha < 1\}$.

Theorem 2. *The following results are valid for the RL of upper one-sided individual Shewhart and EWMA residual charts for the mean and variance of a stationary and invertible ARMA(1,1) process.*

	μ	σ
Shewhart	(3.1) $RL_{S-\mu}(\delta,\theta,\phi,\alpha)\downarrow_{st}$ with ϕ, $(\phi,\alpha)\in G$	(3.5) $RL_{S-\sigma}(\delta,\theta,\phi,\alpha)\downarrow_{st}$ with ϕ, $(\phi,\alpha)\in K$
	(3.2) $RL_{S-\mu}(\delta,\theta,\phi,\alpha)\uparrow_{st}$ with ϕ, $(\phi,\alpha)\in H$	(3.6) $RL_{S-\sigma}(\delta,\theta,\phi,\alpha)\uparrow_{st}$ with ϕ, $(\phi,\alpha)\in L$
	(3.3) $RL_{S-\mu}(\delta,\theta,\phi,\alpha)\downarrow_{st}$ with α, $(\phi,\alpha)\in I$	(3.7) $RL_{S-\sigma}(\delta,\theta,\phi,\alpha)\downarrow_{st}$ with α, $(\phi,\alpha)\in M$
	(3.4) $RL_{S-\mu}(\delta,\theta,\phi,\alpha)\uparrow_{st}$ with α, $(\phi,\alpha)\in J$	(3.8) $RL_{S-\sigma}(\delta,\theta,\phi,\alpha)\uparrow_{st}$ with α, $(\phi,\alpha)\in N$
EWMA	(3.9) $RL_{E-\mu}(\delta,\theta,\phi,\alpha)\downarrow_{st}$ with ϕ, $(\phi,\alpha)\in G$	(3.13) $RL_{E-\sigma}(\delta,\theta,\phi,\alpha)\downarrow_{st}$ with ϕ, $(\phi,\alpha)\in K$
	(3.10) $RL_{E-\mu}(\delta,\theta,\phi,\alpha)\uparrow_{st}$ with ϕ, $(\phi,\alpha)\in H$	(3.14) $RL_{E-\sigma}(\delta,\theta,\phi,\alpha)\uparrow_{st}$ with ϕ, $(\phi,\alpha)\in L$
	(3.11) $RL_{E-\mu}(\delta,\theta,\phi,\alpha)\downarrow_{st}$ with α, $(\phi,\alpha)\in I$	(3.15) $RL_{E-\sigma}(\delta,\theta,\phi,\alpha)\downarrow_{st}$ with α, $(\phi,\alpha)\in M$
	(3.12) $RL_{E-\mu}(\delta,\theta,\phi,\alpha)\uparrow_{st}$ with α, $(\phi,\alpha)\in J$	(3.16) $RL_{E-\sigma}(\delta,\theta,\phi,\alpha)\uparrow_{st}$ with α, $(\phi,\alpha)\in N$

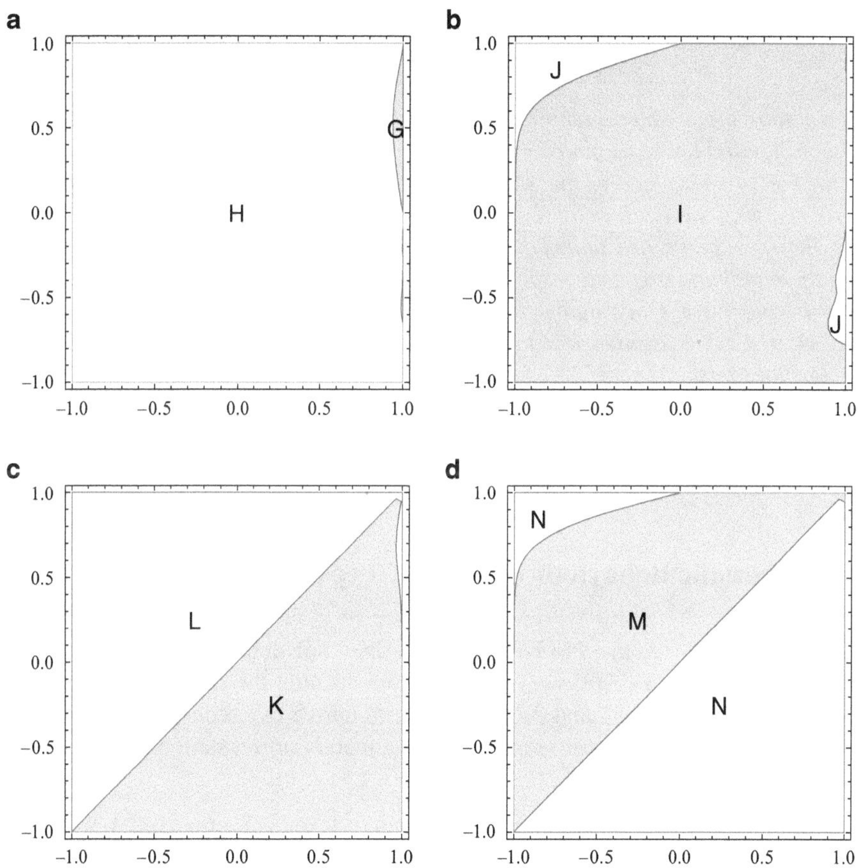

Fig. 2 Decompositions of the stationary region of the stationary and invertible $ARMA(1,1)$ process (the *shaded areas* correspond to sign=1). (**a**) sign $\left(\frac{\partial\xi_{S-\mu}(\delta,\theta,\phi,\alpha)}{\partial\phi}\right)$; (**b**) sign $\left(\frac{\partial\xi_{S-\mu}(\delta,\theta,\phi,\alpha)}{\partial\alpha}\right)$; (**c**) sign $\left(\frac{\partial\xi_{S-\sigma}(\delta,\theta,\phi,\alpha)}{\partial\phi}\right)$; (**d**) sign $\left(\frac{\partial\xi_{S-\sigma}(\delta,\theta,\phi,\alpha)}{\partial\alpha}\right)$

where the subregions H through N are characterized as follows:

Subregion
$G = \{(\phi,\alpha) \in S_{ARMA(1,1)} : \frac{\partial \xi_{S-\mu}(\delta,\alpha,\phi,\alpha)}{\partial \phi} > 0\}$
$H = \{(\phi,\alpha) \in S_{ARMA(1,1)} : \frac{\partial \xi_{S-\mu}(\delta,\alpha,\phi,\alpha)}{\partial \phi} < 0\}$
$I = \{(\phi,\alpha) \in S_{ARMA(1,1)} : \frac{\partial \xi_{S-\mu}(\delta,\alpha,\phi,\alpha)}{\partial \alpha} > 0\}$
$J = \{(\phi,\alpha) \in S_{ARMA(1,1)} : \frac{\partial \xi_{S-\mu}(\delta,\alpha,\phi,\alpha)}{\partial \alpha} < 0\}$
$K = \{(\phi,\alpha) \in S_{ARMA(1,1)} : \frac{\partial \xi_{S-\sigma}(\delta,\alpha,\phi,\alpha)}{\partial \phi} > 0\}$
$L = \{(\phi,\alpha) \in S_{ARMA(1,1)} : \frac{\partial \xi_{S-\sigma}(\delta,\alpha,\phi,\alpha)}{\partial \phi} < 0\}$
$M = \{(\phi,\alpha) \in S_{ARMA(1,1)} : \frac{\partial \xi_{S-\sigma}(\delta,\alpha,\phi,\alpha)}{\partial \alpha} > 0\}$
$N = \{(\phi,\alpha) \in S_{ARMA(1,1)} : \frac{\partial \xi_{S-\sigma}(\delta,\alpha,\phi,\alpha)}{\partial \alpha} < 0\}$

These subregions were identified using once again the Mathematica function RegionPlot and are represented in Fig. 2.
As in the previous figure, the shaded areas represent the subregions where the corresponding derivatives of $\xi_{\mu}(\delta,\theta,\phi,\alpha)$ and $\xi_{\mu}(\delta,\theta,\phi,\alpha)$ of the individual Shewhart residual charts and of the left partial sums defined in Eqs. 5–5, are positive and the remaining areas correspond to the values of (ϕ,α) for which those derivatives are negative. Once again the sign of the derivatives of $\xi_{S-\mu}(\delta,\theta,\phi,\alpha)$ and $\xi_{S-\sigma}(\delta,\theta,\phi,\alpha)$ is symmetrical to the sign of the derivatives of the corresponding left partial sums.

Proof. All the results follow from Lemma 1 and, once more, the simplification of conditions like $\frac{\partial \xi_{S-\mu}(\delta,\alpha,\phi,\alpha)}{\partial \phi} > 0$ proved to be unfeasible.

4 Monotonic Behaviour of PMS of Type IV

As previously noted, the PMS of Type III does not depend on the parameters of the stationary process. However, taking into account the definition of PMS of Type IV and Theorems 1 and 2, we are able to derive monotonicity properties of this performance measure in terms of the parameters of the stationary AR(2) and ARMA(1,1) processes.

Theorem 3. *The following monotonicity properties are valid for the PMS of Type IV of the simultaneous Shewhart and EWMA schemes based on upper one-sided individual residual charts for the mean and variance of a stationary $AR(2)$ process:*

(3.1) $PMS_{IV}(\delta,\phi_1,\phi_2) \uparrow$ *with* ϕ_1 *if* $(\phi_1,\phi_2) \in D$
(3.2) $PMS_{IV}(\delta,\phi_1,\phi_2) \uparrow$ *with* ϕ_2 *if* $(\phi_1,\phi_2) \in B \cap F$
(3.3) $PMS_{IV}(\delta,\phi_1,\phi_2) \downarrow$ *with* ϕ_2 *if* $(\phi_1,\phi_2) \in A \cap E$

where the subregions D, $B \cap F$ and $A \cap E$ are represented in Fig. 3.

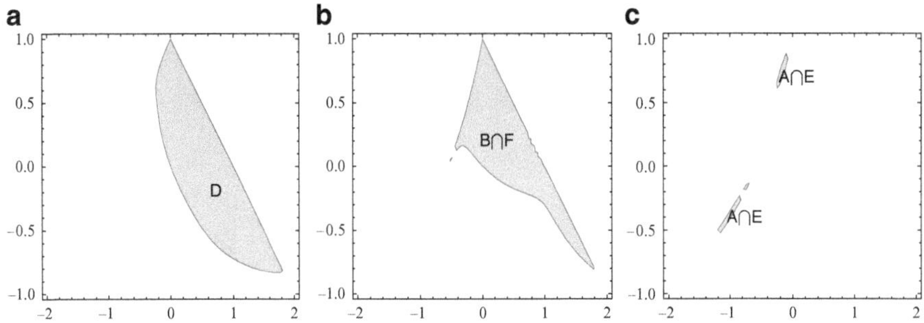

Fig. 3 Monotonicity regions (*shaded areas*) of PMS_{IV} for a stationary AR(2) process. (**a**) $PMS_{IV}(\delta,\phi_1,\phi_2) \uparrow \phi_1$; (**b**) $PMS_{IV}(\delta,\phi_1,\phi_2) \uparrow \phi_2$; (**c**) $PMS_{IV}(\delta,\phi_1,\phi_2) \downarrow \phi_2$

Proof. These results follow directly from the definition of $\text{PMS}_{IV}(\delta,\phi_1,\phi_2)$ and from the monotonicity results in Theorem 1. For example, since for both Shewhart and EWMA schemes,

- $RL_\mu(\delta,\theta,\phi_1,\phi_2) \uparrow_{st}$ with ϕ_1,
- $RL_\sigma(\delta,\theta,\phi_1,\phi_2) \downarrow_{st}$ with ϕ_1 when $(\phi_1,\phi_2) \in D$, and
- $PMS_{IV}(\delta,\phi_1,\phi_2) = P[RL_\sigma(\delta,1,\phi_1,\phi_2) < RL_\mu(\delta,1,\phi_1,\phi_2)]$,

we can conclude that $PMS_{IV}(\delta,\phi_1,\phi_2) \uparrow$ with ϕ_1 when $(\phi_1,\phi_2) \in D$.

As we can see from Fig. 3, the PMS of Type IV increases with ϕ_1 and ϕ_2 when these two parameters are both nonnegative and within the stationarity region. Moreover, from Fig. 3c we can conclude that the subregions where $PMS_{IV}(\delta,\phi_1,\phi_2) \downarrow \phi_2$ have a very small size.

It is important to point out that we were only able to establish a few monotonicity results for the AR(2) process. This suggests that it will be harder to establish monotonicities in more complex processes such as the ARMA(1,1).

Theorem 4. *The PMS of Type IV of the simultaneous Shewhart and EWMA schemes, based on upper one-sided individual residual charts for the mean and variance of a stationary and invertible ARMA(1,1) process, has the following monotonicity properties:*

(4.1) $PMS_{IV}(\delta,\phi,\alpha) \downarrow$ *with* ϕ *if* $(\phi,\alpha) \in G \cap L$
(4.2) $PMS_{IV}(\delta,\phi,\alpha) \uparrow$ *with* ϕ *if* $(\phi,\alpha) \in H \cap K$
(4.3) $PMS_{IV}(\delta,\phi,\alpha) \downarrow$ *with* α *if* $(\phi,\alpha) \in I \cap N$
(4.4) $PMS_{IV}(\delta,\phi,\alpha) \downarrow$ *with* α *if* $(\phi,\alpha) \in J \cap M$

where the subregions $G\cap L$, $H\cap K$, $I\cap N$ *and* $J\cap M$ *are represented in Fig. 4.*

Proof. All these results follow directly from the definition of PMS_{IV} and from the monotonicity results in Theorem 2. For instance, for both Shewhart and EWMA schemes,

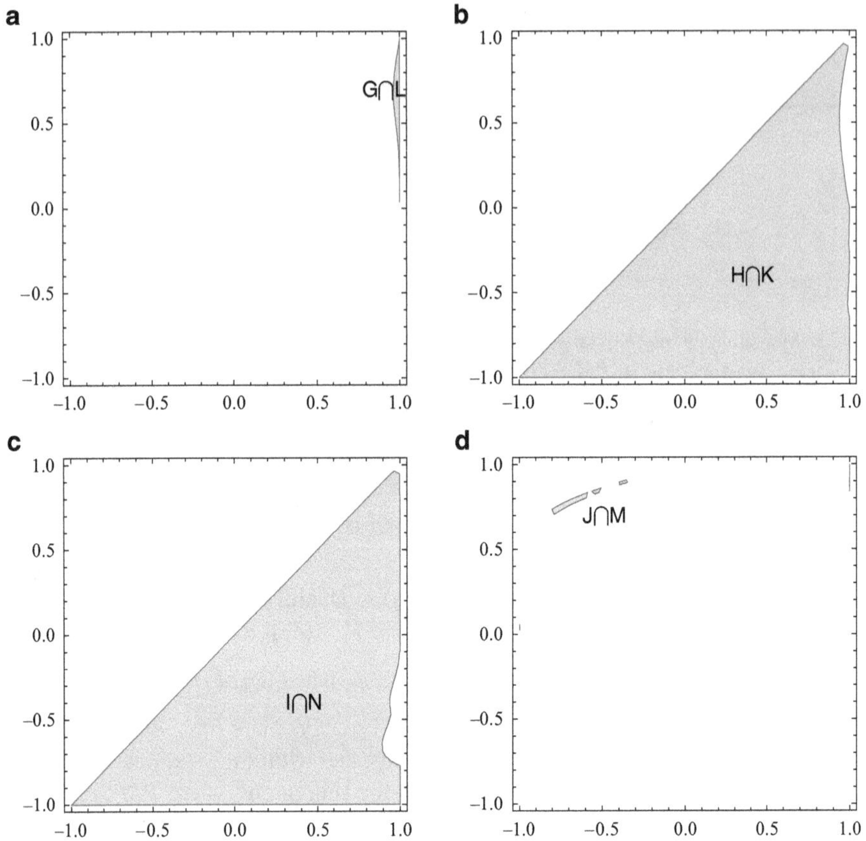

Fig. 4 Monotinicity regions (*shaded areas*) of the PMS_{IV} for a stationary and invertible ARMA(1,1) process. (**a**) $PMS_{IV}(\delta,\phi,\alpha) \downarrow \phi$; (**b**) $PMS_{IV}(\delta,\phi,\alpha) \uparrow \phi$; (**c**) $PMS_{IV}(\delta,\phi,\alpha) \downarrow \alpha$; (**d**) $PMS_{IV}(\delta,\phi,\alpha) \uparrow \alpha$

- $RL_\mu(\delta,\theta,\phi,\alpha) \downarrow_{st}$ with ϕ when $(\phi,\alpha) \in G$
- $RL_\sigma(\delta,\theta,\phi,\alpha) \uparrow_{st}$ with ϕ when $(\phi,\alpha) \in L$ and
- $PMS_{IV}(\delta,\phi,\alpha) = P[RL_\sigma(\delta,1,\phi,\alpha) < RL_\mu(\delta,1,\phi,\alpha)]$,

therefore $PMS_{IV}(\delta,\phi,\alpha) \downarrow$ with ϕ when $(\phi,\alpha) \in G \cap L$.

It is apparent from Fig. 4 that the PMS of Type IV increases with ϕ for most values of (ϕ,α) in the set $\{(\phi,\alpha) \in (-1,1)^2 : \phi < \alpha\}$. We can also add that the subregions where $PMS_{IV}(\delta,\phi,\alpha) \downarrow \phi$ and $PMS_{IV}(\delta,\phi,\alpha) \uparrow \alpha$ have very small size.

We close this subsection by stressing out that all the monotonicity results in Theorems 3 and 4 hold regardless of the value of the smoothing parameter λ, meaning that they are valid for both Shewhart and EWMA residual schemes.

5 Concluding Remarks

The results presented in this paper show that the MS of Type III is not affected by the autocorrelation structure of the stationary process, as previously noted by Antunes (2009) and Knoth et al. (2009) while dealing with stationary AR(1) output.

In addition, the use of stochastic ordering allowed us to make a qualitative assessment of the impact of the presence of autocorrelation on the performance of simultaneous residual schemes. For instance, we have shown that larger nonnegative values of the parameters ϕ_1 and ϕ_2 of the stationary AR(2) process are associated to higher values of PMS of Type IV. We have also proved that the PMS of Type IV increases (resp. decreases) in most cases with the autoregressive parameter ϕ (resp. moving average parameter α) of the stationary and invertible ARMA(1,1) process.

All these results are valid for both simultaneous Shewhart and EWMA residual schemes.

We strongly believe that this paper provides additional insights and answers to the following question: how are PMS of types III and IV influenced by the presence of autocorrelation?

Acknowledgements The first author is supported by grant SFRH/BD/35739/2007 of *Fundação para a Ciência e a Tecnologia* (FCT) and was partially supported by *Centro de Matemática e Aplicações* (CEMAT) and FCT while visiting the Department of Statistics of the European University Viadrina (Frankfurt (Oder), Germany).

References

Antunes, C. (2009). Avaliação do impacto da correlação em sinais erróneos de esquemas de conjuntos para o valor esperado e variância (Assessment of the impact of the correlation on misleading signals in joint schemes for the mean and variance). Master's thesis, Instituto Superior Técnico, Technical University of Lisbon.

Brockwell, P. J., & Davis, R. A. (1991). *Time series: Theory and methods*. New York: Springer.

Brook, D., & Evans, D. A. (1972). An approach to the probability distribution of CUSUM run length. *Biometrika, 59*, 539–549.

St. John, R. C., & Bragg, D. J. (1991). Joint X-bar R charts under shift in mu or sigma. *ASQC Quality Congress Transactions – Milwaukee, 45*, 547–550.

Knoth, S., & Schmid, W. (2002). Monitoring the mean and the variance of a stationary process. *Statistica Neerlandica, 56*, 77–100.

Knoth, S., Morais, M. C., Pacheco, A., & Schmid, W. (2009). Misleading signals in simultaneous residual schemes for the mean and variance of a stationary process. *Communications in Statistics – Theory and Methods, 38*, 2923–2943.

Mathai, A. M., & Provost, S. B. (1992). *Quadratic forms in random variables*. New York: Marcel Dekker.

Morais, M. J. C. (2002). *Stochastic ordering in the performance analysis of quality control schemes*. Ph.D. thesis, Instituto Superior Técnico, Technical University of Lisbon.

Morais, M. C., & Pacheco, A. (1998). Two stochastic properties of one-sided exponentially weighted moving average control charts. *Communications in Statistics – Simulation and Computation, 27*, 937–952.

Morais, M. C., & Pacheco, A. (2000). On the performance of combined EWMA schemes for μ and σ: A Markovian approach. *Communications in Statistics – Simulation and Computation, 29*, 153–174.

Morais, M. C., & Pacheco, A. (2006). Misleading signals in joint schemes for μ and σ. In H. J. Lenz & P. T. Wilrich (Eds.), *Frontiers in statistical quality control* (Vol. 8, pp. 100–102). Heidelberg: Physica-Verlag.

Ramos, P. F., Morais, M. C., & Pacheco, A. (2010). Misleading signals in simultaneous residual schemes for the process mean and variance of AR(1) processes: A stochastic ordering approach (submited for publication).

Reynolds, M. R., Jr., & Stoumbos, Z. G. (2001). Monitoring the process mean and variance using individual observations and variable sampling intervals. *Journal of Quality Technology, 33*, 181–205.

Reynolds, M. R., Jr., & Stoumbos, Z. G. (2004). Control charts and the efficient allocation of sampling resources. *Technometrics, 46*, 200–214.

More on Control Charting Under Drift

Sven Knoth

Abstract The standard task within SPC is the detection of an unforeseen shift in the mean level of the sequence of typically normally distributed random variables. Only some papers deal with a not considerably less common pattern in industrial practice: gradual changes because of tool wear or similar causes. In the small list of the currently available papers both existing control charts for the mean under drift are studied and new ones are created. It is worth noting that except Gan (J Stat Comput Simul 38:181–200, 1991; Statistician 41:71–84, 1992) no convincing numerical algorithms are presented for calculating characteristics for control charts under and for drift. The good message is that mean level control charts are also suited for detecting drifts. This paper provides some more numerical results including a competing method to Gan's algorithm and presents various schemes.

Keywords Statistical process control • Linear drift • Average run length • Numerical methods for calculating *ARL*

1 Introduction

Typically, control charts (surveillance or change point detection schemes) such as Page's CUSUM, Roberts' EWMA, and the Shiryaev-Roberts procedures are setup to detect a step change as quickly as possible while maintaining a low false alarm rate. In practical applications, a change may happen gradually. A linear drift (or trend) would be the first idea to model it. Except Chang and Fricker (1999), most of the literature published about control charting under or for drift considers linear

S. Knoth (✉)
Department of Economics and Social Sciences, Institute of Mathematics and Statistics, Helmut Schmidt University Hamburg, Hamburg, Germany
e-mail: Sven.Knoth@hsu-hh.de

H.-J. Lenz et al. (eds.), *Frontiers in Statistical Quality Control 10*,
Frontiers in Statistical Quality Control 10, DOI 10.1007/978-3-7908-2846-7_4,

drift (even more, linear in the observation number). As usual, these models would already cover a wide range of applications. Thus, this simple linear drift model will be also the model of this paper.

Nevertheless, the numerical treatment of control charts under linear drift, looking for optimal or nearly optimal schemes etc. is less intensively studied than for the popular step change model. While designing a control chart it is not clear, whether a specific drift detection scheme should be set up or the usual step change chart is needed anyway. Thus, one should not only construct new schemes, but also analyze the performance of the classics under drift. Note that a very simple approach for constructing a linear drift detection scheme could rely on the differences of the observed data sequence, which is something like a signed Moving Range statistic. Davis and Woodall (1988) evaluated popular chart examples from the quality circle world – the so-called trend rules. However, they conclude that these charts are ineffective in detecting drifts. Most of the papers deploy Monte Carlo studies to get performance measures such as the Average Run Length. The most accurate numerical algorithm was developed by Gan (1991, 1992). It is surprising that nearly no other author is using his method. Here, his approach and one already used in Knoth (2003) are utilized for one-sided EWMA, CUSUM and Shiryaev-Roberts schemes. Contrary to the step change performance it is quite difficult to evaluate the drift performance of two-sided CUSUM and Shiryaev-Roberts schemes. For all other schemes it is even more complicated so that Monte Carlo studies dominate. The aim of this paper is twofold. First, it illustrates that on the current level of knowledge, the classics could be used for drift detection. Second, it should inspire the SPC community to do more in order to enhance the knowledge about drift detection.

2 Status Quo of Drift Detection with Attention to Numerical Algorithms

One of the first papers is Bissell (1984) who already treated CUSUM charts under drift. His numerical algorithm, a modified Markov chain approach, did not work well (also noted by himself in a correction) and was refined in Asbagh (1985) – presumably it is more a matter of efficient coding than algorithmic difficulties. Some more details follow below. Bissell's conclusions regarding the choice of an appropriate scheme are therefore not reliable, nonetheless they are not wrong. Then, in Davis and Woodall (1988) the Shewhart chart with trend rules was analyzed. The authors clearly discourage from their usage. A more elaborated study was done in Aerne et al. (1991), where based on Markov chain and Monte Carlo methods Shewhart charts with and without runs rules, CUSUM and EWMA were considered. Basically, they finally suggest to apply the competing charts as in the step change case. That is, for small drift coefficients deploy CUSUM or EWMA, otherwise Shewhart charts. In the same year, Gan (1991) published the first time his algorithm to attain high accuracy in calculating Average Run Lengths of classical control charts under drift. His first paper discusses two-sided EWMA charts. In Gan (1992) one-sided CUSUM charts were analyzed. His idea is simple and impressive. It should be sketched here for the two-sided EWMA chart. Beforehand, the general change point model utilized in this paper is described.

Let $X_1, X_2, \ldots$ be a sequence of independent normal random variables with variance 1. Their mean is under risk to change. Specifically,

$$E(X_t) = \begin{cases} \mu_0 = 0, & t < \tau \\ \mu_{t-\tau} = (t - \tau + 1)\Delta, & t \geq \tau \end{cases}, \quad t = 1, 2, \ldots .$$

τ resembles the so-called change point. Note that Gan preferred $\mu_{t-\tau} = (t - \tau)\Delta$ with the special feature $E(X_\tau) = 0 = \mu_0$. Here, the above model is chosen. The parameter Δ is the drift coefficient. Gan considered the case $\tau = \infty$ as in-control scenario and $\tau = 1$, as usual, as representative out-of-control pattern. Imagine now that the change, step or drift, should be detected by an EWMA control chart. Thus, apply Roberts' (1959) classic EWMA chart:

$$\begin{aligned} Z_0 &= z_0 = \mu_0 = 0\,, \\ Z_t &= (1 - \lambda) Z_{t-1} + \lambda X_t\,, \quad \lambda \in (0, 1]\,, \\ L &= \inf \left\{ t \in \mathbb{N} : |Z_t| > c\sqrt{\lambda/(2-\lambda)} =: C \right\} . \end{aligned}$$

The most popular performance measure is the zero-state Average Run Length (*ARL*). It is defined as (denote $E_\tau()$ the expectation for change point τ)

$$ARL = \begin{cases} E_\infty(L), & \text{in-control case: no change at all.} \\ E_1(L), & \text{out-of-control case: change already at the beginning.} \end{cases}$$

It is a simplified measure and already often criticized. In the drift detection area, it is the dominating one. A more suitable notation style is to write the *ARL* as function of the slope Δ, where $\Delta = 0$ corresponds then to $E_\infty(L)$. The so-called steady-state *ARL* is slightly different:

$$\mathcal{D} = \lim_{\tau \to \infty} E_\tau(L - \tau + 1 \mid L \geq \tau)\,,$$

where $\Delta = 0$ and $\Delta \neq 0$ are both involved. $\mathcal{D}$ is calculated only in Table 7.

Now Gan denoted with

$$\mathcal{L}_j(y, \mu_j)\,, \; j = 0, 1, \ldots, m$$

the *ARL* for an EWMA chart starting at $z_0 = y$ and mean sequence $\mu_j = j\Delta$, $\mu_{j+1}, \ldots, \mu_m, \mu_m, \ldots$ He showed in Gan (1991) that for $y \in [-C, C]$

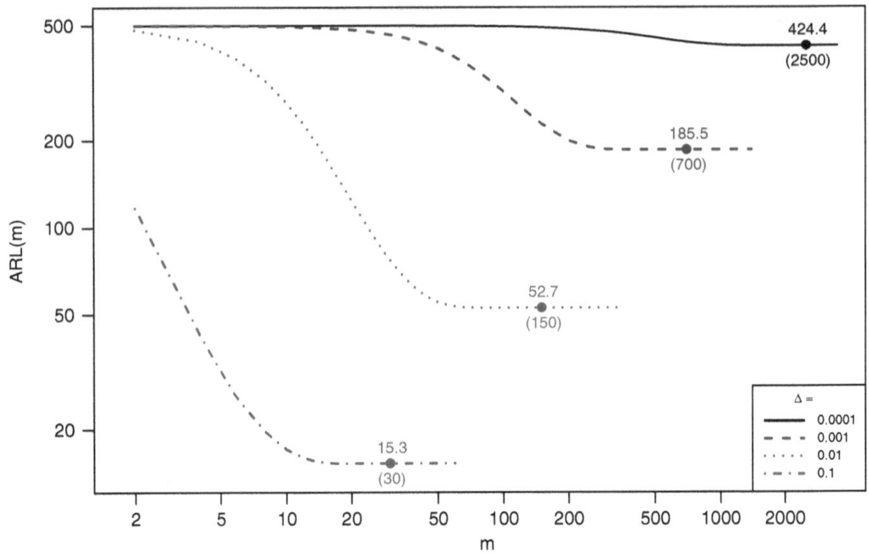

Fig. 1 Illustration of Gan's search for a suitable choice of the constant m – see Gan (1991) for details (EWMA control chart with $\lambda = 0.047$ and $E_\infty(L) = 500$)

$$\mathcal{L}_j(y, \mu_j) \underset{j=0,1,\ldots,m-1}{=} 1 + \int_{-C}^{C} \mathcal{L}_{j+1}(y, \mu_{j+1}) \frac{1}{\lambda} f_{\mu_j}\left(\frac{x - (1-\lambda)y}{\lambda}\right) dx\,,$$

$$\mathcal{L}_m(y, \mu_m) = 1 + \int_{-C}^{C} \mathcal{L}_m(y, \mu_m) \frac{1}{\lambda} f_{\mu_m}\left(\frac{x - (1-\lambda)y}{\lambda}\right) dx\,.$$

For fixed m, the latter integral equation can be solved as usual with the Nyström method utilizing a reasonable quadrature method such as Gauss-Legendre. Afterwards, the first iteration sequence is executed by plugging in the same quadrature as for the integral equation. The final approximation of $\mathcal{L}_1(y = z_0, \mu_1)$ (Gan took $\mathcal{L}_0(y = z_0, \mu_0)$) provides the *ARL* $E_1(L)$. It remains the choice of the index m. A simple idea is to increase m until $\mathcal{L}_0(y = z_0, \mu_0)$ remains constant. It depends heavily on the drift rate Δ. Gan reported for $\Delta = 1, .1, .01, .001$ and $.0001$ the following values for m: 6, 30, 150, 700 and 2,500, respectively. See Fig. 1 for an illustration of the convergence patterns. For implementing Gan's algorithm one should hide the search for sufficiently large m. The resulting procedure is slower than the method described in Knoth (2003). This algorithm could also be used to calculate the *ARL* for drift. Additionally, it could be extended easily for the steady-state *ARL* and *ARL* vehicles for $1 < \tau < \infty$.

In Gan's publication in 1996 of the Algorithm AS 305 for CUSUM an important remark regarding the numerical issues for two-sided CUSUM was given. Gan underlined that the performance of the usual two-sided CUSUM chart – simultaneously running a lower and an upper chart – could not be treated under drift like under a step change. The problem occurs, e.g., by calculating the *ARL* of the lower chart, if an upwards drift is present. The above approach – increase m until the resulting *ARL* does not change anymore – fails because $\mathcal{L}_m(y, \mu_m)^{\text{lower}}$ becomes larger for increasing m. Or to put it in other words, if the lower chart does not signal already during the first drifted values, then it becomes quite unlikely that it will ever signal. To sum up, except Monte Carlo methods no other method is currently available for calculating reliably the *ARL* of two-sided CUSUM schemes under drift.

Some years earlier, Bissell (1984) studied one-sided CUSUM charts (besides Shewhart charts with and without warning limits) and extended the results in Brook and Evans (1972) to non-homogeneous Markov chains. Thus, instead of solving the famous linear equation system to get the *ARL* he considered:

$$ARL = \sum_{m=1}^{\infty} m \cdot P(L = m) \, , \; ARL \approx \sum_{m=1}^{m_{\max}} m \cdot \tilde{P}(L = m) \, ,$$

$$\tilde{P}(L = m) = \tilde{P}(L > m-1) - \tilde{P}(L > m) \, , \; \tilde{P}(L > m) = \underline{z}_0' \left(\prod_{i=1}^{m} {}^{i}\mathbf{P} \right) \underline{1} \, .$$

The transition matrices ${}^{i}\mathbf{P}$ are determined as in Brook and Evans (1972) with slipping means $i\Delta$. $\underline{z}_0$ and $\underline{1}$ are the starting vector and a vector of ones, respectively. Regarding the choice of $m_{\max}$ Bissell mentioned that "for $\Delta \geq 0.01$ at most a few hundred terms need to be evaluated, and for $\Delta \geq 0.1$ only a few tens". The dimension of the matrix was set to 20. For small values of Δ, however, the results in Bissell (1984) are far away from the true ones. In Bissell (1986) the author mentioned that this was mainly caused by rounding errors and compared his previous results with new numbers obtained by a small Monte Carlo study. Before digging into the details one comment has to be made: Bissell and nearly all more recent papers calculated the ARL values for really small values of the slope Δ. This is somehow misleading, especially if one looks at the smallest values used for shifts. However, here (and in the sequel) the original design of the comparison was treated.

See Table 1 for Bissell's results and some up-to-date ones. Note that the columns B_{100} and B_{500} in Table 1 provide numbers based on Bissell's algorithm with $m_{\max} = 100$ and 500. Further increasing of $m_{\max}$ would not change the results anymore. To attain higher accuracy, the matrix dimension has to be increased. It is interesting that they are much better than Bissell's original results. Nowadays it is difficult to identify whether the original computer code was not correct or the accuracy of the matrix vector multiplication in the 1980s was not that high. Bissell (1986) noted that in the Master thesis of Asbagh (1985) the implementation of the above algorithm was an improved one. Eventually, it is highly recommended that

$$ARL \approx \sum_{m=0}^{m_{\max}} \tilde{P}(L > m)$$

Table 1 Bissell's results and corresponding new ones. $\hat{\mathcal{L}}^B_{\text{MC}}$ and the subsequent s. e.^{B} are from Bissell (1986), B from Bissell (1984), B_{100} and B_{500} are based on Bissell's formulas with $m_{\max} =$ 100 and 500, respectively, GoK are calculated with Gan's (or Knoth's) algorithm, and the last two columns provide Monte Carlo results with 10^6 replicates

Δ	$\hat{\mathcal{L}}^B_{\text{MC}}$	s. e.^{B}	B	B_{100}	B_{500}	GoK	$\hat{\mathcal{L}}_{\text{MC}}$	s. e.
0.001	245	23.8	–	95	229	231	231	0.113
0.002	142	12.4	–	92	155	156	156	0.068
0.005	87	7.0	205	81	89	89	89	0.033
0.01	56.4	3.8	101	57.1	57.1	57.2	57.2	0.019
0.02	36.2	2.1	53	36.5	36.5	36.5	36.5	0.010
0.05	18.8	1.2	24	20.4	20.4	20.4	20.4	0.005
0.1	14.3	0.58	14.2	13.3	13.3	13.3	13.3	0.003
0.2	8.4	0.32	9.0	8.8	8.8	8.8	8.8	0.002
0.5	5.2	0.12	5.3	5.3	5.3	5.3	5.3	0.001
1.0	3.40	0.13	3.6	3.60	3.60	3.60	3.60	0.001
2.0	2.44	0.10	2.5	2.50	2.50	2.50	2.50	0.000
3.0	1.96	0.04	2.0	2.01	2.01	2.01	2.01	0.000
# replicates	25						10^6	

should be used, which is simpler to implement and presumably more numerically robust than Bissell's $\ldots m \cdot \tilde{P}(L = m)$.

Already in 1988 Sweet published an algorithm for monitoring data with potential drifts. His coupled EWMA charts directly monitor the slope of the data:

$$
\begin{aligned}
S_0 &= \mu_0 = 0\,, \\
S_t &= (1 - \lambda_S)(S_{t-1} + B_{t-1}) + \lambda_S X_t\,, \\
B_0 &= \Delta_0 = 0\,, \\
B_t &= (1 - \lambda_B) B_{t-1} + \lambda_B (S_t - S_{t-1})\,.
\end{aligned}
$$

The sequence B_t is built for monitoring the trend slope. S_t is, more or less, the usual mean monitor. Sweet's framework allows also to monitor changes from one drift coefficient to another. The whole design stems from the forecasting literature (Holt 1960) and seems to be rarely discussed in SPC literature. Sweet provides some guide lines to design the two control charts. Thereby, he tries to link distributional properties of the sequences to the standard Shewhart 3σ rule. He did not calculate performance measures like the *ARL*. It is not that surprising, because the design of these coupled EWMA charts is not easily treated numerically. Thus, only Monte Carlo studies could be done. Just a quick comparison of Sweet's combined charts and the usual two-sided EWMA control chart is given in Table 2. Note that these numbers are only some first results. More analysis is ongoing. Additionally, a considerably large number of alarms within the Sweet scheme are raised by the mean chart and not by the slope chart. Finally, this setup looks promising for designing control charts that allow to monitor more general in-control processes such as slowly drifting ones.

Table 2 *ARL* results of Sweet's Winter-Holt like EWMA charts (Monte Carlo with 10^6 replicates) and standard mean monitoring EWMA charts (Gan's or Knoth's algorithm). All the EWMA λ values are set to 0.1. The in-control *ARL* is roughly 370

Δ	EWMA	Winter-Holt
0.001	177	211
0.026	28.6	30.4
0.081	14.7	15.4
0.325	6.72	7.08

A further contribution to both the modelling of the change point for potentially drifting data and new monitoring procedures is Chang and Fricker (1999). They look at a broader class of drift patterns: monotonically increasing means. However, they have a maximal value for the mean in mind. The aim of their scheme is not to detect the drift itself, but the exceeding of that maximal value. This sounds reasonable for some practical applications. Chang and Fricker Jr. evaluate one-sided CUSUM and EWMA charts, and a GLR-like scheme. The latter is custom-built for detecting the exceedance of their threshold mean under their more general mean drift model. The basic idea of GLR (generalized likelihood ratio) in change detection is to dispense the explicite knowledge of the post-change parameters. Here, the additional estimation task (under the monotonicity restriction) leads to isotonic regressions. Of course, the complicate implementation of GLR schemes in general and specifically for the model here hampers their application in practice. As side effect it offers a nice on-line estimate of the mean. However, Chang and Fricker Jr. conclude that the classics are not outperformed by the specialized GLR procedure. Thus, there is no reason to switch from the classics in the considered drift change point model. Note that the authors calculated a kind of steady-state *ARL* with Monte Carlo studies. Thus, it is difficult to compare it to any other analysis done for drift monitoring.

More recently, Reynolds and Stoumbos (2001), studied drift behavior in simultaneous monitoring schemes (mean and variance). They allowed both drifts in mean and variance. All was evaluated with Monte Carlo studies. It was concluded, not surprisingly, that EWMA outperforms the Shewhart counterparts for small and moderate drift coefficients. Note that separate analysis of variance control charts under drift would be a interesting and useful task.

Fahmy and Elsayed (2006) introduced a drift detection scheme based on slope estimates on rolling windows. Afterwards, they compared their new chart with one-sided Shewhart, CUSUM and EWMA charts, and drift GLR scheme under drift. They obtained their numbers by Monte Carlo simulation. They concluded that their scheme is better than the classics under drift. Some more reflections about their results will be given in the next section. Given the only slight performance advantages and the higher complexity of the rolling window scheme, in application the classics would outlast.

The most recent paper is Zou et al. (2009). It gives a thorough introduction to the subject and a large comparison study for one-sided EWMA, CUSUM, GEWMA

Table 3 $\mathcal{L}_\Delta^{\text{w/o}}$ and $\mathcal{L}_\Delta^{\text{w/}}$ – *ARL* results of Divoky and Taylor (1995), $\mathcal{L}_\Delta^{\text{w/o*}}$ – *ARL* of Shewhart control chart with tightened limits, and $\mathcal{L}_\Delta^{\text{E}}$ – standard mean monitoring EWMA charts (Gan's algorithm, $\lambda = 0.1$). The in-control *ARL* is roughly 370. The η values are calculated according to the above formula. The larger η, the better the trend detection abilities

Δ	$\mathcal{L}_\Delta^{\text{w/o}}$	type	$\mathcal{L}_\Delta^{\text{w/}}$	$\eta_{\text{w/}}$	$\mathcal{L}_\Delta^{\text{w/o*}}$	$\eta_{\text{w/o*}}$	$\mathcal{L}_\Delta^{\text{E}}$	η_{E}
0	367.61	–	–	–	300	–	370	–
0.003	174.54	15 of 20	163.79	0.867	157.36	0.905	101.73	1.727
0.006	120.70	17 of 24	113.99	0.775	111.68	0.882	68.55	1.772
0.03	43.44	10 of 13	40.87	0.416	41.49	0.855	26.27	1.665
0.06	26.86	9 of 12	25.20	0.236	25.67	0.854	17.45	1.549
0.3	8.12	6 of 8	8.02	0.063	7.90	0.838	7.02	1.164

(the smoothing constant λ is optimized in a certain way), and GLR schemes for the step change and drift. All schemes are analyzed for drift coefficients from very small up to large ones. They deploy only Monte Carlo studies. Based on their measure *RMI* (measures the performance over the whole range of considered drift coefficients) they conclude that the more sophisticated schemes outperform the classics. This will be reviewed in the next section.

Eventually, some notes on detecting drift with so-called trend rules are given. Davis and Woodall (1988) considered rules like "signal if 5 or 6 of consecutive slopes are of the same sign". They demonstrated that simple Shewhart charts and Shewhart charts with traditional runs rules are better in detecting drift without than with additional trend rules. However, in Divoky and Taylor (1995) about 600 different trend rules were studied to find an optimal one. They considered as effectiveness ratio the relation between the reduction in the in-control vs. out-of-control *ARL*, that is,

$$\eta = \frac{\mathcal{L}_0^{\text{w/}} / \mathcal{L}_0^{\text{w/o}}}{\mathcal{L}_\Delta^{\text{w/}} / \mathcal{L}_\Delta^{\text{w/o}}} .$$

Thus, a reasonable trend rule add-on would generate large η values. See Table 3 for their results, a slightly modified Shewhart control chart and a two-sided EWMA chart for mean shifts. From Table 3 it is clear that Divoky and Taylor's best designs, $\mathcal{L}_\Delta^{\text{w/}}$ are the corresponding *ARL* values, are beaten by (1) a simple Shewhart control chart with tightened limits (from 3 to 2.935) as already demonstrated in Davis and Woodall (1988) and (2) by EWMA control charts. Thus, there is only one conclusion: Trend rules should be avoided.

3 New and Old Results

3.1 Control Charts Under Consideration

3.1.1 Standard Charts

This is a short list without any detailed discussion. The charts are described in their one-sided version. It is quite simple (and standard) to get to their two-sided counterparts. GRSR stands for Girshick-Rubin-Shiryaev-Roberts (or more frequently called Shiryaev-Roberts) and GEWMA for generalized EWMA.

$$\text{Shewhart: } L = \inf\{t \in \mathbb{N} : X_t > c_S\},$$

$$\text{CUSUM: } S_t = \max\{0, S_{t-1} + X_t - k\}, \; S_0 = s_0 = 0,$$
$$L = \inf\{t \in \mathbb{N} : S_t > h\},$$

$$\text{EWMA: } Z_t = \max\left\{z_{\text{reflect}}, (1-\lambda)Z_{t-1} + \lambda X_t\right\}, \; \lambda \in (0,1], \; Z_0 = \mu_0 = 0,$$
$$L = \inf\left\{t \in \mathbb{N} : Z_t > c_E \sqrt{\lambda/(2-\lambda)}\right\},$$

$$\text{GRSR: } R_t = (1 + R_{t-1}) \exp(X_t - k), \; R_0 = r_0 = 0,$$
$$L = \inf\{t \in \mathbb{N} : R_t > g\},$$

$$\text{GEWMA: } \tilde{Z}_t(\lambda) = \sqrt{\frac{2-\lambda}{\lambda\left[1-(1-\lambda)^{2t}\right]}} \sum_{i=1}^{t} \lambda(1-\lambda)^{t-i} X_i,$$
$$L = \inf\{t \in \mathbb{N} : \max_{1 \le k \le t} \tilde{Z}_t\big(1/k\big) > c_g\}.$$

For more details on GEWMA see Han and Tsung (2004).

3.1.2 GLR Charts

The generalized likelihood ratio charts allow to get rid of specifying a certain out-of-control value (for the mean or the the drift coefficient). Here, two versions (for shift and for drift) are given.

step change:

$$LR_t(\tau, \mu) = \prod_{i=\tau}^{t} \frac{e^{-[X_i-\mu]^2/2}}{e^{-X_i^2/2}} \to \max_{1 \le \tau \le t, \mu},$$
$$T_t = \max_{1 \le \tau \le t, \mu} LR_t(\tau, \mu) = \max_{1 \le \tau \le t} \max_{\mu} LR_t(\tau, \mu),$$
$$L = \inf\{t \in \mathbb{N} : T_t > h_S\},$$

drift:

$$
\begin{aligned}
LR_t(\tau, \Delta) &= \prod_{i=\tau}^{t} \frac{e^{-[X_i-(i-\tau+1)\Delta]^2/2}}{e^{-X_i^2/2}} \to \max_{1\le\tau\le t,\, \Delta}, \\
\tilde{T}_t &= \text{similar to step change}, \\
L &= \inf\{t \in \mathbb{N} : \tilde{T}_t > h_D\}.
\end{aligned}
$$

Note that it would be more appropriate to write "sup" instead of "max" – τ is a discrete and μ, Δ are continuous parameters so that the above framework is rather sloppy. For the drift case it was simply copied from Zou et al. (2009). In the shift case one should mention that

$$
\arg\max_\mu LR_t(\tau, \mu) = \frac{1}{t-\tau+1} \sum_{i=\tau}^{t} X_i
$$

and $LR_t(\tau, \mu)$ further simplifies.

3.2 *Fahmy and Elsayed (2006)*

These authors introduce a further special drift detection chart. Basically, they consider a rolling window of size w and calculate on each window an OLS (ordinary least square) fit (observation vs. observation number), determine $\hat{\mu}_{wn} = \hat{\alpha}_n + \hat{\beta}_n t_w$ as "final" estimate of the mean, and create

$$
M_n = \frac{(\mu_0 - \hat{\mu}_{wn})^2}{1/w + (t_w - \bar{t})^2/S_{tt}} \quad \text{with} \quad S_{tt} = \sum_{i=1}^{w} (t_i - \bar{t})^2 .
$$

In the in-control case (no drift), M_n follows a χ_1^2 distribution. Finally, Fahmy and Elsayed neglect the autocorrelation of M_n and build a Shewhart like control chart by looking at (here one-sided):

$$
L = \inf\{n \in \mathbb{N} : M_n > c_R\} .
$$

Now, start with a comparison of two-sided EWMA and CUSUM results given in Fahmy and Elsayed (2006) and new ones based on a much more extensive Monte Carlo study and, for EWMA, on Gan's (or Knoth's) algorithm (Tables 4 and 5).

Note that the EWMA results of Fahmy and Elsayed (2006) match quite well to the more recent ones, while there are some slight differences for CUSUM. Fahmy and Elsayed concluded that their rolling regression chart (the resulting statistic is χ^2-distributed) outperform the rest. This is mainly because of their specific setup of the EWMA and CUSUM competitors. Taking different λ values changes the picture, see Table 5. Because the considered drift coefficients are quite large, one

Table 4 Comparison of *ARL* values for two-sided CUSUM and EWMA charts under drift. FE denotes the numbers from Fahmy and Elsayed (2006) based on a MC runs with 10^4 replicates. The remaining numbers stem from MC runs with 10^7 replicates and Gan's procedure (GoK)

	CUSUM		EWMA		
Δ	FE (10^4)	Here (10^7)	FE (10^4)	Here (10^7)	GoK
0	368.333 ± 3.549	368.251 ± 0.111	365.749 ± 3.598	369.021 ± 0.114	368.994
0.10	13.986 ± 0.026	14.086 ± 0.001	12.971 ± 0.029	12.986 ± 0.001	12.986
0.25	8.560 ± 0.014	8.656 ± 0.000	7.738 ± 0.015	7.758 ± 0.000	7.758
0.50	5.946 ± 0.008	6.033 ± 0.000	5.312 ± 0.009	5.318 ± 0.000	5.318
0.75	4.827 ± 0.007	4.898 ± 0.000	4.279 ± 0.007	4.286 ± 0.000	4.285
1.00	4.156 ± 0.006	4.224 ± 0.000	3.680 ± 0.006	3.688 ± 0.000	3.688
2.00	2.950 ± 0.003	2.989 ± 0.000	2.598 ± 0.005	2.616 ± 0.000	2.616

Table 5 Comparison of the *ARL* values of Fahmy and Elsayed (2006) favorite schemes (χ^2) and additional EWMA charts under drift. The *bold* values mark the smallest *ARL* values

	FE χ^2			EWMA			
Δ	$w^* = 3$	$w^* = 5$	$w^* = 20$	$\lambda = 0.1$	$\lambda = 0.2$	$\lambda = 0.3$	$\lambda = 0.5$
0	379.138 ± 3.790	370.048 ± 3.682	373.458 ± 3.546	369	370	370	370
0.10	17.445 ± 0.056	16.047 ± 0.049	12.860 ± 0.035	12.986	**12.747**	13.041	14.136
0.25	8.537 ± 0.032	8.127 ± 0.023	7.623 ± 0.018	7.758	7.304	**7.231**	7.497
0.50	5.027 ± 0.021	4.869 ± 0.013	5.260 ± 0.012	5.318	4.881	4.722	**4.706**
0.75	3.672 ± 0.017	3.673 ± 0.009	4.250 ± 0.009	4.285	3.886	3.715	**3.620**
1.00	**2.939** ± 0.014	3.055 ± 0.007	3.660 ± 0.008	3.668	3.318	3.149	3.023
2.00	**1.816** ± 0.003	2.042 ± 0.000	2.579 ± 0.000	2.616	2.254	2.124	2.005

should apply λ values larger than 0.1. Additionally, in other papers also much smaller drift coefficients are evaluated (see also next subsection). Summing up, the rolling window procedure needs more computational efforts and exhibits not better performance than classical control charts under drift.

3.3 *Zou et al. (2009)*

Zou et al. evaluated one-sided control charts. Among them are the classical EWMA and CUSUM chart, the more sophisticated GEWMA (the smoothing constant of the EWMA sequence is permanently adjusted), and two GLR charts designed for step changes and drifts. All results were calculated by Monte Carlo studies. Based on the measure *RMI*, which evaluates the robustness of the charts to various magnitudes of drifts, they conclude that the GEWMA and the drift GLR outperform the rest. Nevertheless, on smaller ranges of possible drift coefficients, the classics keep up with the newer ones or even beat them. It seems so that mainly the specific dynamic adaption of one of the chart parameters gives the more recent schemes some advance. In Table 6 the EWMA and CUSUM (zero-state) *ARL* results are

Table 6 Zero-state *ARL* values of one-sided EWMA, CUSUM and Shiryaev-Roberts (GRSR) procedures. The schemes are optimized for step changes of size δ. The corresponding λ values are 0.03479, 0.11125, and 0.23052. Additionally, the Monte Carlo results for GEWMA, GLR-S, and GLR-L are taken from Table 1 in Zou et al. (2009). The smallest values are boldly written

	EWMA			CUSUM			GRSR					
Δ	$\delta = 0.5$	$\delta = 1$	$\delta = 1.5$	$\delta = 0.5$	$\delta = 1$	$\delta = 1.5$	$\delta = 0.5$	$\delta = 1$	$\delta = 1.5$	*GEWMA*	*GLR-S*	*GLR-L*
0	1,750	1,747	1,733	1,741	1,742	1,735	1,730	1,730	1,730	–	–	–
0.0005	**318**	378	437	345	412	468	337	399	448	375	381	368
0.001	**215**	254	295	231	276	316	227	267	301	252	257	249
0.005	**83.5**	92.2	106	86.7	98.3	112	85.8	95.7	107	96.2	97.8	95.4
0.01	**55.7**	58.7	66.3	57.0	61.9	69.4	56.6	60.4	66.6	62.1	63.3	62.0
0.05	22.6	**21.1**	22.0	22.6	21.6	22.6	22.7	21.4	22.1	22.4	22.7	22.5
0.1	15.5	**13.9**	**13.9**	15.4	14.0	14.2	15.7	14.1	14.0	14.4	14.6	14.5
0.5	6.65	5.56	**5.09**	6.60	5.54	5.16	6.84	5.76	5.32	5.10	5.23	5.18
1.0	4.67	3.83	3.43	4.63	3.80	3.45	4.86	4.03	3.66	**3.26**	3.38	3.31
2.0	3.21	2.74	2.32	3.17	2.67	2.32	3.42	2.91	2.59	**2.09**	2.16	2.12
3.0	2.86	2.06	1.98	2.79	2.04	1.96	2.97	2.20	2.02	**1.69**	1.75	1.72
4.0	2.14	2.00	1.83	2.10	1.98	1.74	2.39	2.20	1.97	**1.31**	1.37	1.34

Table 7 Steady-state *ARL* ($\mathcal{D}$) values of one-sided EWMA, CUSUM and Shiryaev-Roberts (GRSR) procedures under drift. The schemes are optimized for step changes of size δ. The corresponding λ values are 0.03479, 0.11125, and 0.23052. The smallest values are boldly written

	EWMA			CUSUM			GRSR		
Δ	$\delta = 0.5$	$\delta = 1$	$\delta = 1.5$	$\delta = 0.5$	$\delta = 1$	$\delta = 1.5$	$\delta = 0.5$	$\delta = 1$	$\delta = 1.5$
0.0005	**314**	376	436	340	410	467	333	397	446
0.001	**213**	253	295	228	275	315	224	266	301
0.005	**82.6**	91.8	106	85.4	97.9	112	84.2	95.1	107
0.01	**55.1**	58.4	66.2	55.9	61.6	69.2	55.3	60.0	66.4
0.05	22.3	**20.9**	21.9	21.8	21.4	22.6	21.6	21.1	21.9
0.1	15.4	13.8	13.8	14.8	13.8	14.1	14.7	**13.7**	13.8
0.5	6.59	5.50	**5.05**	6.17	5.36	5.08	6.18	5.38	5.08
1.0	4.62	3.79	**3.40**	4.30	3.65	3.37	4.31	3.68	**3.40**
2.0	3.27	2.66	2.33	2.98	2.53	2.26	3.00	2.58	**2.30**
3.0	2.68	2.13	1.91	2.50	1.99	**1.90**	2.52	2.01	1.92
4.0	2.32	1.90	1.73	2.01	1.89	**1.66**	2.04	1.90	1.73

given now together with Shiryaev-Roberts *ARL* values under drift. To perform the comparison as in Zou et al. (2009) the Monte Carlo results from their Table 1 are added.

Roughly speaking, the classical schemes do not differ heavily under drift in terms of the zero-state *ARL*. Moreover, the EWMA (in it's classical dress and as GEWMA) exhibits the best zero-state *ARL* behavior. Note that the EWMA type schemes start at their mean level, while the other three schemes are evaluated from their worst level.

Therefore, in Table 7 are the related steady-state ARL values (but not ...) given. They are calculated based on the algorithms described in Knoth (2003). Note that it could be done also with Gan's procedure together with the approximated left eigenfunction of the chart transition kernel in the in-control case.

There are only slight differences between the three schemes in both the zero-state and steady-state *ARL* under drift. The EWMA chart (for mean monitoring) remains, however, a good candidate for detecting also drifts.

3.4 Charts for diff(X)

For detecting drift one simple idea would be, naturally, to monitor the differences of the original data by utilizing standard control charts. Denote D_t the first order differences. Hence, for equidistant time points:

$$D_t \underset{t>1}{=} X_t - X_{t-1} \quad \text{and} \quad D_1 = X_1 - \mu_0 = X_1 \,,$$

$$E(D_t) = \begin{cases} 0 & , t < \tau \\ \Delta & , t \geq \tau \end{cases} .$$

The original drift change point model is transformed into a step change model. Additionally, we get

$$Var(D_t) = 2Var(X_t) = 2\sigma_0^2 \,,$$
$$Corr(D_t, D_{t-1}) = -1/2 \,,$$

Thus, the variance is inflated and the differences D_t are negatively correlated. Increased variance means deteriorated detection power, while the negative correlation causes at least two effects: (1) from SPC literature on autocorrelated data it is known that for negative autocorrelation the detection power is increased and (2) it becomes difficult to calculate *ARL*s numerically. Because of (2) Monte-Carlo was used to get some first results – for a more detailed analysis two-dimensional Markov chain (or integral equation algorithms) approximation would be a better choice. Here, we look at a small sample of two-sided EWMA charts applied to D_t and compare it with a classical two-sided EWMA chart. In Table 8 some Monte-Carlo results including the standard error are collected. From Table 8 one concludes that diff(X)-EWMA charting is not useful to detect small drifts. Thus, EWMA charts on the original data or Sweet's coupled charts are better suited to detect drift.

4 Conclusions

Generally speaking, the schemes specifically designed for detecting drifts (instead of a step change) are not really worth the effort. The classical charts as CUSUM, EWMA, and Shiryaev-Roberts for mean surveillance are sufficiently sensitive to detect also drifts, even small ones. Regarding the size of the considered out-of-

Table 8 *ARL*s of two-sided EWMA charts. For diff(X)-EWMA Monte-Carlo with 10^7 rep. was used. The results in the last column were calculated with Gan's (or Knoth's) algorithm

	diff(X)-EWMA				EWMA
Δ	$\lambda = 0.001$	$\lambda = 0.005$	$\lambda = 0.01$	$\lambda = 0.05$	$\lambda = 0.1$
0.000	$499.043_{0.158}$	$499.941_{0.158}$	$500.011_{0.158}$	$499.909_{0.158}$	500
0.001	$352.154_{0.082}$	$446.749_{0.134}$	$480.844_{0.150}$	$498.781_{0.157}$	200.366
0.005	$156.924_{0.024}$	$195.202_{0.037}$	$269.405_{0.068}$	$476.528_{0.150}$	81.377
0.010	$100.950_{0.013}$	$116.058_{0.017}$	$144.803_{0.027}$	$417.469_{0.130}$	53.341
0.050	$32.538_{0.003}$	$33.973_{0.005}$	$36.002_{0.004}$	$79.414_{0.018}$	20.028
0.100	$19.430_{0.002}$	$19.922_{0.002}$	$20.591_{0.002}$	$29.565_{0.004}$	13.343
0.500	$5.703_{0.000}$	$5.742_{0.000}$	$5.790_{0.000}$	$6.227_{0.001}$	5.439
1.000	$3.368_{0.000}$	$3.380_{0.000}$	$3.394_{0.000}$	$3.520_{0.000}$	3.768
2.000	$2.019_{0.000}$	$2.022_{0.000}$	$2.026_{0.000}$	$2.060_{0.000}$	2.688
3.000	$1.537_{0.000}$	$1.538_{0.000}$	$1.540_{0.000}$	$1.553_{0.000}$	2.047
4.000	$1.181_{0.000}$	$1.182_{0.000}$	$1.183_{0.000}$	$1.192_{0.000}$	1.993
5.000	$1.028_{0.000}$	$1.028_{0.000}$	$1.028_{0.000}$	$1.031_{0.000}$	1.927

control slopes Δ one has to note that surprisingly small values are utilized in most of the comparison studies.

It is open how the *ARL* of two-sided CUSUM charts under drift could be numerically calculated. For the Shiryaev-Roberts scheme it is not clear, how the two-sided chart looks like – there are several ideas to create one. Summarizing, statistical drift monitoring is just at its beginning.

References

Aerne, L. A., Champ, C. W., & Rigdon, S. E. (1991). Evaluation of control charts under linear trend. *Communications in Statistics Theory and Methods, 20*, 3341–3349.

Asbagh, N. A. (1985). *Performance of cusum and combined shewhart-cusum charts under linear trend.* Master's thesis, Department of Statistics, University of Southwestern Louisiana, Lafayette.

Bissell, A. F. (1984). The performance of control charts and cusums under linear trend. *Applied Statistics, 33*, 145–151.

Bissell, A. F. (1986). Corrigendum: The performance of control charts and cusums under linear trend. *Applied Statistics, 35*, 214–214.

Brook, D., & Evans, D. A. (1972). An approach to the probability distribution of CUSUM run length. *Biometrika, 59*, 539–549.

Chang, J. T., & Fricker, R. D. Jr. (1999). Detecting when a monotonically increasing mean has crossed a threshold. *Journal of Quality Technology, 31*, 217–234.

Davis, R. B., & Woodall, W. H. (1988). Performance of the control chart trend rule under linear shift. *Journal of Quality Technology, 20*, 260–262.

Divoky, J. J., & Taylor, R. W. (1995). Detecting process drift with combinations of trend and zonal supplementary runs rules. *International Journal of Quality and Reliability Management, 12*, 60–71.

Fahmy, H. M., & Elsayed, E. A. (2006). Detection of linear trends in process mean. *International Journal of Production Research, 44*, 487–504.

Gan, F. F. (1991). EWMA control chart under linear drift. *Journal of Statistical Computation and Simulation, 38*, 181–200.

Gan, F. F. (1992). CUSUM control charts under linear drift. *Statistician, 41*, 71–84.

Gan, F. F. (1996). Average run lengths for cumulative sum control chart under linear trend. *Applied Statistics, 45*, 505–512.

Han, D., & Tsung, F. (2004). A generalized EWMA control chart and its comparison with the optimal EWMA, CUSUM and GLR schemes. *Annals of Statistics, 32*, 316–339.

Holt, C. C. (1960). *Planning production, inventories, and work force*. Englewood Cliffs, NJ: Prentice-Hall.

Knoth, S. (2003). EWMA schemes with non-homogeneous transition kernels. *Sequential Analysis, 22*, 241–255.

Reynolds, M. R. Jr., & Stoumbos, Z. G. (2001). Individuals control schemes for monitoring the mean and variance of processes subject to drifts. *Stochastic Analysis and Applications, 19*, 863–892.

Roberts, S. W. (1959). Control-charts-tests based on geometric moving averages. *Technometrics, 1*, 239–250.

Zou, C., Liu, Y., & Wang, Z. (2009). Comparisons of control schemes for monitoring the means of processes subject to drifts. *Metrika, 70*, 141–163.

Limit Properties of EWMA Charts for Stationary Processes

Manuel Cabral Morais, Yarema Okhrin, and Wolfgang Schmid

Abstract In this paper we consider a general family of EWMA charts for an arbitrary parameter of the target process. Our assumptions on the target process are very weak and they are usually satisfied if it is stationary. We distinguish between the EWMA chart based on the exact variance and the EWMA scheme based on the asymptotic variance. In the case of the EWMA chart with exact variance the in-control variance of the EWMA recursion at time t is used for the decision at time t while in the case of the asymptotic variance at each time point the limit of the in-control variance of the EWMA chart for t tending to infinity is applied. It is analyzed how the distributions of the corresponding run lengths behave if the smoothing parameter tends to zero. We show that the distribution of the run length of the EWMA chart based on the exact variance converges to the distribution of the run length of the repeated significance test while the limit of the EWMA scheme based on the asymptotic variance is degenerate. It is either 0 or 1. This result underlines the weakness of the schemes based on the asymptotic variance if the smoothing parameter is small. Moreover, several properties of the limit chart, i.e. the chart based on the repeated significance test, are presented as well.

Keywords EWMA chart • Stationary process • Statistical process control • Run length

M.C. Morais (✉)
Department of Mathematicss and CEMAT Instituto Superior Técnico, Technical University of Lisbon, Av. Rovisco Pais, 1049-001, Lisboa, Portugal
e-mail: maj@math.ist.utl.pt

Y. Okhrin
Department of Statistics, University of Augsburg, Universitätsstr. 16, 86159, Augsburg, Germany
e-mail: yarema.okhrin@wiwi.uni-augsburg.de

W. Schmid
Department of Statistics, European University Viadrina, Grosse Scharrnstrasse 59, 15230, Frankfurt (Oder), Germany
e-mail: schmid@euv-frankfurt-o.de

H.-J. Lenz et al. (eds.), *Frontiers in Statistical Quality Control 10*,
Frontiers in Statistical Quality Control 10, DOI 10.1007/978-3-7908-2846-7_5,

1 Introduction

Since the introduction of the exponentially weighted moving average (EWMA) chart by Roberts (1959) this scheme has become one of the most discussed control charts in the literature on statistical process control. Because its decision rule is quite simple it has shown to be an attractive control chart for practitioners. Similar to the CUSUM chart of Page (1954) it depends on a further parameter, the smoothing parameter, which regulates the influence of past observations on the present decision. If the smoothing parameter is equal to 1 the chart coincides with the Shewhart chart (cf. Shewhart 1931). The smaller the value of the smoothing parameter the larger is the influence of the preceding observations. Recommendations about the choice of the smoothing parameter were given by several authors. Montgomery (2009) writes that values of λ in the interval $0.05 \le \lambda \le 0.25$ work well in practice. The aim of Lucas and Saccucci (1990) was to find an optimal design procedure. Fixing the desired in-control average run length (ARL) of the EWMA chart they determined the value of the smoothing parameter which minimizes the out-of-control ARL for a given value of the expected shift. In their paper the minimum is determined for values of the smoothing parameter lying within the interval (0.03, 1.0). One reason why a lower bound for the smoothing parameter was chosen probably lies in the fact that the calculation of the average run length of the EWMA chart turns out to be quite complicate (see, e.g., Brook and Evans 1972; Crowder 1987) and numerically instable if the smoothing parameter is very small.

Most of the literature on EWMA charts is concerned with the monitoring of the mean of an independent random sample. A further great advantage of the EWMA approach consists in the fact that it can be easily extended to monitor other parameters of a process and that the structure of the underlying process can be quite general. EWMA charts for the standard deviation of an independent random process have been introduced by Crowder and Hamilton (1992). The extension of the EWMA chart to monitor the mean of a stationary time series process was given by Schmid (1997). EWMA charts for the variance of a stationary process were proposed in Schipper and Schmid (2001) while Rosołowski and Schmid (2003) discussed simultaneous EWMA schemes for the mean, the variances, and the autocovariances. These examples describe only a few applications of EWMA charts.

Many researchers followed the proposal of Lucas and Saccucci (1990) about the determination of the optimal smoothing parameter and they obtained it by minimizing the out-of-control ARL for a fixed in-control ARL. Due to the improvement of the computer power in the last years it is nowadays possible to calculate the ARL for smaller bounds than 0.03 as well. Doing this something interesting happens. It has turned in some cases that the smaller the lower bound the smaller the ARL and thus the optimal smoothing parameter is equal to the lower bound. This problem was described by Chan and Zhang (2000) and it was analyzed in more detail by Frisén and Sonesson (2006); both papers focus on the mean chart and independent samples.

A deeper analysis of the convergence of the distribution of the run length of the one-sided EWMA chart for the mean was provided by Morais et al. (2010). They proved that for a fixed control limit the in-control ARL of the EWMA scheme based on the asymptotic variance is a decreasing function in the smoothing parameter. Moreover, they analyzed the limit of the distribution of the run length as the smoothing parameter turns to zero. The resulting limit chart turns out to be equal to the repeated significance test. These authors also derived several properties of the limit chart.

In the present paper we consider a very general family of EWMA schemes which can be used to monitor an arbitrary real-valued parameter. It is assumed that the statistics to which the EWMA recursion is applied are governed by a stationary process. Because of its generality this approach covers most of the EWMA schemes discussed in literature. We distinguish between two types of decision rules. The first rule is based on a comparison of the deviation of the EWMA statistic and the target parameter with the in-control variance of the EWMA recursion. Because at each time point the variance changes it must be calculated in each step. This procedure is called the EWMA chart based on the exact variance has also been considered by Morais et al. (2010) as well. Because in each step the variance must be calculated most practitioners prefer to work with the asymptotic variance. In that case the distance between the EWMA recursion and the target value is compared with the asymptotic in-control variance of the EWMA recursion. This is our second procedure. It turns out to be much simpler and is used in most papers on EWMA charts.

In Section 2 we introduce the EWMA charts based on the asymptotic and exact variance. Moreover, we analyze the one-sided and the two-sided monitoring problem.

Section 3 deals with the limit distributions of the run lengths of the EWMA recursions. It is shown that the run lengths of the charts based on the exact variance converge to the distribution of the run length of the repeated significance test. This result holds for the in-control and out-of-control case as well. As special cases we consider the monitoring of the mean and the variance of a stationary process. In the second part of this section the limit of the charts based on the asymptotic variance is determined. It is shown to be degenerate, i.e. it is either 0 or 1. This is a clear hint that the charts based on the asymptotic variance should not be used with an extremely small smoothing parameter. In the most realistic case the probability of a signal up to a fixed time point converges to 1 as well in the in-control state as in the out-of-control state. This is a very unpleasant property. Moreover, it shows that the ARL converges to infinity if the smoothing parameter tends to zero.

Section 4 is devoted to the analysis of the limit scheme, i.e. the chart using the repeated significance test. Here we restrict ourselves to independent samples and a one-sided chart. Several properties of the limit scheme are presented which question the usefulness of the ARL as a performance measure for EWMA schemes.

In Section 5 we summarize the main conclusions of our paper.

2 EWMA Charts for Stationary Processes

In what follows $\{Y_t\}$ denotes the target process and $\{X_t\}$ stands for the observed process. Both processes are assumed to be the same up to time point $q-1$, i.e. $X_t = Y_t$ for $t = 1, \ldots, q-1$. At the time point q it is assumed that $X_q \neq Y_q$. The unknown position q is assumed to be a deterministic quantity taking values within the set $\mathbb{N} \cup \{\infty\}$. If $q < \infty$ then a sustained shift has occurred at time $t = q$. The process $\{X_t\}$ is called to be out-of-control. Needless to say, $\{X_t\}$ is said to be in-control if $q = \infty$.

We are interested to monitor a parameter θ of the target process by using the statistic T_t, which we assume is a function of past and present values of the observed process, i.e. $T_t = f_t(X_1, \ldots, X_t)$. The statistic T_t can be interpreted as a point estimator for the parameter θ. Suppose that T_t is an unbiased estimator of θ in the in-control state. If θ is equal to the mean μ of the $\{Y_t\}$ then we can choose, e.g., $T_t = X_t$ (see Schmid 1997), $T_t = \sum_{v=1}^{k} X_{t+1-v}/k$ or $T_t = c_1 med\{X_{t+1-k}, .., X_t\}$. In case θ is the variance of $\{Y_t\}$ possible choices would be, e.g., $T_t = (X_t - \mu)^2$ (e.g., Schipper and Schmid 2001), $T_t = \sum_{v=1}^{k}(X_{t+1-v}-\mu)^2/k$ or $T_t = c_2 \sum_{v=1}^{k} |X_{t+1-v} - \mu|/k$.

The EWMA recursion applied to T_t is given by

$$Z_t = \begin{cases} Z_0, & t = 0 \\ (1-\lambda)Z_{t-1} + \lambda T_t, & t = 1, 2, \ldots, \end{cases} \tag{1}$$

with an initial value Z_0. The parameter λ is a smoothing parameter taking values within the set $(0, 1]$ and it corresponds to the weight given to the most recent observed value. Having in mind that Z_t can be equivalently written as the following moving average

$$Z_t = \lambda \sum_{i=0}^{t-1} (1-\lambda)^i T_{t-i} + (1-\lambda)^t Z_0, \quad t = 1, 2, \ldots, \tag{2}$$

whose weights fall off geometrically, we immediately conclude that a value of λ close to one leads to a short memory EWMA chart – in fact $\lambda = 1$ leads to nothing but a Shewhart chart –, whereas values of λ close to zero lead to EWMA charts that give little importance to the most recent observations.

Since T_t is an unbiased estimator of θ, we can assert that in the in-control state,

$$E_\infty(Z_t) = \theta + (1-\lambda)^t [Z_0 - \theta] .$$

The index "∞" means, throughout the remainder of this paper, that the quantity (an expectation, a variance, a covariance, a probability, etc.) is calculated with respect to the in-control situation.

Now suppose that in the in-control state $\{T_t\}$ is a (weakly) stationary process with mean θ and autocovariance function $\{\gamma_{T,h}\}$. Then

$$\begin{aligned}\mathrm{Var}_\infty(Z_t) &= \lambda^2 \sum_{i,j=0}^{t-1} (1-\lambda)^{i+j} \gamma_{T,|i-j|} \\ &= \frac{\lambda}{2-\lambda}\left[[1-(1-\lambda)^{2t}]\gamma_{T,0} + 2\sum_{v=1}^{t-1}(1-\lambda)^v[1-(1-\lambda)^{2(t-v)}]\gamma_{T,v}\right].\end{aligned} \tag{3}$$

Upward shifts in θ can be detected by upper one-sided EWMA charts which give a signal at the sampling period $t \geq 1$, suggesting that the parameter θ increased, if

$$Z_t > \theta + c\sqrt{\mathrm{Var}_\infty(Z_t)},$$

for some fixed constant critical value c that defines the range of these exact control limits. Note that we make use of the asymptotic mean θ instead of the exact one $E_\infty(Z_t)$. This signal is a valid one, in case the process is out-of-control, and it is called a false alarm, otherwise.

In order to detect an upward or a downward shift, we have to make use of two-sided EWMA charts which trigger a signal whenever

$$|Z_t - \theta| > c\sqrt{\mathrm{Var}_\infty(Z_t)},$$

with $c > 0$.

To address both types of EWMA charts, we define $\mathcal{R}$ as the rejection area and thus a signal is given if

$$\frac{Z_t - \theta}{\sqrt{\mathrm{Var}_\infty(Z_t)}} \in \mathcal{R},$$

where $\mathcal{R} = (c, \infty)$ in the upper one-sided case, and $\mathcal{R} = (-\infty, -c) \cup (c, \infty)$ in the two-sided case.

In practice the asymptotic variance is frequently used instead of the exact one. In that case a signal is given if

$$\frac{Z_t - \theta}{\sqrt{\lim_{t\to\infty} \mathrm{Var}_\infty(Z_t)}} \in \mathcal{R}\,.$$

Note that

$$\lim_{t\to\infty} \mathrm{Var}_\infty(Z_t) = \lambda^2 \sum_{i,j=0}^{\infty} (1-\lambda)^{i+j} \gamma_{T,|i-j|}$$

if $\{T_t\}$ has an absolutely summable autocovariance function $\{\gamma_{T,h}\}$. Another representation of the asymptotic variance based on the spectral density function is given in Schmid (1997).

In what follows, θ, $\text{Var}_\infty(Z_t)$, and $\lim_{t\to\infty}\text{Var}_\infty(Z_t)$ are assumed to be known quantities. We shall not discuss the influence of parameter estimation.

3 Limit Behavior for Stationary Processes

Let $N(\lambda, \mathcal{R})$ denote the run length of control chart based on the exact variance, i.e.

$$N(\lambda,\mathcal{R}) = \inf\left\{t \in \mathbb{N} : \frac{Z_t - \theta}{\sqrt{\text{Var}_\infty(Z_t)}} \in \mathcal{R}\right\},$$

and $N_{asymp}(\lambda, \mathcal{R})$ be the run length of the scheme with the asymptotic variance

$$N_a(\lambda,\mathcal{R}) = \inf\left\{t \in \mathbb{N} : \frac{Z_t - \theta}{\sqrt{\lim_{t\to\infty}\text{Var}_\infty(Z_t)}} \in \mathcal{R}\right\}.$$

We denote the run length of the repeated significance test by

$$N(\mathcal{R}) = \inf\left\{t \in \mathbb{N} : \frac{\bar{T}_t - \theta}{\sqrt{\text{Var}_\infty(\bar{T}_t)}} \in \mathcal{R}\right\},$$

where $\bar{T}_t = \sum_{v=1}^{t} T_v/t$. It is worth mentioning that

$$t\,\text{Var}_\infty(\bar{T}_t) = \gamma_{T,0} + 2\sum_{v=1}^{t-1}(1 - v/t)\gamma_{T,v}, \tag{4}$$

whose limit, when t tends to infinity, exists as long as $\sum_{v=1}^{\infty}|\gamma_{T,v}| < \infty$.

3.1 The EWMA Scheme Based on the Exact Variance

First, we consider the probability of getting no signal up to a fixed time point for the present EWMA scheme if λ tends to zero. It is shown that the limit is equal to the probability of no false signal for the repeated significance test, i.e. the sequential application of the corresponding significance test using the critical value c.

Theorem 1. *Assume that the k-dimensional random vector $(X_1, \ldots, X_k)$ is continuous or discrete and that its distribution does not depend on λ. If $Z_0 = \theta$ then*

$$\lim_{\lambda\to 0+} P[N(\lambda,\mathcal{R}) > k] = P[N(\mathcal{R}) > k],$$

for any fixed $k = 1, 2, \ldots$ and any set $\mathcal{R} \subset IR$ which does not depend on λ.

Proof. The run length of the EWMA scheme with exact variance exceeds k with probability

$$P[N(\lambda,\mathcal{R}) > k] = P\left[\frac{Z_t - \theta}{\sqrt{\mathrm{Var}_\infty(Z_t)}} \notin \mathcal{R} \quad \forall\, t = 1,\ldots,k\right]$$
$$= \int_{A(\lambda)} \cdots \int f_{X_1,\ldots,X_k}(x_1,\ldots,x_k)\,dx_1\ldots dx_k,$$

where $A(\lambda) = \cap_{t=1}^{k} A_t(\lambda)$ with

$$A_t(\lambda) = \left\{(x_1,\ldots,x_t) : \frac{Z_t - \theta}{\sqrt{\mathrm{Var}_\infty(Z_t)}} \notin \mathcal{R}\right\}.$$

Because

$$\frac{Z_t - \theta}{\sqrt{\mathrm{Var}_\infty(Z_t)}} = \frac{\sum_{i=0}^{t-1}(1-\lambda)^i (T_{t-i} - \theta)}{\sqrt{\mathrm{Var}_\infty(Z_t)/\lambda}} \tag{5}$$

and

$$\lim_{\lambda\to 0+} \frac{\mathrm{Var}_\infty(Z_t)}{\lambda^2}$$
$$= \frac{1}{2}\lim_{\lambda\to 0+} \frac{\left[[1-(1-\lambda)^{2t}]\gamma_{T,0} + 2\sum_{v=1}^{t-1}(1-\lambda)^v[1-(1-\lambda)^{2(t-v)}]\gamma_{T,v}\right]}{\lambda}$$
$$= t^2\,\mathrm{Var}_\infty(\bar{T}_t),$$

the use of Eq. 4 leads to

$$\lim_{\lambda\to 0+} A_t(\lambda) = \left\{(x_1,..,x_t) : \frac{\bar{T}_t(x_1,\ldots,x_t) - \theta}{\sqrt{\mathrm{Var}_\infty(\bar{T}_t)}} \notin \mathcal{R}\right\}$$

and

$$\lim_{\lambda\to 0+} \bigcap_{t=1}^{k} A_t(\lambda) = \bigcap_{t=1}^{k} \left\{(x_1,..,x_t) : \frac{\bar{T}_t(x_1,\ldots,x_t) - \theta}{\sqrt{\mathrm{Var}_\infty(\bar{T}_t)}} \notin \mathcal{R}\right\}. \tag{6}$$

Moreover, if we define the set on the right side of Eq. 6 by A, then

$$\lim_{\lambda\to 0+} \int_{A(\lambda)} \cdots \int f_{X_1,\ldots,X_k}(x_1,\ldots,x_k)\,dx_1\ldots dx_k$$
$$= \int_{A} \cdots \int f_{X_1,\ldots,X_k}(x_1,\ldots,x_k)\,dx_1\ldots dx_k,$$

thus proving the result.

Remarkably, the theorem holds both in the in-control and in the out-of-control states. In addition, if we combine the fact that $E[N] = \sum_{k=0}^{\infty} P[N > k]$ and Theorem 1, we can conclude that

$$\lim_{\lambda \to 0+} E[N(\lambda, \mathcal{R})] = E[N(\mathcal{R})].$$

This means that if λ converges to zero then the ARL of the EWMA chart for monitoring θ converges to the ARL of the repeated significance test. We ought to mention that in some cases $E[N(\mathcal{R})] = \infty$; this point is analyzed in more detail in the next section.

Note that the above result is quite general and valid for any EWMA chart whose input statistics $\{T_t\}$ are governed by a stationary process.

Example 1. (a) Monitoring the mean of a stationary process

In this case we have $\theta = \mu$. Following Schmid (1997) we choose $T_t = X_t$. If $\{Y_t\}$ is stationary then in the in-control state $\{T_t\}$ is stationary as well. In the special case that $\{Y_t\}$ is a causal ARMA process (cf. Brockwell and Davis 1991) the autocovariances of $\{Y_t\}$ can be determined recursively by making use of the Yule-Walker equations (e.g., Brockwell and Davis 1991, Chap. 3). If $\{Y_t\}$ is a stationary GARCH process (cf. Tsay 2005, Chap. 3) then the determination of the variance of the EWMA recursion is easier and it holds that

$$\mathrm{Var}_{\infty}(Z_t) = \frac{\lambda}{2-\lambda}[1-(1-\lambda)^{2t}]\gamma_{T,0}$$

(b) Monitoring the variance of a stationary process

Suppose that $\mu = 0$, $\theta = \gamma_0$, $T_t = X_t^2$ (e.g., Schipper and Schmid 2001) and $Y_t = \sum_{i=-\infty}^{\infty} a_i \varepsilon_{t-i}$, where $\{a_i\}$ is absolutely summable. Let $\{\varepsilon_t\}$ be independent and normally distributed with $E(\varepsilon_t) = 0$ and $\mathrm{Var}(\varepsilon_t) = \sigma_\varepsilon^2$ then $\gamma_{T,h} = \gamma_0^2 + 2\gamma_h^2$ (cf. Brockwell and Davis 1991, p. 227), where γ_h stands for the autocovariance function of $\{Y_t\}$. Thus, we get

$$\begin{aligned}\mathrm{Var}_{\infty}(Z_t) = \frac{\lambda}{2-\lambda}\Bigg[\gamma_0^2\Bigg[&[3(1-(1-\lambda)^{2t}]\\ &+\frac{2(1-\lambda)}{\lambda}[1-(1-\lambda)^t][1-(1-\lambda)^{t-1}]\Bigg]\\ &+2\sum_{i=1}^{t-1}(1-\lambda)^i[1-(1-\lambda)^{2(t-i)}]\gamma_i^2\Bigg].\end{aligned}$$

This quantity can be calculated recursively for a stationary ARMA process as described in (a). For the special case of an ARMA(1,1) process $Y_t = \alpha Y_{t-1} + \varepsilon_t + \beta\varepsilon_{t-1}$, we get

$$\begin{aligned}\mathrm{Var}_\infty(Z_t) &= \frac{\lambda}{2-\lambda}\Bigg[\gamma_0^2\Bigg[[3(1-(1-\lambda)^{2t}]\\&+\frac{2(1-\lambda)}{\lambda}[1-(1-\lambda)^t][1-(1-\lambda)^{t-1}]\Bigg]\\&+2(1-\lambda)\gamma_1^2\left[\frac{1-[\alpha^2(1-\lambda)]^{t-1}}{1-\alpha^2(1-\lambda)}-(1-\lambda)^t\frac{(1-\lambda)^{t-1}-\alpha^{2(t-1)}}{1-\lambda-\alpha^2}\right]\Bigg].\end{aligned}$$

since $\gamma_i = \alpha^{i-1}\gamma_1$, for $i \geq 1$, with

$$\gamma_0 = \sigma_\varepsilon^2 \frac{1+2\alpha\beta+\beta^2}{1-\alpha^2} \quad \text{and} \quad \gamma_1 = \sigma_\varepsilon^2 \frac{(1+\alpha\beta)(\alpha+\beta)}{1-\alpha^2}.$$

Note that a stronger result for the mean chart for independent normal variables was shown by Morais et al. (2010). They proved that in the in-control state the probability of a false signal is a decreasing function in $\lambda \in (0,1]$. This result was obtained by using monotonicity results for the multivariate normal distribution (cf. Tong 1990).

3.2 *The EWMA Scheme Based on the Asymptotic Variance*

Next we analyze the run length of the scheme based on the asymptotic variance in the in-control state.

Theorem 2. *Assume that the k-dimensional random vector $(X_1,\ldots,X_k)$ is continuous or discrete and that its distribution does not depend on λ. Let $\mathcal{R}$ be an arbitrary subset of IR which does not depend on λ and suppose that $Z_0 = \theta$.*

(a) If $0 \notin \mathcal{R}$ then

$$\lim_{\lambda\to 0+} P[N_{asymp}(\lambda,\mathcal{R}) > k] = 1,\ k \in \mathbb{N}.$$

(b) If $0 \in \mathcal{R}$ then

$$\lim_{\lambda\to 0+} P[N_{asymp}(\lambda,\mathcal{R}) > k] = 0,\ k \in \mathbb{N}.$$

Proof. We only prove part (a). The proof of part (b) follows immediately.

First, note that

$$\begin{aligned}P[N_{asymp}(\lambda,\mathcal{R}) > k] &= P\left[\frac{Z_t-\theta}{\sqrt{\lim_{t\to\infty}\mathrm{Var}_\infty(Z_t)}} \notin \mathcal{R}, \quad \forall t = 1,\ldots,k\right]\\&= \int_{A_a(\lambda)}\cdots\int f_{X_1,\ldots,X_k}(x_1,\ldots,x_k)dx_1\ldots dx_k,\end{aligned}$$

where $A_a(\lambda) = \cap_{t=1}^{k} A_{a,t}(\lambda)$ with

$$A_{a,t}(\lambda) = \{(x_1, \dots, x_t) : \frac{Z_t - \theta}{\sqrt{\lim_{t\to\infty} \mathrm{Var}_\infty(Z_t)}} \notin \mathcal{R}\}.$$

Because

$$\frac{Z_t - \theta}{\sqrt{\lim_{t\to\infty} \mathrm{Var}_\infty(Z_t)}} = \frac{\sum_{i=0}^{t-1}(1-\lambda)^i (T_{t-i} - \theta)}{\sqrt{\lim_{t\to\infty} \mathrm{Var}_\infty(Z_t)/\lambda}} \tag{7}$$

and

$$\lim_{\lambda\to 0+} \frac{\lim_{t\to\infty} \mathrm{Var}_\infty(Z_t)}{\lambda^2} = \frac{1}{2} \lim_{\lambda\to 0+} \frac{\left[\gamma_{T,0} + 2\sum_{v=1}^{\infty}(1-\lambda)^v \gamma_{T,v}\right]}{\lambda} = \infty,$$

we successively get, for case (a),

$$\lim_{\lambda\to 0+} A_t(\lambda) = \Omega$$

$$\lim_{\lambda\to 0+} \bigcap_{t=1}^{k} A_t(\lambda) = \Omega,$$

and

$$\lim_{\lambda\to 0+} \int_{A(\lambda)} \cdots \int f_{X_1,\dots,X_k}(x_1, \dots, x_k) dx_1 \dots dx_k$$
$$= \int_{\Omega} \cdots \int f_{X_1,\dots,X_k}(x_1, \dots, x_k) dx_1 \dots dx_k = 1,$$

thus proving the result for case (a).

Note that case (a) is the one we usually deal with. It arises if we choose $c > 0$ and take $\mathcal{R} = (c, \infty)$, in the one-sided case, or consider $\mathcal{R} = (-\infty, -c) \cup (c, \infty)$, in the two-sided case. Choosing $c < 0$ means that $0 \in \mathcal{R}$ and that we are dealing with a negative control limit, in the one-sided case, which is rather strange to most practitioners. However, we shall address this case later on.

The result of part (a) is highly desirable in the in-control state because it means that the probability of a false alarm converges to zero as λ tends to zero. However, the result also holds in the out-of-control state, that is, the probability of a signal within the first k samples converges to 1. As a consequence the EWMA scheme with asymptotic variance behaves quite chaotically if λ reaches 0. It also implies that $\lim_{\lambda\to 0+} E[N_{asymp}(\lambda, c)] = \infty$, therefore the behavior of the EWMA control charts with different values of λ cannot be compared by means of the average run length. Let us remind the reader that the in-control ARL is used to determine the control limits but in the present case this cannot be done since this quantity does not exist if λ converges to zero.

For case (b) the scheme shows similar undesirable properties. For instance, the probability of a correct signal converges to 0 in the in-control state when λ converges to zero.

In sum, if λ takes values close to zero the control chart based on the exact variance must be favored: it has at least a reasonable limit behavior – the repeated significance test. For a better understanding of this limit behavior, it is necessary, however, to analyze the properties of the repeated significance test.

4 Some Properties of the Limit Chart

In this section we present some results of the limit chart. We focus on the detection of a change in the mean and choose $T_t = X_t$. Moreover, we restrict ourselves to the one-sided problem, i.e. we choose $\mathcal{R} = (c, \infty)$. Then the limit chart, i.e. the repeated significance test, has run length given by

$$N(c) = \inf\{t \in \mathbb{N} : \frac{\sum_{i=1}^{t}(X_i - \mu)}{\sqrt{t[\gamma_0 + 2\sum_{i=1}^{t-1}(1 - i/t)\gamma_i]}} > c\}.$$

In order to illustrate how the limit scheme may behave, we focus on the case of independent random variables. A detailed analysis of the repeated significance test is given in Morais et al. (2010). In that case we have that

$$N(c) = \inf\{t \in \mathbb{N} : \frac{\sum_{i=1}^{t}(X_i - \mu)}{\sqrt{t\gamma_0}} > c\}.$$

We make use of the following change point model

$$X_t = \begin{cases} Y_t & \text{for } t < q \\ Y_t + a\sqrt{\gamma_0} & \text{for } t \geq q. \end{cases} \tag{8}$$

In what follows we use the symbols $P_{a,q}$, $E_{a,q}$, etc. to denote a probability, expectation, etc. taken with respect to model Eq. 8.

First, we discuss the behavior of the limit scheme in the in-control state.

Theorem 3. *Assume that the random variables $\{Y_t\}$ are independent and identically distributed with mean μ and variance γ_0.*

(a) If $c \geq 0$ then $E_\infty[N(c)] = \infty$.

Table 1 The fraction (in %) of the run lengths falling into the respective interval as a function of c. The computations are based on 10^9 replications of iid Gaussian observations

c	$E[N(c)]$	$N(c) \leq 10$ (in %)	$10 < N(c) \leq 10^4$ (in %)	$10^4 < N(c) \leq 10^7$ (in %)	$10^7 < N(c)$ (in %)
−1.5	1.266	99.64	0.36	3.84e−05	0
−1.4	1.406	99.50	0.51	8.97e−05	0
−1.3	1.614	99.28	0.71	2.05e−04	1e−07
−1.2	2.078	99.02	0.98	4.35e−04	7e−07
−1.1	2.707	98.66	1.34	8.74e−04	4e−07
−1.0	4.342	98.20	1.80	1.83e−03	1.5e−06
−0.9	8.284	97.61	2.38	3.62e−03	6.7e−06
−0.8	21.81	96.87	3.12	7.05e−03	1.53e−05
−0.7	59.14	95.95	4.03	1.34e−02	4.27e−05
−0.6	140.9	94.84	5.15	2.47e−02	1.16e−04

(b) *Suppose that* $P[Y_t = 0] < 1$ *and that the variables* $\{Y_t\}$ *are symmetric around* μ. *If* $c < 0$ *then* $E_\infty[N(c)] < \infty$ *and* $Var_\infty[N(c)] < \infty$.

This result is remarkable. It implies that the in-control ARL of the limit chart is equal to infinity if the control limit is nonnegative. Thus, together with the results of Section 3, it follows that $\lim_{\lambda \to 0+} E_\infty[N(\lambda, c)] = \infty$. Note that practitioners choose a positive control limit and thus the in-control ARL is not finite in the most popular case. Theorem 3 also has an important consequence on the comparison of EWMA charts because it states that EWMA charts should not be compared by means of the average run length, at least when λ is very small.

Table 1 illustrates the divergence of the run length if the critical value tends to zero from the left. Despite of a small in-control ARL, there is a substantial number of run lengths exceeding 10^7. It is assumed for simplicity that the observations are iid and follow Gaussian distribution.

Now let us assume that the control limit c is chosen as a solution of $E_\infty[N(\lambda, c)] = \xi$, with $\xi > 1$, denote $c = c(\lambda, \xi)$ and discuss the behavior of $c(\lambda, \xi)$ as λ tends to 0.

Theorem 4. *Assume that the random variables* $\{Y_t\}$ *are independent and identically distributed to* $\mathcal{N}(\mu, \gamma_0)$. *Then* $\lim_{\lambda \to 0+} c(\lambda, \xi) < 0$.

The control limit of the upper one-sided EWMA chart is, by definition, positive. Theorem 4 shows, however, that there is no positive solution for the limit chart. Thus, it is impossible to compare the behavior of the EWMA chart with that of the limit chart if the control limit is chosen to be equal to a specified constant. Figure 1 shows the behaviour of the control limit c with $\xi = 50$ both for the asymptotic and the exact variances as a function of λ.

Now we investigate the out-of-control behavior of the limit scheme. We ought to begin by noting that contrary to the in-control ARL, the out-of-control ARL is always finite, as stated in the next theorem.

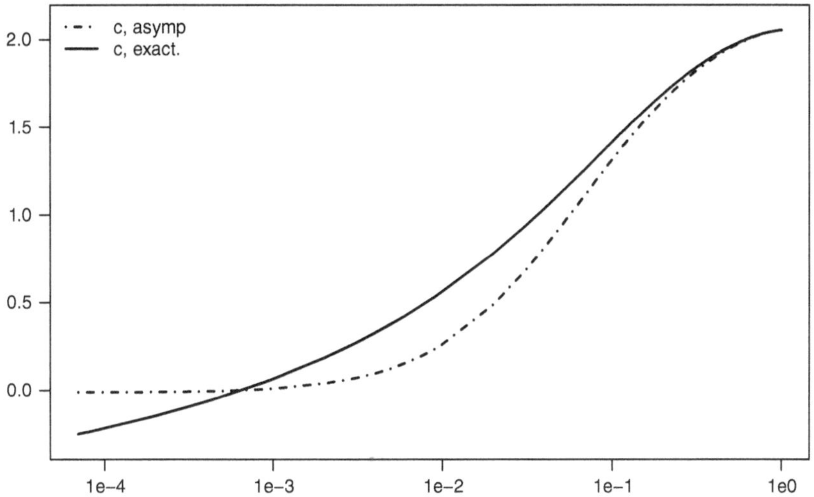

Fig. 1 The critical value c as a function of λ for the EWMA schemes with exact and asymptotic variance, for $\xi = 50$ and iid Gaussian observations. The results are based on a Monte-Carlo study with 10^9 replications

Theorem 5. *Assume that the random variables $\{Y_t\}$ are independent and identically distributed to $\mathcal{N}(\mu, \gamma_0)$. Then $E_{a,q}[N(c)] < \infty$, for all $c \in I\!R$, $q \in \mathbb{N}$, and shifts with magnitude $a > 0$.*

Figure 2 illustrates this result. In fact, it shows that the out-of-control ARL converges to a finite value as $\lambda \to 0+$ for several upward shifts with magnitude a.

We shall now discuss the average delay of the limit scheme, another frequently used performance measure. Recall that, when dealing with control chart with run length N, the average delay is given by

$$E_{a,q}(N(c) - q + 1 | N(c) \geq q) = \sum_{k=q}^{\infty} \frac{P_{a,q}[N(c) \geq k]}{P_{a,q}[N(c) \geq q]}.$$

Note that Theorem 5 allows us to assert that the average delay exists for $q < \infty$ and $a > 0$. The next theorem refers to the behavior of the average delay when the magnitude of the shift, a, tends to zero.

Theorem 6. *Assume that the random variables $\{Y_t\}$ are independent and identically distributed to $\mathcal{N}(\mu, \gamma_0)$ and let $q \in \mathbb{N}$.*

(a) If $c > 0$ then $\lim_{a \to 0+} E_{a,q}[N(c) - q + 1 | N(c) \geq q] = \infty$.
(b) If $c < 0$ then $\lim_{a \to 0+} E_{a,q}[N(c) - q + 1 | N(c) \geq q] < \infty$.

Part (a) from Theorem 6 reads as follows: the average delay of the limit chart has an undesirable behavior when a positive control limit is at use. Moreover, since this

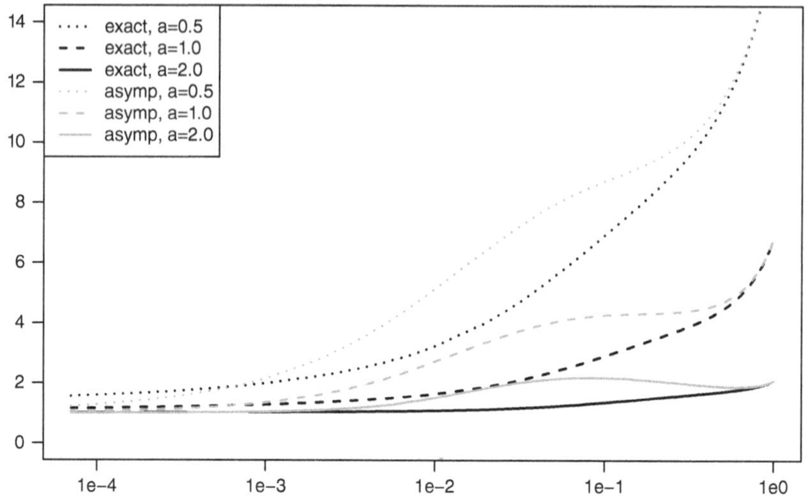

Fig. 2 The out-of-control ARL as a function of λ for the EWMA schemes based on the exact and the asymptotic variance, for $\xi = 50$ and iid Gaussian observations. The results are based on a Monte-Carlo study with 10^9 replications

result implies that the EWMA scheme behaves similar as λ tends to zero, we can conclude that all criteria based on the first moment of the run length are not suitable to assess or compare the performance EWMA charts with small values of λ.

5 Concluding Remarks

This paper essentially provides a thorough study on the behaviour of the run length of EWMA charts with exact and asymptotic control limits when λ converges to zero. We ought to stress that the results are quite general and refer to the control of any parameter of a stationary process. For instance, we proved that, when the smoothing parameter λ tends to zero:

- The run length of EWMA charts based on the exact variance has the same behavior as the run length of a chart based on a repeated significance test, what we called the limit chart;
- The out-of-control run length of EWMA charts based on the asymptotic variance is infinite if the rejection area $\mathcal{R}$ includes the origin.

Finally, this study also brought to light a few useful results concerning the run length of the limit chart, namely for the mean independent and identically distributed processes:

- Its in-control ARL is not finite if the control limit c in nonnegative.

In summary, our results permit the following conclusions:

1. The EWMA scheme with exact variance must be preferred over the EWMA scheme with asymptotic variance.
2. The EWMA scheme with asymptotic variance should not be applied if the smoothing parameter is small (about $\lambda \leq 0.1$).
3. The average run length is no suitable performance measure for EWMA charts if the smoothing parameter is very small.
4. One-sided EWMA charts should make use of a reflecting boundary.

References

Brockwell, P. J., & Davis, R. A. (1991). *Time series: Theory and methods*. New York: Springer.

Brook, D., & Evans, D. A. (1972). An approach to the probability distributions of CUSUM run length. *Biometrika, 3*, 539–549.

Chan, L. K., & Zhang, J. (2000). Some issues in the design of EWMA charts. *Communications in Statistics – Simulation and Computation, 29*, 207–217.

Crowder, S. V. (1987). A simple method for studying run-length distributions of exponentially weighted moving average charts. *Technometrics, 29*, 401–407.

Crowder, S. V., & Hamilton, M. D. (1992). EWMA for monitoring a process standard deviation. *Journal of Quality Technology, 24*, 12–21.

Frisén, M., & Sonesson, C. (2006). Optimal surveillance based on exonentially weighted moving averages. *Sequential Analysis, 25*, 379–403.

Lucas, J. M., & Saccucci, M. S. (1990). Exponentially weighted moving average control schemes: Properties and enhancements. *Technometrics, 32*, 1–12.

Montgomery, D. C. (2009). *Introduction to statistical quality control* (6th ed.). New York: Wiley.

Morais, M. C., Okhrin, Y., & Schmid, W. (2009). *On the limiting behaviour of EWMA charts with exact control limits*. Discussion Paper 272 of the Faculty of Business Administration and Economics, EUV, Frankfurt (Oder).

Page, E. S. (1954). Continuous inspection schemes. *Biometrika, 41*, 100–114.

Roberts, S.W. (1959). Control charts tests based on geometric moving averages. *Technometrics, 1*, 239–250.

Rosołowski, M., & Schmid, W. (2003). EWMA charts for monitoring the mean and the autocovariances of stationary Gaussian processes. *Sequential Analysis, 22*, 257–285.

Schmid, W. (1997). On EWMA charts for time series. In H. J. Lenz & P.-Th. Wilrich (Eds.), *Frontiers of statistical quality control* (Vol. 5, pp. 115–137). Heidelberg: Physica.

Schipper, S., & Schmid, W. (2001). Sequential methods for detecting changes in the variance of economic time series. *Sequential Analysis, 20*, 235–262.

Shewhart, W. A. (1931). *Economic control of quality of manufactured product*. Princeton: Van Notrand.

Tong, Y. L. (1990). *The multivariate normal distribution*. New York: Springer.

Tsay, R. S. (2005). *Analysis of financial time series* (2nd ed.). Hoboken, NJ: Wiley.

Economic Control Chart Policies for Monitoring Variables When There Are Two Components of Variance

Erwin Saniga, James Lucas, Darwin Davis, and Thomas McWilliams

Abstract When controlling a process mean one can achieve optimal performance in terms of the criterion of average run length (ARL) by using a CUSUM control chart rather than a Shewhart control chart, although for very large shifts the Shewhart control chart is equivalent to a CUSUM chart. Using cost as a criterion, several authors have shown that the ARL dominance of the CUSUM chart does not translate to a cost dominance unless the fixed cost of sampling is very small and some other configurations of the input parameters are met. Additionally, because of the simplicity of the Shewart chart in terms of user training, ease of design and ease of use it may be preferable to a CUSUM chart in these situations. Here, using a large experiment, we investigate the cost advantages of the CUSUM chart versus a common Shewhart control chart, the $\overline{X}$ chart, in the situation when one is monitoring a process mean and there are two components of variance. Our results are similar to the single component of variance results in that there are predictable regions where there is a large cost advantage to using CUSUM charts and there are also predictable regions where one can use an $\overline{X}$ without incurring any large increase in cost.

Keywords Control charts • Economic design • CUSUM charts • Shewhart charts • Two components of variance

E. Saniga (✉) · D. Davis
Department of Business Administration, University of Delaware, Newark, DE 19716, USA
e-mail: saniga@udel.edu; dd@udel.edu

J. Lucas
J. M. Lucas and Associates, 5120 New Kent Road, Wilmington, DE 19808, USA
e-mail: James.Lucas@verizon.net

T. McWilliams
Department of Decision Sciences, Bennett S. LeBow College of Business, Drexel University, 3141 Chestnut St., Philadelphia, PA 19104, USA
e-mail: tmcwilliams@drexel.edu

H.-J. Lenz et al. (eds.), *Frontiers in Statistical Quality Control 10*,
Frontiers in Statistical Quality Control 10, DOI 10.1007/978-3-7908-2846-7_6,

1 Introduction

If a control chart is to be used to monitor a stable process where the variable of interest is continuous one must choose between several different types of charts and policies for using these charts. For example one can employ the usual recommendation (e.g. see Montgomery 2001) to employ a CUSUM chart if the expected shift is small. Moustakides (1986) proved that the CUSUM chart is optimal in the sense that the average run length (ARL) of the CUSUM is minimum for detecting any particular shift for a fixed in control ARL. We note that for large expected shifts the $\overline{X}$ chart is a special case of a CUSUM chart so the ease of use and design and lack of abstractness of this chart might lead to its employment in this case. Lucas (1982) and Hawkins and Olwell (1998) recommend that one can use both simultaneously; e.g. one can use a CUSUM chart to monitor small shifts and a Shewhart chart such as the $\overline{X}$ to monitor large shifts.

In a large study comprised of a large range of shifts and other conditions, Reynolds and Stoumbos (2004) have compared Shewhart charts to CUSUM charts as well as other charts. Their general recommendation is that one should employ CUSUM or EWMA charts with samples sizes of $n = 1$ because of their strong performance overall.

Statistical performance is a sound criterion of design in a practical sense because one limits false searches and unnecessary process adjustments (which may lead to more variability) as well as guaranteeing high quality because assignable causes of poor quality are detected rapidly. From a theoretical perspective, statistical performance meshes well with the theory of hypothesis testing expressed in articles by Neyman and Pearson (1928, 1933a,b).

Another performance criterion of much practical importance is cost, and Neyman and Pearson (1928) emphasize the point that alternative criteria should be used in any decision. Additionally, one may prefer to use a Shewhart chart rather than a CUSUM chart because of its ease of design and use and lack of abstractness.

Some authors including Goel (1968), Von Collani (1987) and Saniga et al. (2006a,b) have compared the CUSUM chart to the $\overline{X}$ chart in terms of cost under the assumption that each was designed in an economically optimal fashion.

Some of the conclusions of the former three studies are that the cost advantages of the CUSUM chart over the $\overline{X}$ chart is small and that the optimal $\overline{X}$ chart's design was much like that of the CUSUM chart in that the reference value k is relatively large and the decision interval h is small (the $\overline{X}$ chart is a special case of the CUSUM with $k = 3$ and $h = 0$).

The Saniga et al. (2006b) study, which was conducted by running a very large experiment, contradicts some of the findings of some of the previous research on the subject. The differences are easily explained because the Saniga et al. (2006b) results were derived from a very wide configuration of problems considered in an experimental design. The four major conclusions were as follows: (1) The existence of fixed sampling costs make the choice of an $n = 1$ CUSUM far from economically optimal in most cases; (2) There are identifiable regions where a CUSUM chart is economically advantageous and there are regions where the use of an $\overline{X}$ chart may be preferred because of its simplicity and because there is no substantial cost disadvantage to using it; (3) CUSUM charts that have an economic advantage over $\overline{X}$ charts are dissimilar to the $\overline{X}$ chart in that k is small and h is large; and (4) There are regions with small shifts where a CUSUM chart has no economic advantage over an $\overline{X}$ chart.

This paper is an extension of the Saniga et al. (2006b) paper that considers the same questions except for the fact that we consider the employment of the CUSUM and $\overline{X}$ chart in a situation where there are two components of variance rather than one. For example, in process industries there are often two components of variance where there is between-samples variance that cannot be removed.

In the next section we discuss the problem in detail and present the experiment. Section 3 defines standardized shifts. Section 4 describes the experiment and Sect. 5 contains an analysis of the results. Some conclusions are drawn in Sect. 6.

2 The Cost Model

Lorenzen and Vance (1986) presented a general model for control chart design based on the initial model of Duncan (1956). In their model, the expected cost per hour is defined as C, and additional parameters are as defined below.

n = sample size
h = hours between samples
L = number of standard deviations from control limits to center line for the $\overline{X}$ chart
k = reference value for the CUSUM chart
h = decision interval for the CUSUM chart
g = intersample interval
δ = number of standard deviations slip when out of control
E = time to sample and chart one item
Y = cost per false alarm
A = fixed cost per sample
C_0 = quality cost per hour while the process is in control
C_1 = quality cost per hour while the process is out of control
T_0 = expected time to search for a false alarm
T_1 = expected time to discover the assignable cause
T_2 = expected time to repair the process
a = fixed cost per sample
b = variable cost per sample
λ = 1/(mean time the process is in control)

$\delta_1 = 1$ if production continues during searches, 0 if production ceases during search
$\delta_2 = 1$ if production continues during repair, 0 if production ceases during repair
W = cost to locate and repair the assignable cause

In this model, the standard assumptions in economic control chart design are made. These assumptions are that the process time in control follows the negative exponential distribution, the size of the shift is known, and the cost and system parameters are fixed and known.

Economic design involves finding the control chart parameters that minimize C. These control chart parameters are n, L and g for the $\overline{X}$ chart and n, g, k and h for the CUSUM chart. We use this model to find the optimal economic designs for the $\overline{X}$ chart and for the CUSUM chart. The solution procedure is described in Saniga et al. (2006b).

3 Definitions of Standardized Shifts

Suppose X is the quality variable of interest; we assume that the in control distribution of X is $X \sim N(\mu, \sigma^2)$ where μ and σ are respectively the in control mean and standard deviation of the process. To derive the standardized shift for the case where there is a single component of variance we define the out of control distribution of X as $X \sim N(\Delta\sigma + \mu, \sigma^2)$, where Δ is the magnitude of the process shift. Note that the distribution of the $\overline{X}$s are the same as above with the exception that the variance is $(\frac{\sigma}{\sqrt{n}})^2$. The standardized difference between the two $\overline{X}$s is $\frac{(\mu+\Delta\sigma)-\mu}{\sigma/\sqrt{n}} = \Delta\sqrt{n}$, which is the usual definition for the magnitude of the shift in the mean.

Now, with two components of variance such as the case where we have a between-samples variance, σ_b^2, in addition to the within-samples variance, σ_w^2, we have the in control distribution of X as $X \sim N(\mu, \sigma_b^2 + \sigma_w^2)$. The out of control distribution is $X \sim N(\Delta\sigma + \mu, \sigma_b^2 + \sigma_w^2)$ where $\sigma = (\sigma_b^2 + \sigma_w^2)^{1/2}$. Also, note that $\overline{X} \sim N(\mu, (\sigma_b^2 + \frac{\sigma_w^2}{n}))$ is the in control distribution of $\overline{X}$ where X_i are i.i.d. random variables and $\overline{X} \sim N(\Delta\sigma + \mu, (\sigma_b^2 + \frac{\sigma_w^2}{n}))$ is the out of control distribution of $\overline{X}$. Standardizing this difference we get the standardized shift as $Z = \frac{\Delta\sigma+\mu-\mu}{(\sigma_b^2+\sigma_w^2/n)^{1/2}} = \Delta\frac{(\sigma_b^2+\sigma_w^2)^{1/2}}{(\sigma_b^2+\sigma_w^2/n)^{1/2}}$.

Note that if $\sigma_b^2 = 0$ this becomes $Z = \Delta\sqrt{n}$. In other words the two models will give equal results.

To run experiments we set $\sigma_w^2 = 1$ and let $\sigma_b^2 = 0, 0.25, 0.5, 1, 2$.

4 The Experiments

We ran many configurations of input variables in three experiments in which we find the economically optimal design and cost of a CUSUM chart and an economically optimally design and cost of an $\overline{X}$ chart.

The three experiments are defined by two well know example problems in the literature. In the first experiment we found designs for 7,680 configurations of cost and system parameters based upon Chiu's (1974) first example. In this experiment we set

$C_0 = 0$
$C_1 = 100, 500, 1{,}000$
$b = 0.1, 1$
$a = 0.5, 10$
$\lambda = 0.01, 0.05$
$\delta = 0.5, 1, 2, 3$
$\delta_1 = \delta_2 = 0, 1$
$Y = W = 75, 500$
$t_0 = t_1 = 0.1, 0.5$
$t_2 = 0.2, 0.5$

The other parameter is $E_0 = 0$. Additionally, we set $\sigma_b^2 = 0, 0.25, 0.5, 1, 2$. Hereafter, we call this Experiment $A(E = 0)$.

The second experiment uses the same configuration of parameters as the first experiment with the exception that $E_0 = 0.5$. We call this Experiment $A(E = 0.5)$.

The third experiment is based upon an example proposed by Lorenzen and Vance (1987). Here, we found optimal designs for both the CUSUM and $\overline{X}$ chart for 17,280 configurations where we set

$C_0, C_1 = (0, 835)(114.2, 949)$
$b = 0.1, 0.5$
$a = 0, 10, 50, 200$
$\lambda = 0.02, 0.05$
$\delta = 0.5, 0.86, 1.5$
$Y = W = (200, 200)(977, 977)(1500, 1500)$
$t_0 = t_1 = (0.0833, 0.0833)(0.5, 0.5)$
$t_2 = 0.75, 1.5$
$E_0 = 0.0833, 0.5$

As in Lorenzen and Vance (1987) we set $\delta_1 = 1, \delta_2 = 0$. Again, we set $\delta_b^2 = 0, 0.25, 0.5, 1, 2$. We call this Experiment B. In each experiment we calculate the ratio of the cost of the optimally designed $\overline{X}$ chart to the cost of the optimally designed CUSUM chart. Note that we assume that all input costs are assumed to be the same for the CUSUM charts and the $\overline{X}$ charts; in practice it is usually true that the CUSUM charts cost more in terms of implementation and use which implies that our results are biased, if bias exists, towards the CUSUM chart. Note also that the ratio of costs

Table 1 Quantiles of the distribution of (Cost of an optimal $\overline{X}$ design)/(Cost of an optimal CUSUM design)

		Experiment		
Quantiles (%)		$A(E=0)$	$A(E=0.5)$	B
100.00	Maximum	2.3840	2.2013	3.5024
99.50		2.1456	2.0511	2.6610
97.50		1.7782	1.7461	2.1238
90.00		1.3391	1.3791	1.5114
75.00	Quartile	1.1141	1.1538	1.1886
50.00	Median	1.0045	1.0345	1.0179
25.00	Quartile	1.0002	1.0007	1.0000
10.00		1.0000	1.0000	1.0000
2.50		1.0000	1.0000	1.0000
0.50		0.9999	1.0000	1.0000
0.00	Minimum	0.9999	0.9537	0.9819
	N	7,680	7,680	17,280
	Mean	1.1033	1.9603	1.1613

must be greater than one since the ARL of the CUSUM chart is always less than or equal to the $\overline{X}$ chart.

In some cases the costs of using a control chart are larger than the cost of using a regular search policy; those cases were not reported in the results.

5 Analysis of Results

In Experiment $A(E = 0)$ we can fix $\sigma_b^2 = 0$. Note that here we are performing an experiment similar to Saniga et al. (2006), where there is only one component of variance. In this case, the maximum ratio of optimal $\overline{X}$ cost to optimal CUSUM cost is 1.0052. Thus, the maximum cost disadvantage of using an $\overline{X}$ chart over a CUSUM chart is slightly more than 0.5%. Compare this case to the case reported in Table 1, which gives the quantiles of the distribution of the ratio of the cost of an optimal $\overline{X}$ design to the cost of an optimal CUSUM design when σ_b^2 varies. Notice that for experiment $A(E = 0)$, the existence of two components of variance drastically changes the quantiles. Here, the advantage of the CUSUM is at a maximum 238% and in at least 25% of the cases the advantage is 11.4%. On the other hand, one may look at these results from the other perspective and see that the median advantage of the CUSUM control chart over the optimal $\overline{X}$ chart is very small, with advantages of 0.5%, 3.5% and 1.8% respectively in the three experiments. Generally, there are cases in which one should employ a CUSUM chart if there are two components of variance to gain the cost advantage (along with the ARL advantage, but not the advantage of simplicity) and there are cases in which one may employ the optimal $\overline{X}$ chart to take advantage of its simplicity without incurring an opportunity cost.

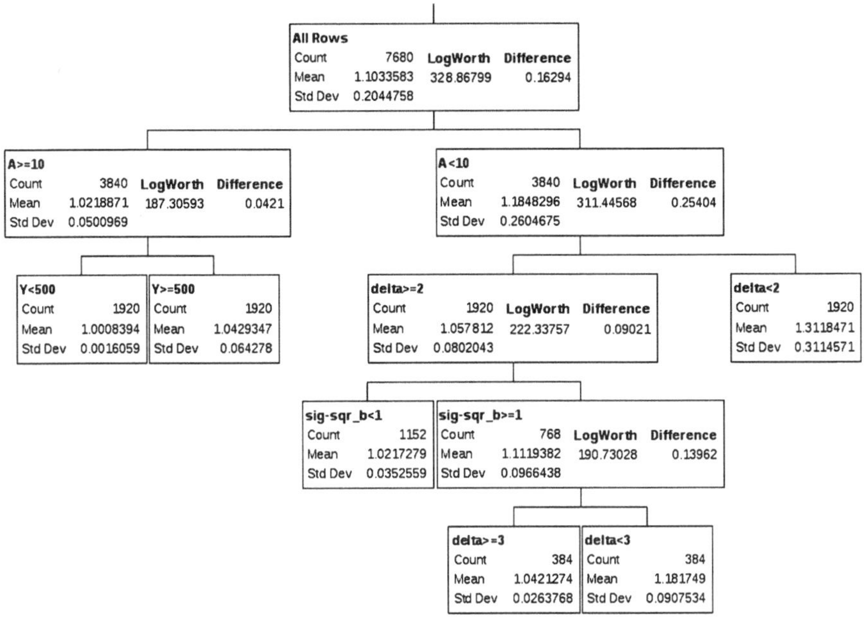

Fig. 1 Regression tree of (Cost of an optimal $\overline{X}$ design)/(Cost of an optimal CUSUM design) for experiment $A(E = 0)$

We investigate the situations in which one should use one chart versus the other with the aid of regression trees.

The regression tree for experiment $A(E = 0)$ is given in Fig. 1. In each cell the mean is defined as the average ratio of the cost of the $\overline{X}$ chart to the cost of the CUSUM chart. Note that the lower bound on this ratio is 1 since the CUSUM chart is an optimal procedure. Here, it is seen that if the fixed cost of sampling is high, say $A \geq 10$, then an optimal $\overline{X}$ chart has a mean disadvantage in cost of only 2%. This accounts for about half of the 7,680 cases we investigated in this experiment. If one considers that the standard deviation is 0.05 the implication is that the 99th percentile of the distribution of the cost ratios indicates a less than a 17% advantage to the CUSUM chart when $A \geq 10$. Also, if the cost of a false alarm is small, $Y < 500$, the 99th percentile of the distribution of the cost ratios is about 0.5%, which means that the optimal $\overline{X}$ chart should always be used due to its simplicity.

If we follow the regression tree for the smaller fixed costs of sampling one notes that for large shifts with a single component of variance ($\delta \geq 2$, $\sigma_b^2 < 1$) the 99th percentile of the distribution of the cost ratios is about an 11% cost disadvantage to the $\overline{X}$ chart, and further, for the largest shifts ($\delta \geq 3$) the 99th percentile of the distribution of the cost disadvantage is again 11% for the $\overline{X}$ chart even for the largest levels of σ_b^2. Generally, though, in this experiment there are substantial cost advantages to the CUSUM chart when $\sigma_b^2 \geq 1$.

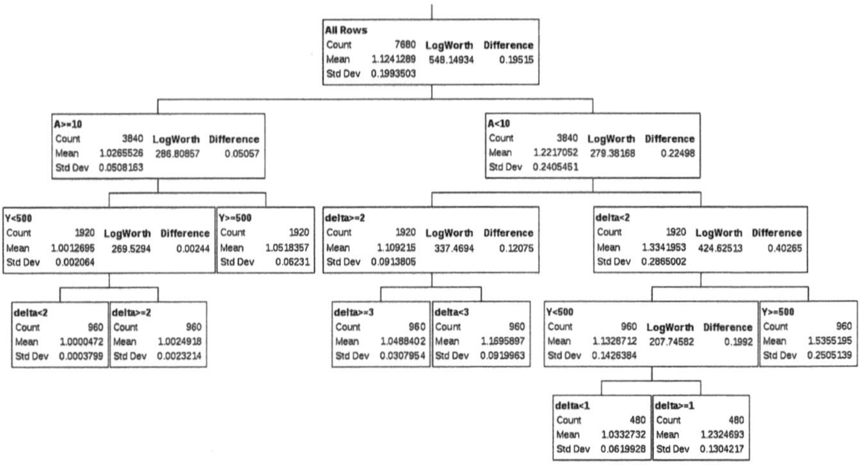

Fig. 2 Regression tree of (Cost of an optimal $\overline{X}$ design)/(Cost of an optimal CUSUM design) for experiment $A(E = 0.5)$

Experiment $A(E = 0.5)$ yields very similar results for the left side of the regression tree (see Fig. 2). Here, we see that for the larger fixed costs of sampling and smaller cost of searching for a false alarm that the 99th percentile of the distribution of the cost ratio indicates a less than 1% advantage in cost of the CUSUM chart. For the $A < 10$ side we find that the CUSUM is preferred other than if the shift is very large ($\delta \geq 3$) where the 99th percentile of the distribution of the cost ratios is about a 13% cost advantage of the CUSUM chart. A counterintuitive result is that when $Y < 500$, we find that the performance of the CUSUM is relatively worse when $\delta = 0.5$ when compared to when $\delta \geq 1$.

Experiment B, which is the largest experiment with 17,280 designs determined for each chart, yields very similar results in that if the fixed cost of sampling is large ($A \geq 50$) one should employ an $\overline{X}$ chart and if the cost is small one should employ a CUSUM chart (see Fig. 3).

One surprising finding gleaned from the regression trees (other than for one tree on one experiment) is that the effect of the second component of variance does not seem to make a difference unless one is dealing with a situation in which the fixed costs of sampling are small and the expected shift is small. To look at this a little closer we present Table 2, which shows the relationship of σ_b^2 on the cost ratios for the three experiments. This table indicates the increasing dominance of the CUSUM chart as σ_b^2 increases. Still, the regression tree analysis we presented earlier points to the determining factor being invariably the fixed cost of sampling.

In the interest of space, we do not present results here showing the actual designs of the charts for various scenarios. These are available upon request. But we wish to mention that optimal CUSUM chart designs, when there is a cost advantage to them, are characterized by small samples taken very frequently (when compared to the usual recommendation of taking a sample every $g = 1$ h). This is expected because

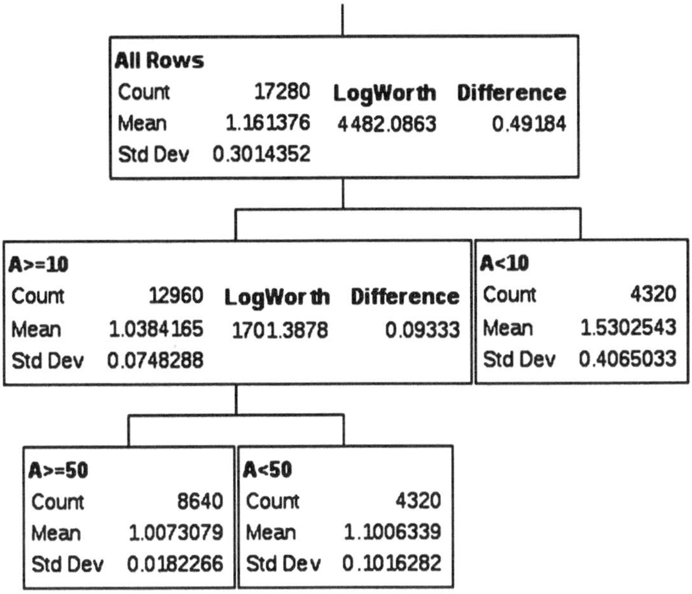

Fig. 3 Regression tree of (Cost of an optimal $\overline{X}$ design)/(Cost of an optimal CUSUM design) for experiment B

Table 2 Average value of (Cost of Optimal $\overline{X}$/Cost of CUSUM) by level of σ_b^2

σ_b^2	Exp $A(E = 0)$	Exp $A(E = 0.5)$	Exp B
0	1.0009	1.1037	1.1205
0.25	1.1028	1.1388	1.2478
0.5	1.1232	1.1456	1.2837
1	1.1642	1.1558	1.3217
2	1.1669	1.1655	1.3545

we have shown that the CUSUM is preferred when the fixed cost of sampling is very small which would indicate one would take smaller samples more frequently because of cost. We also note that the "average" CUSUM design is quite different than the design with $k = 3$ and $h = 0$ where the CUSUM mirrors the $\overline{X}$ chart.

We also have data to support the fact that although $n = 1$ is not generally an economically optimal sample size for the CUSUM the frequency of occurrence of optimal $n = 1$ increases as σ_b^2 increases. Recall that Reynolds and Stoumbos (2004) make the recommendation to use $n = 1$ CUSUMs in general.

6 Conclusions

The usual recommendation is to use a CUSUM chart if δ is small. Here, in these experiments where we consider a situation characterized by two components of variance we find that if one uses cost as a criterion rather than ARL or an equivalent, a more important factor to use in making the decision between using a CUSUM or an $\overline{X}$ chart is the fixed cost of sampling. If this cost is large one may find it advantageous to employ an $\overline{X}$ chart because of its ease of use and relatively equal costs. We have also shown that the larger the between-samples variance is, the more the CUSUM chart has an advantage in cost, that $n = 1$ is not necessarily an optimal economic design sample size for a CUSUM chart, and that optimally designed CUSUM charts do not mirror an $\overline{X}$ chart.

Surprisingly, the single component of variance results reported by Saniga et al. (2006b) are very similar in the sense that if the fixed cost of sampling is high, an $\overline{X}$ chart may be employed without incurring an opportunity cost. Moreover, even if the fixed cost of sampling is lower, there are cases usually characterized by large expected shifts in the process mean where again, an $\overline{X}$ chart may be employed without incurring an opportunity cost. As in the two component of variance case, it was shown that $n = 1$ is not generally an optimal sample size for the CUSUM and in cases where the CUSUM is an optimal design economically, it is dissimilar to an $\overline{X}$ chart in terms of k and h.

References

Chiu, W. K. (1974). The economic design of cusum charts for controlling normal means. *Applied Statistics, 23*, 420–433.

Duncan, A. J. (1956). The economic design of X bar charts used to maintain current control of a process. *Journal of the American Statistical Association, 51*, 228–242.

Goel, A. L. (1968). *A comparative and economic investigation of X bar and cumulative sum control charts*. Unpublished Ph.D. dissertation, University of Wisconsin.

Hawkins, D. M., & Olwell, D. H. (1998). *Cumulative sum charts and charting for quality improvement*. New York: Springer.

Lorenzen, T. J., & Vance, L. C. (1986), The economic design of control charts: A unified approach. *Technometrics, 28*, 3–10.

Lucas, J. M. (1982). Combined Shewhart-CUSUM quality control schemes. *Journal of Quality Technology, 14*, 51–59.

Montgomery, D. C. (2001). *Introduction to statistical quality control* (4th ed.). New York: Wiley.

Moustakides, G. V. (1986). Optimal stopping times for detecting changes in distributions. *The Annals of Statistics, 14*, 1379–1387.

Neyman, J., & Pearson, E. S. (1928). On the use and interpretation of certain test criteria for the purposes of statistical inference. *Biometrika, 20*, 175–240.

Neyman, J., & Pearson, E. S. (1933a). On the testing of statistical hypotheses in relation to probabilities a priori. *Proceedings Cambridge Philosophical Society, 29*.

Neyman, J., & Pearson, E.S. (1933b). On the problem of the most efficient tests of statistical hypotheses. *Philosophical Transactions, Series A, 231*, 289–337.

Reynolds, M., Jr., & Stoumbos, Z. (2004). Control charts and the efficient allocation of sampling resources. *Technometrics, 46*, 200–214.

Saniga, E., McWilliams, T., Davis, D., & Lucas, J. (2006a). Economic advantages of CUSUM control charts for variables. In H.-J. Lenz, & P.-T. Wilrich (Eds.), *Frontiers in statistical quality control* (Vol. 8). Heidelberg: Physica.

Saniga, E., McWilliams, T., Davis, D., & Lucas, J. (2006b). Economic control chart policies for monitoring variables. *International Journal of Productivity and Quality Management, 1*(1), 116–138.

Von Collani, E. (1987). Economic process control. *Statistica Neerlandica, 41*, 89–97.
Woodall, W. H. (1986a). The design of CUSUM quality control charts. *Journal of Quality Technology, 18*, 99–102.
Woodall, W. H. (1986b). Weaknesses of the economic design of control charts, (Letter to the Editor). *Technometrics, 28*, 408–410.

Process Monitoring Using an Online Nonlinear Data Reduction Based Control Chart

Issam Ben Khediri and Claus Weihs

Abstract Recent advances of multivariate Statistical Process Control (SPC) show that the introduction of Principal Component Analysis (PCA) methods for reduction of process data is a promising area in system monitoring and fault diagnosis. The advantage of these techniques is to identify sets of variables which describe the key variations of the operating data and which allow process handling and control based on a reduced number of charts. However, because the basic PCA method stipulates that relationships between process characteristics are linear, the application of such techniques to nonlinear systems that undergo many changes has been limited in many real cases. In order to overcome this issue, some recent studies suggested the use of nonlinear adaptive PCA methods in order to track process variation and detect abnormal events at early stages. For this reason, this study develops and analyses an online Kernel PCA chart as a key technique to model nonlinear systems and to monitor the evolution of non-stationary processes. Results based on an analysis of a simulated process show that the control chart is robust and provides a reduced rate of false alarms with high fault detection abilities.

Keywords Statistical process control • Kernel principal components analysis • Online monitoring

1 Introduction

Recent industrial developments through introduction of high technology in production processes has led several companies to adopt condition monitoring strategies to allow significant improvements in plant efficiency. In order to improve productivity,

I.B. Khediri (✉) · C. Weihs
Department of Statistics, TU Dortmund University, 44221 Dortmund, Germany
e-mail: issam.ben@tu-dortmund.de; claus.weihs@tu-dortmund.de

H.-J. Lenz et al. (eds.), *Frontiers in Statistical Quality Control 10*,
Frontiers in Statistical Quality Control 10, DOI 10.1007/978-3-7908-2846-7_7,

to reduce faults and to provide safer systems, successful control procedures usually require efficient modelling of complex systems. Indeed, with the development of measurement and data storage equipment, handling of large and complex information is strongly required.

To reduce complexity and to extract only simple but useful information from large process data, Principal Component Analysis (PCA) method may be used Kano et al. (2001). This technique allows dealing with high dimensional, noisy and correlated data through projection of data into a lower dimensional space that contains most of the information characterizing the process. However, application of the PCA model stipulates that the relationship between process variables is linear and that the process is stationary. In fact, the PCA sometimes shows quite a poor monitoring performance for nonlinear processes (Dong and McAvoy 1996). This fact restricts the use of the PCA method, since usually industrial systems are non-linear non-stationary continuous systems that tend to drift due to various phenomena as the process may undergo changes.

Because of this limitation many researchers looked for tools in order to be able to handle nonlinear systems. One of the most innovative and frequently used techniques is the Kernel Principal Component Analysis (KPCA) method. This technique is applied by Lee et al. (2004) as a new nonlinear process monitoring strategy. Results showed that the KPCA approach is effective in capturing nonlinear relationships in process variables and that it has a superior process monitoring performance compared to linear PCA. Hoffmann (2007) investigates the use of KPCA for novelty detection and demonstrated that it has a competitive performance. Cui et al. (2007) improved KPCA for fault detection by applying a feature vector selection scheme plus Fisher Discriminant Analysis to improve the fault detection performance of KPCA. Their simulation results show the effectiveness of these improvements for fault detection performance. This method has many advantages since it allows learning the particular structure of a model from data and can handle nonlinear relationships.

Recently, to better suit online monitoring and to construct online control charts able to handle non-stationary systems, Liu et al. (2009) proposed application of adaptive KPCA for online process monitoring and showed that applying such models can provide good detection results. However, in order to train KPCA continually, the adopted approach allows introduction and elimination of only one observation at a time. Because of this fact the window size of KPCA is assumed to be constant. However, in many practical situations not only one new observation is provided but a block of new data is present. Moreover sometimes it is of interest to freeze the model for a certain time or to eliminate a number of observations that do not characterize the process states.

In order to overcome these limitations, this paper proposes an adaptive KPCA based control chart that can test the state of a group of data at one time. A monitoring procedure with a variable window size model that can provide a flexible control strategy is investigated. Also, an algorithm that allows a recursive calculation of both window size and chart control limits is proposed. Finally, comparisons between the developed Adaptive KPCA, Batch KPCA, Adaptive PCA and Batch PCA control charts are performed.

This paper is organized as follows: Sect. 2 presents the principle of the Kernel Principal Components Analysis method. In Sect. 3, the proposed online KPCA based control chart is introduced. Section 4 provides analyses of the adopted chart for a fermentation process of a penicillin production. Then, Sect. 5 resumes the results and proposes future research.

2 Kernel Principal Component Analysis

Kernel PCA is a method introduced by Scholkopf et al. (1998) that has the advantage of estimating nonlinear relationships between variables. The basic idea of KPCA is to first map the input space into a feature space via nonlinear mapping and then to compute the PCs in that feature space. This principle supposes that, using a mapping of the original space into a higher-dimensional space, we can find a space where the data can vary linearly. As a result, KPCA performs a nonlinear PCA in the input space. Figure 1 provides the principle of this method.

Suppose we have the covariance matrix of the transformation data $x_i \in R^m$, $i = 1, .., n$, defined as,

$$K = \frac{1}{n} \sum_{i=1}^{n} \phi(x_i)\phi(x_i)^T, \tag{1}$$

where m is the number of variables, n the number of observations and it is assumed that $\sum_{i=1}^{n} \phi(x_i) = 0$ and $\phi(.)$ is a nonlinear mapping. To find the principal components, one has to solve the eigenvalue problem in the feature space such that

$$\lambda\mu = C\mu, \tag{2}$$

where eigenvalues $\lambda \geq 0$, μ a vector of eigenloadings and there must exist coefficients γ_i, $i = 1, .., n$, such that

$$\mu = \sum_{i=1}^{n} \gamma_i \phi(x_i). \tag{3}$$

Equation 2 is equivalent to

$$\lambda < \phi(x_k), \mu >=< \phi(x_k), C\mu > \quad k = 1, \ldots, n. \tag{4}$$

Combining Eqs. 2–4, we obtain for $k = 1, \ldots, n$,

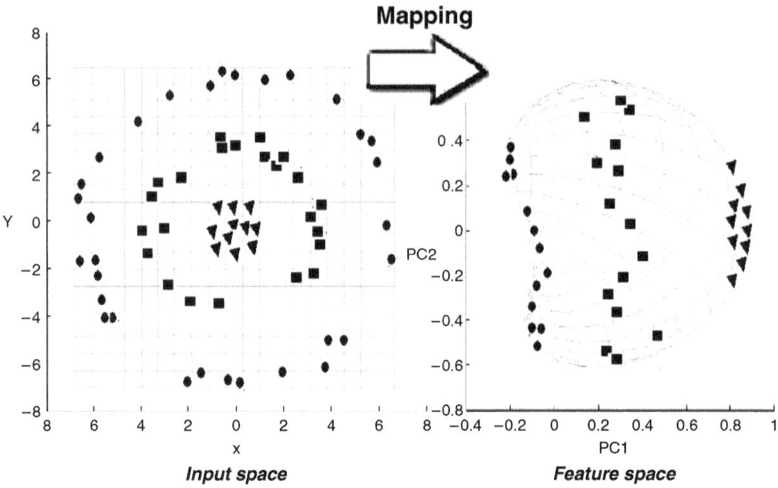

Fig. 1 Basic idea of kernel PCA

$$\lambda \sum_{i=1}^{n} \gamma_i < \phi(x_i), \phi(x_k) >= \frac{1}{n} \sum_{i=1}^{n} \gamma_i \left\langle \phi(x_k), \sum_{j=1}^{n} \phi(x_j) < \phi(x_j), \phi(x_i). > \right\rangle \quad (5)$$

Then, the inner product $< \phi(x_i), \phi(x_j) >$ is changed by the kernel function $K(x_i, x_j)$ and abbreviated by K_{ij}. Equation 5 can be expressed as,

$$\lambda \sum_{i=1}^{n} \gamma_i K_{ik} = \frac{1}{n} \sum_{i=1}^{n} \gamma_i \left(\sum_{j=1}^{n} K_{kj} K_{ji} \right) \quad k = 1, \ldots, n. \quad (6)$$

Equation 6 can be written as follows,

$$\lambda K V = \frac{1}{n} K^2 V, \quad (7)$$

$$\lambda V = \frac{1}{n} K V, \quad (8)$$

where $V = [\gamma_1, \ldots, \gamma_n]^T$ and K is an $(n \times n)$ matrix defined by K_{ij}.

Now, performing PCA is equivalent to solving the eigen-problem of Eq. 8. This yields eigenvectors $V^1, \ldots, V^n$ with eigenvalues $\lambda_1 \geq \ldots \geq \lambda_n$. In order to insure the normality of $u^1, \ldots, u^n$, (Eq. 2), the corresponding vectors $V^1, \ldots, V^n$ should be scaled such that

$$< u^l, u^l >= 1, \quad \forall l = 1, .., n \quad (9)$$

Using Eq. 3 this translates to

$$\sum_{i=1}^{n}\sum_{j=1}^{n}\gamma_i^l\gamma_j^l\left\langle\phi(x_i),\phi(x_j)\right\rangle = 1, \quad l = 1,..,n \tag{10}$$

$$\sum_{i=1}^{n}\sum_{j=1}^{n}\gamma_i^l\gamma_j^l K_{ij} = 1, \quad l = 1,..,n \tag{11}$$

$$\left\langle V^l, K V^l\right\rangle = 1, \quad l = 1,..,n \tag{12}$$

Using Eq. 8, $V^1,\ldots,V^n$ should be normalized such that

$$n\lambda_l\left\langle V^l, V^l\right\rangle = 1, \quad l = 1,..,n \tag{13}$$

$$\left\langle V^l, V^l\right\rangle = \frac{1}{n\lambda_l}, \quad l = 1,\ldots,n \tag{14}$$

The first p principal components (t_z) of a test vector x are then extracted by projecting x into eigenvectors $V^1,\ldots,V^p$, where,

$$t_z = \sum_{i=1}^{n} V_i^z K(x_i, x). \tag{15}$$

Because the kernel function is known as a measure of similarity, KPCA can work very well for process monitoring issues since the goal is to distinguish aberrant observations from others. One of the most used kernel functions is the radial function which is expressed as follows:

$$K(x, x\prime) = \exp\left(\frac{-\|x - x\prime\|^2}{\sigma^2}\right), \tag{16}$$

where $\sigma \in R$ is determined beforehand.

3 Online Adaptive KPCA Based Control Chart

In order to monitor multivariate processes based on the PCA method, this study uses the Squared Prediction Error (SPE) as a statistic to test the state of the non-stationary systems. This is also known as the Q statistic and it has the ability to be updated to the condition on which the system is operating. The KPCA-based monitoring method is similar to that using PCA in that the Q statistic in the feature space can be interpreted in the same way. For a new observation x_{new} of size $(1 \times m)$, where m is the number of variables, using linear PCA, the Q statistic is defined as follows,

$$Q = e^T e = x_{new} x_{new}^T - (x_{new} P)(x_{new} P)^T \tag{17}$$

where e represents the residuals vector of the reconstructed data and P the matrix of PCA eigenloadings. The Q statistic for KPCA is defined by Choi et al. (2005) as follows

$$Q = |\bar{k}(x_{new}, x_{new}) - \hat{k}(x_{new}, x_{new})|, \tag{18}$$

$$= |\bar{k}(x_{new}, x_{new}) - tt^T|, \tag{19}$$

where $\bar{k}$ is the scaled kernel product, $\hat{k}$ the projection of $\bar{k}$ into KPCA model obtained from Eq. 8 where

$$t = [\bar{k}(x_{new}, x_1), .., \bar{k}(x_{new}, x_n)] V_{np}, \tag{20}$$

where $V_{np} = [V^1, \ldots, V^p]$.

where $\bar{k}$ is the scaled kernel product and $\bar{k}_{x_{new}} = [\bar{k}(x_{new}, x_1), .., \bar{k}(x_{new}, x_n)]$. The scaling is based on mean centering and variance scaling.

The Q statistic indicates the extent to which each sample conforms to the PCA model. It is a measure of the amount of variation not captured by the principal component model. The upper limit for the Q statistic is given by,

$$Q_{limit} = \frac{\theta_2}{2\theta_1} \chi^2_\alpha \left(\frac{2\theta_1^2}{\theta_2} \right), \tag{21}$$

where θ_1 and θ_2 are the sample mean and variance of Q values and α is the risk level (Nomikos and MacGregor 1995).

For slowly time-varying processes, the upper limit for detection indices changes with time, making adjustment of this limit necessary for online monitoring. For the Q statistic, parameter values of θ_1, θ_2 are recursively updated using the p largest eigenvalues after each new data block making Q_{limit} time-varying. Also, as it is done for training of KPCA, influence of the oldest Q values from the mean and variance values θ_1 and θ_2 are excluded.

In order to have an efficient adaptive control chart, the definition of the window size which gives information about the training data is also an important fact. A use of constant window size is very restrictive and can sometimes provide poor performance because it can imply a use of corrupted training samples especially for processes that undergo several changes. For certain applications it is of interest to use a variable window size for the training of the KPCA model. The idea here is to train the model in regions that characterize well the actual state of the process. By this way the window size can grow or decrease depending on the states of the process. An approach to determine this window size H is to use the information contained in the Q statistic. If certain old observations imply an increase of the standard deviation of the Q statistic then this would mean that these old observations differ from the actual process state and therefore it may be better to eliminate their

Table 1 Algorithm to determine KPCA window size

$H \leftarrow H+$ size(new data)
for $h = 1 : k$
    $SD(h) \leftarrow std(Q(w(h-1)+1 : H))$
end
$h \leftarrow Index(\min(SD))$
$H \leftarrow H - (h-1) \times w$
if $H > H_{max}$
    $H \leftarrow H_{max}$
elseif $H < H_{\min}$
    $H \leftarrow H_{\min}$
end

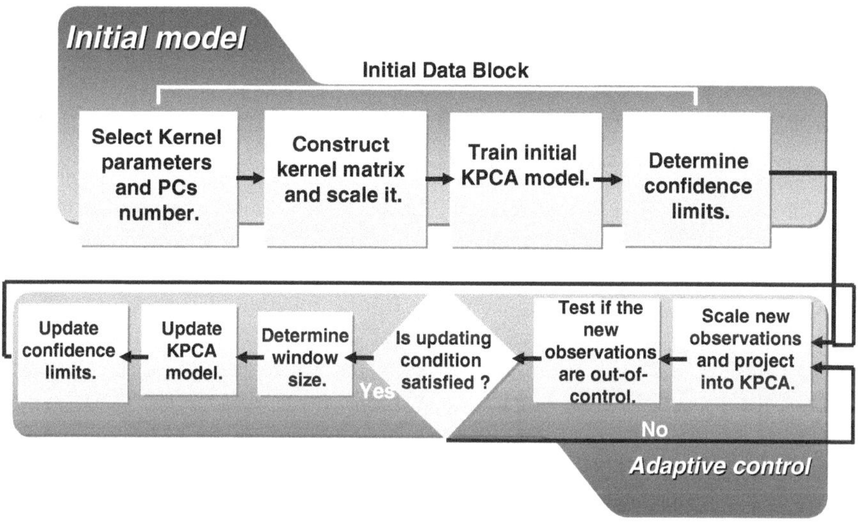

Fig. 2 Adaptive online kernel KPCA control chart

effect from the model. We also note that other indexes than the standard deviation of Q can be used. The estimated varying window size can be implemented by the algorithm of Table 1. Let H be the actual window size, H_{max} and H_{min} the maximal and the minimal window sizes that can be used, w , h and k are the size, the number of samples and the allowed maximal number of samples to be eliminated.

In addition to the window size, to obtain an adaptive chart, not affected by integration of out-of-control variables, a condition for updating the KPCA model and Q_{limit} value is introduced. This condition lets the model avoid contamination by observations which could make the model insensitive to faults. Moreover, because the model can produce false alarms, specially in the case of a non-stationary process, where sometimes an out-of-control signal can characterize a change in the relationship between variables and not a proper fault, a margin of acceptability of observations is introduced in the Q_{limit} value. Thus the adjustment condition is activated only if the Q value of the observation does not exceed the value of ηQ_{limit}, where $\eta > 1$. In this paper a value of η is taken to be equal to 1.3. The main algorithmic step of the proposed adaptive window kernel principal component chart is shown in Fig. 2.

4 Control Chart Analysis for a Fed-Batch Fermentation Process

In order to evaluate the proposed adaptive KPCA chart with respect to the current PCA control strategies, this study applies different procedures to a simulated benchmark case study of Fed-batch fermentation process of penicillin production. Evaluation is performed by reporting the degree of accuracy of each method in detecting the true out-of-control situations and avoiding false alarms. Performance evaluation of the accuracy can be reported by using the false alarm rate, the detection rate and the Run Length (RL) criterion. The first statistic gives information about the robustness of the adopted method against normal system changes while the second and third statistic give information about the sensitivity and efficiency of detecting faults. This study proposes to compare the performance of Adaptive online KPCA (AKPCA) with Batch KPCA, Batch PCA and Adaptive PCA. In order to assess these strategies, a simulator, supplied by the Illinois Institute of Technology, can be used. The program is called PenSim 2 and can be found in the following website address: http://216.47.139.198/software.html. The simulation time of this process is 100 h with a sampling time equal 0.05 h which gives a good description of the system. The initial condition as well as the set points parameters used in this study present values recommended by Birol et al. (2002).

In this study, the monitoring procedure is applied to the state variables of the process which are Substrate concentration, Biomass concentration, Culture volume, Dissolved Oxygen saturation, Penicillin concentration and CO_2 concentration. Figure 3 illustrates the dynamics of these variables for normal operating condition.

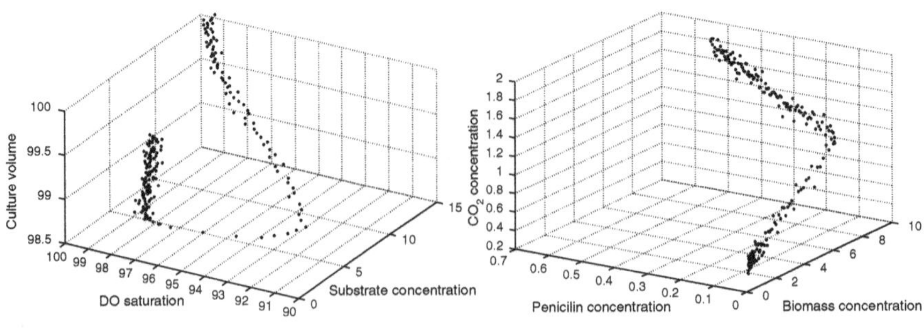

Fig. 3 Illustration of penicillin fermentation state variables

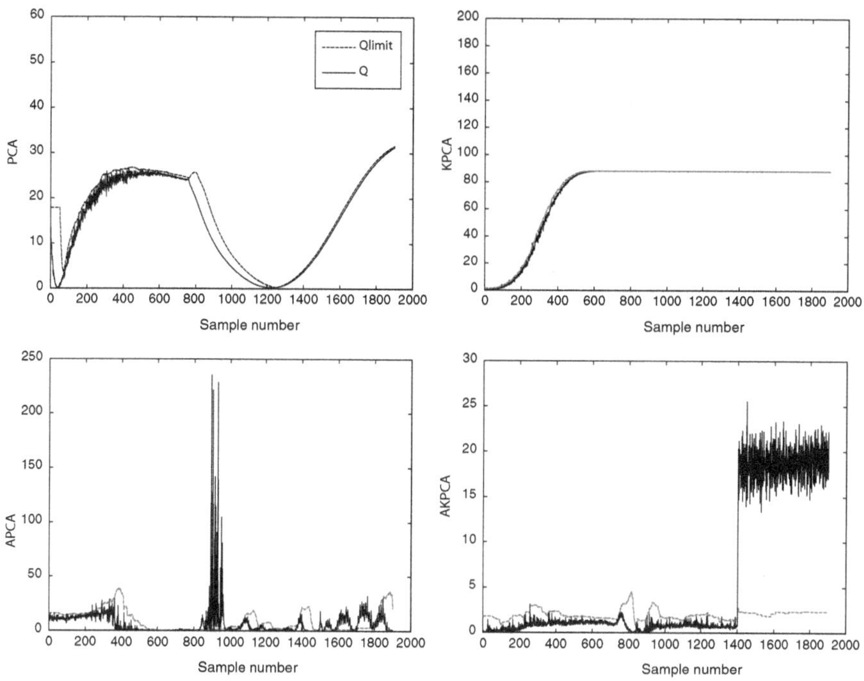

Fig. 4 PCA monitoring charts using Fault 1 of Fed-batch process. The *black* and *grey* lines represent respectively the value of Q and Q_{limit}

To control this process, for PCA and Adaptive PCA, three PCs that explain 99% of the total variance are selected. For KPCA and AKPCA, a sigma value equal to 5 and a number of PCs equal to 3 are selected and they explain more than 99% of the total variance. The training data consists of 150 samples with 100 samples used to train initial PCA models and 50 samples used for Q computation. The window size for AKPCA can vary between 80 and 120. Introduction of observations into the AKPCA is made by blocks of size 5. To investigate the sensitivity of the proposed control strategy against abnormal conditions, three faults are introduced in the process starting from time 1,500. Fault 1 presents an increase in Dissolved Oxygen saturation by 3%, Fault 2 presents a decrease of the Culture volume by 2% and Fault 3 presents an increase of the Biomass concentration by 4%.

For the risk level α fixed at a 1% level, Fig. 4 illustrates the obtained results of the different control charts using simulated data with Fault 1 and Table 2 resumes the analysis of the different strategies for all types of faults.

As shown in Table 2, Batch KPCA and Batch PCA provide poor robustness results in detecting abnormalities for all faults with detection rates that do not exceed respectively 40% and 74%. Whereas for Adaptive PCA, even though the control chart exhibits good performance in terms of RL as concerns Fault 2 and Fault 3, the chart in question is unable to detect Fault 1 and provides bad results in terms of

Table 2 Analysis of PCA, KPCA, APCA and AKPCA charts for several faults

		Fault 1		Fault 2		Fault 3	
	False alarm	Detection	RL	Detection	RL	Detection	RL
PCA	0.018	0	500	0.74	1	0	500
APCA	0.129	0.51	92	0.03	1	0.05	1
KPCA	0.043	0.40	51	0.31	1	0.25	1
AKPCA	0.005	1	1	1	1	1	1

Type II error rates that exceed 49%. As far as the analysis concerns AKPCA, results show that the procedure has a good sensitivity to faults with an RL equal to 1 for all out-of-control conditions with a good stability indicated by Detection rate and better robustness abilities than the other charts with a False alarm rate of 0.5%. Figure 4 makes clear that AKPCA allows a better control strategy especially than Batch and linear PCA charts.

5 Conclusion

Recent industrial advances made by introduction of information technology has opened the gate for the development of innovative monitoring procedures. One of the most often used techniques that have taken advantages of this development is the data reduction method PCA. Following this new research area, this study proposes a variable window size adaptive Kernel PCA modelling approach that allows estimation of nonlinear processes integrated with an online smoothing procedure to track normal operating drifts. In order to perform such a procedure, an algorithm that allows determining the process state and the need of updating is presented. Also, because this study allows a variable window size, a method that allows calculating the optimal window size is proposed. To investigate the performance of this chart, we conducted a comparison of the adaptive Kernel PCA with various PCA based procedures, named Batch PCA, Batch KPCA and Adaptive PCA. The obtained results show that first, Batch PCA and KPCA are unable to adequately control non-stationary processes since they are based on the use of a single data block for monitoring highly dynamic processes, whereas Adaptive PCA models are unable to monitor non-linear systems as they are based very restrictive assumptions such as normality and linearity of the relationship between variables. However, Adaptive KPCA with variable window size overcomes these shortcomings by providing excellent ability of detecting out-of-control conditions with a reduced false alarm rate. But, there are still some issues that could be investigated. Indeed, a fast procedure for updating the KPCA model should be developed. Also, a monitoring procedure for a recursive determination of the number of principal components and kernel parameters should be the object of further research.

References

Birol, G., Undey, C., & Cinar, A. (2002). A modular simulation package for Fed-batch fermentation: Penicillin production. *Computers and Chemical Engineering, 26*, 1553–1565.

Choi, S. W., Lee, C., Lee, J. M. P., Park, J. H., & Lee, I. B. (2005). Fault detection and identification of nonlinear processes based on KPCA. *Chemometrics and Intelligent Laboratory Systems, 75*, 55–67.

Cui, P., Li, J., & Wang, G. (2007). Improved kernel principal component analysis for fault detection. *Expert Systems with Applications, 36*, 1423–1432.

Dong, D., & McAvoy, T. J. (1996). Nonlinear principal components analysis based on principal curves and neural networks. *Computers and Chemical Engineering, 20*, 65–78.

Hoffmann, H. (2007). Kernel PCA for novelty detection. *Pattern Recognition, 40*, 863–874.

Kano, M., Hasebe, S., Hashimoto, I., & Ohno, H. (2001). A new multivariate statistical process monitoring method using principal component analysis. *Computers and Chemical Engineering, 25*, 1103–1113.

Lee, J. M., Yoo, C., Choi, S. K., Vanrolleghemb, P. A., & Lee, I. (2004). Nonlinear process monitoring using kernel principal component analysis. *Chemical Engineering Science, 59*, 223–234.

Liu, X., Xie, L., Kruger, U., Littler, T., & Wang, S. (2009). Moving window kernel PCA for adaptive monitoring of nonlinear processes. *Chemometrics and Intelligent Laboratory Systems, 96*, 132–143.

Nomikos, P., & MacGregor, J. F. (1995). Multivariate SPC charts for monitoring batch processes. *Technometrics, 37*, 41–59.

Scholkopf, B., Smola, A., & Muller, K. B. (1998). Nonlinear component analysis as a kernel eigenvalue problem. *Neural Computation, 10*, 1299–1319.

On the Integration of SPC and APC: APC Can Be a Convenient Support for SPC

Ken Nishina, Masanobu Higashide, Hironobu Kawamura, and Naru Ishii

Abstract This paper is developed from Higashide et al. (Front Stat Qual Control 9:71–84, 2010). Automatic process control (APC) is frequently used in the semiconductor manufacturing process; however, statistical process control (SPC) is also needed to control the APC controller. This is an earlier paradigm on the integration of SPC and APC. Our viewpoint is different from the earlier one as follows:

(a) APC reinforces SPC.
(b) SPC complements APC.

Through case studies on the semiconductor manufacturing process, the remarks above are discussed. Our proposals for the integration of SPC and APC are as follows:

(a) The process rate is used as the control characteristic to control the between-subgroup variation.
(b) Principal component analysis is applied to control the within-subgroup variation.

These proposals can lead to developments of the traditional $\bar{X} - R$ charts.

Keywords Semiconductor manufacturing process • Process rate • Principal component analysis • Traditional $\bar{X} - R$ chart

K. Nishina (✉) · N. Ishii
Nagoya Institute of Technology, Gokiso-cho, Showa-ku, Nagoya, 466-8555, Japan
e-mail: nishina@nitech.ac.jp; naru@nitech.ac.jp

M. Higashide
Renesas Electronics Co., 1753 Shimonumabe, Nakahara-ku, Kawasaki, 211-8669, Japan
e-mail: masanobu.higashide.wt@renesas.com

H. Kawamura
Tokyo Science University, 2641 Yamazaki, Noda, 278-8510, Japan
e-mail: kawamura@ia.noda.tus.ac.jp

H.-J. Lenz et al. (eds.), *Frontiers in Statistical Quality Control 10*,
Frontiers in Statistical Quality Control 10, DOI 10.1007/978-3-7908-2846-7_8,

1 Introduction

Statistical process control (SPC) and automatic process control (APC) have a common aim of reducing process variation, but the approach to achieve it differs between SPC and APC, which originated in different industries, the parts industries and the process industries, respectively (Box and Kramer 1992). SPC can be achieved by detecting assignable causes and taking action against them. SPC improves and maintains the process in the long run. Control charts are frequently used as statistical tools. On the other hand, in APC, it can be done by process adjustment by correcting the manipulated variable. APC is a repeated action applied corresponding to the manual. In many cases feedback control is used.

Control charts could be built into the APC system to decide when to adjust the process. But it is not appropriate for process adjustment and may unnecessarily increase variation about the target value (Box and Kramer 1992). SPC and APC should complement each other very effectively.

The origin of the integration is to monitor the APC controller using control charts. It is useful to extend the idea of common causes and special causes to APC schemes (Box and Kramer 1992). SPC could be used for analyzing APC controller performance and as a diagnostic tool for the APC controller (MacGregor 1988; MacGregor and Harris 1990).

As mentioned above, the earlier paradigm of the integration of SPC and APC is, if anything, that SPC should be introduced into the process with APC. In this paper we develop the earlier paradigm into new remarks according to which APC reinforces SPC and SPC complements APC. The new remarks can also lead to developments of the traditional control charts.

We consider semiconductor manufacturing, in which feedback control is frequently used because it is very difficult to detect and eliminate assignable causes due to including some 100 steps downstream and environmental effects.

2 Traditional $\bar{X} - R$ Charts Are Not Available in the Semiconductor Manufacturing Process

A chemical mechanical polish (CMP) process is considered. The monitoring output is the remaining film thickness on the wafer surface after polishing. The function of this CMP process is polarization of an uneven wafer surface by polishing. But practically it can not be realized, and the wafer surface is a little uneven. It is a within-wafer variation.

Observations of the remaining film thickness are obtained at nine measurement points on a wafer surface as shown in Fig. 1. The data sheet for charts is also shown in Fig. 1. The between-wafer variation is controlled by a $\bar{X}$ chart and the within-wafer variation by a R chart. In such cases, traditional $\bar{X} - R$ charts are not

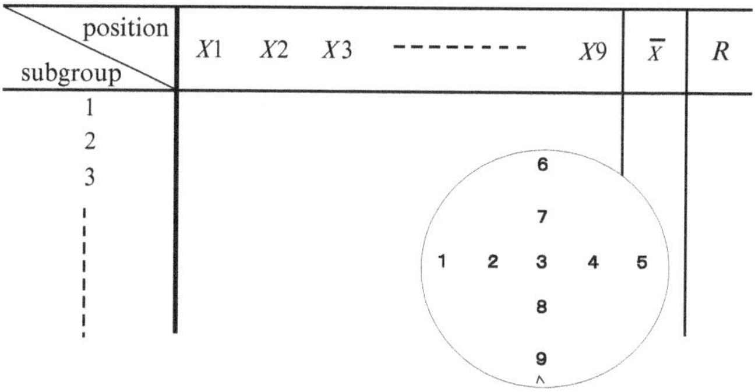

Fig. 1 Data sheet of $\bar{X} - R$ and measurement positions on wafer

appropriate because the between-wafer variation includes the compensated impact by the manipulated variable, and there are some systematic patterns in the within-wafer variation due to the deterioration of some parts of the CMP process.

3 APC Reinforces SPC

3.1 Interaction Involving the Manipulated Variable

Let y_t and y_0 be the observation at time t and the target value, respectively. When the relationship between y and the manipulated variable, w, is

$$y = a + bw \tag{1}$$

and the EWMA with the parameter λ is used as the one-step-ahead predictor, the value of the manipulated variable at time t, w_t, is determined as follows (see Montgomery 2001):

$$w_t = \frac{1}{b}(y_0 - y_{t-1}) \times \lambda + w_{t-1}. \tag{2}$$

We focus on the slope b. If an interaction between the manipulated variable and a factor of the manufacturing process exists, the slope b can change as the factor is varied. Then the feedback control following Eq. 2 can lead to more process variation as shown in Fig. 2.

Kawamura et al. (2008b) verified the existence of the interaction involving the manipulated variable in the photo resist process, in which the exposure time is used as the manipulated variable and the output is the critical dimension. They showed the interaction between the exposure time and the focus of lens, which is a factor of this process, and moreover, the interaction between the exposure time and the photo

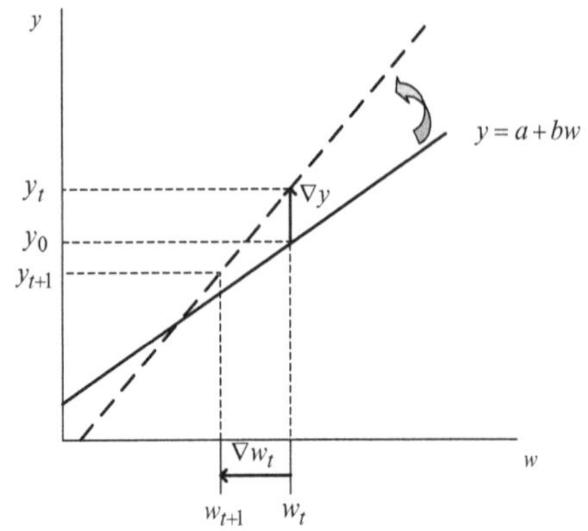

Fig. 2 Inflation of process variation by interaction

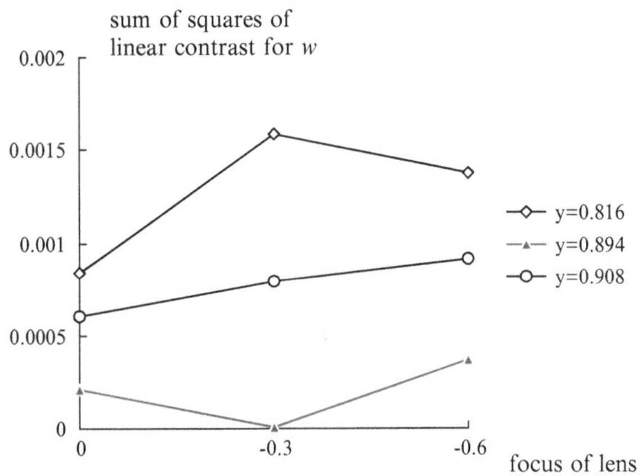

Fig. 3 Interaction between the manipulated variable w and the focus of lens

resist thickness, which is the output of the previous process. Figure 3 indicates these interactions. It can be seen that the sum of squares due to the linear contrast is varied with the focus of the lens and the photo resist thickness. In many another manufacturing processes interactions involve a manipulated variable as shown in Fig. 3.

3.2 Properties of the Manipulated Variable

The manipulated variable needs to be operated easily and shall have a simple relationship to the output of process like Eq. 1. In addition, as a result of the discussion in Sect. 3.1 a factor with no interaction with another factor in the process should be selected as the manipulated variable . In practice, however, the manipulated variable may indeed have an interaction.

In the semiconductor manufacturing process, most of the manipulated variables represent the input energy for processing in a process such as the polishing time in CMP process, the exposure time in the photo resist process and so on. It means that the APC scheme originally consists in the input–output relationship. The manipulated variable can be regarded as not only the input variable of APC scheme but also the input of the manufacturing process.

3.3 From Passive Control to Active Control

As mentioned in Sect. 1, the integration of SPC into APC is obtained by monitoring the APC schemes using control charts. From the discussion in Sect. 3.1, the disturbance of the input–output relationship can be caused by the interaction between the manipulated variable and a factor of the manufacturing process. Therefore, monitoring the input–output relationship in APC scheme is not confined to "controlling the controller" but leads to "controlling the manufacturing process".

Originally, process monitoring by traditional control charts is passive control because only the output is observed. On the other hand, the control chart of applied to the process with APC can be considered as an active control because both the input value and the output value can be observed, and then the input value is varied. APC can realize such an active control by control charts. APC can thus provide convenient support for SPC.

Next, we consider what should be monitored in the manufacturing process with APC. Some related proposals have been presented as follows:

- Amount of correction in the manipulated variable necessary to keep the process at target (MacGregor and Harris 1990; Capilla et al. 1999, etc.)
- One-step-ahead prediction errors (Harris and Ross 1991; Montgomery and Mastrangelo 1991, etc)
- Both of the process outputs and the manipulated inputs (using bivariate SPC) (Tsung et al. 1999).

In addition to the proposals above, we proposed the process rate y_t/w_t as a control characteristic (see Kawamura et al. 2008a). In monitoring the interaction between the manipulated variable and another factor, $\nabla y/\nabla w$ (where ∇y is the difference between the observed value y and the target value y_0 and ∇w is the amount of change in the manipulated variable, $w_t - w_{t-1}$, shown in Fig. 2) could be a better control characteristic. However, when process adjustment is not performed, the denominator ∇w is zero. In many cases there is an adjustment band and no adjustment may occur. Therefore, in practice the process rate y_t/w_t is preferred to $\nabla y/\nabla w$. If there is no adjustment band, $\nabla y/\nabla w$ is appropriate.

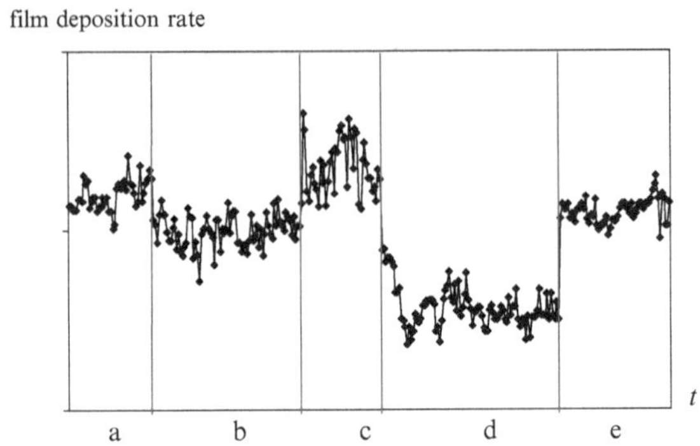

Fig. 4 Film deposition rate of LPCVD (Higashide et al. 2010)

3.4 Example

Higashide et al. (2010) presented an example in which a control chart was applied to the LPCVD (low pressure chemical vapor deposition) process. In the LPCVD process a thin polysilicon film is formed on a single-crystal silicon wafer. The output of LPCVD is the thickness of the measurement wafer. The between-batch variation is caused by the accumulated reaction by-product in the chamber. A feedback control following Eq. 2 is performed to reduce between batch variation. The manipulated variable of the process is the deposition time.

As mentioned in Sect. 3.3, the process rate is considered as the control characteristic. For the LPDVD process we use the film deposition rate corresponding to the slope of the correction formula in Eq. 1.

Figure 4 shows the time series data of the film deposition rate. Five periods (a–e) stand for the intervals between the maintenances. The film deposition rate data are autocorrelated. Examination of the autocorrelation structure can be represented by an EWMA model as follows:

$$z_t = 0.519 y_t + (1 - 0.519)\, z_{t-1}$$

where y_t stands for the film deposition rate at time t. The parameter 0.519 of the EWMA was estimated for the data of "period a".

We use the residual EWMA control chart. Figure 5 shows an example of the residual EWMA control chart for the data of "period d". The residual control chart uses the control limits determined for the data of "period a". The residual control chart shown in Fig. 5 indicates that there is a run with length nine.

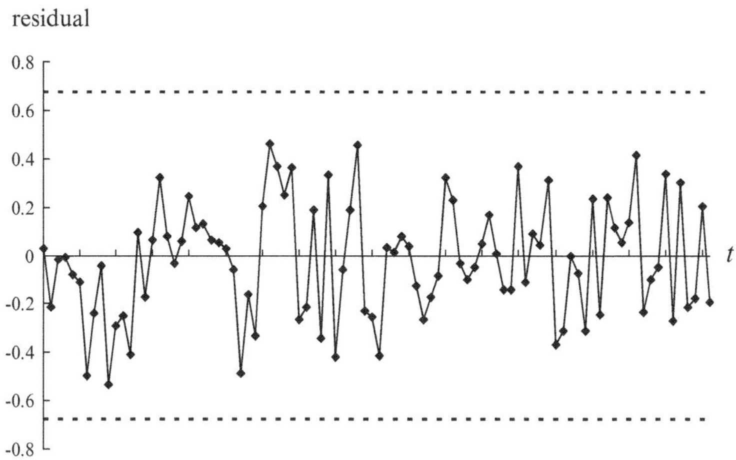

Fig. 5 Residual EWMA control chart for the period *d*

4 SPC Complements APC

4.1 Systematic Within-Wafer Variation

We consider again the CMP process in which the between-wafer variation is reduced by APC (feedback control); however, the within-wafer variation, that is, the uneven wafer surface, can not be controlled by APC. The uneven wafer surface after polishing indicates a systematic variation. In such a case, the within-wafer variation must not be controlled from the viewpoint of the amount of variation but rather the pattern of variation.

Knowing how to quantify the pattern is necessary to monitor and control the pattern. Principal component analysis (PCA) is very useful for the quantification. Some case studies explain applications of PCA to process monitoring, for example, Roes and Does (1995) and Gonzales and Sanchez (2008).

In our case, the CMP process, PCA is also applied to monitoring an uneven wafer surface. See again the data sheet for $\bar{X} - R$ charts shown in Fig. 1. The data can be regarded as a set of multivariate time series data. Double centralized transformation x_{ij} (see Eq. 3) is useful as preprocessing to identify a wafer with a different remaining film pattern by eliminating the between-wafer variation, which is adjusted by feedback control (Higashide et al. 2010). Then the data set x_{ij} is analyzed using the Mahalanobis distance and PCA:

$$x_{ij} = y_{ij} - \bar{y}_{\cdot j} - \bar{y}_{i\cdot} + \bar{\bar{y}} \tag{3}$$

$$\bar{y}_{i\cdot} = \frac{\sum_{j=1}^{9} y_{ij}}{9}, \quad \bar{y}_{\cdot j} = \frac{\sum_{i} y_{ij}}{\text{sample size}}, \quad \bar{\bar{y}} : \text{total average.}$$

4.2 Two Distances for Multivariate Data

It is well known that the Mahalanobis distance D can be decomposed as follows:

$$D_i^2 = \mathbf{x}_i^T \hat{\Sigma}^{-1} \mathbf{x}_i = (1/\lambda_1) z_{1i}^2 + (1/\lambda_2) z_{2i}^2 + (1/\lambda_3) z_{3i}^2 + \ldots + (1/\lambda_p) z_{pi}^2$$

where

$\mathbf{x}$: vector of observations x_{ij}
$\hat{\Sigma}$: estimate of the variance covariance matrix using reference data,
$\lambda_j \quad (j = 1, 2, \ldots, p)$: Eigenvalues of $\hat{\Sigma}$,
$z_j \quad (j = 1, 2, \ldots, p)$: the jth principal component score.

Jackson and Mudholkar (1979) proposed a residual statistic associated with PCA,

$$Q = \sum_{j=k+1}^{p} z_j^2.$$

Jackson and Mudholkar (1979) concluded that Q may be quite useful for the detection of outliers.

Now, it should be noted that Q is not a standardized distance but an Euclidean distance. The Mahalanobis distance D for future data may be unstable when the variance covariance matrix Σ is not estimated efficiently. One of the reasons is that some Eigenvalues and the corresponding Eigen vectors are unstable (Jackson and Hearne 1973). Therefore, Q as a non standardized distance is more stable.

The other statistic T^2 is standardized as follows:

$$T^2 = \sum_{j=1}^{k} \frac{z_j^2}{\lambda_j}.$$

We apply two distances, T^2 and Q, for monitoring the uneven wafer surface. T^2 is applied to monitor the systematic variation. On the other hand, Q is applied to detect any departure from the systematic variation. If an individual principal comportment (PC) score, which composes T^2, has an engineering meaning, it should be monitored.

Jackson and Hearne (1973) showed a transformation for the sake of normal approximation. However, the log transformation can be sufficiently approximated by a normal distribution.

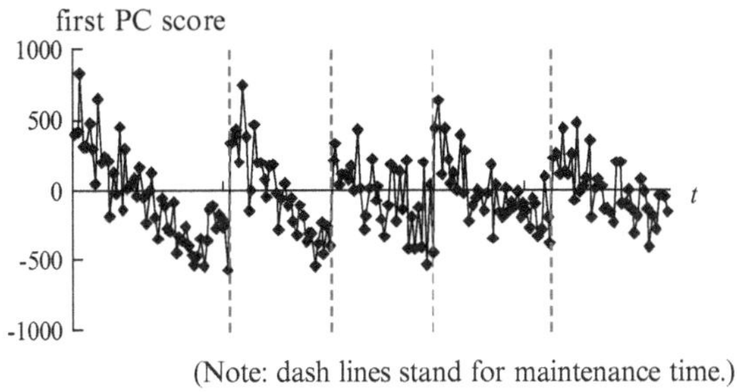

Fig. 6 Trend in the first principal component score

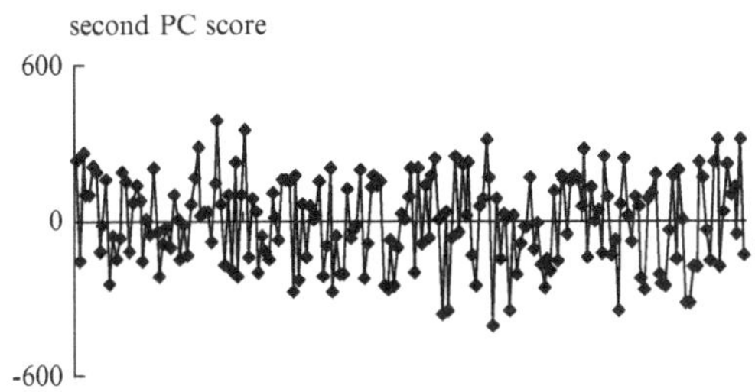

Fig. 7 Second principal component score

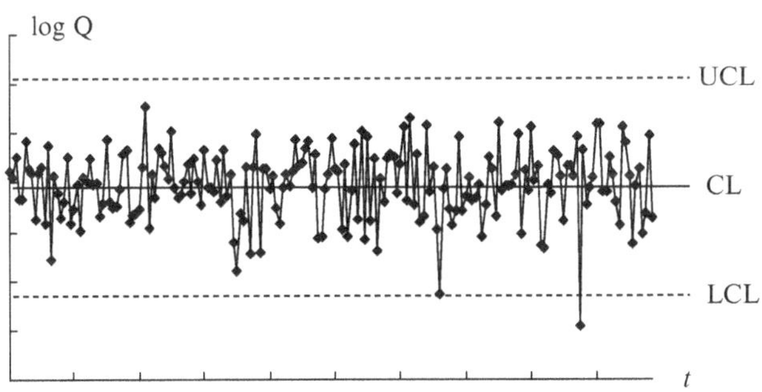

Fig. 8 log Q control chart

4.3 *Example*

PCA is performed for the uneven wafer surface data of the CMP process following the procedure shown in Sect. 4.1. As the results of PCA, the first principal component explains 51.7% of the overall variation. The first PC scores are plotted in Fig. 6. The broken lines stand for the times of maintenance, for example, parts replacement. Trends can be seen in Fig. 6. The trends repeat in every period between maintenances. It means that the first principal component has an engineering meaning and can be monitored for process maintenance. Figure 7 shows a plot of the second PC scores. The second PC explains 19.4% of the overall variation but no systematic variation can be seen in Fig. 7. So in this case, only the first PC score represents a systematic variation of the within-wafer variation.

Figure 8 shows a control chart of the log Q. One point is beyond the three sigma control limit. But no unusual effect occurs at this time. It may be a false alarm.

5 Conclusive Remarks

In this paper, by some case studies of the semiconductor manufacturing process, the integration of SPC and APC is discussed from a different viewpoint than in earlier studies. The conclusions can be expressed by the following two remarks:

First, APC can be a convenient support for SPC. Originally, the control chart approach is a passive control using the output of the process under study; however, in a process with APC, the approach can be changed to an active control using the process rate as the control characteristic.

Second, SPC can complement APC. Generally, APC cannot control the within-subgroup variation; however, both the systematic variation and the departure from it in the within-subgroup variation can be monitored by using SPC, for example, PCA and Q charts.

These two remarks can be extended to other processes than the semiconductor manufacturing process.

References

Box, G., & Kramer, T. (1992). Statistical process monitoring and feedback adjustment – a discussion. *Technometrics, 34*(3), 251–285.

Capilla, C., Ferrer, A., & Romero, R. (1999). Integration of statistical and engineering process control in a continuous polymerization process. *Technometrics, 41*(1), 14–28.

Gonzales, I., & Sanchez, I. (2008). Principal alarms in multivariate statistical process control. *Journal of Quality Technology, 40*(1), 19–30.

Harris, T. J., & Ross, W. H. (1991). Statistical process control procedures for correlated observations. *The Canadian Journal of Chemical Engineering, 69*, 48–57.

Higashide, M., Nishina, K., Kawamura, H., & Ishii, N. (2010). Statistical process control for semiconductor manufacturing process. *Frontiers in Statistical Quality Control, 9*, 71–84.

Jackson, J. E., & Hearne, F. T. (1973). Relationships among coefficients of vectors used in principal components. *Technometrics, 15*(3), 601–610.

Jackson, J. E., & Mudholkar, G. S. (1979). Control procedures for residuals associated with principal component analysis. *Technometrics, 21*(3), 341–349.

Kawamura, H., Nishina, K., & Higashide, M. (2008a). Discount factors and control characteristics in the semiconductor manufacturing process. *Proceedings of the 6th Asian Network Quality Congress* 2008, CD. Bangkok: ANQ 2008.

Kawamura, H., Nishina, K., Higashide, M., & Shimazu, K. (2008b). Integrating SPC and APC in semiconductor manufacturing process. *Quality, 38*(3), 99–107 (in Japanese).

MacGregor, J. F. (1988). On-line statistical process control. *Chemical Engineering Process, 84*, 21–31.

MacGregor, J. F., & Harris, T. J. (1990). Exponential weighted moving average control schemes: Properties and enhancement – Discussion. *Technometrics, 32*(1), 23–26.

Montgomery, D. C. (2001). *Introduction to statistical quality control* (4th ed.). New York: Wiley.

Montgomery, D. C., & Mastrangelo, C. M. (1991). Some statistical process control methods for autocorrelated data. *Journal of Quality Control, 23*(3), 179–204.

Roes, K. B., & Does, R. J. M. M. (1995). Shewhart-type charts in nonstandard situations. *Technometrics, 37*(1), 15–40.

Tsung, F., Shi, J., & Wu, C. F. J. (1999). Joint monitoring of PID-controlled processes. *Journal of Quality Control, 31*(3), 275–285.

Process Adjustment Control Chart for Simultaneous Monitoring of Process Capability and State of Statistical Control

Hironobu Kawamura, Ken Nishina, and Tomomichi Suzuki

Abstract In the production of chemicals, a process adjustment such as feedback control is frequently used to reduce process variability. It is very important to judge whether or not the adjustment should be done automatically because an automatic process control (APC) system requires a large capital investment.

This paper presents the determination of the adjustment timing on the basis of the process capability, and control charts combining information about the state of statistical control and process capability are also presented for the judgment of adjustment timing. Practitioners can assess both the adjustment interval and the number of adjustments by simulation or trial using the presented method. Moreover, the information is very useful for judging whether or not the automatic adjustment system should be introduced.

Keywords SPC • APC • Dead band schemes • Adjustment timing • Control chart

1 Introduction

In the production of chemicals, a process adjustment such as feedback control is frequently used to reduce process variability. It is very important to judge whether or not the adjustment should be done automatically because an automatic process control (APC) system requires a large capital investment. Engineering process control

H. Kawamura (✉) · T. Suzuki
Tokyo University of Science, 2641 Yamazaki, Noda, Chiba, 278-8510, Japan
e-mail: kawamura@ia.noda.tus.ac.jp; suzuki@ia.noda.tus.ac.jp

K. Nishina
Nagoya Institute of Technology, Gokiso-cho, Showa-ku, Nagoya, Aichi, 466-8555, Japan
e-mail: nishina@nitech.ac.jp

H.-J. Lenz et al. (eds.), *Frontiers in Statistical Quality Control 10*,
Frontiers in Statistical Quality Control 10, DOI 10.1007/978-3-7908-2846-7_9,

(EPC) studies deal with the cost of adjustments, e.g., Box et al. (2009), but the process engineer and process manager are typically interested in the cost of introducing an APC system. In order to examine the advantages and disadvantages of an APC system, information for engineers on whether or not automatic adjustment should be introduced is needed. Thus, to provide such information, we consider a method for determining adjustment timing, so that practitioners can assess both the adjustment interval and the number of adjustments by simulation or trial.

This paper presents the adjustment timing determination on the basis of process capability. The basic concept of statistical process control (SPC) is to attain the required process capability early and maintain it. Therefore, the purpose of process adjustment is to achieve process capability predetermined by the product designer or the manufacturing engineer.

The determination method presented in this study comprises three steps, including an application of a time-series model, assessment of process capability, and judgment of the adjustment operation. We model process disturbances by an integrated moving average (IMA(1, 1)), and consider the usefulness of the exponentially weighted moving average (EWMA) as a time-series model, process capability indices for non-stationary processes, and appropriate adjustment timing. Finally, control charts combining information about the state of statistical control and the process capability are presented for the judgment of adjustment timing.

2 A Case Study on Low Pressure Chemical Vapor Deposition

The low pressure chemical vapor deposition (LPCVD) process is outlined in Fig. 1. The wafer is introduced into a reaction chamber called a silicon carbide tube and heated to several hundred degrees under reduced pressure, after which a polysilicon raw material gas is flowed in. A polysilicon film then forms through a chemical reaction by heating. In this process, multiple lots composed of single lots with a maximum of 25 silicon wafers are processed simultaneously. This work unit is called a batch. The thickness of the polysilicon film formed is a quality characteristic.

In this process, between-batch variation is generated by the accumulated reaction byproducts. As a result, the average film thickness of the batch gradually becomes thinner without adjustment (Fig. 2). APC is used to reduce such between-batch variation.

The control rule used in this process is

$$w_t = \frac{1}{b} \times (T - y_{t-1}) \times \beta + w_{t-1}, \tag{1}$$

where w_t is the manipulated variable that we should set at time t, b is a constant called the process gain by Montgomery (2005) and is a regression coefficient that

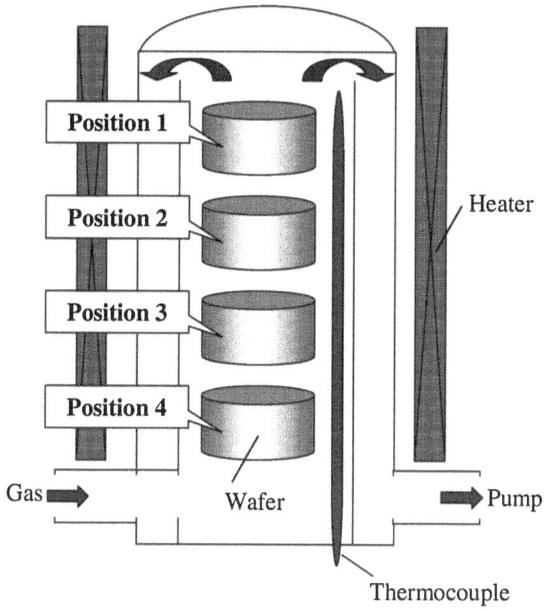

Fig. 1 LPCVD process (Higashide et al. 2010)

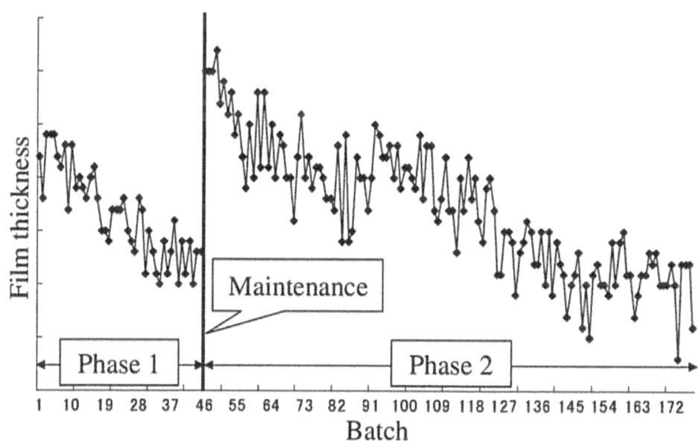

Fig. 2 Film thickness dataset

relates the magnitude of a change in w_t to a change in y_t, T is the target value of the average film thickness, y_{t-1} is the average of the measured quality characteristic at time $t-1$, β is the damping factor called by Taguchi (1993), and w_{t-1} is the manipulated variable that we set at time $t-1$. Moreover, the process gain b is determined by regression analysis of experimental data when the process is designed. This type of process adjustment is called integral control. The damping factor in this process is currently determined by the views of engineers.

3 Process Capability Assessment

It is necessary to evaluate the process capability for product design, process design, process improvement, and process control. The process capability indices (PCI) have been widely used as a means of assessing process capability. For instance, these indices are

$$C_p = \frac{USL - LSL}{6\sigma}, \tag{2}$$

$$C_{pk} = min\left\{\frac{USL - \mu}{3\sigma}, \frac{\mu - LSL}{3\sigma}\right\}, \tag{3}$$

where μ and σ are the mean and the standard deviation of the process, to be estimated from sampled data, and USL and LSL are the upper and the lower specification limits, respectively.

These indices are based on several assumptions; one of the most essential of them is that observations are statistically independent. However, there are many processes, particularly in the chemical industry, where the data are inherently correlated. Therefore, these indices may lead to inappropriate judgment because the correlation effect is not taken into account.

When the subgroup size is n, the measure of process capability for dynamic processes proposed by Spiring (1991) is

$$\hat{C}_{pm} = \frac{min\{USL - T, T - LSL\}}{3\sqrt{MSE_t + \frac{n}{n-1}(\bar{z}_t - T)^2}}, \tag{4}$$

where USL, LSL and T are the usual upper specification, lower specification, and target, respectively, used in assessing process capability, while $\bar{z}_t$ represents the average and MSE_t is the variation of the process at time t. This measure of process capability considers only the proximity to the target value T, and the variation associated with random causes as the linear effect of the tool wear is effectively removed using

$$MSE_t = \frac{\sum_{i=1}^{n}(z_{t_{a_i}} - \hat{z}_{t_{a_i}})^2}{n-2} \tag{5}$$

of sequentially selected points (i.e., $t_{a_1}, t_{a_2}, t_{a_3}, \cdots$) rather than the sample variance. In Eq. 5, MSE_t is the mean square error associated with the regression equation $\hat{z}_{a_i} = \alpha_a + \beta t_{a_i}$ and where t_{a_i} is the sequence number of the sampling unit.

The measure of process capability without a subgroup is needed to assess the LPCVD process. Therefore, we use the measure given by

$$\hat{C}^*_{pmk} = \frac{min\{USL - \hat{z}_{t+1}, \hat{z}_{t+1} - LSL\}}{3\sqrt{MSE}}, \tag{6}$$

Table 1 Results of Shapiro-Wilk test

Model	W statistic	P-value
Regression model	0.980	0.049
EWMA model	0.990	0.507

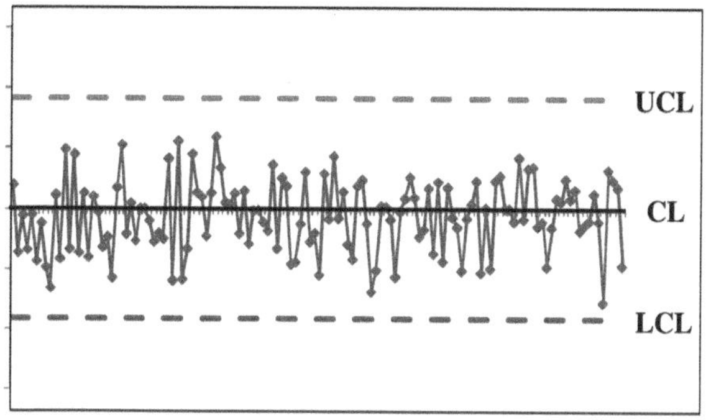

Fig. 3 Residual chart applied to film thickness

where $\hat{z}_{t+1}$ and MSE are the one-step-ahead prediction and the variance of the residual calculated by EWMA. Our main purpose in this study is to investigate the adjustment timing determination on the basis of the process capability. Since it is desirable to judge the adjustment timing when the estimates of process capability indices are less than the needed value $\hat{z}_{t+1}$ is used to calculate the one-step-ahead $\hat{C}^*_{pmk}$. This measure of process capability considers an approach to the specification limits of the process average and the variation associated with random causes (considered the systematic effect of the accumulated reaction byproducts) is removed by EWMA. The EWMA statistic is

$$\hat{z}_t = \lambda z_{t-1} + (1 - \lambda)\hat{z}_{t-1} \,, \tag{7}$$

where $0 < \lambda \leq 1$, z_{t-1} is an observation, and the initial value $\hat{z}_0$ is the target value.

The estimates of C^*_{pmk} are similar to that of C_{pk}, which has the assumption that the observations are normally distributed with constant variance. Since this measure uses the residual of EWMA, the residual should be approximately normally distributed and have constant variance. After applying both regression and the EWMA model to the data in Phase 2 of Fig. 2, we examine whether or not the residual of each model is normally distributed by means of a test. Table 1 shows the results of the Shapiro-Wilk tests for residuals. This test result seems a strong evidence suggesting that selecting the EWMA model is more appropriate than selecting the regression model.

Similarly, Fig. 3 shows the residual chart when EWMA is applied to the data in Phase 2 of Fig. 2. It is found that the residual of EWMA has constant variance.

4 Determination Method of Adjustment Timing

There are several approaches to determine adjustment timings. Dead band schemes, also called bounded adjustment schemes, were initially proposed by Box and Jenkins (1963) to deal with process adjustment applications wherein an important cost is incurred every time the process is adjusted. Taguchi (1981) also approached the process adjustment problem using a loss function. Later, Box and Kramer (1992) generalized the dead band schemes approach to include sampling costs. Luceño (2000) provided general methods to evaluate the long-run average adjustment interval and mean-squared deviation from target corresponding to dead band adjustment schemes under very general assumptions concerning both the process disturbances and the process dynamics.

In all of the above methods, adjustment timings are determined on the basis of costs. We propose a method based on the process capability because it is typically difficult to estimate the costs (e.g., off-target cost per squared deviations from target) in practice. The estimates of C^*_{pmk} are calculated whenever a characteristic value is observed, and the adjustment is made when the estimates are less than the desired value (e.g., set generally to 1.33).

The wandering behavior of the process, which comes from its inherent disturbances, is often described by an IMA(1, 1) model (see Box and Kramer 1992; Vander 1996; Park 2007). Several low-order ARIMA models with parameters estimated in Phase1 of Fig. 2 are assessed by the Akaike information criterion (AIC). As shown in Table 2, it is found that the IMA(1,1) model (i.e., ARIMA(0, 1, 1)) is most adequate in the LPCVD process.

The IMA(1, 1) model is expressed as:

$$z_t = z_{t-1} + a_t - \theta a_{t-1} \ , \tag{8}$$

where a_t is a sequence of white noise with variance σ_a^2 and θ is a smoothing constant with $0 \leq \theta < 1$. Muth (1960) showed that the minimum mean square error forecast of the IMA(1, 1) is the EWMA of the actual observations z_t with the weight $\lambda = 1 - \theta$, which ranges from $(0, 1]$.

We describe the calculation for the upper boundary line L_u for adjustment timing in terms of the (USL). The lower boundary line L_l is also similarly calculated.

Table 2 Results of AIC

Model	AIC
ARIMA(1, 0, 0)	98.545
ARIMA(2, 0, 0)	88.112
ARIMA(1, 1, 0)	85.902
ARIMA(2, 1, 0)	84.795
ARIMA(1, 0, 1)	86.027
ARIMA(0, 1, 1)	71.348
ARIMA(2, 0, 1)	87.173

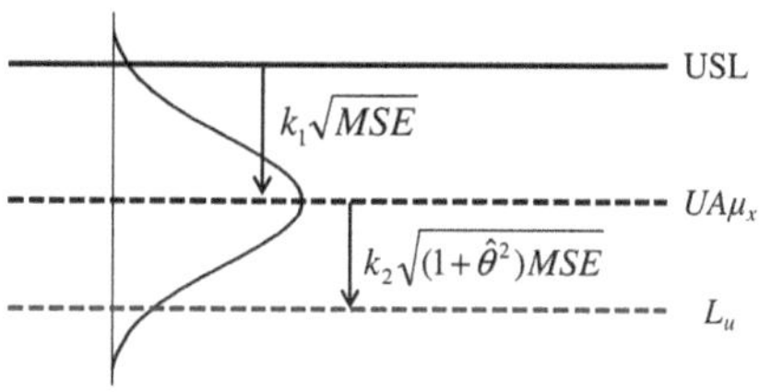

Fig. 4 Determination of *boundary line* L_u for adjustment

First of all, the maximum allowable value for the process average ($UA\mu_x$) can be determined as

$$UA\mu_x = USL - k_1\sqrt{MSE}\ , \tag{9}$$

where k_1 is a constant that determines the probability of producing material above the USL. The k_1 is calculated as shown below:

$$k_1 = 3PCI\ , \tag{10}$$

where the PCI value is set to the desired value for the process in the same way as C_p or C_{pk}. If the PCI is set to 1.33, C^*_{pmk} becomes 1.33 when the process average is $UA\mu_x$.

The upper boundary line L_u then would be

$$L_u = UA\mu_x - k_2\sqrt{(1+\hat{\theta}^2)MSE}\ , \tag{11}$$

where k_2 is a constant corresponding to the probability of prediction error for the process average, and $\hat{\theta}$ is set to $1-\hat{\lambda}$ from the relationship between IMA(1, 1) and EWMA, and $\sqrt{(1+\hat{\theta}^2)MSE}$ is the standard deviation of the prediction error because the distance of $UA\mu_x$ and L_u is predicted by the MA(1) model.

Combining Eqs. 9–11 gives the upper boundary line L_u for adjustment as

$$L_u = USL - (k_1 + k_2\sqrt{1+\hat{\theta}^2})\sqrt{MSE}\ . \tag{12}$$

Figure 4 graphically displays the calculation of Eq. 12.

5 Control Charts for Adjustment Timing

A description of acceptance control charts was published by Freund (1957), and Holmes and Mergen (2000) proposed the EWMA acceptance chart. Since these charts are used to relax the level of surveillance provided by the standard control

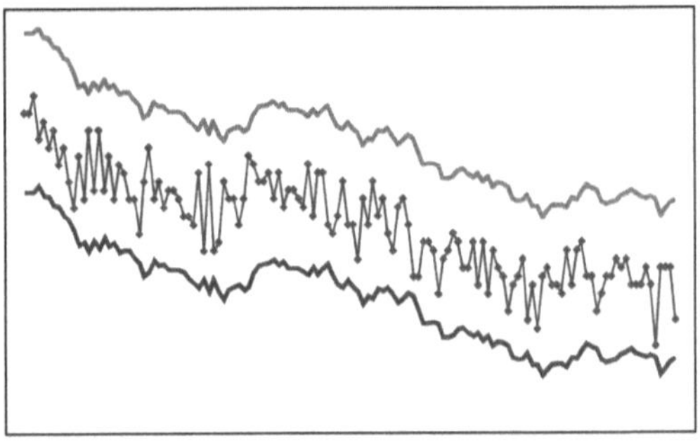

Fig. 5 Moving *center-line* EWMA control chart applied to film thickness

chart, the process does not need to be in a state of statistical control. However, it is desirable for the process that has process adjustment to be in a state of statistical control in many cases.

Our presented determination method resembles the acceptance control chart in appearance but not in precondition. We assume that the process is in a state of statistical control because the state of statistical control is needed when the process capability is evaluated using C_p or C_{pk}. Therefore, it is useful to combine information about the process capability and the state of statistical control in one chart. We propose a control chart that visualizes both types of information by modifying a moving center-line EWMA control chart.

The moving center-line EWMA control chart was proposed by Montgomery and Mastrangelo (1991). They point out that it is possible to combine information about the state of statistical control and process dynamics in one control chart. If one assumes that the model residuals are normally distributed, then the usual three-sigma control limits on the control chart on these residuals satisfy the following probability statement:

$$P[\hat{z}_t(t-1) - 3\sigma \leq z_t \leq \hat{z}_t(t-1) + 3\sigma] = 0.9973\,, \tag{13}$$

where σ is the standard deviation of residuals. This control chart uses $\hat{z}_t$ as the center line on a control chart with upper and lower control limits at

$$UCL = \hat{z}_t(t-1) + 3\sigma \tag{14}$$

and

$$LCL = \hat{z}_t(t-1) - 3\sigma \tag{15}$$

and the observation z_t would be compared to these limits to test for statistical control. Figure 5 shows the moving center-line EWMA control chart applied to the data in Phase 2 of Fig. 1.

When $k_1 = 4$ and $k_2 = 3$, Fig. 6 shows our presented control chart for adjustment timing adapted to the data in Phase 2 of Fig. 1, with parameters $MSE, \hat{\lambda}, \hat{\theta} = 1 - \hat{\lambda}$ estimated by the data in Phase 1 of Fig. 2. This chart has the boundary lines

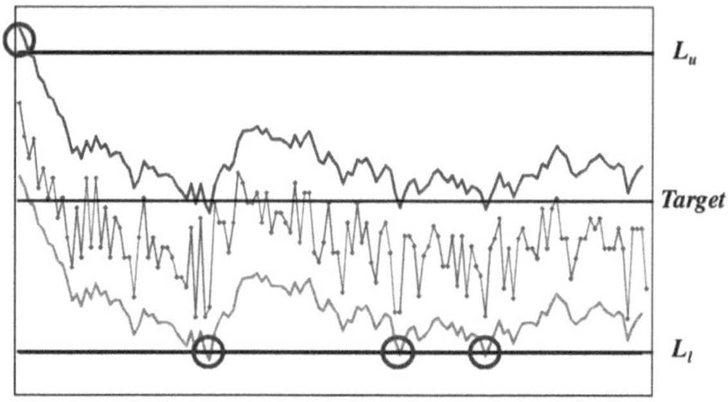

Fig. 6 Control chart for adjustment timing

(i.e., L_u and L_l) indicated in Sect. 4 and control limits of the moving center-line EWMA control chart. Moreover, the specification limits, which are used for calculating L_u and L_l in this case, are different from the actual specification limits.

In this control chart, we can visualize the process capability by the nearest distance of the boundary line and the control limit of the moving center-line EWMA control chart (i.e., L_u and UCL, or L_l and LCL). That is, when $LCL \leq L_l$ or $UCL \geq L_u$, the adjustment is made since it is expected that the process capability will be less than the desired value at the next time. The open circles indicate $LCL \leq L_l$ or $UCL \geq L_u$, and then the adjustment is made, cf. Fig. 6. That is, the adjustment is made to cancel out the deviation from target only when $LCL \leq L_l$ or $UCL \geq L_u$. Moreover, the control rule, which is indicated in Eq. 1, is used to illustrate the control chart shown.

The advantage of using this control chart is that we can judge adjustment timing, assessing the process capability after confirming the state of statistical control.

6 Conclusion

This paper has focused on the methods for determining adjustment timing in order to judge whether or not an APC system should be introduced. We proposed a method for determining adjustment timing using the boundary lines calculated on the basis of the process capability. It is natural that the adjustment timing is determined on the basis of the process capability in terms of statistical process control. The usefulness of EWMA as a time-series model was indicated using real data obtained from the LPCVD process.

We also presented a control chart for adjustment timing by combining the boundary lines and the moving center-line EWMA control chart. This chart is very useful because engineers can visually judge the adjustment timing while assessing both the process capability and the state of statistical control. Moreover, if the disturbance is not IMA(1,1), the presented control chart may not well perfom because EWMA is not the best forecast. But even when it is not, it often does quite well in predicting nonstationary series (see Box et al. 2009).

Acknowledgements This work was supported by KAKENHI (22710151). The authors acknowledge the Renesas Electronics Corporation Group for their constructive comments and provision of data.

References

Box, G. E. P., & Jenkins, G. M. (1963). *Further contributions to adaptive optimization and control: Simultaneous estimation of dynamics: Non-zero costs* (pp. 943–974). Ottawa: Bulletin of the International Instiitute, 34th Session.

Box, G. E. P., & Kramer, T. (1992). Statistical process monitoring and feedback adjustment – A discussion. *Technometrics, 34*(3), 251–267.

Box, G. E. P., & Luceño, A., & Paniagua-Quiñones, M. C. (2009). *Statistical control by monitoring and adjustment*. New York: Wiley.

Freund, R. A. (1957). Acceptance control charts. *Industrial Quality Control, 14*(4), 13–23.

Higashide, M., Nishina, K., Kawamura, H., & Ishii, N. (2010). Statistical process control for semiconductor manufacturing process. *Frontiers in Statistical Quality Control, 9*, 71–84.

Holmes, D. S., & Mergen, A. E. (2000). Exponentially weighted moving average acceptance charts. *Quality and Reliability Engineering International, 16*, 139–142.

Luceño, A. (2000). Minimum cost dead band adjustment schemes under tool-wear effects and delayed dynamics. *Statistics & Probability Letters, 50*, 165–178.

Montgomery, D. C. (2005). *Introduction to statistical quality control*. New York: Wiley

Montgomery, D. C., & Mastrangelo, C. M. (1991). Some statistical process control methods for autocorrelated data. *Journal of Quality Technology, 23*(3), 179–193.

Muth, J. F. (1960). Optimal properties of exponentially weighted forecasts of time series with permanent and transitory components. *Journal of the American Statistical Association, 55*, 299–306.

Park, C. (2007). An algorithm for the properties of the integrated process control with bounded adjustments and EWMA monitoring. *International Journal of Production Research, 23*(1), 5571–5587.

Spiring, F. A. (1991). Assessing process capability in the presence of systematic assignable cause. *Journal of Quality Technology, 23*(2), 125–134.

Taguchi, G. (1981). *On-line quality control during production*. Tokyo: Japanese Standard Association.

Taguchi, G. (1993). *Taguchi on Robust Technology Development*. New York: ASME.

Vander Wiel, S. A. (1996). Monitoring processes that wander using integrated moving average models. *Technometrics, 38*, 139–151.

Adaptive Threshold Methods for Monitoring Rates in Public Health Surveillance

Linmin Gan, William H. Woodall, and John L. Szarka

Abstract We examine one of the methods implemented by the U.S. Centers for Disease Control and Prevention's (CDC) BioSense program. The program uses data from hospitals and public health departments to detect outbreaks using the Early Aberration Reporting System (EARS). The EARS W2r method allows one to monitor the proportion of counts of a particular syndrome at a facility relative to the total number of visits. We investigate the performance of the W2r method with negative binomial inputs designed using an empirical recurrence interval (RI). An adaptive threshold monitoring method is studied based on estimating the underlying negative binomial distributions, then converting the current counts to a Z-score through a p-value. We study the effect of the input distributions on the upper thresholds required for both Shewhart and exponentially weighted moving average (EWMA) versions of the W2r and adaptive threshold methods. We simulate 1-week outbreaks and compare the outbreak detection properties of the methods.

Keywords Biosurveillance • Exponentially weighted moving average chart • Negative binomial distribution • Outbreak detection • Recurrence interval

1 Introduction

The Centers for Disease Control and Prevention (CDC) established the BioSense program with the intent of providing real-time biosurveillance for early disease outbreak detection (CDC 2007). The primary purpose of Early Aberration Reporting

L. Gan (✉) · W.H. Woodall
Virginia Polytechnic Institute and State University, 406 Hutcheson Hall, Blacksburg, VA 24061-0439, USA
e-mail: gan@vt.edu, bwoodall@vt.edu

J.L. Szarka
W. L. Gore and Associates, Inc., 501 Vieve's Way, Elkton, MD 21922, USA
e-mail: jszarka@wlgore.com

H.-J. Lenz et al. (eds.), *Frontiers in Statistical Quality Control 10*,
Frontiers in Statistical Quality Control 10, DOI 10.1007/978-3-7908-2846-7_10,

System (EARS) within BioSense is to provide national, state, and local health departments with several alternative aberration detection methods that have been developed for syndromic surveillance by CDC and non-CDC epidemiologists. Currently, hundreds of hospitals and public health departments across the United States provide data to BioSense, where the EARS methods are used for determining whether or not syndromic outbreaks have occurred (CDC 2008). For additional information on EARS methods and BioSense, see Hutwagner et al. (2003) and http://www.cdc.gov/biosense/index.html.

There are two methodologies EARS uses for detecting outbreaks. The W2 count (W2c) method focuses on the number of cases of a particular syndrome at a facility on a given day. The W2rate (W2r) method is based on the proportion of visits corresponding to a particular syndrome, which accounts for the total number of daily visits. The W2 statistics are based on 7-day moving windows. The short baseline is intended to accumulate recent information on a given syndrome. A 2-day lag is also incorporated in the calculation of the statistics, meaning the previous 2 days are not included in the baselines. If the current day's syndromic rate is large relative to the baseline data, this will result in a large W2 statistic. If a W2 value exceeds a specified threshold, an alarm is given.

The W2 statistics are calculated separately for weekdays and weekends. This is done because many health care facilities have fewer visits during weekends. We examine the case where weekday and weekend counts follow negative binomial distributions. The number of cases of a syndrome given the total number of daily visits follows a conditional binomial distribution for Poisson inputs and follows what we refer to as a "conditional negative binomial distribution" for negative binomial inputs. Comparisons of methods with the Poisson input data streams were given by Gan (2010) and Szarka et al. (2011).

An adaptive threshold method proposed by Lambert and Liu (2006) for computer network monitoring is also considered in our study. Using the baseline data, the parameters of the negative binomial distributions are fit using method of moments estimators. The current day's syndromic count yields an upper-tail p-value calculated from the estimated distribution conditioned on the total number of visits for the day. A Z-score is computed by taking the inverse standard normal cumulative distribution function (CDF) of one minus the p-value, giving an approximately standard normal statistic when there is no outbreak. The successive Z-scores are used for process monitoring.

The W2 methods have thresholds based on an empirical recurrence interval (RI) metric. Kleinman et al. (2005) explained that if monitoring of a process continues without interruption after any alarm, the RI is the fixed number of time periods for which the expected number of false alarms is one. Table 3 of the CDC's Hospital User Guide gives the W2r thresholds associated with a range of RI values from 10 to 2,000 when the length of the baseline is $n = 7$ days. For a given threshold, the empirical RI is the reciprocal of the proportion of days for which the threshold is exceeded. Using simulation, we computed our own empirical RI threshold curves and compared these to the values in the table given by BioSense. We also compared the RI threshold functions of the adaptive threshold and the W2r methods across different parameter sets and baseline lengths. Since a single upper threshold value is used once a specified RI value is selected, it is important that the non-outbreak performance of the method not depend too much on the characteristics of the input data. In general, however, we do not recommend the use of the recurrence interval metric. It is used in our paper because the CDC used this metric in their design of the W2 methods. For more information on the limitations of the RI metric, see Fraker et al. (2008).

We evaluated the various methods using baselines of $n = 7$, 14, and 28 days. These baseline lengths were used in Tokars et al. (2009), but with no more than 56 days of historical data being used. Therefore for weekends, only 8 weeks of data were available, leading to only 16 days of data in their baseline. We do not consider this restriction. The current baseline of $n = 7$ used by BioSense is a short baseline that in many instances is insufficient for reliable estimation. However, a baseline that is too long will mitigate the ability of the statistic to adjust to seasonal variation. This can lead to a decreased chance in signaling an outbreak. The W2r method is a Shewhart-type approach that signals an outbreak if the current day's statistic exceeds a threshold based on the desired value of the RI. In our study we also used the exponentially weighted moving average (EWMA) method with both the W2r and adaptive threshold approaches.

A simulation study was used to analyze the ability of the W2r and adaptive threshold methods to detect outbreaks, i.e., a power or sensitivity analysis. This was performed by generating samples from a reference distribution for several weeks, then systematically injecting a specified increase in the average number of syndrome counts. The outbreaks are assumed to last for 7 days. It is of interest to determine how frequently the various methods signal, given different magnitudes of shifts and various baseline window sizes. Both the Shewhart and EWMA approaches for detecting outbreaks are considered.

The W2 methods are reviewed in Sect. 2. In Sect. 3, we introduce the adaptive threshold method for the conditional negative binomial distribution. The performance evaluation with negative binomial inputs is presented and discussed in Sect. 4. Our conclusions are given in Sect. 5.

2 The W2 Methods

Let X_t be the count of a specific syndrome for day t at a particular facility. The baseline data for day t is dependent on its day of the week. The W2c value for day t is

$$W2c(t) = \frac{x_t - \bar{x}_t}{s_t}, \tag{1}$$

where $\bar{x}_t$ and s_t are the sample mean and standard deviation from the baseline period. These values are expressed as

$$\bar{x}_t = \frac{1}{n}\sum_{j=1}^{n} y_{jt}, \qquad s_t = \sqrt{\frac{1}{n-1}\sum_{j=1}^{n}(y_{jt} - \bar{x}_t)^2}, \tag{2}$$

where y_{jt}, $j = 1, 2, \ldots, n$, correspond to the eligible baseline data for day t. If s_t is less than one, it is reassigned a value of one CDC (2007). The relative performance of the W2c method and an adaptive threshold method was reported by Szarka et al. (2011) and Szarka (2011).

For day t, we let X_{1t} represent the syndrome count; X_{2t} the non-syndrome count; and $D_t = X_{1t} + X_{2t}$ the total number of visits to a facility, $t = 1, 2, \ldots$, with observed value d_t. The corresponding counts and numbers of visits for the baseline days are Y_{jt} and D_{jt}, $j = 1, 2, \ldots, n$. We let BLS_t and BLV_t represent the total number of syndromic counts and facility visits over the baseline period. Thus, the average rate of syndrome counts over this period is equal to BLS_t / BLV_t. The W2r value for day t is

$$W2r(t) = \frac{x_{1t} - \hat{\mu}_t}{MAR_t}, \tag{3}$$

where the expected value for day t is a function of the average rate, and the denominator is the mean absolute residual (MAR), i.e.,

$$\hat{\mu}_t = \frac{d_t BLS_t}{BLV_t} \quad \text{and} \quad MAR_t = \frac{1}{n}\sum_{j=1}^{n} \left| y_{it} - \hat{\mu}_{jt} \right|, \tag{4}$$

where $\hat{\mu}_{jt}$ refers to the estimated mean count for day j in the baseline period. If MAR_t is less than one, it is assigned a value of one CDC (2007).

Tokars et al. (2009) reported that use of the W2r method produces a more accurate expected count value and lower residuals than with use of the W2c method. They used real CDC daily syndrome counts as baseline data and assessed the power of the rate algorithm to detect injected outbreaks.

Tokars et al. (2009) proposed four algorithm modifications to address shortcomings in the previously used C2 algorithm by the CDC, which does not stratify baselines by weekends and weekdays. Those modifications included stratifying the baseline days into weekdays versus weekends, lengthening the baseline period, adjustment for total daily visits (i.e., use of the W2r algorithm), and increasing the minimum value for the estimated standard deviation to unity from 0.2.

3 The Adaptive Threshold Method

An adaptive threshold method used by Lambert and Liu (2006) for computer network monitoring has led to an alternative to the W2 methods. We consider two independent negative binomial distributions for modeling count data for the W2r method. The probability mass function (pmf) for the count X_{1t} or X_{2t} is

$$f(x_{it}|r_i, p_i) = \binom{r_i + x_{it} - 1}{x_{it}} p_i^{r_i}(1-p_i)^{x_{it}}, \tag{5}$$

where $x_{it} = 0, 1, 2, \ldots, r_i > 0$, and $0 < p_i < 1$ for $i = 1, 2$. Note that r_1 and p_1 are the negative binomial parameters for syndrome counts, and r_2 and p_2 are the negative binomial parameters for non-syndrome counts. For the remainder of our paper, we note the indexing for i will always be for values of 1 and 2. The mean and variance of the negative binomial distributions are $\mu_i = \frac{r_i(1-p_i)}{p_i}$ and $\sigma_i^2 = \frac{r_i(1-p_i)}{p_i^2}$, respectively. Conditional on the total number of visits for day t, the syndrome count X_{1t} is distributed as what we refer to as a conditional negative binomial random variable. The probability mass function (pmf) for the count X_{1t} conditioned on $d_t = \nu$ is

$$f(x|\nu, r_1, p_1, r_2, p_2) = I_{x\in(0,1,\ldots,\nu)} \frac{\binom{r_2+\nu-x-1}{\nu-x}\binom{r_1+x-1}{x}\left(\frac{1-p_1}{1-p_2}\right)^x}{\sum_{x=0}^{\nu}\binom{r_2+\nu-x-1}{\nu-x}\binom{r_1+x-1}{x}\left(\frac{1-p_1}{1-p_2}\right)^x}. \tag{6}$$

The conditional negative binomial distribution's parameters were estimated using the method of moments (MOM) estimators. These estimators for the negative binomial parameters are

$$\hat{p}_{it} = \frac{\bar{x}_{it}}{\hat{\sigma}_{it}^2} \quad \text{and} \quad \hat{r}_{it} = \frac{\bar{x}_{it}{}^2}{\hat{\sigma}_{it}^2 - \bar{x}_{it}}, \tag{7}$$

where $\bar{x}_{it} = \sum_{j=1}^{n} y_{ijt}$, $\hat{\sigma}_{it}^2 = \frac{1}{n}\sum_{j=1}^{n}\left(y_{ijt} - \bar{x}_{it}\right)^2$, and y_{ijt} correspond to the eligible baseline data for day t. The domain of these parameters of the fitted negative binomial distributions is violated if $\bar{x}_{it} > \hat{\sigma}_{it}^2$, so we must account for this.

The values of Z-scores for the conditional negative binomial distribution are determined by the p-values, whereas the p-values depend on six components: the input data x_{1t}, x_{2t}, and the parameter estimates $\hat{r}_1$, $\hat{r}_2$, $\hat{p}_1$, and $\hat{p}_2$. In order to improve the performance of the parameter estimators for our model, we used the following algorithm to compute the Z-score values:

1. If a large syndromic count x_{1t} or a non-syndromic count x_{2t} is beyond the 0.9999 quantile of the fitted negative binomial distribution, then it is replaced with a random count beyond the 99th percentile of the corresponding fitted distribution.
2. If $\bar{x}_{it} > \hat{\sigma}_{it}^2$, we set $\hat{\sigma}_{it}^2 = 1.05\bar{x}_{it}$.
3. We compute the pmf given in Eq. 6 for the syndromic count x_{1t} given d_t using the estimators from Eq. 7.
4. We calculate the p-value. If the p-value is less than 10^{-6}, then it is replaced with a random value between 10^{-6} and 10^{-5}. A Z-score is computed by taking the inverse standard normal CDF of one minus the p-value.

Lambert and Liu (2006) also used a technique of outlier removal by replacing an extremely large outlier that is beyond 0.9999 quantile of the estimated distribution with a randomly generated count beyond the 0.99 quantile of the fitted distribution. The second step of the algorithm is a technique implemented by Watkins et al. (2008). This is an ad-hoc method to give valid estimators. For an incoming observed count x_{1t} at time t, the proposed adaptive threshold method with a Shewhart approach signals an outbreak when $Z_t \geq h_{Z_{AT}}$, where $h_{Z_{AT}} > 0$ is a specified threshold value.

While a Shewhart decision rule relies on using one observation at a time, the EWMA statistic incorporates information using past observations with observations closer to the current time point given larger weights than those further back in time. For standardized variables, say v_t, the EWMA statistics E_t are

$$E_t = \alpha v_t + (1 - \alpha) E_{t-1}, \tag{8}$$

where $t = 1, 2, \ldots$, α is the weight given to the current observation, and $E_0 = 0$. When $\alpha = 1$, the EWMA method reduces to a Shewhart chart. Montgomery (2009, p. 423) recommended using weights between 0.05 and 0.25 for EWMA charts. Smaller values of α are recommended for detecting smaller shifts quickly, and larger values are recommended for larger shifts. We used $\alpha = 0.20$ in our simulation studies.

In most industrial applications, a two-sided EWMA chart is used, signaling for abnormally low or large values of the EWMA statistic. However, we are only concerned with outbreaks in our applications, so a one-sided chart is used. The one-sided EWMA statistics are expressed as

$$E_t = \max[0, \alpha v_t + (1 - \alpha) E_{t-1}], \tag{9}$$

for $t = 1, 2, \ldots$. A signal is given if $E_t \geq h_{ET}$, where $h_{ET} > 0$ is a specified threshold. The reflecting barrier at zero is used so that the statistic does not become very small. If this is not done and an outbreak occurred when the statistic is very small, it would be more difficult to signal. Failure to use a reflecting barrier in a one-sided EWMA chart can lead to serious inertial problems, a topic discussed by Woodall and Mahmoud (2005). Lambert and Liu (2006) recommended using a one-sided EWMA chart, but did not use the reflecting barrier at zero.

In traditional quality control applications, the EWMA statistic is reset to zero after a signal. This happens as a result of stopping a process, taking a corrective action, and then resuming the process. However, the EWMA statistic will not be reset after a signal in our applications because the monitoring statistics are not usually reset following a signal in public health surveillance.

4 Performance Evaluation

In this section, we report the results of a simulation study for the adaptive threshold method and the W2r method when the input data streams are assumed to have a negative binomial distribution. The negative binomial distribution is often more realistic than the Poisson distribution for public health data since counts are often overdispersed compared to the Poisson distribution. We examine the performance of these methods in terms of the RI threshold function analysis and a power analysis for both Shewhart and one-sided EWMA chart approaches.

We assumed weekday and weekend counts each follow independent negative binomial distributions. More precisely, we assumed the syndrome counts in weekdays follows a negative binomial distribution with the parameters (r_1, p_1), the non-syndrome counts in weekdays follows a negative binomial distribution with the parameters (r_2, p_2), the syndrome counts in weekends follows a negative binomial distribution with the parameters (r_1', p_1'), and the non-syndrome counts in weekends follows a negative binomial distribution with the parameters (r_2', p_2'). For simplicity, however, we let $r_1 = r_1'$, $r_2 = r_2'$, $p_1 = p_1'$, and $p_2 = p_2'$. Eight cases of parameter combinations considered are listed in Table 1 for the negative binomial inputs. For some cases, we see extreme overdispersion. The variance-to-the-mean ratio ranges from 2 to 10 for these 8 cases.

In Fig. 1 (left), we see the RI threshold functions for the adaptive threshold method for the Shewhart approach with the negative binomial parameters assumed to be known for $n = 7$, 14, and 28, and Fig. 1 (right) shows the corresponding RI threshold functions for the W2r Shewhart method. The W2r threshold functions are close to the one given in the CDC manual, which were obtained using authentic data. The W2r threshold functions become less variable with a longer baseline. From Fig. 2 (left), the adaptive threshold method using MOM estimators for the conditional negative binomial distribution tends to outperform the W2r method by giving more consistent threshold results across the eight cases for each baseline window length.

For outbreak detection, 10 weeks of in-control baseline data were simulated, and then an outbreak lasting 7 days was injected. This process was repeated

Table 1 Negative binomial distributions used

Case	(r_1, p_1)	(μ_1, σ_1^2)	(r_2, p_2)	(μ_2, σ_2^2)
1	100, 0.2	400, 2000	50, 0.1	450, 4500
2	150, 0.3	350, 1167	50, 0.1	450, 4500
3	150, 0.3	350, 1167	100, 0.2	400, 2000
4	150, 0.3	350, 1167	150, 0.3	350, 1167
5	200, 0.2	800, 4000	200, 0.2	800, 4000
6	150, 0.5	150, 300	50, 0.1	450, 4500
7	150, 0.5	150, 300	100, 0.2	400, 2000
8	150, 0.5	150, 300	150, 0.5	150, 300

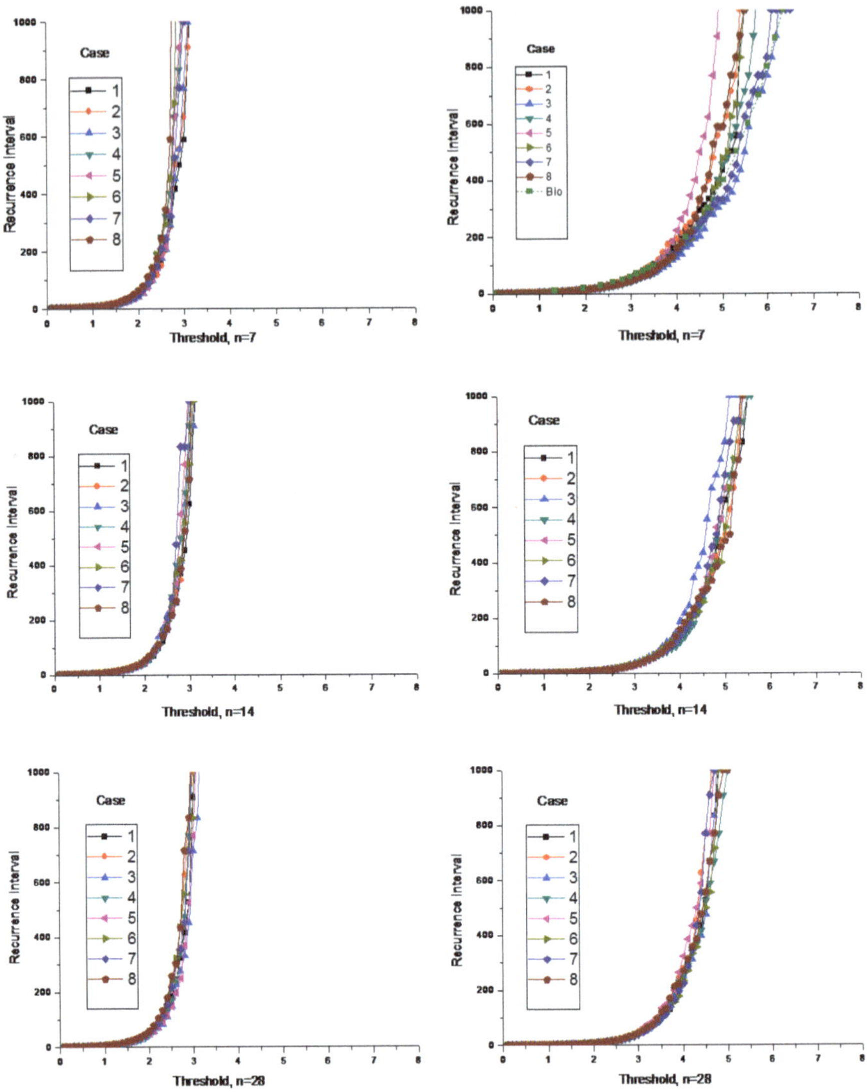

Fig. 1 RI Thresholds for adaptive threshold method with known parameters (*left*) and W2r method (*right*) for different baselines – Shewhart approach

100,000 times for each parameter combination considered where the in-control RI = 500. For each of these transient shifts, we determined the proportion of times the various methods signaled during the outbreak.

Table 2 shows the percentage increases of the power values of the adaptive threshold method compared to the W2r method. The values of the percentage increases in Table 2 are computed by using the formula

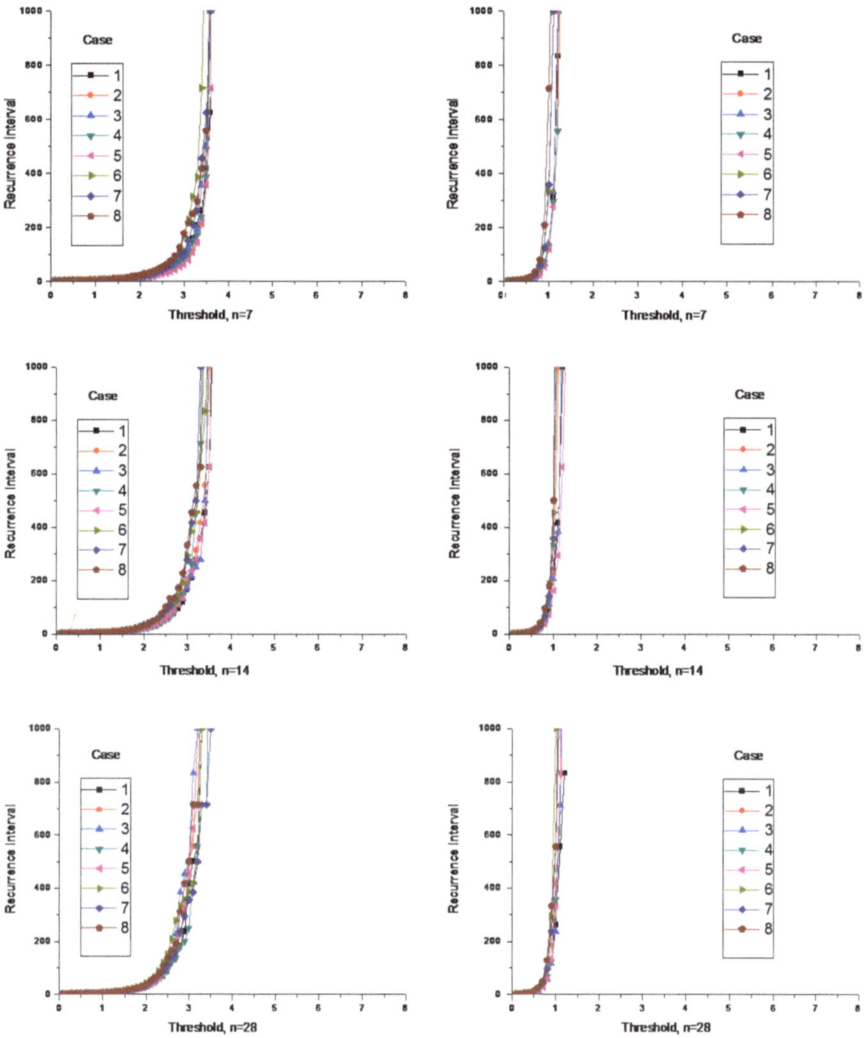

Fig. 2 RI thresholds for adaptive threshold method using MOM for different baselines – Shewhart (*left*) and EWMA (*right*) methods

$$\frac{\text{Power}_{\text{ATM}} - \text{Power}_{\text{W2r}}}{\text{Power}_{\text{W2r}}} \times 100\%. \tag{10}$$

We observed that for a given increase $\delta(100)\%$ in r_1 and a given baseline n, the adaptive threshold method works better than the W2r method with only a few exceptions for larger shifts. Gan (2010) provided the actual power values.

In Table 3, we show the percentage increase in power of the adaptive threshold EWMA approach relative to the EWMA approach used with the W2r statistics. It is

Table 2 Percentage increase of the power values for adaptive threshold method using MOM estimators compared to W2r method – Shewhart

		Case							
δ	n	1	2	3	4	5	6	7	8
	7	0.84	0.74	1.83	1.63	1.53	3.25	2.31	1.07
0.1	14	3.52	2.18	3.87	2.13	3.26	3.93	2.45	1.82
	28	4.52	5.25	3.29	6.61	8.17	4.37	5.48	4.71
	7	11.01	10.91	10.49	10.16	10.98	10.86	11.51	11.59
0.2	14	7.83	9.85	9.05	8.99	9.22	9.66	7.64	9.87
	28	8.64	11.29	8.06	10.20	7.88	2.96	8.90	7.23
	7	22.63	41.20	25.90	13.60	9.20	38.30	21.66	22.47
0.5	14	6.27	6.55	0.75	0.37	0.45	9.95	4.83	4.58
	28	5.45	0.77	0.18	0.13	0.62	1.27	0.84	0.25
	7	2.47	3.33	0.35	−3.08	−0.66	6.75	1.15	2.60
1	14	−2.83	0.02	−1.98	−3.00	−3.00	0.30	−0.50	−5.18
	28	−1.00	0.00	0.00	−1.50	0.00	0.00	0.00	−1.50
	7	0.00	0.00	0.00	−4.00	0.00	0.00	0.00	−0.35
2	14	0.00	0.00	0.00	−4.00	0.00	0.00	0.00	0.00
	28	0.00	0.00	0.00	0.00	0.00	0.00	0.00	0.00

Table 3 Percentage increase of the power values for adaptive threshold method using MOM estimators compared to W2r method – EWMA

		Case							
δ	n	1	2	3	4	5	6	7	8
	7	10.94	8.51	7.92	7.22	10.84	6.11	6.76	7.73
0.1	14	9.26	14.83	15.36	2.05	10.87	12.49	9.65	7.06
	28	11.74	13.33	9.41	0.27	3.74	16.63	13.11	5.76
	7	21.08	17.03	7.32	5.83	5.49	18.27	15.00	14.19
0.2	14	21.51	27.87	20.25	5.06	2.77	34.36	18.00	26.29
	28	27.19	27.04	4.49	3.26	0.20	47.44	33.31	10.51
	7	4.69	1.01	0.23	0.14	0.11	11.46	6.69	10.44
0.5	14	0.40	0.79	0.13	0.22	2.75	3.56	0.04	0.27
	28	0.35	0.06	0.73	0.00	0.18	1.35	0.02	0.06
	7	−0.95	−0.15	−0.03	0.02	−0.01	0.20	−0.10	0.19
1	14	−0.90	−0.20	−0.02	0.00	0.00	0.00	−0.02	−0.05
	28	−0.65	−0.02	0.00	0.00	0.00	0.00	0.00	−0.02
	7	−0.02	−0.02	0.00	0.00	0.00	0.00	0.00	0.00
2	14	0.00	0.00	0.00	0.00	0.00	0.00	0.00	0.00
	28	0.00	0.00	0.00	0.00	0.00	0.00	0.00	0.00

clear that the adaptive threshold method has better performance. The incorporation of the EWMA approach also led to higher power with both the adaptive threshold and W2r methods. Again, Gan (2010) provided the actual power values.

Figure 3 shows the power values for Case 6 for both the Shewhart (on left) and EWMA (on right) variations of the W2r and adaptive threshold methods. The adaptive threshold method generally has higher power for a given baseline length.

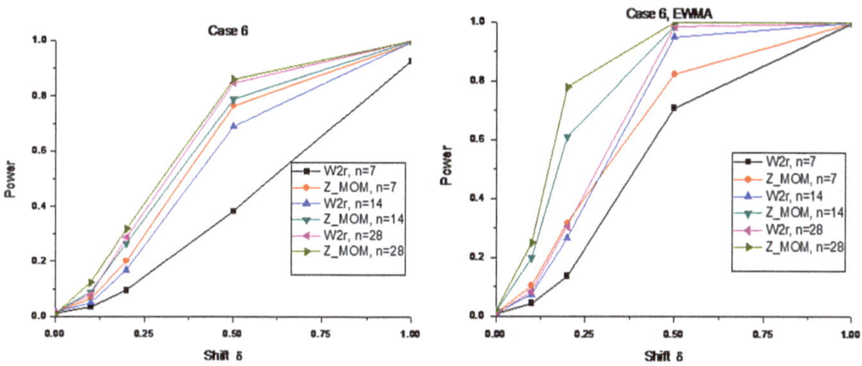

Fig. 3 Power of the W2r and adaptive threshold methods for varying baseline lengths n

The power values generally increase for all approaches as the length of the baseline increases. The EWMA methods have higher power that the Shewhart methods. These results hold for the other negative binomial cases studied as well, as reported in Gan (2010), and for Poisson input data streams.

5 Conclusions

Our results show that the adaptive threshold method has better performance than the W2r method. The corresponding RI threshold functions are more tightly grouped and the power to detect an outbreak is higher. All of the methods studied perform better with a baseline longer than 7 days. The use of the EWMA approach leads to improved performance for both the W2r and adaptive threshold approaches. The aspect of the W2r statistic that seems to lead to its relatively poor performance is the fact that the estimator of the variation of the syndromic count on a given day does not depend on the total number of visits on that day.

Acknowledgements This work was supported in part by NSF Grant CMMI-0927323 and a grant from Merck & Co., Inc.

References

Centers for Disease Control and Prevention (2007). Real-time hospital data user guide: Application version 2.11. http://www.cdc.gov/biosense/files/CDC_BioSense_BioSense_Hospital_Data_User_Guide_V2.11.pdf. Accessed May 2009.

Centers for Disease Control and Prevention (2008). Biosense technical overview of data collection, analysis, and reporting. http://www.cdc.gov/BioSense/files/BioSense_Techn_Overview_102908_webpage.pdf. Accessed May 2009.

Fraker, S. E., Woodall, W. H., & Mousavi, S. (2008). Performance metrics for surveillance schemes. *Quality Engineering, 20*(4), 451–464.
Gan, L. (2010). *Adaptive threshold method for monitoring rates in public health surveillance*. Ph.D. thesis, Virginia Polytechnic Institute and State University.
Hutwagner, L., Thompson, W., Seeman, G. M., & Treadwell, T. (2003). The bioterrorism preparedness and response early aberration reporting system (EARS). *Journal of Urban Health: Bulletin of the New York Academy of Medicine, 80*(2(Suppl. 1)), i89–i96.
Kleinman, K. P., Abrams, A., Mandl, K., & Platt, R. (2005) Simulation for assessing statistical methods of biologic terrorism surveillance. *Morbidity and Mortality Weekly Report, 54*(Suppl), 101–108.
Lambert, D., & Liu, C. (2006). Adaptive thresholds: Monitoring streams of network counts. *Journal of the American Statistical Association, 101*(473), 78–88.
Montgomery, D. C. (2009). *Introduction to statistical quality control* (6th ed.). Hoboken, NJ: Wiley.
Szarka III, J. L. (2011). *Surveillance of negative binomial and bernoulli processes*. Ph.D. thesis, Virginia Polytechnic Institute and State University.
Szarka III, J. L., Gan, L., & Woodall, W. H. (2011). Comparison of the early aberration reporting system (EARS) W2 methods to an adaptive threshold method. *Statistics in Medicine, 30*(5), 489–504.
Tokars, J. I., Burkom, H., Xing, J., English, R., Bloom, S., Cox, K., & Pavlin, J. A. (2009). Enhancing time-series detection algorithms for automated biosurveillance. *Emerging Infectious Diseases, 15*(4), 533–539.
Watkins, R. E., Eagleson, S., Veenedaal, B., Wright, G., & Plant, A. (2008). Applying cusum-based methods for the detection of outbreaks of Ross River virus disease in Western Australia. *BMC Medical Informatics and Decision Making, 8*, 37.
Woodall, W. H., & Mahmoud, M. A. (2005). The inertial properties of quality control charts. *Technometrics, 47*(4), 425–436.

Spatiotemporal Bio Surveillance Under Non-homogeneous Population

Sung Won Han, Wei Jiang, and Kwok-Leung Tsui

Abstract Motivated by the applications in healthcare surveillance, this paper discusses the spatiotemporal surveillance problem of detecting the mean change of Poisson count data in a non-homogeneous population environment. Through Monte Carlo simulations, we investigate several likelihood ratio-based approaches and compare them under various scenarios depending on four factors (1) the population trend, (2) the change time, (3) the change magnitude, and (4) the change coverage. Most literature of spatiotemporal surveillance evaluated the performance based on the average run length if a change occurs at the beginning of surveillance, which is often noted by ARL_1. On the other hand, our comparison is based on the average run length after the time when a change occurs later. Our simulation study shows that no method is uniformly better than others in all scenarios. It is found that the difference between generalized likelihood ratios (GLR) approach and weighted likelihood ratios (WLR) approach depends on population trend and change time, not the change coverage or change magnitude.

S.W. Han (✉)
Department of Biostatistics and Epidemiology, University of Pennsylvania, Philadelphia, PA 19104, USA
e-mail: hansungw@mail.med.upenn.edu

W. Jiang
Department of Operations Management, Shanghai Jiaotong University, Shanghai, P.R. China
e-mail: jiangwei08@gmail.com

K.-L. Tsui
Department of Manufacturing Engineering and Engineering Management, City University of Hong Kong, Tat Chee Avenue, Hong Kong, P.R. China

School of Industrial and Systems Engineering, Georgia Institute of Technology, Atlanta, GA 30332, USA
e-mail: ktsui@isye.gatech.edu; kltsui@cityu.edu.hk

H.-J. Lenz et al. (eds.), *Frontiers in Statistical Quality Control 10*,
Frontiers in Statistical Quality Control 10, DOI 10.1007/978-3-7908-2846-7_11,

Keywords Spatiotemporal surveillance • Generalized likelihood ratios • Weighted likelihood ratios • Non-homogeneous Poisson • Change point detection • Clusters • Detection delay • Scan statistics

1 Introduction

In health care and bio surveillance as well as manufacturing SPC, the problem of detecting a change in the mean or occurrence rate in a monitored process is important. Examples include the detection of an increase in the infection rate of diseases such as Avian flu or H1N1, or in the incidence rate of chronic diseases such as cancer. Kulldorff (2001), and Rogerson and Yamada (2004) applied their spatiotemporal surveillance method to detect clusters of increased rates in male tyroid cancers and breast cancers, respectively. Such detection problems have a spatial structure and the sample size (often, the population size) is not homogeneous. In these problems, the main goal is to detect the rate increase in some regions (or cluster) at an unknown time point by taking into account the population size. For the detection, we consider several likelihood ratio based methods for variable sample sizes devised for spatiotemporal surveillance. We evaluate their performance at both an initial change time and a later change time because the performance of distinct change times differs due to non-homogeneous baselines.

Several temporal surveillance methods for non-homogeneous sample sizes have been developed and studied. Ryan and Woodall (2009) reviewed the existing detection methods for non-homogeneous populations based on likelihood ratios. Montgomery (2005) recommended the Shewhart u-chart to handle non-homogeneous sample sizes. Several researchers modified the cumulative sum (CUSUM) statistic to deal with non-homogeneous sample sizes. Yashchin (1989), Hawkins and Olwell (1998), and Mei et al. (2009) developed weighted CUSUM methods derived from likelihood ratios for variable sample sizes. Mei et al. (2009) studied the theoretical properties of the weighted CUSUM chart under the assumption that the mean of Poisson data is proportional to the sample size. Sparks et al. (2009) also studied the weighted CUSUM method under a more general case of the Poisson mean.

Comparisons of the CUSUM and weighted CUSUM methods for temporal surveillance with non-homogeneous sample sizes have been thoroughly investigated by Ryan and Woodall (2009) and Mei et al. (2009). By assuming randomly generated populations, Ryan and Woodall (2009) compared the initial average run length (ARL) of different surveillance methods and found that the CUSUM chart is the most efficient. However, Mei et al. (2009) showed that the initial ARL does not provide a complete and fair comparison between different methods since the population changes over time and the ARL measure calculated after the change time strongly depends on the actual population size after that time point. In addition, the non-homogeneous population is randomly generated from a uniform distribution in their simulations, which does not reflect the reality in many real cases as demonstrated later. In fact, the random effect of population changes may mask the comparison based on the ARL. One may expect that the performance under such a non-homogeneous population setting is similar to the case with a constant population whose size equals to the mean of the uniform distribution. For example, the ARL performance under the population size following a uniform distribution $U[5, 15]$ is close to that under the constant population size of 10. Mei et al. (2009) compared three detection methods based on likelihood ratios, particularly the CUSUM methods based on counts and incidence rates. They used Lorden's criteria for performance comparison under monotonously increasing or decreasing population patterns. However, Lorden's criteria is the performance measure for the worst detection delays and may not be appropriate for practical applications.

For spatiotemporal surveillance, several methods have been proposed with based on likelihood ratio statistics without consideration of temporally non-homogeneous populations. Tsui et al. (2009) discussed a general framework of spatiotemporal surveillance that includes many well-known methods as special cases. A basic approach for spatiotemporal surveillance is to take the maximum of the CUSUM statistics over regions or potential clusters. Woodall and Ncube (1985) and Tartakovsky and Veeravalli (2004) suggested monitoring the maximum of multiple CUSUM statistics for each single region. Mei (2009) proposed using the summation of multiple CUSUM's instead of the maximum. Beyond a single region, Raubertas (1989) considered potential clusters of nearest neighbor regions and constructed multiple CUSUM charts for each potential cluster before taking the maximum over all potential nearest neighbor clusters. Sonesson (2007) suggested multiple CUSUM charts with variable regions, the idea of which came from the generalized likelihood ratio approach proposed in Kulldorff (2001). These methods can be extended to temporally non-homogeneous population cases because the population information can also be incorporated into the likelihood ratio statistics in each CUSUM chart of single regions or clusters. We thus call this approach the generalized likelihood ratio (GLR) method.

Due to the discussions of temporal surveillance methods for non-homogeneous populations, weighted CUSUM methods can be easily generalized to spatiotemporal surveillance. In contrast to the GLR method, we call these generalizations weighted likelihood ratio (WLR) method in this paper. In particular, we will discuss different CUSUM methods for spatiotemporal surveillance with variable temporal sample sizes. As Mei (2009) suggested, we will also compare the maximum and summation of different CUSUM statistics over potential regions or clusters to understand the efficient way for aggregating spatial information and evidence along time. We will compare the run length performance of these surveillance methods assuming temporally non-homogeneous Poisson data with monotonously increasing or decreasing population trends. The effect of change point time will also be investigated in the comparisons.

The rest of this paper is organized as follows. In Sect. 2, we discuss the spatiotemporal surveillance problem as a sequential change point detection problem and discuss performance measures. In Sect. 3, we elaborate GLR- and WLR-based surveillance methods. We compare the proposed methods through Monte Carlo simulations in Sect. 4 and discuss conclusions in Sect. 5.

2 Problem Formulation

In spatiotemporal surveillance, we are interested in detecting an increase of the occurrence rate in some regions. Suppose we observe a sequence of independent random vectors over time, $\{\mathbf{Y}_1, \mathbf{Y}_2, \ldots\}$. The random vector $\mathbf{Y}_t$ has an entry $[Y_{ct}]$, where t represents time and c represents the index of spatial regions as shown in Fig. 1. We denote a spatial location c by $c = (m, n)$, where $m = 1, 2, \ldots, M$, $n = 1, 2, \ldots, N$, and (m, n) represents the position coordinates in space.

We assume that Y_{ct} follows a Poisson distribution with mean $\lambda_{ct} = n_{ct} \times p_{ct}$, where n_{ct} is population size at region c at time t, and p is the individual risk. Under the normal condition, $p_{ct} = p_0$ for all region c's at any time. After an unfavorable event such as an outbreak occurs at an *unknown* time ν, p_{ct} changes from p_0 to p_1 at some adjacent region c's. In other words, for some $\nu \geq 1, \mathbf{Y}_1, \ldots, \mathbf{Y}_{\nu-1}$ are Poisson random vectors with mean $\lambda_{ct} = [n_{ct} \times p_0]$ for all region c's, whereas $\mathbf{Y}_\nu, \mathbf{Y}_{\nu+1}, \ldots$ are Poisson random vectors, which have means of $\mathbf{n}_{ct} \times p_1$ in some region c's and $\mathbf{n}_{ct} \times p_0$ for other regions. The goal is to detect an increase in the individual risk from p_0 to p_1 as soon as possible after an unfavorable event occurs.

This problem can be formulated as the following sequential hypothesis testing problem:

$$H_0 : \lambda_{ct} = n_{ct} \times p_0 \quad \text{for } t \geq 1 \text{ and any } c$$

against the composite alternative hypothesis

$$H_1 : \lambda_{ct} = \begin{cases} n_{ct} \times p_0, & \text{for all } c\text{'s if } 1 \leq i \leq \nu - 1 \\ n_{ct} \times p_1, & \text{for some } c\text{'s if } t \geq \nu. \end{cases}$$

This hypothesis test (H_0 versus H_1) is conducted at each time based on the sequence of independent Poisson random vectors. To simplify the problem, we assume that the outbreak coverage is a circle. We define an outbreak position by $c = (i, j)$ and outbreak coverage at the given position by A_c^r, where $A_c^r = \{d = (m, n) \big| \|d - c\| \leq r\}$. r is a radius that indicates outbreak coverage, and $d = (m, n)$ is a coordinate that indicates the position inside of the coverage. See Fig. 2 for illustrations.

In the sequential change point detection problem, we apply a detection statistic or algorithm S to the data when new data are available. If the detection statistic S is greater than a pre-determined threshold h, then an alarm is raised. The alarm time is given by $T = \min\{t | S_t > h\}$, where S_t is the detection statistic at time t. The performance of a detection method can be evaluated by the following two criteria: ARL_0 (average run length under the in-control state) and *detection delay* under the out-of-control state. A more detailed explanation of performance measures can be found in Han et al. (2010). The ARL_0 is defined by $E[T|\nu = \infty]$, where $\nu = \infty$ indicates that no change occurs. In the out-of-control state, we consider the *conditional expected delay*, defined as $CED(\nu) = E[T - \nu + 1 | T \geq \nu]$.

1	2	3	4	5	6
7	8	9	10	11	12
13	14	15	16	17	18
19	20	21	22	23	24
25	26	27	28	29	30
31	32	33	34	35	36

(1,1)	(1,2)	(1,3)	(1,4)	(1,5)	(1,6)
(2,1)	(2,2)	(2,3)	(2,4)	(2,5)	(2,6)
(3,1)	(3,2)	(3,3)	(3,4)	(3,5)	(3,6)
(4,1)	(4,2)	(4,3)	(4,4)	(4,5)	(4,6)
(5,1)	(5,2)	(5,3)	(5,4)	(5,5)	(5,6)
(6,1)	(6,2)	(6,3)	(6,4)	(6,5)	(6,6)

Fig. 1 Vector and coordinate expression of regional data ($p = 36$)

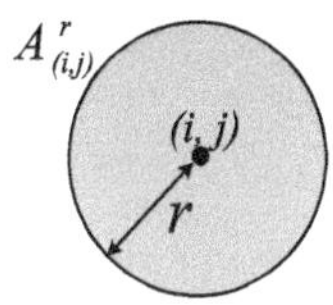

Fig. 2 Outbreak position $c = (i, j)$, radius r, and coverage $A^r_{(i,j)}$

When comparing spatiotemporal surveillance methods, we propose to use $CED(\nu = 1)$, $CED(\nu = ARL_0/2)$, and $CED(\nu = ARL_0)$ for each method. $CED(\nu = 1)$ indicates the initial average run length and $CED(\nu = ARL_0)$ is similar to the value of the steady-state average run length. We will focus on comparing several surveillance methods based on $CED(\nu)$ subject to

$$ARL_0 \geq \gamma, \tag{1}$$

where γ is the lower bound of the target ARL_0.

3 Spatiotemporal Surveillance Methods

In this section, we will discuss spatiotemporal surveillance methods based on generalized likelihood ratios. These methods correspond to the likelihood ratio methods under homogeneous populations. Suppose that the observation in the position $d = (m, n)$ at time s, Y_{ds}, follows $f(\cdot | n_{ds}, p)$ which is the Poisson distribution with the mean $n_{ds} p$. The log-likelihood ratio statistic of this observation follows

$$G_{ds} = \log \frac{f(Y_{ds} | n_{ds} p_1)}{f(Y_{ds} | n_{ds} p_0)}.$$

For each position c, we can derive cumulative log-likelihood ratios of a potential cluster A^r_c within time window $[\nu, t]$ as follows,

$$U_{cv}^{rt} = \sum_{s=v}^{t} \sum_{d \in A_c^r} G_{ds}.$$

If we focus on a single region, this can be simplified as

$$U_{cv}^{t} = \sum_{s=v}^{t} G_{cs}.$$

We now discuss different methods based on the information of shift coverage.

3.1 *GLR-Based Methods*

3.1.1 GLR for Single Regions

If the shift coverage is a single region, i.e., $r = 0$, we can obtain the following statistic for surveillance,

$$M1GLR_t = \max_c \left\{ \max_{1 \le v \le t} \left\{ U_{cv}^t, 0 \right\} \right\}. \quad (2)$$

This corresponds to running individual CUSUM statistics for each single region and taking the maximum over all CUSUM statistics. Note that $M1GLR_t$ can be recursively obtained as

$$M1GLR_t = \max_c \{ S_{ct} \},$$

where

$$S_{ct} = \max\{0, S_{c(t-1)} + G_{ct}\}.$$

If n_{ct} remains constant over time and space, $M1GLR_t$ reduces to the multivariate CUSUM (or multiple CUSUM) in Tartakovsky and Veeravalli (2004) and Woodall and Ncube (1985).

3.1.2 GLR with Unknown Shift Coverage

Sonesson (2007) assumed that the shift coverage is unknown and suggested the following statistic,

$$MVGLR_t = \max_c \left\{ \max_r \left\{ \max_{1 \le v \le t} \left\{ U_{cv}^{rt}, 0 \right\} \right\} \right\}, \quad (3)$$

where A_c^r can include at most 50% of the entire area from the center. It also has a recursive form as

$$MVGLR_t = \max_c \left\{ \max_r \left\{ S_{ct}^r \right\} \right\},$$

where

$$S_{ct}^r = \max \left\{ 0, S_{c(t-1)}^r + \sum_{d \in A_c^r} G_{dt} \right\}.$$

3.1.3 Sum of GLR for Single Regions

Corresponding to the GLR method for single regions, Mei (2009) suggested using summation instead of the maximum across spatial regions in order to combine evidence of outbreaks. The test statistic is

$$S1GLR_t = \sum_c \left\{ \max_{1 \le \nu \le t} \left\{ U_{c\nu}^t, 0 \right\} \right\}. \tag{4}$$

If the population is homogeneous, $S1GLR_t$ is equivalent to the sum of multiple CUSUM's in Mei (2009). The characteristics between the maximum and the summation have been explained in Tsui et al. (2009) using distance arguments.

3.2 WLR-Based Methods

The above spatiotemporal surveillance statistics are derived based on the likelihood ratio statistics. When these methods signal an outbreak, it is difficult to unmask signals due to a real incidence rate change or population changes. Now we discuss modified spatiotemporal surveillance methods based on weighted likelihood ratios. The following methods adjust likelihood ratios by the Kullback information number, which is defined by $I(n_{ds}, p_1, p_0) = E_{p_1}\left[\log \frac{f(Y_{ds}|n_{dt} p_1)}{f(Y_{ds}|n_{dt} p_0)}\right]$. Thus the weighted likelihood ratio follows

$$F_{ds} = \frac{1}{I(n_{ds}, p_1, p_0)} \log \frac{f(Y_{ds}|n_{ds} p_1)}{f(Y_{ds}|n_{ds} p_0)},$$

where s is time and $d = (m, n)$ is a regional index. Since $I(n_{ds}, p_1, p_0) = n_{ds} \times I(p_1, p_0)$, where $I(p_1, p_0) = p_1 \log(p_1/p_0) - (p_1 - p_0)$, we can simply divide the likelihood ratios by n_{ds} instead of $I(n_{ds}, p_1, p_0)$. Similar to generalized likelihood ratio, for each position c, we define aggregated WLR statistics for possible nearest neighbor clusters A_c^r and limited window of time $[\nu, t]$ by

$$V_{cv}^{rt} = \sum_{s=v}^{t} \sum_{d \in A_c^r} F_{ds}.$$

If we focus on a single region, i.e., r = 0, the notation V_{cv}^{rt} is simplified as

$$V_{cv}^{t} = \sum_{s=v}^{t} F_{cs}.$$

3.2.1 WLR for Single Regions

Similar to the GLR statistic for single regions, we suggest the following statistic to run multiple CUSUM charts for the WLR statistics,

$$M1WLR_t = \max_c \left\{ \max_{1 \le v \le t} V_{cv}^t, 0 \right\}. \tag{5}$$

It has a recursive form as

$$M1WLR_t = \max_c \{S_{ct}\},$$

where $S_{ct} = \max\{0, S_{c(t-1)} + F_{ct}\}$.

3.2.2 WLR with Unknown Shift Coverage

If the shift coverage is unknown, we can apply Sonesson's framework (Eq. 3) to the weighted likelihood ratios and the monitoring statistic follows

$$MVWLR_t = \max_c \left\{ \max_r \left\{ \max_{1 \le v \le t} \left\{ V_{cv}^{rt}, 0 \right\} \right\} \right\}, \tag{6}$$

where A_c^r can include at most 50% of the entire area from the center. The recursive form is

$$MVWLR_t = \max_c \left\{ \max_r \left\{ S_{ct}^r \right\} \right\},$$

where

$$S_{ct}^r = \max \left\{ 0, S_{c(t-1)}^r + \sum_{d \in A_c^r} F_{dt} \right\}.$$

3.2.3 Sum of WLR for Single Regions

We can also apply the summation principle to weighted likelihood ratios over all regions and obtain the surveillance statistic as

$$S1WLR_t = \sum_c \left\{ \max_{1 \le v \le t} \left\{ V_{cv}^t, 0 \right\} \right\}. \quad (7)$$

4 A Simulation Comparison

We conducted a simulation study to compare the detection ability of the above methods for spatiotemporal surveillance with non-homogeneous population. We explore their performances under various population trends and different time of outbreak. Our simulation is motivated by male thyroid cancer data in New Mexico (Kulldorff 2001; Mei et al. 2009; Tsui et al. 2009).

4.1 Experiment Settings

We first use the increasing population trend, which is based on the following logistic models derived from the New Mexico data,

$$n_t = \frac{\hat{\phi}_1}{1 + \exp[-(t - \hat{\phi}_2)/\hat{\phi}_3]},$$

where n_t is the population in 100,000. Mei et al. (2009) estimated the parameters based on the population time series as $\hat{\phi}_1 = 13.8065$, $\hat{\phi}_2 = 11.8532$, and $\hat{\phi}_3 = 26.4037$. Note that if t goes to infinite, n_t converges to $\hat{\phi}_1$. Another alternative population trend is a decreasing pattern,

$$n_t = \frac{\hat{\phi}_1}{1 + \exp[(t - \hat{\phi}_2)/\hat{\phi}_3]} + \hat{\phi}_1,$$

which decreases and converges to the same population size as the increasing case. The third population trend is a constant population, and the population size is $\hat{\phi}_1$. Figure 3 plots the three different cases of the population trends for our simulations.

For the spatial structure in the simulation, we use a 6×6 regular map in Sonesson (2007) and Tsui et al. (2009). As shown in Fig. 4, we consider three outbreak patterns centered around the cell (4,3): S-1, S-5, and S-13. The number after 'S-' indicates the number of affected regions. Our simulation assumes that the data in each region follow a Poisson distribution with parameter $p_0 n_t/36$ under the normal state, but $p_1 n_t/36$ under the outbreak. In this experiment, p_0, the baseline risk per

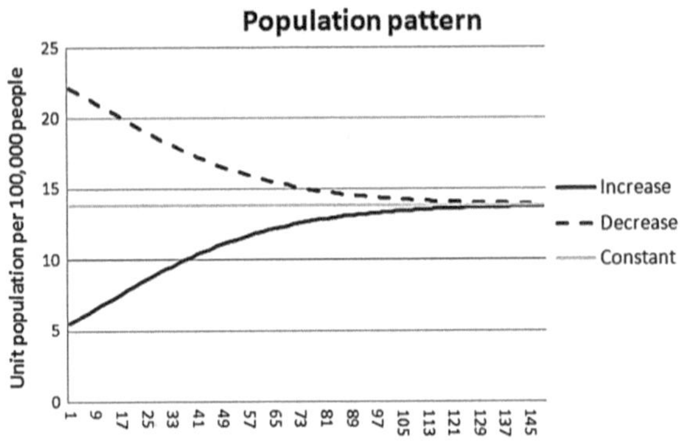

Fig. 3 Population trends

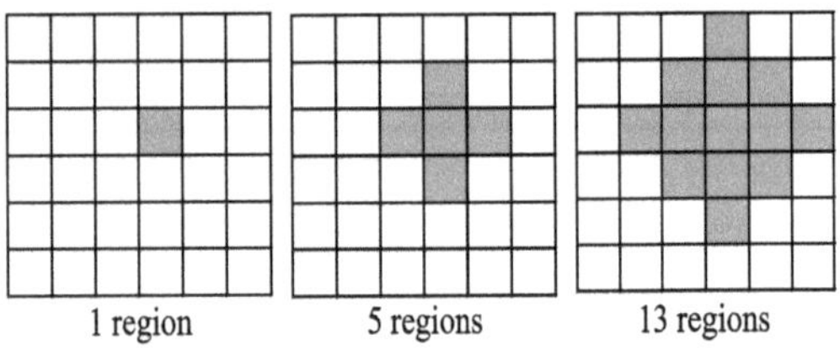

Fig. 4 Three outbreak patterns

Table 1 The thresholds for the detection methods for $p_1 = 2.7$

	Decrease	Constant	Increase
M1GLR	2.535	2.3494	2.132
M1WLR	5.502	6.126	7.42
S1GLR	22.33	20.84	19.17
S1WLR	48.78	54.33	66.2
MVGLR	4.687	4.4895	4.207
MVWLR	10.415	11.705	14.182

100,000 people, is estimated as 2.4 and the post-change risk is $p_1 = 2.7$ for small change magnitude or 3.6 for a large magnitude.

In our simulation, we set ARL_0 as close to 100 as possible. In other words, for all detection methods, we select the threshold so that each method generates a false alarm about once every 100 time periods under the baseline risk. We searched the thresholds for the targeted ARL_0 based on 100,000 replications. Table 1 shows the thresholds for the detection methods for $p_1 = 2.7$. We simulated $CED(\nu, p_1)$ for different change time such as $\nu = 1, 50$, and 100 using 50,000 replicates.

4.2 Performance Comparison

We compared the performance of the aforementioned methods under the simulation settings. We first investigate the performance of the detection methods under the constant population case followed by those under the non-homogeneous population case.

Table 2 *CED* under constant population

	$\nu = 1$			$\nu = 50$			$\nu = 100$		
	S-1	S-5	S-13	S-1	S-5	S-13	S-1	S-5	S-13
					$p_1 = 2.7$				
M1GLR	93.2*	75.9*	61.9*	53.8	43.5*	35.4*	38.2	31.9	25.6*
M1WLR	93.2*	75.9*	61.9*	53.8	43.5*	35.4*	38.2	31.9	25.6*
S1GLR	93.3*	73.7	54.0	**48.8**	**36.2**	24.4	**32.1**	**23.6**	**15.6**
S1WLR	93.3*	73.7	54.0	**48.7**	**36.2**	24.4	**31.9**	**23.6**	**15.6**
MVGLR	**86.8**	**50.6**	**29.1**	66.7*	39.0	**21.9**	64.0*	38.1*	21.1
MVWLR	**86.7**	**50.6**	**29.1**	66.6*	39.1	**21.9**	64.1*	38.1*	21.1
					$p_1 = 3.6$				
M1GLR	**39.6**	21.4	15.9*	**30.6**	16.0*	11.2*	**30.0**	15.8*	11.0*
M1WLR	**39.6**	21.4	15.9*	**30.6**	16.0*	11.2*	**30.0**	15.8*	11.0*
S1GLR	50.5*	22.1*	11.9	34.6	12.9	5.8	34.3	12.7	5.7
S1WLR	50.5*	22.1*	11.9	34.6	12.9	5.8	34.2	12.7	5.7
MVGLR	42.0	**11.7**	**5.2**	37.2*	**10.4**	**4.6**	37.2*	**10.3**	**4.6**
MVWLR	41.7	**11.6**	**5.2**	36.9*	**10.4**	**4.6**	36.8*	**10.2**	**4.6**

4.2.1 Performance Under the Constant Population

Under the constant population, it is easy to see that the GLR- and WLR-based methods are equivalent since one is scaled by a constant from the other (Mei et al. 2009). Table 2 shows the *CED* values of the aforementioned surveillance methods under the constant population in the case of $p_1 = 2.7$ and 3.6. It is easy to see that the *CED*'s of M1GLR and M1WLR (S1GLR and S1WLR, or MVGLR and MVWLR) are identical or within sampling errors ($0.2 \sim 0.4$).

Tsui et al. (2009) compared these methods in the case of independent normal distributions with mean of 0 and variance of 1. They showed that for the initial change time, M1GLR and S1GLR perform similarly for single region changes (S-1), while S1GLR is better than M1GLR for multiple region changes (S-5, S-13). If the outbreak occurs at later time (steady-state case), they showed that S1GLR is uniformly better than M1GLR for any change coverage. In addition, they also compared M1GLR and MVGLR and found that M1GLR is better than MVGLR for single region changes (S-1), but vice versa in the case of multiple region changes (S-5, S-13).

This paper investigates the detection schemes under Poisson data. The main conclusion is similar to that of Tsui et al. (2009) with subtle differences. First, comparing M1GLR and S1GLR in Table 2, we found that the *CED*'s of M1GLR and S1GLR are similar for S-1 (93.2 and 93.3, respectively) for $p_1 = 2.7$ with the change time $\nu = 1$ but the *CED* of latter (54.0) is significantly smaller than that of the former (61.9) for S-13. If $\nu = 100$, the *CED*'s of S1GLR are uniformly better than those of M1GLR for any change coverage. This result is identical to that in Tsui et al. (2009). However, in the case of large shift magnitude ($p_1 = 3.6$), the *CED* of M1GLR for S-1 is smaller than that of S1GLR at all change time $\nu = 1$, 50,

2.1%	2.4%	2.1%	2.5%	2.5%	2.5%
2.3%	2.2%	2.3%	2.2%	2.4%	2.3%
2.4%	2.1%	2.3%	19.8%	2.5%	2.2%
2.3%	2.4%	2.4%	2.0%	2.3%	2.3%
2.3%	2.2%	2.4%	2.2%	2.6%	2.1%
2.5%	2.2%	2.1%	2.6%	2.1%	2.3%

2.8%	3.0%	3.2%	3.7%	2.9%	3.9%
2.1%	2.8%	2.5%	3.2%	3.3%	2.9%
2.9%	2.5%	2.9%	3.2%	3.3%	3.9%
2.6%	2.1%	2.5%	2.9%	2.9%	3.2%
1.9%	3.0%	1.8%	2.3%	2.5%	2.7%
2.4%	1.9%	2.4%	3.0%	2.3%	2.7%

Fig. 5 Signal distribution (%) during outbreak: The *left plot* shows the signal distribution of M1GLR, and the *right plot* shows that of MVGLR

and 100. This indicates that large change magnitude favors M1GLR for the single coverage (S-1). On the other hand, S1GLR outperforms M1GLR for large coverage (S-13) at any change time. This result is consistent with the analysis in Tsui et al. (2009).

When comparing the performance between M1GLR and MVGLR, it is found that MVGLR is uniformly better than M1GLR for any change coverage when $p_1 = 2.7$ at $\nu = 1$. If $\nu = 100$, M1GLR is better than MVGLR for S-1, but vice versa for S-13. When $p_1 = 3.6$, the performance pattern is the same as that in the case of $p_1 = 2.7$ except that for S-1, M1GLR is better than MVGLR if $\nu = 1$. The result for $p_1 = 3.6$ is essentially consistent with that in Tsui et al. (2009).

One may expect that M1GLR is optimally designed for a single region change (S-1) and should outperform MVGLR in this case. However, as seen in Table 2, the CED of MVGLR (86.8) is smaller than that of M1GLR (93.2) for S-1 at $\nu = 1$ in the case of $p_1 = 2.7$. This is interesting and may be partially explained by the trade-off between detection speed and cluster identification. Even though the CED of MVGLR is smaller than that of M1GLR for detecting small change magnitudes, MVGLR gives more false identification than M1GLR. In other words, under out-of-control state, M1GLR gives high percentage of correct signals on the outbreak region, but MVGLR gives signals (in terms of center of detected areas) mostly outside of the outbreak region. Figure 5 shows the signal distribution (%) of M1GLR and MVGLR during outbreaks when $p_1 = 2.7$. Under the small shift magnitude, both the noise and signal affect MVGLR detection and make the CED of MVGLR smaller than that of M1GLR. However, the correctness of the detection of MVGLR is often sacrificed. Nonetheless, M1GLR outperforms MVGLR when $p_1 = 3.6$ in S-1.

4.2.2 Performance Under the Non-homogeneous Population

Under temporally non-homogeneous population such as the increasing population, the population size is small at initial time, but large at later time. The threshold value incorporates the information of both small population at initial time and large

Table 3 *CED* under decreasing population

	$\nu = 1$			$\nu = 50$			$\nu = 100$		
	S-1	S-5	S-13	S-1	S-5	S-13	S-1	S-5	S-13
					$p_1 = 2.7$				
M1GLR	90.1	68.6	52.8	55.6	44.5*	34.6*	46.2	37.2	29.2*
M1WLR	93.4	76.9*	63.5*	51.1	42.5	34.3*	30.8	26.1	21.4
S1GLR	91.0	66.2	44.7	49.9	**35.3**	22.1	38.9	27.6	17.9
S1WLR	94.6*	76.5*	56.8	**48.4**	37.1	25.4	**25.2**	**19.3**	**13.3**
MVGLR	**83.4**	**42.6**	**22.0**	72.5*	39.7	21.2	73.6*	41.5*	22.8
MVWLR	89.2	55.4	30.8	56.6	**35.4**	**20.4**	47.0	29.5	17.6
					$p_1 = 3.6$				
M1GLR	**30.4**	15.2	11.2	29.9	14.8*	10.3	31.5	16.4*	11.3*
M1WLR	40.0	21.8	16.7*	**28.6**	15.2*	10.8*	23.5	12.9	9.0
S1GLR	42.2	16.4	8.5	34.7	12.1	5.3	35.9	12.9	5.7
S1WLR	59.9*	26.0*	12.9	32.5	13.3	6.1	**22.7**	9.6	4.6
MVGLR	33.3	**8.2**	**3.6**	35.8*	**9.5**	**4.1**	38.4*	10.3	4.6
MVWLR	48.2	11.9	5.1	32.5	**9.7**	**4.3**	26.2	**8.5**	**4.0**

population at later time. Table 1 shows thresholds for the detection methods. Thus the population at different time affects the GLR and WLR methods differently. Under the out-of-control-state (outbreak), the time with large populations may favor GLR, whereas the time with small populations may favor WLR, as observed in Mei et al. (2009).

Tables 3 and 4 present the *CED* values of different surveillance methods under decreasing and increasing trend of populations respectively. Under the decreasing trend of population, the GLR approach (M1GLR, S1GLR, and MVGLR) outperforms the WLR approach (M1WLR, S1WLR, and MVWLR, respectively) if $\nu = 1$, but vice versa if $\nu = 100$. For example, when $p_1 = 2.7$ at $\nu = 1$, the *CED*'s of M1GLR are 90.1, 68.6, and 52.8 for S-1, S-5 and S-13, respectively. The corresponding *CED*'s of M1WLR are 93.4, 76.9, and 63.5. Thus M1GLR uniformly ourperforms M1WLR if $\nu = 1$. This performance comparison agrees with the analysis in Mei et al. (2009) that under the decreasing population, the population size at $\nu = 1$ is larger than that at $\nu = 100$, so the GLR performs better (or worse) than the WLR at $\nu = 1$ (or $\nu = 100$). On the other hand, under the increasing trend of population, the WLR approach outperforms the GLR approach at $\nu = 1$, but vice versa at $\nu = 100$.

Generally, the performance comparison between the maximum and summation operations under the constant population is similar to that under the non-homogeneous population except some minor differences for S-1 depending on population trend, change magnitude, and WLR/GLR. In most cases, the maximum operation is as good as the summation operation for S-1 if $\nu = 1$, except under the increasing population for $p_1 = 3.6$. On the other hand, the summation operation is generally as good as the maximum operation for S-1 if $\nu = 100$, except under the decreasing population for $p_1 = 3.6$.

Table 4 *CED* under increasing population

	$\nu = 1$			$\nu = 50$			$\nu = 100$		
	S-1	S-5	S-13	S-1	S-5	S-13	S-1	S-5	S-13
					$p_1 = 2.7$				
M1GLR	95.7*	83.4*	71.8*	50.7	42.9	35.3	30.0	25.4	21.1
M1WLR	91.4	71.6	56.7	63.3	49.6	38.2*	59.2	46.4	36.3*
S1GLR	95.6*	81.0	64.7	**48.5**	38.1	27.2	**25.7**	**19.6**	**13.5**
S1WLR	89.7	65.4	45.8	57.0	38.3	23.3	55.9	37.7	22.0
MVGLR	90.4	61.7	41.2	59.5	**36.9**	**21.8**	53.5	32.8	19.2
MVWLR	**80.3**	**41.1**	**24.2**	108.2*	50.5*	26.4	125.2*	57.8*	29.8
					$p_1 = 3.6$				
M1GLR	53.8	33.9*	26.7*	**30.6**	16.9	11.9	**27.2**	14.5	10.1
M1WLR	31.2	18.0	12.9	37.7	18.7*	12.8*	44.8	22.4*	15.7*
S1GLR	61.9*	33.0	20.2	34.6	13.9	6.4	30.6	12.0	5.5
S1WLR	25.2	13.9	9.1	59.9*	15.7	6.4	81.2*	22.1*	8.8
MVGLR	53.0	20.0	10.0	37.6	**11.5**	**5.3**	35.1	**10.1**	**4.6**
MVWLR	**21.3**	**8.4**	**4.6**	54.7	13.1	5.7	63.7	14.6	6.2

5 Conclusions

In this paper, we discussed several outbreak detection schemes for spatiotemporal surveillance with variable sample sizes. Using Monte Carlo simulations, we compared these methods under various scenarios depending on four factors (1) the population trend, (2) the change time, (3) the change magnitude, and (4) the change coverage. We showed that the *CED* performance of the methods depend on these factors, and no detection method is uniformly the best regardless of these four factors.

With temporally non-homogeneous population, the GLR approach outperforms the WLR approach in the time period with smaller population, but vice versa in the case with larger population. It means that in the out-of-control state, the smaller (or larger) population favors the GLR (or the WLR). This performance result holds true under any change coverage and change magnitudes.

For large change coverage (S-13), the summation operation performs better than the maximum operation at any change times and population trends. For the case S-1, the maximum operation is usually better than the summation if $\nu = 1$, and vise versa if $\nu = 100$. However, under non-homogeneous population, the performance comparison for S-1 depends on the population trend, the change magnitude, and WLR/GLR.

Similarly, for large change coverage (S-13), the variable radius approach performs better than the approach based on single regions regardless of the change times and population trends. On the other hand, for small change coverage (S-1), the latter approach performs better than the former if $\nu = 100$ or $\nu = 1$ when the change magnitude is large ($p_1 = 3.6$), but vise versa if $\nu = 1$ when the change magnitude is small ($p_1 = 2.7$). Finally, M1GLR is always better than S1GLR and MVGLR for S-1 if $p_1 = 3.6$ and $\nu = 1$.

References

Basseville, M., & Nikiforov, I.V. (1993). *Detection of abrupt changes – Theory and application*. Englewoood Cliffs, NJ: Prentice-Hall.

Crosier, R.B. (1988). Multivariate generalizations of cumulative sum quality-control schemes. *Technometrics, 30*, 291–303.

Han, S.W., Tsui, K.-L., Ariyajunyab, B., & Kim, S.B. (2010). A comparison of CUSUM, EWMA, and temporal scan statistics for detection of increases in poisson rates. *Quality and Reliability Engineering International, 26*, 279–289.

Hawkins, D.M., & Olwell, D.H. (1998). *Cumulative sum charts and charting for quality improvement*. New York: Springer.

Jiang, W., Han, S.W., Tsui, K.-L., & Woodall, W.H. (2010). Spatiotemporal bio surveillance in presence of spatial correlations. *Statistics in Medicine*. doi:10.1002/sim.3877.

Kulldorff, M. (2001). Prospective time periodic geographical disease surveillance using a scan statistic. *Journal of the Royal Statistical Society, Series A, 164*, 61–72.

Mei, Y.J. (2009). Robust scalable schemes for monitoring multiple data streams. submitted for publication.

Mei, Y.J., Han, S.W., & Tsui, K.-L. (2009). Early detection of a change in Poisson rate after accounting for population size effects. *Statistica Sinica* (in press).

Montgomery, D.C. (2005). *Statistical quality control* (5th ed.). Wiley: New York.

Raubertas, R.F. (1989). An analysis of disease surveillance data that uses the geographic locations of the reporting units. *Statistics in Medicine, 8*, 267–271.

Rogerson, P.A., & Yamada, I. (2004). Monitoring change in spatial patterns of disease: Comparing univariate and multivariate cumulative sum approaches. *Statistics in Medcine, 23*, 2195–2214.

Ryan, A.G., & Woodall, W.H. (2009). Control charts for Poisson count data with varying sample sizes. submitted for publication.

Sparks, R., Carter, C., Graham, P.L., Muscatello, D., Churches, T., Kaldor, J., Turner, R., Zheng, W., & Ryan, L. (2009). Understanding sources of variation in syndromic surveillance for early warning of natural or intentional disease outbreaks. submitted for publication.

Sonesson, C. (2007). A CUSUM framework for detection of space-time disease clustering using scan statistics. *Statistics in Medicine, 26*, 4770–4789.

Tartakovsky, A.G., & Veeravalli, V.V. (2004). Change-point detection in multichannel and distributed systems with applications. In N. Mukhopadhyay, S. Datta, & S. Chattopadhyay (Eds.), *Applications of Sequential Methodologies* (pp. 331–363). New York: Marcel Dekker, Inc.

Tsui, K.-L., Han, S.W., Jiang, W., & Woodall, W.H. (2009). Likelihood ratio methods for spatial and spatiotemporal surveillance. submitted for publication.

Woodall, W.H., & Ncube, M.M. (1985). Multivariate CUSUM quality contrl procedures. *Technometrics, 38*, 291–303.

Yashchin, E. (1989). Weighted cumulative sum technique. *Technometrics, 31*, 321–338.

Monitoring Hospital-Associated Infections with Control Charts

Christina M. Mastrangelo and Anna M. Gillan

Abstract Hospital-associated infections are a major concern in hospitals due to the potential loss of life and increased treatment costs. Monitoring the incidences of infections is an established part of quality maintenance programs for infectious disease departments in hospitals. However, traditional methods of analysis are often inadequate since the incidences of infections occur at relatively low rates. The g-type control chart is ideal for use since it monitors days between infections. However, users of the control charts find the g-type chart counter-intuitive and would prefer to use a u-chart or even a control chart for individuals. In this paper, we investigate g-type chart alternatives and how these charts may be applied to infection control surveillance data from Seattle Childrens Hospital.

Keywords G-type control charts • Disease monitoring • Negative binomial control chart

1 Introduction

Hospital-associated infections (HAIs) can affect any organ or organ system and manifest themselves in a multitude of ways. The National Nosocomial Infection Surveillance System (NNIS) and Pediatric Prevention Network data reveal that bloodstream infections are the most common occurrence of HAIs and account for 32–53% of infections, while respiratory, gastrointestinal, and urinary tract infections are reported less frequently (Agency for Healthcare Research and Quality 2001; Health Protection Agency 2008; Healthcare Infection Control Practices Advisory Committee 2009). The problem of HAIs is quite significant in terms of affecting

C.M. Mastrangelo (✉) · A.M. Gillan
Industrial & Systems Engineering, University of Washington, Box 352650, Seattle, WA 98195-2650, USA
e-mail: mastr@uw.edu; amgillan@uw.edu

H.-J. Lenz et al. (eds.), *Frontiers in Statistical Quality Control 10*,
Frontiers in Statistical Quality Control 10, DOI 10.1007/978-3-7908-2846-7_12,

patient lives, adding to the economic cost of the health care, and putting additional strain on the hospital resources. Effective monitoring of infection rates can alert clinicians to a change of infection rates, prompt the quality improvement teams to identify causes behind the abnormal increase, and stimulate efforts to look for effective interventions to reduce them. A control chart is an effective tool for this (Benneyan 1998c; Carey 2002; Montgomery 2009).

The use of control charts is increasingly being suggested for a variety of applications in healthcare in an effort to improve the quality of healthcare delivery. Components of variability exhibited by healthcare data make them attractive candidates to apply control charting techniques (Matthes et al. 2007). Woodall (2006) and Sonneson and Bock (2003) summarize various types of control charts in healthcare monitoring and in public health surveillance, as well as discussing the issues related to these charts.

The use of control charts is also widely used for monitoring infections in an effort to improve patient safety. Benneyan (1998a,b) reason the use of SPC in other fields; that is, understanding current process performance, achieving a consistent level of process quality, monitoring for process deterioration and reducing process variation, are very much applicable to the case of monitoring infections as well. In order to address some of the concerns of traditional control charts in this setting, alternate charts have been suggested. Gustafson (2000) suggests the use of risk-adjusted control charts based on a standardized infection ratio calculated by dividing the observed number of infections by the expected number of infections during a particular period. Benneyan (2001a,b) develop the g-type and h-type control charts based on inverse sampling from geometric and negative binomial distributions (further discussed in the next section) for evaluating the number of cases or the number of days between HAIs as they can exhibit greater detection power over conventional binomial-based approaches. Morton et al. (2009) demonstrate use of counted-data EWMA and CUSUM charts for monitoring of hospital-associated infections.

Limaye et al. (2008) demonstrate three control chart techniques for monitoring infection surveillance data in the pediatric ICU of Seattle Childrens Hospital. The u-chart, counted data CUSUM chart and the g-type chart were compared. Note that while these charts monitor infection data in general, the plotted point differs: the u-chart monitors the number of infections per 1,000 patient days; the counted data CUSUM chart monitors the number of infections per month; the g-type control chart monitors the number of days between infections. Each chart has a set of advantages and disadvantages (Kaminsky et al. 1992; Radaelli 1998; Shore 2000). The g-type control chart is simple to construct, and it can quickly indicate long periods having no infection occurrences. However, the g-type charts are not very helpful in detecting increased rates of infection because the lower control limit may be zero. See the next section for this discussion. If the lower control limit is zero, there would be no signal to indicate when an increase in the number of infections occurred.

The goal of this research is to develop a modified g-type chart to yield a more intuitive control chart with a precise interpretation of days between infections and to evaluate it utilizing historical infection data from Seattle Childrens Hospital. Several alternatives were identified. The next question asked was would the control chart alternatives be able to detect a change in infection rate and how effectively? Comparisons between charts are made using the Average Time Between Signal Events (ATBSE), an alternate performance metric to the Average Run Length (see Fraker et al. 2008 and the references therein). Control chart designs most appropriate for modeling days-between-infection data will be presented.

2 Concerns Regarding g-Type Chart Use in Infection Control

The g-type (geometric) control chart monitors time between events. Figure 1 is a g-type chart applied to 5 years of infection surveillance data (from 2003 to 2006) from the pediatric ICU of Seattle Childrens Hospital (Limaye et al. 2008). The data represents the number of HAIs in the unit for five of the most common types of infections: central venous catheter associated bloodstream infections, ventilator associated pneumonia, urinary tract infections, respiratory virus infections and rotavirus infections. These 5 infections account for 75% of total infections in the unit.

For the g-type chart shown in Fig. 1, each infection occurrence is an event, and a count of infection-free days before each event is the plotted point on the chart. These charts are particularly useful when the data are low-count events. Infection control may be hesitant to use the g-type control chart as a monitoring tool because interpreting the chart is counter-intuitive when compared to more commonly used control charts. For example, as the geometrically-distributed days between infections increase, lower infection events occur (hence a lower rate); yet these points plot above the upper control limit in Fig. 1. When infection events occur more frequently, they plot toward the bottom of the control chart.

In contrast, an upward (or lower) trend in infections per month plotted on a u-chart (Fig. 2) which would indicate higher (or lower) infection rates. Figure 2 shows that, on average, there are 11.2 HAIs per 1,000 patient days. Currently, u-charts are used for reporting to hospital administration; while at the same time, signage displaying the number of Days Without Infection are posted at the unit level. The u-chart is used here because each patient day is an "area of opportunity" in which one or more infections could occur. If the data had been recorded as the number of patient days with one or more infections, a p-chart would be more appropriate. P-charts also work best if the rate of non-conforming is greater than 0.05% (Montgomery 2009). The HAI rates are typically smaller than that (e.g., 0.009 per day).

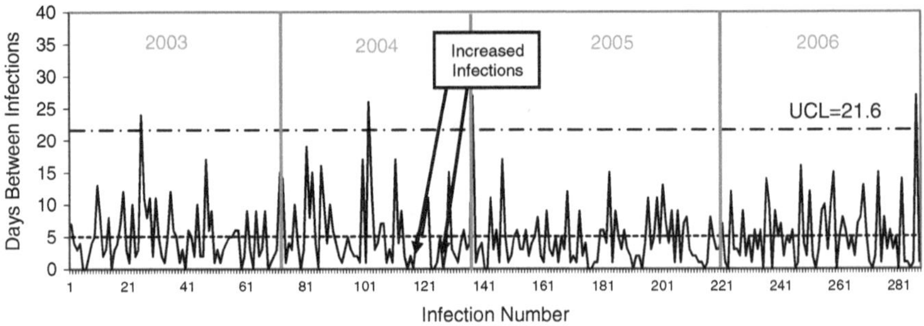

Fig. 1 G-type control chart of the number of days between HAIs (CL = 5.1)

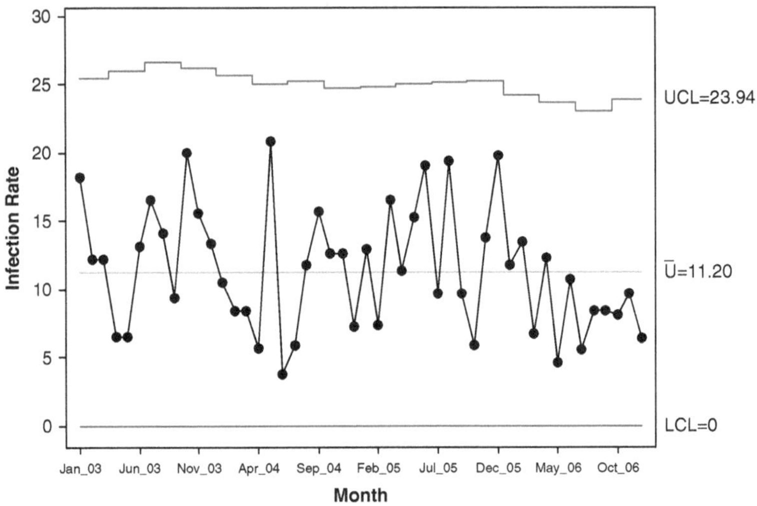

Fig. 2 A monthly *U* chart of the number of infections per 1,000 patient days

"Days between" infection data is universally understood. Monitoring geometrically-distributed days between infections data allows the chart to be interpreted once every new infection occurs, as opposed to waiting for the completion of a specified sampling period (as in the u-chart). In a healthcare environment, where timely intervention is essential to preventing further loss, continuous updating is a significant advantage and enables more effective use of out of control action plans. Figures 1 and 2 do show a downward trend starting in 2006 when an intensive hospital-wide initiative to reduce blood-stream infections began. In 2004, changes in the type of bone marrow transplants occurred. The yearly variation in the data is not unsurprising since hospitals continuously strive to improve the quality of their services with new procedures, policy and technology to reduce the possibility of infection events.

Benneyan (2001b) distinguishes between two counting schemes for collecting geometric data: Type I and Type II. Type I refers to the number of days until an

Table 1 Control limits for the g-type control chart where $\bar{x}$ is the average number of days between infections, p is the rate of infection (if known), and k is the control limit constant

Chart type	UCL	CL	LCL
G-type chart *(Infection rate known)*	$\frac{1-p}{p} + k\sqrt{\frac{1-p}{p^2}}$	$\frac{1-p}{p}$	$\frac{1-p}{p} - k\sqrt{\frac{1-p}{p^2}}$
G-type chart *(Infection rate estimated)*	$\bar{x} + k\sqrt{\bar{x}(\bar{x}+1)}$	$\bar{x}$	$\bar{x} - k\sqrt{\bar{x}(\bar{x}+1)}$

infection, beginning the count the day *after* the previous infection and including the day of the next infection. Using Type I data means that the smallest data value is 1. Type II refers only to the number of days between infections, excluding all infection day occurrences from the count. For example, if an infection is found on day 5 and the next infection occurs on day 10, then the Type II count would be 4 (whereas the Type I count would be 5). This would appear trivial except for the case when an infection occurs more frequently or even daily. For example, say infections were detected on two consecutive days. The Type I count would be 1, and the Type II count would be 0. If you were to ask a practitioner how many days between infections there were for two consecutive days, they would say 0 implying the Type II count. However, as the next section will demonstrate, Type II data is not amenable to g-type chart alternatives, so Type I count data were used.

The control limits in Fig. 1 were calculated using the equations in Table 1. In Table 1, $\bar{x}$ is the average number of days between infections, p is the rate of infection (if known), and k is the control limit constant. In this application, p is unknown, so $\bar{x}$ is used. Note that when the lower limit is negative, it is rounded up to zero. In practice, the g-type chart is quite asymmetrical, as in Fig. 1, even when the days between infections are quite long because of the large standard deviation. The concern over the traditional g-chart is the inadequate capability of the calculated LCL to be used as a monitoring tool and enabling detection of true changes in infection rates.

Benneyan (2001b) notes that the g-type chart exhibits "little-to-no power to detect increases in the infection rates due to the lower control limit equal to zero". Xie et al. (2002) draw similar conclusions, stating that "the LCL will always be less than zero making it useless in practice." We found this to be the case as well. Changing the k-sigma limits of the chart in Fig. 1 failed to provide a positive LCL until $k = 1$ (or $\alpha = 0.32$) which is unreasonable. Probability control limits may be a means to remedy the LCL problem and maintain the statistical nature of the Shewhart control limits (Liu et al. 2007; Benneyan 2001a,b).

3 Modified g-Type Charts

Several control chart modifications were studied including transformations, the EWMA g-type chart, and charts with non-parametric limits. This section presents the two preferred methods in terms of performance and non-zero lower control limit.

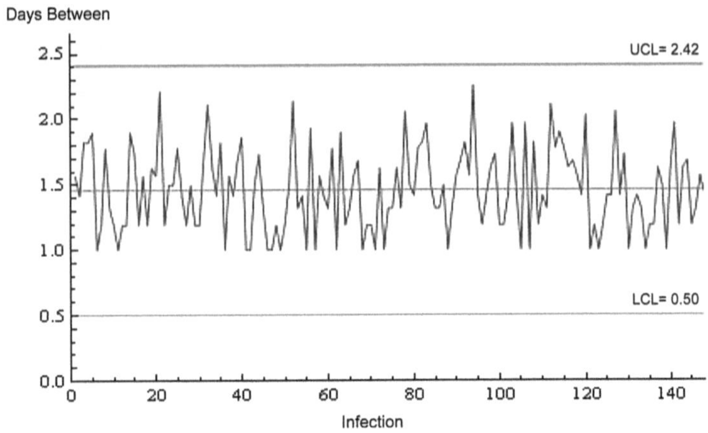

Fig. 3 G-type control chart with the Nelson transformation

Multiple modifications were initially considered but found to be inappropriate for application. For example, a g-type EWMA chart did not address concerns as it had the same issues as the g-type chart (i.e. LCL = 0). In addition by transforming the data, the g-type EWMA hides essential characteristics (i.e. actual days between) of the data that could benefit the hospital when plotted directly. A modified g-type chart devised from an empirical reference distribution failed to provide improved control limits with the limited amount of data (Willemain and Runger 1996).

3.1 Geometric Data Transformations

Transformation of the geometrically-distributed days-between data is essentially a means of normalizing the data (Liu et al. 2007). Normalized data allows accurate k-sigma limits to be used in lieu of probability limits. We refer to the following transformation as the Nelson Weibull transformation (Nelson 1994): $y = \sqrt[3.6]{x}$. Figure 3 is an example of this chart.

The second transformation is the Kittlitz transformation (Kittlitz 1999): $y = \sqrt[4]{x}$, and an example of this chart is shown in Fig. 4. Note that these transformations are similar in terms of the exponent on the transformation and the control chart limits, so there is little practical difference between the methods.

The transformations do results in lower control limit values that are greater than 0.0. However, the problem with the transformation approach is immediately apparent: Type I data will never have a value less than 1 after transformation. As such, these charts meet the primary goal of a non-zero LCL, however, they will never signal an increased infection rate. Hence, they are not considered in the next section. Note that if Type II data was used, the gap between the LCL and data would not occur. However, the LCL = 0 and this too fails to detect an increased infection rate.

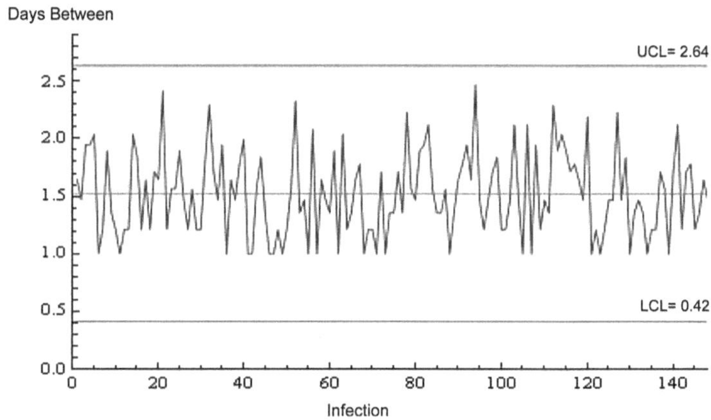

Fig. 4 G-type control chart using the Kittlitz transformation

3.2 Negative Binomial Control Chart

Various charts exist which plot data modeled by the negative binomial distribution (Albers 2010; Xie et al. 2002; Schwertman 2005; Tang and Cheong 2004). The negative binomial models the number of successes for a given number of failures. The designated number of failures is represented by r, and in this application, r represents the event of an infection. The negative binomial control chart is an extension of the Cumulative Count of Conforming (CCC) chart and is denoted as CCC-r. This control chart was explored because it has the potential to have a non-negative LCL. In addition, a situation where a HAI occurs successively does not necessarily indicate an increased infection rate. For example, every HAI in the ICU is documented. This documentation includes patient condition, diagnosis and health scores, and a chronically and seriously ill child could have two different HAIs in a short period of time (in other words, data that is 1, 2, 1 days). So configuring a negative binomial to detect 2 or more successive infection events could be mechanism to detect an increased infection rate. Note that the geometric distribution is simply the negative binomial distribution with $r = 1$.

By increasing the value of r, the LCL becomes greater than zero. However, a trade off occurs at higher values of r, because the time necessary for amassing r data points can become limiting, detracting from the possibility of prompt detection of a rate increase. Values of r between 2 and 5 are most commonly used for low count data, and this range is used here.

Probability control limits (Xie et al. 2002) were calculated using:

$$F(UCL, r, p) = \sum_{i=r}^{UCL} \binom{i-1}{r-1} p^r (1-p)^{i-r} \approx 1 - \frac{\alpha}{2}$$

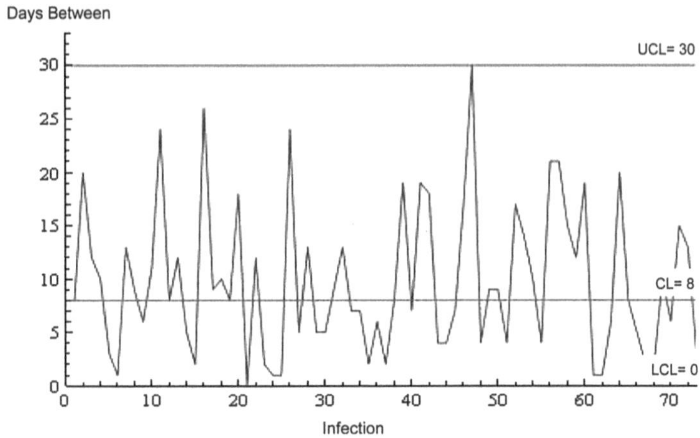

Fig. 5 CCC-r control chart for $r = 2$

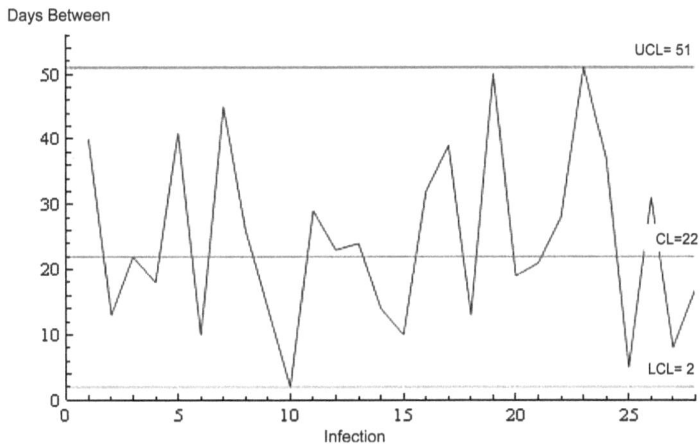

Fig. 6 CCC-r control chart for $r = 5$

$$F(CL, r, p) = \sum_{i=r}^{CL} \binom{i-1}{r-1} p^r (1-p)^{i-r} \approx \frac{1}{2}$$

$$F(LCL, r, p) = \sum_{i=r}^{LCL} \binom{i-1}{r-1} p^r (1-p)^{i-r} \approx \frac{\alpha}{2}$$

An advantage of the negative binomial chart is that it utilizes geometric data (versus, say Poisson data in a u-chart) and has intuitive appeal in a clinical setting. Figures 5 and 6 are examples of the chart with r-values of 2 and 5, respectively.

4 Methodology

To compare the performance of the g-type and negative binomial charts, 10 years of geometrically-distributed days-between-infection data were randomly generated to reflect a steady-state condition. Control limits for each control chart type were calculated using this data, and then applied to an additional 10 years of generated infection data. Five years of historical infectious disease data were used to calculate a value of p, and $p = 0.173$. This represents a baseline scenario. An alternative is to use the number of infections per patient day ($p = 0.009$) which provides a lower p value (Albers 2010). However, the use of calendar days is more intuitive and faster to plot.

Probability limits are used for the g-type chart and the negative binomial chart. A value of 0.1 establishes upper and lower control limits at the 95th and 5th percentile, respectively, of the g-type chart and the negative binomial charts (adjusted for r). For the g-type chart, probability limits given in Xie et al. (2002) were used:

$$LCL = \gamma_\alpha \frac{\ln(1-\alpha/2)}{\ln(1-p_0)}$$

$$UCL = \gamma_\alpha \frac{\ln(\alpha/2)}{\ln(1-p_0)}$$

$$\gamma_\alpha = \frac{\ln\left[\frac{\ln(1-\alpha/2]}{\ln(\alpha/2)]}\right]}{\ln\left[\frac{\alpha/2}{1-\alpha/2}\right]}.$$

An example of this chart is shown in Fig. 7. The process of creating 10 years of data, calculating control limits, and applying control limits to the subsequent 10 years of data was repeated a total of 10,000 times. The average of the performance metrics for all runs was calculated. In the event that a 10 year span of data fails to signal entirely, the run is excluded from the final average.

To reflect an increase in the infection rate, the value of p of the geometrically-distributed data was shifted upward. The second 10 year span of data was sectioned into two 5 year periods. The first 5 year period was calculated using the initial in control value of p, whereas the second 5 year period was calculated using a value of p that increased by a multiple of 0.025. The values of 0.198, 0.223, and 0.248 are used for the entire 5 year period. Control chart response to the shifts is calculated as percentage change in performance metric between the two 5 year periods.

The Average Time Between Signal Events (ATBSE) was used to evaluate chart performance in place of the Average Run Length (ARL). Note that in this control scheme the control limits are not reset after a signal (Fraker et al. 2008). The term 'signal' denotes a point below the LCL. The ATBSE was calculated as follows: after the infection rate shift, every signal is recorded and the time between signals calculated. These times are then averaged. The ATBSE results are given in

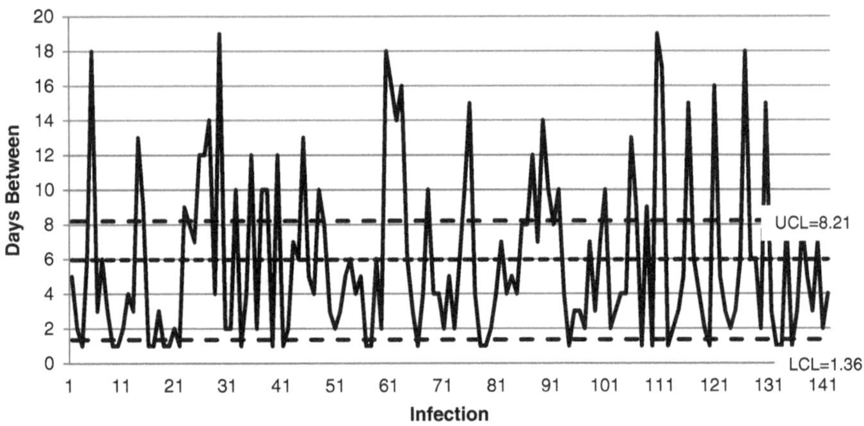

Fig. 7 G-type control chart with Type I data

Table 2 ATBSE values for the various control charts. The standard deviation of the ATBSE is given in parentheses

	Value of p			
	0.173	0.198	0.223	0.248
g-Type chart	31.1 (4.2)	27.9 (4.8)	25.2 (5.6)	22.9 (6.3)
CCC-2	229. 2 (105.5)	209.5 (96.9)	186.5 (89.3)	165.7 (86.4)
CCC-3	446.1 (276.6)	421.4 (262.6)	373.3 (238.9)	326.7 (224.6)
CCC-4	514.2 (325.9)	470.9 (299.4)	409.1 (271.9)	351.9 (256.6)
CCC-5	675.1 (421.3)	651.5 (411.3)	592.4 (382.8)	517.9 (359.4)

Table 2. Note that the first column uses the historical value of p (and $\alpha = 0.1$) in the respective control chart calculations. It does not 'tune' the control limits to a common ATBSE value. Compare the ATBSE values to the historical value of $p = 0.173$. As expected, the ATBSE declines as the infection rate increases. However, the practical importance is questionable. The historical scenario will signal, on average, every 31 days, and when the infection rate increases by two thirds (to $p = 0.248$) the ATBSE only declines by about 8 days. With respect to the ATBSE, the g-type control chart has the 'best' performance. Again this is not surprising as the negative binomial charts only signal upon successive events.

5 Conclusion

The g-type and negative binomial control charts with probability limits and Type I data are relatively straightforward to construct and have non-zero LCLs. Another perspective in selecting between the two control charts is the average time between signal events (ATBSE). Based on that metric, the g-type chart performs the best.

The g-type control chart is a more logical tool for monitoring infection rates because days-between infections (versus, say, days-between three successive infections) is universally understood in hospital environments. An advantage of the g-type chart is its ability to continuously update the control chart upon the occurrence of each infection. Note that the specific focus of this work was to determine if a control chart for geometrically-distributed data could be developed that supported detecting an increase in infection rate. There are other methods, such as a CUSUM, that could be used in this application. However, it was not studied here as a result of the findings of Limaye et al. (2008). See Szarka and Woodall (2010) for a comprehensive discussion of monitoring high-quality processes.

References

Albers, W. (2010). The optimal choice of negative binomial charts for monitoring high-quality processes. *Journal of Statistical Planning and Inference, 140*, 214–225.

Benneyan, J. C. (1998a). Statistical quality control methods in infection control and hospital epidemiology, part i: Introduction and basic theory. *Infection Control and Hospital Epidemiology, 19*(3), 194–214.

Benneyan, J. C. (1998b). Statistical quality control methods in infection control and hospital epidemiology, part ii: Chart use, statistical properties, and research issues. *Infection Control and Hospital Epidemiology, 19*(4), 265–283.

Benneyan, J. C. (1998c). Use and interpretation of statistical quality control charts. *International Journal for Quality in Healthcare, 10*(1), 69–73.

Benneyan, J. C. (2001a). Number- between g-type statistical quality control charts for monitoring adverse events. *Health Care Management Science, 4*, 305–318.

Benneyan, J. C. (2001b). Performance of number-between g-type statistical control charts for monitoring adverse events. *Health Care Management Science, 4*, 319–336.

Carey, R. G. (2002). How do you know that your care is improving? part ii: Using control charts to learn from your data. *Journal of Ambulatory Care Management, 25*, 78–88.

Agency for Healthcare Research and Quality. (2001). *Making health care safer: A critical analysis of patient safety practices* (Technical Report, Publication No. 01-E058). Rockville, MD.

Fraker, S. E., Woodall, W. H., & Mousavi, S. (2008). Performance metrics for surveillance schemes. *Quality Engineering, 20*, 451–464.

Gustafson, T. L. (2000). Practical risk-adjusted quality control charts for infection control. *American Journal of Infection Control, 28*(6), 406–414.

Health Protection Agency. (2008). *General information healthcare associated infections*. Accessed August 2008.

Healthcare Infection Control Practices Advisory Committee (HICPAC). (2009). *Guidelines for prevention of catheter-associated urinary tract infections* (Centers for Disease Control and Prevention). Accessed July 2010.

Kaminsky, F. C., Benneyan, J. C., Burke, R. J., & Davis, R. D. (1992). Statistical control charts based on a geometric distribution. *Journal of Quality Technology, 24*(2), 63–69.

Kittlitz, R. G. (1999). Transforming the exponential for spc applications. *Journal of Quality Technology, 31*(3), 301–308.

Limaye, S. S., Mastrangelo, C. M., & Zerr, D. M. (2008). A case study in monitoring hospital associated infections with count control charts. *Quality Engineering, 20*, 404–413.

Liu, J. Y., Xie, M., Goh, T. N., & Chan, L. Y. (2007). A study of ewma chart with transformed exponential data. *International Journal of Production Research, 45*(3), 743–763.

Matthes, M., Ogunbo, S., Pennington, P., Wood, N., Hart, M., & Hart, R. (2007). Statistical process control for hospitals: Methodology, user education, and challenges. *Quality Management in Health Care, 16*, 205–214.

Montgomery, D. C. (2009). *Introduction to Statistical Quality Control* (6th ed.). Hoboken: Wiley.

Morton, A., Whitby, M., McLaws, M., Dobson, A., McElwain, S., Looke, D., Stockelroth, J., & Sartor, A. (2009). The application of statistical process control charts to detection and monitoring of hospital-acquired infections. *Journal of Quality in Clinical Practise, 21*(4), 112–117.

Nelson, L. S. (1994). A control chart for parts-per-million nonconforming items. *Journal of Quality Technology, 26*(3), 239–240.

Radaelli, G. (1998). Planning time-between-events Shewhart control charts. *Total Quality Management, 9*(1), 133–140.

Schwertman, N. C. (2005). Designing accurate control charts based on the geometric and negative binomial distributions. *Quality and Reliability Engineering International, 21*, 743–756.

Shore, H. (2000). General control charts for attributes. *IIE Transactions, 32*, 1149–1160.

Sonneson, C., & Bock, D. (2003). A review and discussion of prospective statistical surveillance in public health. *Journal of the Royal Statistical Society: Series A, 166*, 5–21.

Szarka, J. L., & Woodall, W. H. (2010). A review and perspective on control charting for high quality Bernoulli processes. *Quality and Reliability Engineering International, 27*, 735–752.

Tang, L. C., & Cheong, W. T. (2004). Cumulative onformance count chart with sequentially updated parameters. *IIE Transactions, 36*, 841–853.

Willemain, T.R., & Runger, G.C. (1996). Designing control charts using an empirical reference distribution. *Journal of Quality Technology, 28*, 31–38.

Woodall, W. H. (2006). The use of control charts in health-care and public health surveillance. *Journal of Quality Technology, 38*(2), 89–104.

Xie, M., Goh, T. N., & Kuralmani, V. (2002). *Statistical models and control charts for high quality processes*. Norwell: Kluwer Academic Publishers.

Design and Implementation of Systems for Monitoring Lifetime Data

Emmanuel Yashchin

Abstract We discuss the problem of monitoring data streams corresponding to a sequence of lifetime tests with censoring. The parameters of the lifetime distribution undergo abrupt changes of unknown magnitude at some unknown points in time. Situations of this type are common in the areas of ongoing reliability tests (ORT) and analysis of warranty data. The main technical difficulty of the detection problem is related to the dynamic nature of the data acquisition process, as the information collected at consecutive points in time generally affects data for a broad range of lifetime tests in progress. We discuss issues and approaches related to monitoring such types of data and give examples related to reliability monitoring of computer components. The emphasis is on a Cusum – based approach and issues related to its implementation in systems for warranty data monitoring deployed in the IBM Personal Systems Division.

Keywords Change-points • Cusum • Sequential analysis • SPC • Reliability • Warranty • Wearout • Weibull models

1 Introduction

The problem of monitoring lifetime data arises in a number of industrial settings. For example, consider a manufacturing line for an electronic component. From the manufacturer's perspective, one of the challenging tasks is to ensure that the product stream meets some given reliability requirements. The reliability of a component is usually described by some statistical model and the lifetime characteristics are functions of the model parameters. Typically, one cannot count on the model

E. Yashchin (✉)
IBM, Thomas J. Watson Research Ctr., Box 218, Yorktown Heights, NY 10598, USA
e-mail: yashchi@us.ibm.com

H.-J. Lenz et al. (eds.), *Frontiers in Statistical Quality Control 10*,
Frontiers in Statistical Quality Control 10, DOI 10.1007/978-3-7908-2846-7_13,

parameters to remain stable over time: rather, these parameters are subject to changes of various types (shifts, drifts, etc.) – and such changes can potentially have a harmful effect on the lifetime of a device. Therefore, it is customary to have a system of ongoing reliability tests (ORT) that are based on (a) periodic selection of product samples, (b) application of stresses (such as temperature, voltage, humidity) that accelerate failures, (c) estimation of lifetime model parameters for selected samples and (d) monitoring the model parameters using some form of control charting. In the literature, a one can find a number of cases of this type (e.g., see Wu and Meeker 2002).

The example described above presents a number of technical challenges. These are primarily related to accounting for the relevant failure modes, modeling the device lifetime, and understanding the acceleration factors, establishing the stress factors, test parameters and censoring policies. However, situations of this type tend to be relatively "mathematically clean", since many of the key factors tend to be controlled by a relatively small and focused organization. Among other things, this typically ensures a good quality of lifetime data.

In this paper we will focus on situations that present special challenges from the perspective of design and deployment. One of such situations is related to warranty management of electronic systems. In this context, one usually needs to deal with data collection systems that are not exclusively centered on monitoring of component lifetimes. Databases related to warranty management are typically multi-purpose; they are, for example, used by finance specialists in order to obtain warranty cost projections, by technical support organizations to establish warranty entitlement; and by quality assurance personnel and procurement organization to keep track of the early field fallout rates. Accordingly, these databases tend to fall under the administration of different organizations, which makes development of lifetime data monitoring system a difficult task that requires compromises from the parties involved, including quality and reliability specialists.

Even with data quality issues resolved, numerous technical challenges are due to the special nature of control schemes required for monitoring of this kind of data; most of them are related to the dynamic nature of observations that serve as a basis for control charting. In particular, the charts are generally based on a sliding window of data and, as new information gets accrued, a wide range of the points on the charts undergo changes that necessitate continuous "on the fly" re-adjustment of alarm thresholds.

In recent years, a number of articles and books have been published that deal with various aspects of monitoring lifetime data. For example, likelihood ratio methods for monitoring parameters of lifetime distributions in the standard Cusum setting were discussed in Biswas and Kalbfleisch (2008), Olteanu and Vining (2009) and Sego et al. (2009). Several methods for monitoring warranty data by using Shewhart-type procedures are discussed in Wu and Meeker (2002). Steiner and McKay (2000, 2001) discuss methods and applications related to monitoring of type I censored data. This type of data (in conjunction with an EWMA monitoring procedure and Weibull observations) was considered in Zhang and Chen (2004). Analysis of warranty claims data is discussed in Blischke and Murthy

(2000), Doganaksoy et al. (2006), Kalbfleisch et al. (1991), Lawless (1998) and Lawless and Kalbfleisch (1992). Methods for analysis of failure data based on marginal counts of warranty claims (and under incomplete information about items introduced into the field) are discussed in Karim et al. (2001). Methods based on change-point analysis for hazard curves have also been considered by a number of authors, e.g., Patra and Dey (2002).

In the next section we will describe the basic parts of a system for monitoring lifetimes of computer system components based on warranty data. In Sect. 3 we discuss the key issues in the problem of design, analysis and deployment of such systems, with emphasis on warranty data. In Sect. 4 we discuss the basic approach to the problem of lifetime data monitoring with dynamically changing observations. In Sect. 5 we focus on the problem of detecting multiplicative changes in hazard curves and changes in wearout conditions that are of special importance in the context of warranty data analysis.

2 Time-Managed Lifetime Data

We will focus on a particular system for detection of unfavorable conditions in warranty data; we will refer to it as the Early Detection Tool (EDT, e.g. see Dubois et al. 2008). The goal of this tool is to provide a timely alert that reliability of some of the components might be unacceptable, and facilitate diagnostics and corrective actions. To simplify the presentation, we will refer to the manufactured systems as PCs (though in practice a much wider range of computers share common components and thus participate in a common monitoring program). We will also refer to the components as FRUs (*Field – Replaceable Units*). For example, a particular 200 gig hard drive with specified speed and other characteristics would be a FRU in the sense that a PC repair involves replacing such a drive with an equivalent one. A particular PC has a *Machine Type* (MT) that, for example, identifies this PC as a laptop of a certain type.

The schematics of EDT are shown in Fig. 1. In particular, one can see that construction of data streams in this case required pulling information from two databases. The first database, named *Ship DB*, serves primarily for the purpose of establishing warranty entitlement of customers requesting repairs. This database contains a record for every shipped PC, specifying its serial number and providing information on its components. The components (FRUs) are generally bar-coded. The barcode uniquely identifies the component and thus enables the company to validate that a PC undergoing a warranty replacement of, say, a hard drive indeed contains the original hard drive that was shipped with the machine (warranty entitlement validation). However, barcodes contain a trove of information that is useful for quality and reliability assurance. In particular, it identifies the FRU manufacturer, the FRU manufacturing date and lot number.

The second database useful for lifetime data monitoring is the *Service DB*. This database records acts of warranty repairs: PC serial number, type of repair, component(s) replaced, cost and customer information. This DB is typically used

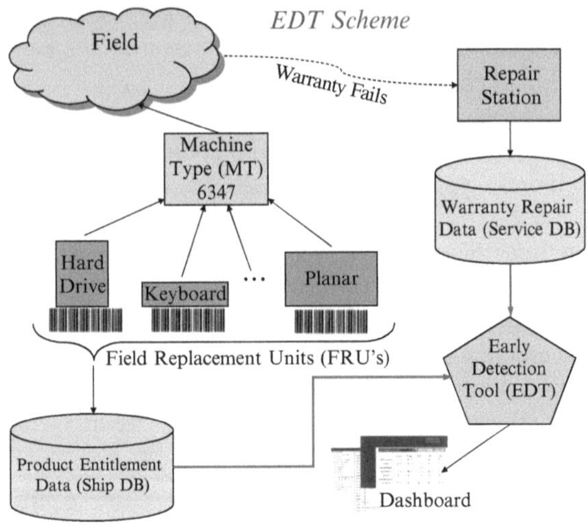

Fig. 1 Schematics of an early detection tool (EDT) for detection of changes in lifetime data

by finance specialists; for example, one can produce monthly reports of actual and projected warranty costs. Once again, this type of DB is of great use for the purpose of lifetime data monitoring, because, in conjunction with the Ship DB we can produce a data structure that is amenable to reliability modeling. In particular, let us consider a particular FRU. Based on the two DBs, we can reconstruct the data stream from the perspective of the FRU manufacturer. For example, for this FRU we can create a table the rows of which correspond to consecutive FRU manufacturer vintages. For every vintage (row) we can give information on lifetimes of FRUs of this vintage. Some FRUs fail (providing failure times), others survive till the time when the table is compiled (e.g. present time) and are thus right-censored. There can also be FRUs that are censored at some earlier points in time: for example, if a machine had only a 1-year warranty. One example of such a table is in Fig. 2. This table corresponds to data summary as of Oct 30, 2001. The columns M1, M2, M3 contain pairs corresponding to the first 3 months of service. For example, row #3 is interpreted as follows: On Aug 20, the FRU manufacturer produced 160 units; of these, 149 units already "saw" their first month of service and suffered no failures in this month. Also, the same 149 units entered into their second month of service and saw no failures in the second month either. None of these FRUs entered their third month of service. In essence, Fig. 2 represents a discretized version of a sequence of lifetime tests.

Data structures of type shown in Fig. 2 can serve as a basis of a control chart (or a set of charts), where points on a chart correspond to FRU manufacturing vintages. Such charts serve the usual purpose of "detecting smoke before the fire breaks out", and they could be tuned towards detecting specific unfavorable conditions related to the FRU manufacturing process. One will have to establish the usual trade-offs between rate of false alarms and sensitivity – however, the conventional "Run Length" framework for doing that is no longer suitable: the particular nature of this

VINTAGES	VOLUME	M1		M2		M3	
20010817	2	2	0	2	0	0	0
20010820	14	14	0	13	0	0	0
20010824	160	149	0	149	0	0	0
20010901	380	349	1	349	1	0	0
20010904	102	102	0	0	0	0	0
20010907	140	136	0	0	0	0	0
20010908	501	473	1	0	0	0	0
20010912	202	191	1	0	0	0	0
20010912	1	1	0	0	0	0	0
20010913	252	235	0	0	0	0	0
20010913	4	4	0	0	0	0	0
20010914	431	406	1	0	0	0	0
20010915	172	172	0	0	0	0	0

Fig. 2 A typical data table for a series of component (FRU) manufacturing vintages

type of lifetime data requires a specialized approach to the monitoring process and design of charts. The key feature of warranty lifetime data is that it is typically *time-managed*. Consider again data in Fig. 2. As time goes by, one will see new vintages (rows) added to the stable, and the table will maintain its characteristic "triangular" structure: early vintages will tend to see more months of service than the later ones. However, in practice, the data corresponding to earlier vintages will tend to get discarded as the corresponding PCs exit from the "warranty horizon". Of course, one can argue for benefits of maintaining a long window of history related to old vintages – however, the fact is that the owners of constituent databases (like Ship DB or Service DB) typically have no interest in them, and one can count on these vintages to disappear from the view. In all fairness, these old vintages are of limited use anyway, because they cannot serve as a reliable source of lifetime data beyond the warranty span. So, their primary use would be to enhance inference pertaining to the older vintages that still have items under warranty (for example, they would be helpful in detecting wearout earlier) – but this argument carries only a limited weight, depending on the data ownership.

Another important feature of the data is its dynamic structure. In contrast to conventional control charting, where data once observed remain static in the course of the charting process, control schemes for lifetime data tent to involve regular updating of previously observed points on the chart. For example, a point on a control chart corresponding to vintage #3 of Fig. 2 is likely to be modified next time the chart is compiled: we can expect to see new FRUs entering service, additional failures and some FRUs entering their third month of service. We refer to monitoring procedures corresponding to this type of data as *control schemes with dynamically changing observations (DCO)*.

In the process of control scheme design, of key importance is sorting of data in accordance with types of unfavorable changes (or unfavorable conditions; in both cases we will refer to them as UCs in what follows) that one intends to detect. As noted earlier, sorting of type shown in Fig. 2 is especially useful for detection of changes at the FRU manufacturers (such changes would be also graphically apparent

SHIP VINTAGES	VOLUME	M1		M2		M3	
20010925	221	206	1	186	1	0	0
20010926	287	261	1	161	0	0	0
20010927	350	316	0	117	0	0	0
20010928	385	381	1	49	0	0	0
20011001	506	492	0	0	0	0	0
20011002	612	578	1	0	0	0	0

Fig. 3 Data for a given FRU, as summarized in accordance with PC ship vintages

in DCO Cusum-type charts corresponding to this sorting). However, FRUs can also be replaced because of problems in the PC assembly process. In order to detect such problems, one may want to sort the data in accordance with the PC *Ship vintages*, see Fig. 3. The interpretation of this data is the same as the one described above; it is derived from the same combination of databases and pertains to the same FRU. So, the total volumes of manufactured FRUs (column 2) in Figs. 2 and 3 will generally be the same. The total number of FRUs that saw their first (or second) month of service will also be the same, and the number of failures in different months of service will also match (five failures in each table). One can apply DCO schemes to a series of PC vintages in Fig. 3; these will tend to reveal any bad conditions or regimes related to PC assembly process.

Other types of sorting and data segmentation involve, for example, sorting by calendar time: for every day, we compile a record of the number of machines "at risk" (in every age category) and the number of failures observed. The resulting table looks similar to ones in Fig. 2. This type of sorting helps one to detect changes related to calendar time, for example those due to introduction of a new system software, demand-related stresses or seasonal effects. Vintage-by-vintage lifetime data is usually analyzed separately for various causes of component replacement. Separate sets of charts are also maintained for different geographies (as electrical grid parameters and type of usage are typically geography-specific), and for various customer groups and even some individual customers. The reason for the latter is related to the necessity to prevent a sequence of "bad" vintages from having too strong an impact on an individual customer; furthermore, one could anticipate that at least some failure modes might be influenced by ways in which the customer uses the product.

3 Key Issues in the Design of a Monitoring System

A monitoring system for massive volumes of lifetime data generally needs to satisfy requirements of a wide range of customers, including management, procurement, brands, quality and reliability teams. This necessitates deployment of analytical and graphical tools that are capable of satisfying a wide range of users. For example, a high level manager may want a dashboard and summary that emphasizes the global picture and does not focus on details of a particular component, say, showing signs of reliability degradation for some narrow class of machines. On the other hand, a reliability engineer responsible for hard drive qualification and monitoring would need tools that not only provide alarms, but also facilitate diagnostics activity that helps in establishing the root cause and taking corrective actions. At this level, one would typically seek confirmation that the incoming process of lifetimes conforms to some expected (or "target") behavior. Once the dashboard (or other system of automatic electronic notification) indicates that this is not the case for a particular combination of Machine Type and FRU, one would like a quick answer to questions of type:

1. What is the likely source of the problem (supplier's process? assembly/configuration process? calendar events?)
2. What is the geographic extent of the problem (e.g., is it limited to a single geography?)
3. Is the problem limited to an individual machine type or to a relatively narrow family of machine types?
4. Is the problem FRU – specific? Is it limited to a subset of FRUs?
5. How many production lots (of FRUs and machine types) are affected?
6. What is the range of customers affected?
7. Is there any evidence of presence of unexpected failure types (e.g., related to early fails)
8. Has the process of failures been stable (albeit at an unacceptably high level), or there is an evidence of change-points?
9. Is there any evidence of increase in failure rate (wearout)?
10. What is the current state of the failure rate process?
11. What is the projected impact of the problem? What severity (or priority) should it be assigned?

Detecting UCs and addressing the above types of questions is especially challenging in the environment of massive data streams. For a PC manufacturing process, one could easily reach a condition where hundreds of thousands of MT/FRU combinations are monitored simultaneously, while the number of people to handle the related alarms remains very limited.

Of special importance is the issue of *false alarms*. Even in a well-designed massive data monitoring system one could expect to have a few hundred false alarms, so it is important that the users of a system be aware of this phenomenon. A requirement to eliminate them completely would be too stringent and it could lead to an overly conservative system that would not be appealing to users. However, it is good practice to err on the side of having fewer false alarms. Excessive alarms can quickly destroy credibility of a monitoring system as they will lead to its signals being ignored, setting the stage for a potentially damaging oversight of conditions that are really important to detect early (e.g., see Brown 2010).

One can reduce the intensity of false alarms (while maintaining acceptable detection capability) by following a careful design procedure that emphasizes *practical significance* of the detected conditions as opposed to their *statistical significance*. One important point is to ensure that every monitored sequence has a well defined acceptable envelope. In the lifetime data setting, this can frequently be achieved by specifying some "target" hazard curve and then specifying what constitutes acceptable and unacceptable deviations from this curve. Thus, the area of *target setting* is of key importance in the process of monitoring. This process can generally be (at least partially) automated by using a combination of engineering knowledge about statistical behavior of monitored components, data available for this component and similar components (including its predecessors) and technical requirements based on the so called *service cost estimates*. These type of requirements will ensure that target curves are not merely driven by historic performance and engineering knowledge, but also take into account business requirements for reliability improvement.

To achieve a high level of statistical power, it is also important for the monitoring process to be based on modern statistical tools. In our experience, we found the tools based on the statistical theory of sequential analysis, likelihood-ratio tests and change-point theory to be especially effective in this regard. We found the procedures based on modified *weighed Cusum-Shewhart (CS) schemes* (e.g. see Yashchin 1993; Hawkins and Olwell 1998) to be especially useful; these types of schemes formed the backbone of the monitoring system deployed in the IBM Personal Systems Division. These procedures provide high statistical power as measured by standard Key Performance Indicators (KPI) used in business management. One of these indicators is the so called Mean Time to Detect (MTTD) – and one can usually make a case for acceptance of Cusum techniques on this basis. Furthermore, the weighted Cusum – Shewhart methodology offers an especially appealing design process based on very few design parameters and provides graphical instrumentation that greatly facilitates diagnostics by helping the user to visually identify regimes and points of change. As we will see later, however, in the presence of time-managed lifetime data the CS method alone is not sufficient and one needs to use supplemental tests in order to obtain good statistical performance.

To illustrate some of the above points, consider the opening screen of the Early Detection Tool (EDT), see Fig. 4. This screen is providing a summary of the Latin America FRUs, with vintage ordering corresponding to machine ship vintages (see Fig. 3). This view could be of value, for example, to management responsible for overall Reliability Assurance (RA) or to a manager responsible for support of Latin America. Several cells are showing signs of reliability-related problems; one of them corresponds to Hard Drives for Power Server systems. Some information pertaining to the cell is available immediately (one MT/FRU combination out of 55 analyzed has been flagged). Additional information can be made available via special properties of the interface: we found "tool tips" (pop-ups appearing upon mouse hovering over the star) to be especially effective. For example, this supplemental information could provide estimates for the most severe condition or to alert the user that a condition exists that has not been seen before.

Figure 5 shows the result of the drill-down obtained by clicking on the red star in Fig. 4 (level-2 nested view). This view would also be of special use to the team

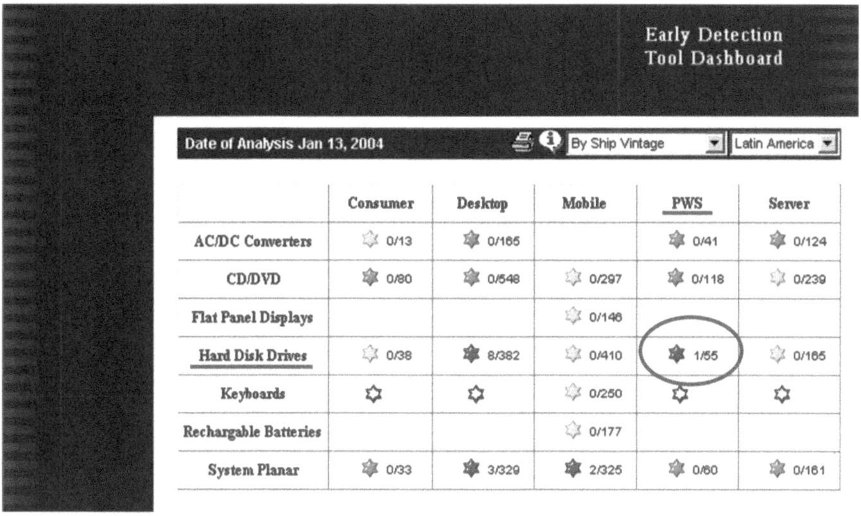

Fig. 4 A *top* level of the multi-layer dashboard for the Early Detection Tool (EDT) representing the Jan 13, 2004 view for Latin America. The *rows* correspond to commodities (types of FRU) and the *columns* – to Brands. One can see that reliability of some of the Hard Disk Drive types for Desktop – type models and for Power Servers (PWS) is considered problematic. The numbers next to the *red star* indicate the total number of analyses for this Brand – Commodity combination, Geography and type of ordering (55) and the number of combinations flagged (1). Clicking on a *red star* leads to the level-2 nested view (dashboard) providing a more detailed view, see Fig. 5. Colors are used to provide a summary of nested conditions: for example, *Green* = *Grey* = OK, *Yellow* = *Light Gray* = no conditions flagged, but some of the data is still missing, *Red* = *Dark Grey* = flagged conditions present

responsible for reliability of hard drives. It shows that a particular FRU (we will refer to it as XXXX) has been flagged for the first time (as signified by our use of triangle instead of a star). Once again, one can get some supplemental information about this condition by taking advantage of the interface properties. In particular, a further drill-down would lead on to the level-3 nested view containing some plots and reports; we will show some examples in the next section.

An effective monitoring system will generally offer a range of post-alarm activities to help its users in handling the newly detected UCs. One of them includes support for *alarm prioritization*. Engineering experience suggests that simply "piling-on" alarms is not an effective strategy, as the users usually have more alarms on their plates than they can handle. Having to handle additional alarms that are of lower importance than those already in the system can be quite irritating to users. It is, therefore, necessary to provide users with some degree over control of the intensity of alarms that goes beyond the control of false alarm rates. Accordingly, in the EDT we provided a sub-system that characterizes alarms in terms of several features that facilitate prioritization, depending on the objectives of this or another group of users. One of the features is the *severity index* that is a function of several statistics computed from the DCO charts: in the CS framework, such a function

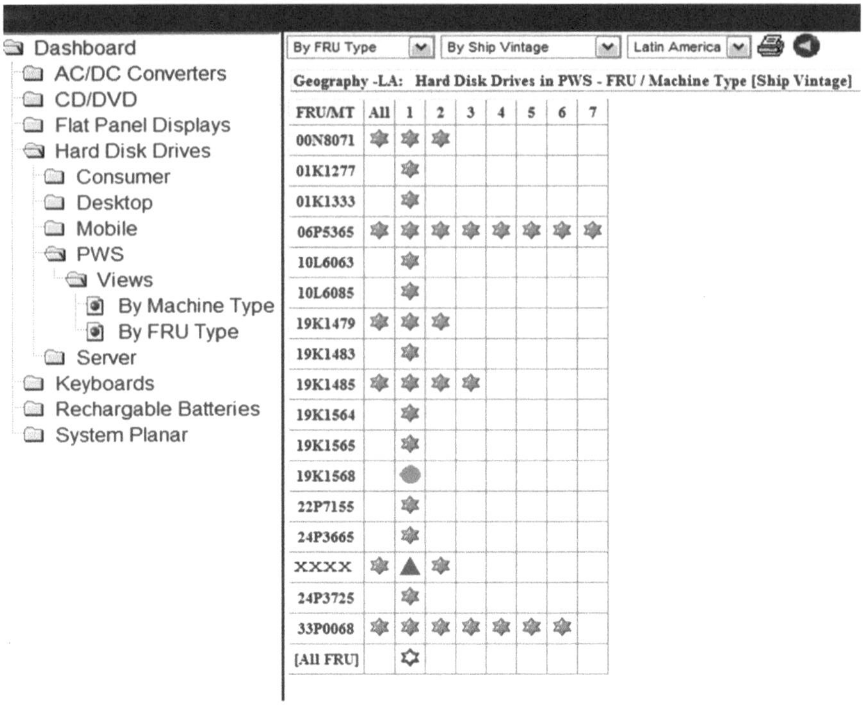

Fig. 5 The level-2 nested dashboard obtained by clicking on the *red star* in Fig. 4. The *rows* correspond to a specific FRUs (Hard Disk Drives). The *columns* are row-dependent, as any given FRU fits only a subset of Machine Types (MT). The *red triangle* (instead of *star*) indicates that this condition has not been seen in the previous analysis. Clicking on the *red triangle* leads to next drill-down layer, including a report and a set of DCO charts

could take into account the degree of threshold violation, closeness of detected UCs to the current point in time and the degree of engineering intervention that is feasible with respect to the flagged component. While prioritization of alarms by severity makes sense for users whose primary responsibility is to track and minimize the impact of existing UCs it may be of limited value to other groups of users. For example, users who are primarily interested in early detection of new UCs may be more interested in knowing how "fresh" the flagged conditions are. So, another feature that we refer to as *recentness* is typically useful in differentiating between newly emerging and pre-existing UCs.

In most cases, flagged conditions with high severity indices correspond to conditions that have been around for some time; on the other hand, newly emerging reliability problems will tend to have a relatively low severity index. With a sufficiently high index of recentness, however, such an alarm could be prioritized high due to the fact that the detected UCs would be recognized as ones meriting special engineering attention. One example of application of such an index can be seen in Fig. 5, where the detected UC has been classified as "new". In addition to prioritization-related alarm features, the users can benefit from many other estimates and indices that can be derived based exclusively on the data at hand. In particular, of special use is the list of data regimes and change-points, with data interpretation that identifies, for example, a given regime as a shift or drift in a parameter related to flagged lifetime data. Providing filtered estimates of the current level of monitored parameters and related quantities is also useful in many situations, especially those where these parameters can be directly adjusted so as to improve the reliability of the new product entering the field.

In many cases, however, one will have to deal with situations in which newly detected UCs pertain to vintages that were produced, say, 1 year earlier and it is already known that the product currently entering the field has different reliability properties. In such situations one would benefit from supplemental automated analysis of affected vintages and forecasts related to increase in fallout rates (due to the newly flagged UC) that are expected in future periods of time.

In design and deployment of systems for monitoring massive amounts of lifetime data one will need to overcome a number of computational challenges spawned by the sheer scope of the analysis, as well as by complexity of the DCO models (as compared to conventional non-DCO CS models, which can usually be designed and analyzed by using Markov Chains). Of special importance here are Monte-Carlo techniques, including resampling analysis and rare event simulation related to alarm threshold derivation and estimation of severity indices. These challenges are discussed in Yashchin (2010).

4 Basic Approach to Analysis of DCO Schemes

In the process of design and analysis, we first of all rely on engineers and managers providing an adequate description of targets, acceptable and unacceptable behaviors of the hazard curves governing reliability of various components. We can typically count on obtaining a form of a *reference hazard* curve $h_0(t)$. This curve can be given in terms of a formula that could reflect, for example complex behavior including infant mortality, maturity and late-life behavior. More often, this curve is given via month-by-month setpoints. In practical terms, one can usually fit a continuous curve to such setpoints – this enables one to reduce the number of parameters in the underlying target-setting system; so, for the sake of simplicity we could think of $h_0(t)$ as a positive continuous curve.

To define acceptable and unacceptable behaviors of hazard curves on a massive scale, it is convenient to work in terms of the time scale transformation

$$Y(t) = H_0(t) = \int_0^t h_0(z)dz, \quad (1)$$

where $H_0(t)$ is the cumulative hazard curve corresponding to $h_0(t)$. The main advantage of handling failures on the Y-scale is related to our ability to efficiently parameterize the data and thus define the acceptable and unacceptable regions in terms of very few parameters that are relatively easy to explain to users and administrators. In particular, if it happens that the underlying hazard curve for a particular FRU is proportional to $h_0(t)$, then the failures on the Y-scale follow an exponential distribution; by defining acceptable and unacceptable regions for the parameter of this distribution we obtain a scheme for detection of multiplicative changes in the hazard curve. Furthermore, one can see that a power-type change in a cumulative hazard curve from $H_0(t)$ to

$$H_1(t) = aH_0^c(t) \tag{2}$$

leads to change in distribution of Y from exponential to Weibull with shape parameter c and scale parameter $\beta = a^{-1/c}$. So, if the data and engineering knowledge support the assumption that unfavorable changes in reliability of primary interest are of type (Eq. 2), then we can work within the convenient framework of Weibull laws. Otherwise, one might need to work within a different (hopefully, parametric) framework – however the methodology similar to one described below should still be applicable.

In what follows, we will assume that all the necessary transformations have been applied and we will proceed (without loss of generality) under the assumption that the lifetimes are Weibull. The basic monitoring strategy calls for decomposing of the data stream into set of *control sequences* of statistics (one control sequence per monitored parameter). For example, in the case of warranty data monitoring one could define control sequences (corresponding to vintages sorted in a particular way) for (a) scale parameter of the underlying Weibull law (b) shape parameter or (c) negative log-probability of a FRU to survive 3 years (i.e., $[H_0(3)/\beta]^c$). One should generally try, where possible, to formulate acceptable and unacceptable regions in terms of parameters that are meaningful to users and to use control sequences that provide *unbiased* estimates of these parameters for the individual vintages. The reason for this preference is related to graphical interpretation of CS charts: changes in process levels appear as changes in slope of Cusum trajectories, and thus the process levels are visually estimated via these slopes; since slopes correspond to linear functions of control sequences, the estimates obtained in this way are consistent only when the individual terms are unbiased. Of course, one needs to be aware of the fact that linear estimation procedures are not the most efficient statistically – however, in many practical situations one can get good statistical power, at least within the framework of Weibull models.

Once the list of monitored parameters has been established and the control sequences defined, one needs to establish the criteria for statistical performance of the corresponding charts. Unfortunately, because of the DCO property (in conjunction with time-managed data), we cannot take advantage of the customary concept of the Average Run Length (ARL) used in the conventional control charting. The fact is that the chart is re-computed from scratch at every time point where new information becomes available – and this information spreads over a wide range of available vintages. Even if no data for a new vintage is introduced, there is still a need to update the alarm threshold(s). In EDT, we control the rate of false alarms in terms of *probability of flagging* of a parameter. For example, one can set this probability to 0.001 for all Hard Disk Drives, and then validate that the sensitivity with respect to unacceptable deviations is satisfactory.

For example, consider again the problem of monitoring the mean rate λ of FRU replacements corresponding to data of type shown in Fig. 2. A fragment of one of the summary reports available from the EDT is shown in Fig. 6. One can see that the data table for the 13 vintages can be directly computed from data in Fig. 2. In this case the control sequence corresponds to vintage-by-vintage replacement rates (number of replacements per Machine-Months (MM) of service, see column "RATES"). Associated with this control sequence is the sequence of *weights* corresponding to the total number of MM for FRUs in consecutive vintages. For this particular component, the target behavior was represented by a constant hazard curve, and the acceptable replacement rate was $\lambda \leq \lambda_0 = 0.001$ replacements per MM. The unacceptable region was $\lambda > \lambda_1 = 0.003$. For the DCO chart in Fig. 7, the false alarm probability (i.e., probability of flagging for $\lambda = 0.001$) was set to 0.01. The top chart shows the replacement rates, with spaces between points corresponding to weights. The bottom chart gives the corresponding weighted Page's scheme (see Yashchin 1993; a generalized procedure will be described later). The Evidence trajectory has the property that it tends to "stick" to the bottom when $\lambda \leq k$, where k is the so called *reference value* given by

$$k = \frac{\lambda_1 - \lambda_0}{\ln \lambda_1 - \ln \lambda_0} \approx (\lambda_0 + \lambda_1)/2, \tag{3}$$

in line with likelihood-ratio test formulation on which such charts are based. On the other hand, the Evidence (i.e., weighted Page's) trajectory tends to "float up" towards the threshold when $\lambda > k$.

One can see that for the data in Fig. 6 some vintages have unacceptable failure rates and the evidence chart of Fig. 7 is indeed moving up – however, as of Oct 30, 2001 the evidence for flagging this condition was insufficient. The header of Fig. 6 lists the basic facts about this MT/FRU combination that are made available to the user. In particular, the severity index of the data set was 0.6. In general, severity exceeding 0.99 is needed to flag the condition, in light of the desired probability 0.01 of protection against false alarms. The wearout severity is reported at 0.3. The formal definition of the severity used in the EDT is given below. Figure 8 shows the charts for the same FRU compiled on Nov 30, 2001. At this point we had enough evidence to flag this component (data for this view is omitted). The shape of the Evidence trajectory indicates that the reliability of this FRU was unacceptable from the very beginning – though it appears as if the midway vintages had a slightly lower replacement rate. Note that points on the chart tend to cluster at the right end, reflecting the fact that "younger" vintages tend to have fewer MM of service. In the top chart this fact leads to higher variability of replacement rates – however, this variability is not present on the bottom chart because the corresponding points are assigned lower weights.

```
Brand-Commod=XXXX, Mtype=XXXX, FRU=XXXXXXX, RunDate=01-10-30
FRU Vintages 01-08-17/01-09-15, Tested 2235, Repl 5, Rate/MM 0.002
Accept/Unaccept levels: 0.001 0.003. Severity: 0.6. Sevwear: 0.3
```

OBS	VINTAGES	WMONTHS	WFAILS	RATES	TESTED	1	2	3
1	20010817	4	0	0	2	0	0	*
2	20010820	27	0	0	14	0	0	*
3	20010824	298	0	0	149	0	0	*
4	20010901	698	2	0.0029	349	1	1	*
5	20010904	102	0	0	102	0	*	*
6	20010907	136	0	0	136	0	*	*
7	20010908	473	1	0.0021	473	1	*	*
8	20010912	191	1	0.0052	191	1	*	*
9	20010912	1	0	0	1	0	*	*
10	20010913	235	0	0	235	0	*	*
11	20010913	4	0	0	4	0	*	*
12	20010914	406	1	0.0024	406	1	*	*
13	20010915	172	0	0	173	0	*	*

Fig. 6 A type of summary table derived from data shown in Fig. 2

Though plots and analyses shown in Figs. 7 and 8 are definitely informative, there is a *need for supplemental flagging criteria* in order for the approach to become fully useable. The main problem with the described approach is that it relies on the fixed probability of producing a false alarm as its main performance criterion. As a consequence, the alarm thresholds will tend to go up as larger volumes of data are accrued for a given MT/FRU combination. This will make the chart insensitive to sudden onset of a very large change in the replacement rate. As noted above, recent vintages come with relatively low associated MM (weights) – and, therefore, their impact is lower than that for the older vintages. Supplemental tests address this problem by focusing specifically on *recent* UCs. To this end, in the EDT we use the concept of an *active component* defined in terms of a threshold of D_a days (in many cases, $D_a \approx 60$ days is a good choice). A FRU is considered *active* if there were vintages present in data within the last D_a days from the current point in time T. Supplemental tests are applied to active FRUs only. Absence of vintages within the last D_a days usually indicates that there is little one can do about preventing this FRU from entering the field. So, even if we detect a problem pertaining to recent vintages, we are not in a position to capitalize on it. On the other hand, inactive components are still under warranty and thus need to be monitored, especially for signs of emerging wearout. Since any supplemental tests dilute the detection power with respect to conditions that we are really afraid of, there is little point of introducing them for inactive components.

We will now give a brief summary of main and supplemental tests for monitoring replacement rate and other control sequences. For a particular lifetime distribution parameter (say, λ), we will denote by $(X_i, w_i), i = 1, 2, \ldots, N$ the corresponding control sequence and weights. For example, X_i and w_i could represent the estimate of the Weibull scale parameter based on the vintage i data and the inverse of variance of this estimate, respectively. As shown in Yashchin (1989), choice of weights in this manner leads to good statistical performance in the sequential setting,

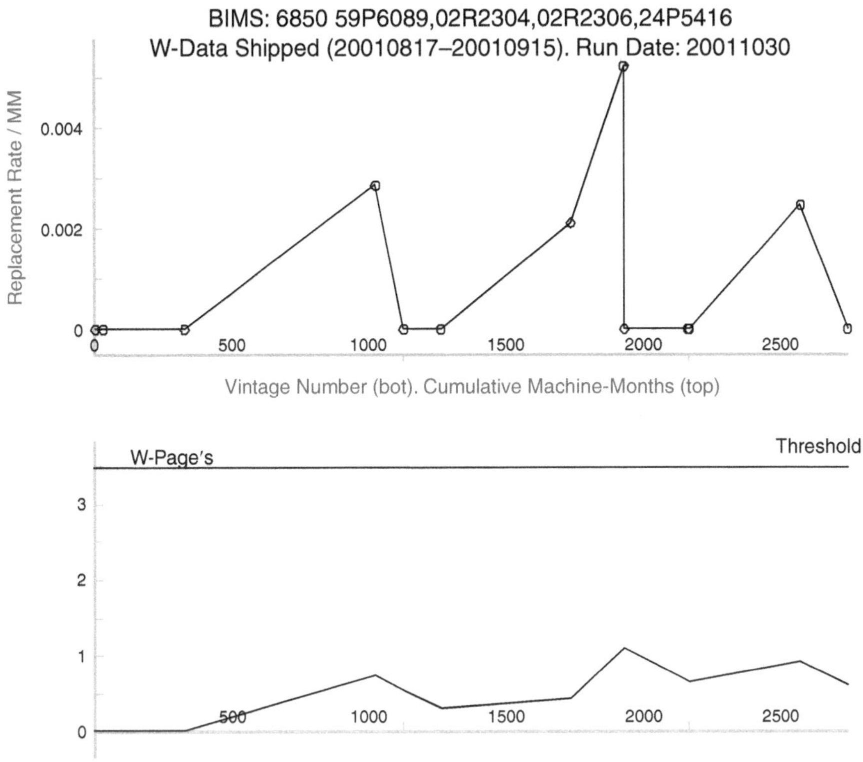

Fig. 7 A DCO control chart corresponding to data in Figs. 2 and 4 (view of Oct 30, 2001). The threshold on the *bottom* (Evidence) chart is chosen so as to ensure the desired level of protection against false alarms. The *top* X-scale corresponds to cumulative MM of service. Note that the data set of this type could be flagged by *supplemental tests* even if the plot does not show threshold violation. In such cases we mark the end of the evidence trajectory by a sign (e.g. "+") corresponding to particular test(s) that caused flagging

and the performance generally tends to be good in the DCO context. In practice, performance of the procedures is not too sensitive to moderate deviations from this principle; for example, in many cases weights simply proportional to sample sizes (or, sometimes, to MM) lead to similar statistical properties.

Next, for every monitored parameter we convert this sequence to values of a control scheme $S_i, i = 1, 2, \ldots, N$ by using a version of the weighed Cusum algorithm. For example, for λ corresponding to replacement rate one can use the *Weighted Geometric Cusum* defined by

$$S_0 = 0, S_i = max[0, \gamma S_{i-1} + w_i(X_i - k)], i = 1, 2, \ldots, N, \quad (4)$$

where γ is typically chosen in [0.7, 1] and the *reference value* k is given by Eq. 3. For schemes with DCO we define $S = max[S_1, S_2, \ldots, S_N]$. The FRU is flagged

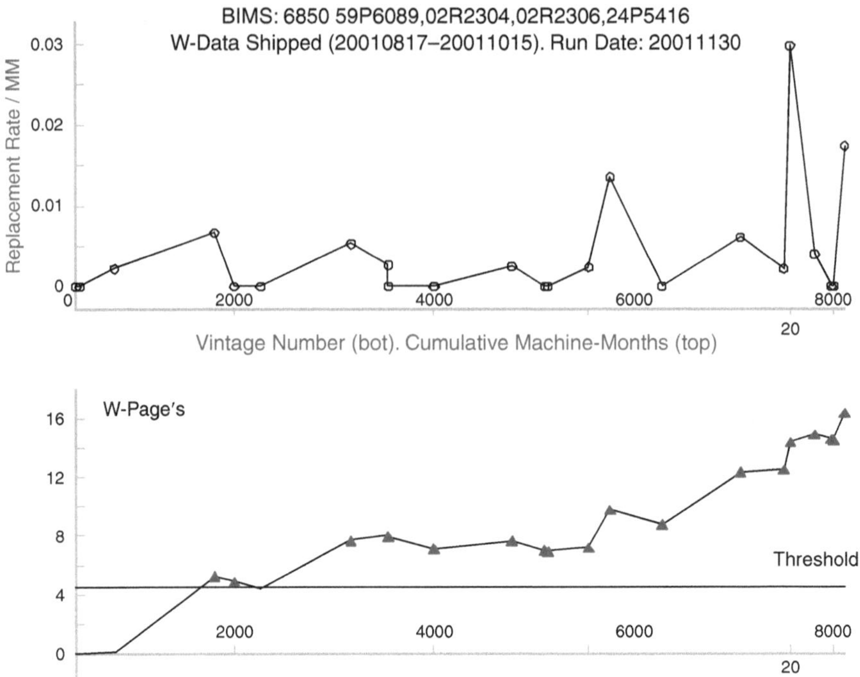

Fig. 8 A DCO chart for the FRU of Figs. 2 and 4 (view of Nov 30, 2001)

at the current point in time T if $S > h$, where the threshold h is chosen based on the required degree of protection against false alarms:

$$Prob[S > h|N, \lambda = \lambda_0] = \alpha_0. \tag{5}$$

The value of α_0 is typically chosen in the interval $(0.01, 0.001)$ – but in some cases it could be even lower. In situations where UCs are related to an abrupt (step) change in the monitored parameter, choice of $\gamma = 1$ generally works well. However, when drifts in parameters are expected, use of $\gamma < 1$ offers performance advantages as it enables one to enhance sensitivity of the detection procedure with respect to drifts in the lifetime distribution parameter (at the expense of somewhat lower sensitivity with respect to shifts). These advantages are similar, in both nature and magnitude, to those observed for non-DCO schemes, e.g., see Yashchin (1989). Using $\gamma < 1$ also helps to enhance sensitivity with respect to newly emerging UCs – however, the need for supplemental tests is still there.

The procedure (Eq. 4) is computationally convenient for applications involving massive data streams because its recursive nature enables easy parallelization of resampling tests. Note however that in the DCO setting the above scheme is non-Markovian: the whole control sequence (and weights), and then the control scheme are re-computed at every new point in time. This leads to necessity of special Monte

Carlo algorithms, e.g., of type discussed in Yashchin (2010). One also needs to make sure that on-line recalculation of control limits does not lead to loss of sensitivity by preventing adaptation of a chart to worsening quality. It is therefore essential to derive control limits based on fixed on-target conditions: the only adaptation allowed is related to the DCO nature of the procedure, but not to movement of parameters themselves. For example, when computing a threshold for the replacement rate λ, the Monte Carlo replications are (a) based on the fixed acceptable level λ_0 and (b) conditioned on the values of MM observed up to the present point in time, which are assumed to be *ancillary* for λ. Because of that, the computations are unaffected by the actual underlying levels of λ or its estimates. However, if there are reasons to believe that the weights (on which we condition in simulated replications) are affected by changes in the underlying monitoring parameter, one will need to take special precautions to ensure that the generated thresholds conform to the nominal levels of protection against false alarms and to the required sensitivity.

Supplemental Tests. In EDT, we use two supplemental criteria. The first criterion returns (for an active FRU part) the p-value from the test of a hypothesis that the data observed within the last D_a days conforms to an underlying replacement rate not exceeding λ_0. For Weibull populations, the corresponding p-value can be computed numerically. The criterion flags the FRU if this p-value is smaller than a pre-specified threshold.

The second criterion is based on the final value S_N of the scheme (Eq. 4); high values indicate that the recent vintages are not conforming to an acceptable set of lifetime parameters. P-values of this test can be computed by using the combination of Monte Carlo simulation and asymptotic theory of Brownian Motion process, see Yashchin (2010).

A combination of MT/FRU is flagged if at least one of the main test (Eq. 4) and supplemental tests produces a p-value that is below a threshold that is tuned to provide the target protection against false alarms. The complement of the computed p-value was found to be a suitable *index of severity*; it can be approximated by a function of the individual p-values of these three tests, $\psi(P_1, P_2, P_3)$ (upper case is used to emphasize that p-values are random variables). In most practical situations, correlation between the supplemental test statistics and S is negligible and can be ignored. However, the supplemental tests do tend to be correlated among themselves, substantially complicating the problem of severity evaluation. In EDT, this is done via failure process simulation under the assumption that the parameters are at the edge of the acceptable region.

5 Monitoring of Shape and Scale Parameters

Providing a framework for monitoring the replacement rate is usually one of the key engineering requirements. Our experience with EDT suggests that after some training, the users feel quite comfortable with charts of type shown in Figs. 7 and 8 and the accompanying reports (Fig. 6). However, it is also highly desirable to have charts for monitoring Weibull parameters. Of course, if it happens that the hazard curve of the MT/FRU combination of interest is proportional to the target hazard curve $h_0(t)$, then the Weibull shape is $c = 1$ and the charts presented above provide a complete picture. However, in practice one can expect changes in hazard curves that bring us, after the time transformation (Eq. 1), into the Weibull domain with $c > 1$ – and this possibility (i.e., onset of wearout) is of great concern. Failure to detect in time presence of wearout conditions can cause phenomenal warranty losses, and even lead to demise of an otherwise sound manufacturing company within a very short period of time. This is especially important for companies involved in mass manufacturing, as illustrated by well publicized events in the electronic industry (e.g., see Vance 2010) and automotive industry.

One can see that presence of wearout is, in fact, detectable on charts for monitoring replacement rate. What we will see upon onset of a wearout is a "bulge" on the chart, especially for early vintages (or a bulge in the middle of the chart, if onset of wearout occurs in midstream). Before too long, the bulge will grow high enough to cause threshold crossing on the evidence (bottom) chart. At this point, one could ask the question: did the threshold crossing occur because the vintages in question simply have an unacceptable (but steady, i.e., no wearout) fallout rate – or is it manifestation of wearout? The reports of type shown in Fig. 6 help to answer this question: some clues appear in the last columns of the report that give the month of service in which individual failures have occurred. Roughly speaking, in the presence of wearout one will usually see failures clustering toward last months of service. In many situations, however, it may be difficult to assess visually whether such clustering is indeed taking place, especially in highly censored situations, when vintages contain data corresponding to several warranty policies. It is, therefore, highly desirable to have a separate chart for monitoring wearout. As will be illustrated in Fig. 9, such a chart can be presented on the same plot as the chart for monitoring replacement rate. The severity of wearout condition can be given on the same charts and reports; in fact, one can find them in Figs. 6 and 9.

Monitoring procedures for c can be developed by using strategy similar to that described in the previous section. For this parameter, it is generally unadvisable to have daily summaries on the charts; it is better (from the perspective of statistical power, visual interpretation and computational burden) to have the points computed on a sparser grid of *macro-vintages*. In EDT, the wearout estimates are computed based on monthly data (e.g., the rows of Fig. 2 are consolidated so as to provide just one row per month). Note, however, that the charts for monitoring c are still updated daily, along with other charts – however, we use a sparser control sequence to detect changes and evaluate the state of c for various macro-vintages. The last point could correspond to partial data from the last month (if we use calendar months for computing the control sequence) – but its role in the detection process is typically insignificant: there is practically no information about wearout in the most recent vintages.

Following the monitoring strategy described above, one can specify the acceptable and unacceptable levels as c_0 and $c_1 > c_0$ (in many practical situations, $c_0 = 1$ is a good choice). We then compute the control sequence consisting of

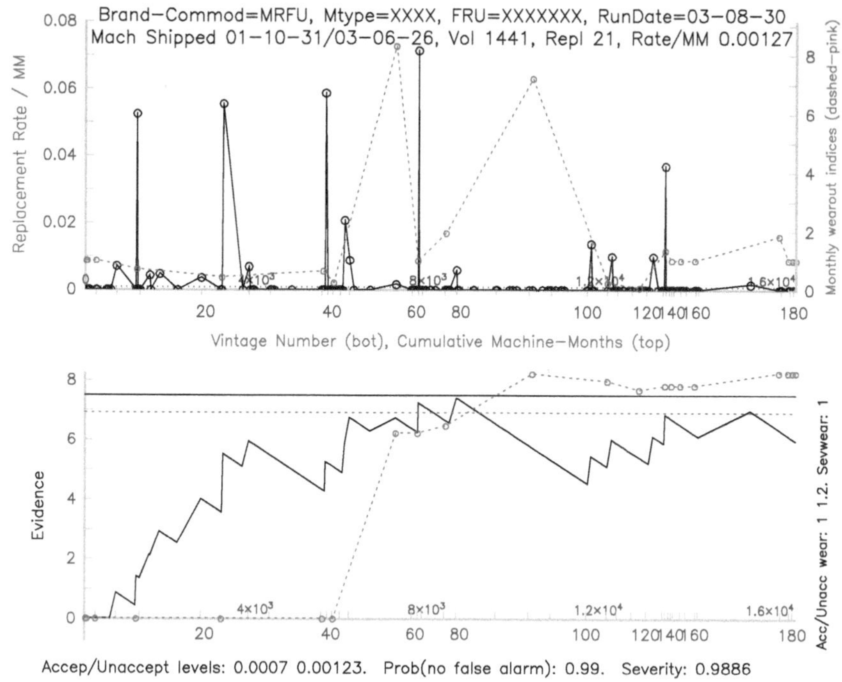

Fig. 9 A DCO chart for simultaneous monitoring of replacement rate and Weibull shape parameter c (wearout index). The horizontal *dashed line* on the *top plot* corresponds to the acceptable level of replacement rate. The *dashed lines* on the *bottom plot* give the chart for c based on month-by-month estimates (updated daily) and the corresponding horizontal threshold. Both charts have the probability of a false alarm 0.01. The bottom scale symbols correspond to vintage indices. Note that the severity of the detected wearout condition is 1 (maximal possible value) and most of the evidence for wearout comes from data in a four consecutive months (7 to 10, corresponding to vintages 37 to 91). On the upper plot, we also show the actual sequence of the estimates $\hat{c}_1, \hat{c}_2, \ldots$ (dashed line with circle symbols), with a separate right axis

consecutive *unbiased* estimates $\hat{c}_1, \hat{c}_2, \ldots, \hat{c}_M$, where M is the number of months (i.e., macro-vintages) for which data is available. These values are then used in the Geometrically Weighted Cusum test:

$$S_{0w} = 0, S_{iw} = max[0, \gamma_w S_{i-1,w} + w_{iw}(\hat{c}_i - k_w)], i = 1, 2, \ldots, M, \quad (6)$$

where γ_w is the geometric parameter (typically in [0.7, 1]) and the reference value is $k_w \approx (c_0 + c_1)/2$. In the above notation the letter "w" stands for "wearout". It is because of unbiasedness of the control sequence that we can use a fixed reference value in Eq. 6 and produce Cusum charts for c with a fixed centering constant (examples of such charts can be found in Yashchin 1993).

The weight w_{iw} can be chosen, for example, as the number of failures f_i observed for the macro-vintage i (and not the overall number of MM for the macro-vintage i, as in Figs. 7 and 8). This choice can be motivated by the fact that the scale parameter can vary from one macro-vintage to another, and thus it is the number of failures and the corresponding lifetimes that carry information about c. The theory of weighted schemes suggests using type-1 weights w_{iw} that are inversly proportional to the variance of the corresponding members of control sequence – and our simulation studies show that the variance of $\hat{c}_i$ indeed tends to be roughly proportional to f_i.

For the DCO scheme, the decision statistic is $S_w = max[S_{1w}, S_{2w}, \ldots, S_{Mw}]$ and the FRU is flagged at time T if $S_w > h_w$, where h_w is chosen from the equation:

$$Prob[S_w > h_w | M, c_0, f_1, f_2, \ldots, f_M] = \alpha_0. \tag{7}$$

As usual, the threshold h_w and p-value of the test are derived by using Monte Carlo simulation. The details can be found in Yashchin (2010). The latter paper also discusses a *supplemental test* for active components similar to the second supplemental test mentioned above that calls for flagging a component if the last value S_{Mw} of Eq. 6 becomes too large.

We will now discuss briefly computational aspects of the DCO control charting for c. Of special importance in the decision-making process is the bias-correction procedure for deriving the sequence of estimates $\hat{c}_i$. In the context of Maximum Likelihood estimation, this procedure is greatly simplified by the fact that the relative bias of the Weibull shape MLE estimator is primarily a function of the sample size and the number of uncensored failures; its dependence on c or β is negligible. We will refer to this property as the *Property A*. This property can be proven theoretically only in the case of ungrouped data with type-2 censoring (see Johnson et al. 1994; McCool 1970): it follows from the fact that there exists a data transformation (namely, log-lifetime) that turns the Weibull family into a location-scale family with the scale parameter $1/c$. Under such conditions, one can prove even a stronger statement, namely, that the ratio $\hat{c}/c$ is pivotal.

Our experiments indicate that in the type-1 censoring environment Property A still holds to a degree suitable for practical application. One can motivate this statement by the fact that type-1 censored data with r observed failures yields a likelihood function that can generally be well approximated by the type-2 censored likelihood function based on the observed r failures. In particular, our experience with MLE-based charts for c indicates that this property continues to yield substantial reductions in the bias and mean square error (MSE) of the shape estimators for the case of grouped failures, at least in cases when group boundaries are not spread too widely.

The EDT data structure is even more complex: not only is this data type-1 censored and grouped by month of service, but it can also be multiply-censored. Such censoring can occur, for example, when some of the items of a given vintage are shipped as part of machines that carry a 1-year warranty, while the remaining items are shipped with machines under a 3-year warranty. Even in cases when multiple censoring is not present in the original data, such type of censoring can occur as a result of consolidating data by macro-vintages. In light of such data complexity, substantial amounts of experimentation was necessary in order to establish adequacy of Property A for this type of data and models of interest.

Moreover, in the context of the EDT data we also found that Property A holds quite reliably for estimators of c that are obtained by (a) rounding the failure times to the midpoint of the corresponding service interval and (b) proceeding to obtain MLEs based on these rounded times as if they were the actually observed failure times. In what follows, let us denote the corresponding sequence of MLE estimates of c and β by $\{\tilde{c}_i\} = \tilde{c}_1, \tilde{c}_2, \ldots, \tilde{c}_M$ and $\{\tilde{\beta}_i\} = \tilde{\beta}_1, \tilde{\beta}_2, \ldots, \tilde{\beta}_M$, respectively. Let $\{n_i, f_i, A_i\}$ be the total number of items on test, the total number of failed items observed and the total length (in months) of service period for the i-th macro-vintage. For example, if the longest period of time for which an item of macro-vintage i was observed is 5 months (including the current month of service), then $A_i = 5$. For EDT-specific levels of time discretization we found that the sequence $\{\tilde{c}_i\}$ delivers statistical power that is comparable to power of the MLEs $\{\hat{c}_i\}$ based on grouped data, at a lower processing cost (the cost of computation is of importance in applications like EDT because we rely heavily on resampling techniques in the process of bias reduction and threshold derivation).

Next, for every macro-vintage i we compute the MLE estimate $\tilde{\beta}_{i0}$ of the scale parameter β_i under the assumption that $c = c_0$. Under this assumption, the probabilities of a failure falling in service months $1, 2, \ldots, A_i$ (given that a failure was observed) can be approximated by the vector $\mathbf{p}_i = \{p_{ij}\},\ j = 1, 2, \ldots, A_i$, where

$$p_{ij} = \left\{exp(-(j-1)\tilde{\beta}_{i0}^{-1})^{c_0} - exp(-j\tilde{\beta}_{i0}^{-1})^{c_0}\right\} / \left\{1 - exp(-A_i\tilde{\beta}_{i0}^{-1})^{c_0}\right\} \quad (8)$$

Now we produce B replications of $\tilde{c}_i$ by (a) randomly re-distributing the observed f_i failures to A_i groups based on multinomial distribution with the probabilities p_{ij} (typically, B = 200 is sufficient for our purposes) and (b) re-computing the MLE estimates $\{\tilde{c}_i, \tilde{\beta}_i\}$ for the re-sampled set of failure times. In essense, this amounts to a *parametric bootstrap* procedure (e.g., see Good 2005) with additional conditioning on the observed number of failures. Denote the resampled sets by $\{\tilde{c}_i^{(b)}, \tilde{\beta}_i^{(b)}, b = 1, 2, \ldots, B\}$. Since for the given macro-vintage i the set $\{\tilde{c}_i^{(b)}\}$ is generated conditionally on $c = c_0$ and on the overall number of uncensored failures, we can rely on Property A to develop a bias-correction factor to be applied to $\tilde{c}_i$:

$$u_c = (Bc_0) / \sum_{b=1}^{B} \tilde{c}_i^{(b)}. \quad (9)$$

The bias-corrected sequence of observations for Eq. 6 is then $\{u_c\tilde{c}_i\}$. The bias-corrected bootsrap replications are then used to evaluate the threshold h_w of Eq. 6. In particular, we construct B replications of the scheme (Eq. 6):

$$S_{0w}^{(b)} = 0, S_{iw}^{(b)} = max[0, \gamma_w S_{i-1,w}^{(b)} + f_i(u_c\tilde{c}_i^{(b)} - k_w)], i = 1, 2, \ldots, M, \quad (10)$$

and observe the replicated set of maxima $S_w^{(b)} = max[S_{1w}^{(b)}, S_{2w}^{(b)}, \ldots, S_{Mw}^{(b)}]$. The value of h_w is then selected as the empirical $(1-\alpha_0)$-th quantile of the set

$\{S_w^{(b)}, b = 1, 2, \ldots, B\}$. Substantial reduction in the simulation effort can be achieved by utilizing the property that the right tail of the $S_w^{(b)}$ distribution is decreasing exponentially (see Yashchin 2010). At the present point, however, we do not have a formal proof of this property. It is also recommended to apply an additional correction of h_w upward to account for the fact that the bootstrap replications $\{\tilde{c}_i^{(b)}\}$ slightly under-represent the variability of the underlying population of $\{\tilde{c}_i\}$ (corresponding to some fixed simulation setup with $c = c_0$) because the replications assume that the underlying β is equal to its estimated value.

Monitoring the scale parameter. In practice, it is also useful to provide charts for the Weibull scale parameter β. As in the case of c, it is quite sufficient to use monthly summaries as a basis for control sequence. Individual monthly estimates can be shown on the basic charts (like in Fig. 9), along with the replacement rate and wearout charts. It is also convenient to plot the sequence of estimated Weibull rates $\hat{\beta}_1^{-1}, \hat{\beta}_2^{-1}, \ldots, \hat{\beta}_M^{-1}$ on the upper plot because its scale (i.e., measurement units) coincides with the scale for the replacement rate, and so these charts can share a common left axis. Note that the control sequence $\hat{\beta}_i^{-1}$ is a by-product of computing the sequence $\hat{c}_i$. In practical applications, it is highly recommended to bias-correct the control sequence $\{\tilde{\beta}_i^{-1}\}$ before using it in the weighted procedure of type

$$S_{0s} = 0, S_{is} = max[0, \gamma_s S_{i-1,s} + w_{is}(\tilde{\beta}_i^{-1} - k_s)], i = 1, 2, \ldots, M, \tag{11}$$

for reasons mentioned above (in the above formula the letter "s" stands for "scale". In our experience, substantial reduction in bias of $\{\tilde{\beta}_i^{-1}\}$ (as an estimate of $1/\beta$) can be obtained by simply estimating its value under the assumption $c = u_c\tilde{c}_i$. Choice of the weights w_{is} corresponding to the total number of MM for the respective macro-vintages is typically adequate. The reference value can be chosen midway between the acceptable and unacceptable levels of $1/\beta$, the geometric factor γ_s is chosen based on standard considerations, and the decision threshold h_s for this scheme can be obtained by simulation. Note, however, that in some cases one may want to simply produce the evidence plot of the resulting chart (if the monitoring procedure for the replacement rate has already been deployed) and rely on the chart (Eq. 11) for diagnostic purposes only.

6 Conclusions

Because of the nature of the data in many systems for lifetime data (time-managed data, DCO), the problem of monitoring its characteristics becomes quite challenging. Some of the problems are of theoretical nature (design and analysis of DCO control schemes), others are related to practical implementation (multi-level dashboards, handling engineering input, alarm management, diagnostics, corrective actions), some computational (efficient use of Monte Carlo techniques for computing thresholds, severities on a massive scale). We illustrated how some of the challenges can be addressed in a particular type of a monitoring environment related to warranty data. There is no doubt that data from other industries (like automotive) has enough idiosyncrasies to warrant whole classes of new approaches; we hope, however, that some basic ideas presented above could still be useful for a broader range of situations, at least in areas related to reliability assurance and warranty management.

There are a number of issues affecting degree of feasibility of the discussed methods – however, we did not run into real "show-stoppers" in the case of IBM Personal Systems warranty data. This type of data appears to be among the cleanest massive-scale sets that an industrial practitioner can realistically hope to encounter. There were, of course, some data quality issues – but it turned out that we could introduce imputations and approximations that our users felt comfortable with. For example, the Service DB was required to contain the id of a FRU that was used as a replacement – but not the FRU being replaced. It turned out that, after some automated search, one could recover this information for most of the cases (the Ship DB proved to be useful in this respect). Moreover, a number of ship dates were incorrect – but we managed to detect a large number of them and obtain a useful estimate for these dates.

Our work was focused on detecting UCs – however, similar methods could be used for detecting improvements in the process (and some were indeed detected, based on periodic or informal dashboard exploration). In general, it appears that detection of improvements does not have the same urgency as detection of UCs, and so it might be a good strategy to use periodic surveys for this purpose – however, we also saw a number of cases where green swaths on a dashboard were due to poor target-setting.

We found that the "repeated weighted Cusum-Shewhart" approach was well received by teams working in various business areas – though some amount of training was a must. The level of familiarity with this approach is still too low, as most engineers associate monitoring and SPC with the Western Electric system of decision making. We decided not to use EWMA for monitoring – mostly because of its poor inertial properties (see Yashchin 1993; Woodall and Mahmoud 1995) – however, some modifications of this technique are useful in post-alarm activities.

The effectiveness of the warranty monitoring process is somewhat limited by the problem of multiple replacements related to difficulties by warranty personnel in identifying the failed component. In many cases, it is actually cheaper to replace several components at once than risk the necessity of a return visit. The information in databases, however, will typically contain information on every replaced part – so, sometimes it is not possible to tell whether the replacement action was indeed a replacement of a failed component. In the industry this problem is known as *masking*, e.g., see Yashchin (2008). Some of the methods for handling masked data could be quite useful in this setting.

Though the analysis of schemes with DCO can be handled within the framework of continuous time, in this paper we emphasized an approach that uses discretization of the time scale. This approach was driven by pragmatic factors, mostly related to preferences and constraints of the Early Detection Tool users and to properties of the available data collection and processing tools.

Finally, the proposed methodology has proven quite effective for prediction of future fallout rates and related costs. By segmenting the control charts and identifying regimes and change-points, one can predict the contribution of every vintage to the future overall rate of replacements.

Acknowledgements I would like to thank David L. Jensen, Ildar Khabibrakhmanov and Prince Stanley (IBM Thomas J. Watson Research Center) for help in development of the Early Detection Tool. I am also thankful to Andrew Dubois, Stephen Restivo, Christopher Ross (IBM Personal Systems Division) for help with implementation and for their valuable input, and to Vaugh Robert Evans (IBM Server Group) for his generous feedback and advice that helped to enhance the power and appeal of both tool and methodology. I am especially grateful to Sara Basson (IBM Research) and the anonymous referee whose comments and suggestions led to substantial improvement of the paper.

References

Blischke, W. R., & Murth, D. N. P. (2000). *Reliability modeling, prediction and optimization.* New York: Wiley.

Biswas, P., & Kalbfleisch, J. (2008). A risk adjusted CUSUM in continuous time based on the Cox model. *Statistics in Medicine, 27*, 3382–3406.

Brown, R. (2010, July 24). Siren on oil rig was kept silent, technician says. *New York Times*, A1–A11.

Doganaksoy, N., Hahn, G. J., & Meeker, W. Q. (2006). Improving reliability through warranty data analysis. *Quality Progress, November*, 63–67.

Dubois, A., Evans, V. R., Jensen, D., Khabibrakhmanov, I., Restivo, S., Ross, C., & Yashchin, E. (2008). System and method for early detection of system component failure. US *Patent 7401263*. Washington, DC: US Patent Office.

Good, P. (2005). *Permutation, parametric, and bootstrap tests of hypotheses* (3rd ed.). New York: Springer.

Hawkins, D., & Olwell, D. (1998). *Cumulative sum charts and charting for quality improvement.* New York: Springer.

Johnson, N., Kotz, S., & Balakrishnan, B. (1994). *Continuous univariate distributions* (Vol. 1). New York: Wiley.

Kalbfleisch, J. D., Lawless, J. F., & Robinson, J. A. (1991). Methods for the analysis and prediction of warranty claims. *Technometrics, 33*, 273–285.

Karim, M. R., Yamamoto, W., & Suzuki, K. (2001). Statistical analysis of marginal count failure data. *Lifetime Data Analysis, 7*, 173–186.

Lawless, J. F. (1998). Statistical analysis of product warranty data. *International Statistical Review, 66*, 41–60.

Lawless, J. F., & Kalbfleisch, J. D. (1992). Some issues in the collection and analysis of field reliability data. In J. P., Klein, & P. K., Goel (Eds.), *Survival analysis: State of the art* (pp. 141–152). Netherlands: Kluwer Academic.

Mccool, J. I. (1970). Inference on Weibull percentiles and shape parameter from maximum likelihood estimates. *IEEE Transactions on Reliability, R-19*(1), 2–9.

Olteanu, D., & Vining, G. G. (2009). CUSUM charts for censored lifetime data. *Paper at the 2009 joint statistical meetings*, Washington, DC.

Patra, K., & Dey, D. K. (2002). A general class of change point and change curve modeling for lifetime data. *Annals of the Institute of Statistical Mathematics, 54*(3), 517–530.

Sego, L., Reynolds, M., & Woodall, W. (2009). Risk-adjusted monitoring of survival times. *Statistics in Medicine, 28*, 1386–1401.

Steiner, S. H., & Mackay, R. J. (2000). Monitoring processes with highly censored data. *Journal of Quality Technology, 32*, 199–208.

Steiner, S. H., & Mackay, R. J. (2001). Detecting changes in the mean from censored lifetime data. In H. J., Lenz, & P. T., Wilrich (Eds.), *Frontiers in statistical quality control* (Vol. 6, pp. 275–289). Heidelberg: Physica.

Vance, A. (2010, June 29). Suit over faulty computers highlights Dell's decline. *New York Times, Business Day Section*, B1–B2.

Woodall, W. H., & Mahmoud, M. A. (2005). The inertial properties of quality control charts. *Technometrics, 47*, 425–436.

Wu, H., & Meeker, W. Q. (2002). Early detection of reliability problems using information from warranty databases. *Technometrics, 44*(2), 120–133.

Yashchin, E. (1989). Weighted cusum technique. *Technometrics, 31*, 321–338.

Yashchin, E. (1993). Statistical control schemes: Methods, applications and generalizations. *International Statistical Review, 61*, 41–66.

Yashchin, E. (2008). Masked failure data. In F., Ruggeri, R. S., Kenett, & F. W., Faltin (Eds.), *Encyclopedia of statistics in quality and reliability* (pp. 1037–1042). Wiley, New York.

Yashchin, E. (2010). Computational and Monte-Carlo aspects of systems for monitoring reliability data, In Y., Lechevallier, & G., Saporta (Eds.), *Proceedings of the COMPSTAT 2010* (pp. 253–262). Paris: Springer.

Zhang, L., & Chen, G. (2004). EWMA charts for monitoring the mean of censored Weibull lifetimes *Journal of Quality Technology, 36*(3), 321–328.

A Robust Detection Procedure for Multiple Change Points of Linear Trends

Seiichi Yasui, Hidehisa Noguchi, and Yoshikazu Ojima

Abstract A flexible manufacturing system (FMS) enables the production of multiple-items with short production run. By using an automatic measurement system, it is possible to observe a large amount of items in a short time. The observations from the FMS include some variation patterns and outliers, thereby making it difficult to implement a conventional statistical process control. In this study, a retrospective analysis of such a process dataset is proposed. Our procedure detects multiple change points for a dataset with outliers and variation patterns such as shifts and trends. The locally weighted scatter plot smoothing and the jump/roof/valley detection procedure based on a local polynomial kernel smoothing are useful to develop our procedure. We modify these procedures and propose a robust procedure for detecting multiple change points.

Keywords Change points • Robustness • Local polynomial kernel smoothing

1 Introduction

A flexible manufacturing system (FMS) enables the production of multiple-items with short production run. Although the production run is short, it is easy to obtain adequate data for monitoring quality characteristics by an automatic measurement system. However, such a sophisticated process makes it difficult to implement a conventional statistical process control that is based on control charts and PDCA cycle.

The flexibility develops a flexible production schedule. Even if the production volume is large, the production is intermittent because the change to other

---

S. Yasui (✉) · H. Noguchi · Y. Ojima
Tokyo University of Science, 2641, Yamazaki, Noda, Chiba, 278-8510, Japan
e-mail: yasui@rs.noda.tus.ac.jp; j7410702@ed.noda.tus.ac.jp; ojima@rs.noda.tus.ac.jp

H.-J. Lenz et al. (eds.), *Frontiers in Statistical Quality Control 10*,
Frontiers in Statistical Quality Control 10, DOI 10.1007/978-3-7908-2846-7_14,

production types is easy. Production line is not defined as an exclusive process of a particular product, since other products are in pipeline after a production run is complete. A rigid production schedule does not provide time to improve the process or take corrective action. Hence, an adjustment is made to maintain the expected quality level of the products.

Since it is practically difficult to implement on-line process control, a retrospective analysis is useful and important to control the quality of future products. Sullivan (2002) and Harnish et al. (2009) proposed detection methods for multiple change points from clustering observations when multiple shifts are present. However, a set of observations from the FMS includes outliers and some variation patterns such as trends and jumps.

Qiu and Yandell (1998) proposed a detection procedure for finding jumps in a regression curve and in first- or higher-order derivatives based on local polynomial fitting. Joo and Qiu (2009) proposed a jump/roof/valley detection method using local polynomial kernel smoothing. A jump is a discontinuous point of a regression curve. A roof/valley is an indifferentiable point of a regression curve (i.e. a jump in the first-order derivative of the regression curve).

This method can detect change points for datasets with multiple patterns such as shifts and trends. However, it cannot be directly applied to a process that uses a FMS because such systems have outliers in the observations. Thus, it is necessary to add robustness to the jump/roof/valley detection method. Therefore, we propose a robust jump/roof/valley detection procedure.

Cleveland (1979) proposed a robust locally weighted regression for a polynomial model known as Locally weighted satter plot smoothing (LOWESS). In this study, by modifying Cleveland (1979) and combining it with Joo and Qiu (2009), we propose a procedure for detecting multiple change points for datasets including some linear trends and outliers.

In Sect. 2, the underlying idea for detecting jumps and roofs/valleys is described. Our robust change point detection procedure is introduced in detail in Sect. 3. In Sect. 4, the performance evaluation through a simulation study is shown. Section 5 present the concluding remarks.

2 Underlying Idea for Detecting Jumps and Roofs/Valleys

The underlying idea for detecting jumps and roofs/valleys is described in Joo and Qiu (2009). First, the principle of jump detection is introduced. A step shift of the process mean is one example of the jumps.

The first-order derivative is a spike at the position where the step shift occurred and it is flat (zero in case of a step shift) in the neighborhood of a jump position. The second-order derivative is zero at the position where the step shift occurred and it changes its sign from positive (negative) to negative (positive) in the neighborhood

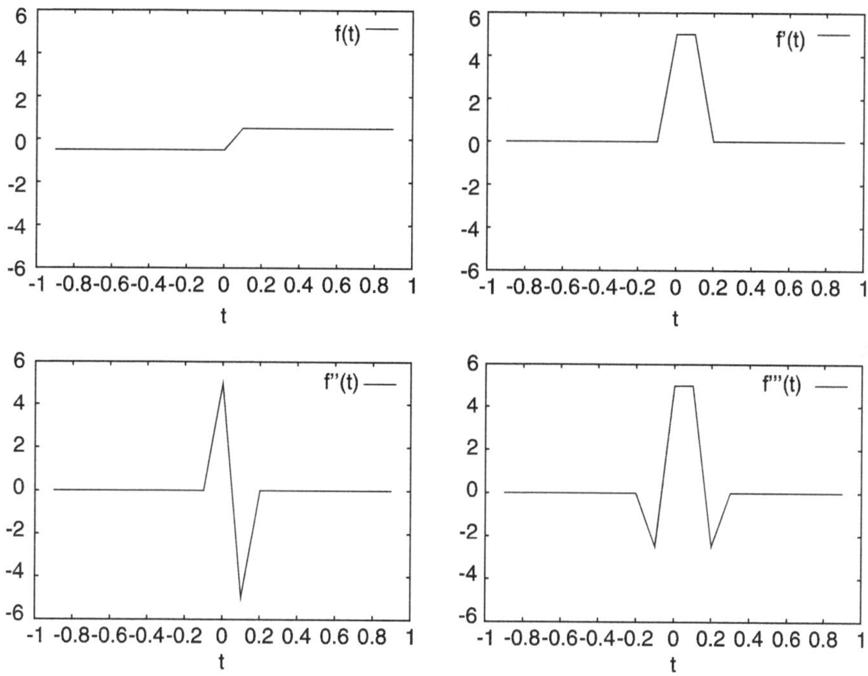

Fig. 1 A step function and its first-, second- and third-order discrete derivatives

of a jump position. This feature is called the *zero-crossing* property in Joo and Qiu (2009).

Next, the principle of roof/valley detection is introduced using the above assumption about the observations. When we suppose that observations consist of only linear functions with two different slopes (e.g. $|x|$), the first-order derivative is a step function in the neighborhood of the roof/valley position. The second-order derivative is a spike at the position where the roof/valley occurred and it is zero (generally, flat) in the neighborhood of the roof/valley position. The third-order derivative is zero at the position where the roof/valley occurs and it changes its sign from positive (negative) to negative (positive) in the neighborhood of the roof/valley position. The first-, second- and third-order discrete derivatives of the step function are shown in Fig. 1. From the features of each derivative, we develop simple criteria for detecting jumps and roofs/valleys compared to the complex criteria proposed by Joo and Qiu (2009). Three derivative coefficients at each data point are estimated by local cubic kernel smoothing with robustness. These estimated derivatives are reasonable approximations of the true derivatives.

3 Robust Change Points Detection Procedure

3.1 Estimation of Derivatives at Each Point

Let $\{y_t | t = 1, \ldots, n\}$ be a sequence of observations from the process, in which some of the observations are outliers. The data model without outliers is given as

$$y_t = f(t) + \varepsilon_t, \tag{1}$$

where $f(t)$ is the unknown function with jumps and/or roofs/valleys at unknown positions and some of $\{\varepsilon_t | t = 1, \ldots, n\}$ have an independent normal distribution with mean 0 and unknown variance σ^2. To obtain estimates of the derivatives of $f(t)$ at a given point t, we utilize the robust local cubic kernel smoothing procedure.

First, we implement the following local cubic kernel smoothing procedure:

$$\min_{a,b,c,d} \sum_{i=1}^{n} \{y_i - [a + b(i-t) + \frac{c}{2}(i-t)^2 + \frac{d}{6}(i-t)^3]\}^2 \times K\left(\frac{i-t}{m_h}\right) \tag{2}$$

where $K(u)$ is a kernel function with support $[-1, 1]$ and m_h is a bandwidth parameter. Let m_h be the number of observations in the neighborhood of a given point t. The solutions to a, b, c and d of the objective function (Eq. 2) are denoted as $\hat{a}_t, \hat{b}_t, \hat{c}_t$ and $\hat{d}_t$, respectively. The $\hat{a}_t$ is the estimate of $f(t)$, and the $\hat{b}_t, \hat{c}_t$ and $\hat{d}_t$ are non-robust estimates of the derivative of $f(t)$.

Then, to obtain robust estimates, robustness weights are calculated. Let $N_h(t)$ be the neighborhood $[t - m_h, t + m_h]$ of the point t. We calculate residuals $e_j = y_j - \hat{y}_j$ for $j \in N_h(t)$, where

$$\hat{y}_j = \hat{a}_t + \hat{b}_t(j-t) + \frac{\hat{c}_t}{2}(j-t)^2 + \frac{\hat{d}_t}{6}(j-t)^3. \tag{3}$$

The robustness weights are defined as

$$\delta_j = B\left(\frac{e_j}{6M}\right),$$

where M is the median of $\{|e_j| : j \in N_h(t)\}$,and

$$B(u) = \begin{cases} (1-u^2)^2 & |u| \leq 1 \\ 0 & |u| > 1 \end{cases} \tag{4}$$

is the bisquare weight function.

Next, by replacing the kernel function in Eq. 2 to the $\delta_j \times K(u)$, we compute new $\hat{a}_t, \hat{b}_t, \hat{c}_t$ and $\hat{d}_t$ that are candidates for robust estimates at a given point t. The $\hat{a}_t, \hat{b}_t, \hat{c}_t$ and $\hat{d}_t$ are obtained as the solutions of the minimization problem (Eq. 2).

Finally, by iterating the re-weighting and minimization for R times, we obtain robust estimates of the derivatives of $f(t)$ at each point t.

The solution of the minimization problem (Eq. 2) is given by the weighted least squares. Thus, the robust estimators of $f(t)$ and its derivatives are given by

$$\hat{\boldsymbol{\beta}}_t = (X'W_tX)^{-1}X'W_t\boldsymbol{y}, \tag{5}$$

where $\hat{\boldsymbol{\beta}}_t = (\hat{a}_t, \hat{b}_t, \hat{c}_t, \hat{d}_t)'$, W_t is the diagonal matrix of the robustness weights after an iteration, X is the design matrix of a cubic polynomial function and $\boldsymbol{y}$ is an observation vector. Although elements of W_t are the function of $\{|e_j| : j \in N_h(t)\}$, we consider elements of W_t as fixed variables. Then, the expectation and variance of $\hat{\boldsymbol{\beta}}_t$ are

$$E[\hat{\boldsymbol{\beta}}_t] = \boldsymbol{\beta}_t + (X'W_tX)^{-1}X'W_t\boldsymbol{d}_t \tag{6}$$

$$V[\hat{\boldsymbol{\beta}}_t] = \sigma^2(X'W_tX)^{-1}(X'W_t)(W_tX)(X'W_tX)^{-1} \tag{7}$$

where $\boldsymbol{\beta}_t = (f(t), f'(t), f''(t), f'''(t))'$ and $\boldsymbol{d}_t = (f(1), \ldots, f(n))' - X\boldsymbol{\beta}_t$. The formulae (6) and (7) are used for developing the change point detection procedure.

3.2 Detection of Multiple Change Points

The points at which jumps and/or roofs/valleys occur are change points. Assume that $f(t)$ is a single linear function in the range between change points, such that the second- and third-order derivatives at each point, except for change points, are zero. Thus, from the expectation (Eq. 6) and the variance (Eq. 7) of $\hat{\boldsymbol{\beta}}_t$, if neither jumps nor roofs/valleys exist in the neighborhood of the point t, then

$$\hat{c}_t \sim N(0, \sigma(3,3)\sigma^2) \tag{8}$$

$$\hat{d}_t \sim N(0, \sigma(4,4)\sigma^2) \tag{9}$$

where $\sigma(3,3)$ and $\sigma(4,4)$ are $(3,3)$ and $(4,4)$ elements of $V[\hat{\boldsymbol{\beta}}_t]/\sigma^2$, respectively.

If a jump of a linear trend occurs at a point t, the *zero crossing* of $\hat{c}_t$ occurs in the neighborhood of the point t. On the basis of this characteristic of $\hat{c}_t$, jump points of the linear trend are detected by the following procedure :

1. Sequentially check the condition $|\hat{c}_t| \geq C_v(c_t)\sqrt{\sigma(3,3)}\hat{\sigma}$ for $t = 1$ to n ;
2. If the above condition is true at the point $t = k$, then for points $t \geq k_1 + 1$, sequentially check whether each point satisfies the condition $|\hat{c}_t| < C_v(c_t)\sqrt{\sigma(3,3)}\hat{\sigma}$

until this condition is false. Let k_2 be the point at which $|\hat{c}_t| \geq C_v(c_t)\sqrt{\sigma(3,3)}\hat{\sigma}$ is true ;

3. If $\hat{c}_{k_1}\hat{c}_{k_2} < 0$ and $0 < k_2 - k_1 - 1 \leq m$, then there is a jump in the interval $[k_1, k_2]$. Then the change point is estimated as $\arg_t \min\{|\hat{c}_{k_1+1}|, \ldots, |\hat{c}_{k_2-1}|\}$. Otherwise, there is no jump in the interval $[k_1, k_2]$.

The roof/valley detection procedure is developed by replacing $\hat{c}_t$ with $\hat{d}_t$ in the jump detection procedure, because the roof/valley of a linear trend is a jump of the second-order derivative (c_t).

The standard deviation of observations $\hat{\sigma}^2$ is estimated through two one-sided local linear kernel smoothings and the robust estimator of these residuals was introduced by Lenth (1989). Two one-sided local linear kernel smoothing were utilized by Joo and Qiu (2009). Two linear regression functions are fitted in each area $[t - m_h, t)$ and $(t, t + m_h]$ with linear regression functions denoted as $\gamma_{0-}(t) + \gamma_{1-}(t)(i - t)$ and $\gamma_{0+}(t) + \gamma_{1+}(t)(i - t)$, respectively. These regression coefficients are obtained as the solutions of the following minimization problems:

$$\min_{\gamma_{0-}(t),\gamma_{1-}(t)} \sum_{i\in[t-m_h,t)} \{y_i - [\gamma_{0-}(t) + \gamma_{1-}(t)(i - t)]\}^2 \times K\left(\frac{i-t}{m_h}\right), \tag{10}$$

$$\min_{\gamma_{0+}(t),\gamma_{1+}(t)} \sum_{i\in(t,t+m_h]} \{y_i - [\gamma_{0+}(t) + \gamma_{1+}(t)(i - t)]\}^2 \times K\left(\frac{i-t}{m_h}\right). \tag{11}$$

The residuals are

$$e_-(i) = y_i - (\gamma_{0-}(t) + \gamma_{1-}(t)(i - t)), i \in [t - m_h, t), \tag{12}$$

$$e_+(i) = y_i - (\gamma_{0+}(t) + \gamma_{1+}(t)(i - t)), i \in (t, t + m_h]. \tag{13}$$

By Lenth's pseudo standard error (*PSE*), the two standard deviations $\hat{\sigma}_-$ and $\hat{\sigma}_+$ are calculated for each side $[t - m_h, t)$,$(t, t + m_h]$. Hence, the estimator of the standard deviation of observations on $[t - m_h, t)$ is given as

$$\hat{\sigma}_- = 1.5\, median_{|e_-(i)|<2.5s_0} |e_-(i)|, \tag{14}$$

where the *median* is computed among the $|e_-(i)|$ with $e_-(i) < 2.5s_0$ and

$$s_o = 1.5\, median\, |e_-(i)|. \tag{15}$$

The estimator of the standard deviation of observations on $(t, t+m_h]$ is also obtained in the same manner. The overall estimator of the standard deviation of observations on $[t - m_h, t + m_h]$ is determined as the smallest between $\hat{\sigma}_-$ and $\hat{\sigma}_+$.

4 Simulation Study

In this section, the proposed change point detection procedure is evaluated through the Monte Carlo method. The observations are generated from the following structure :

$$y_t = \sum_{s=1,2,\cdots} \delta I(t > 100s) - \frac{\delta}{100}(t-1) + \varepsilon_t, t = 1, \ldots, n, \tag{16}$$

where ε_t's are independent random errors and $I(t > s_j)$ is the index function. We focus on the process with outliers. Thus, some of the random errors are distributed to $N(0, 1^2)$ with a probability p, and the rest of the random errors are distributed to a t-distribution with degrees of freedom ϕ as outliers.

The evaluation is carried out under the following conditions:

1. Total number of observations:$n = 1,000$,
2. Amount of jumps:$\delta = 2, 3$,
3. Parameters of outliers:$\phi = 3$, $p = 0.2$,
4. The kernel function : the tricube weight function $K(u) = (1 - |u|^3)^3 I(|u| \leq 1)$ and the Epanechnikov kernel function $K(u) = 0.75(1 - u^2)$,
5. Parameters of the detection procedure:$m_h = 21$(bandwidth for the tricube weight function), $m_h = 55$ for the Epanechnikov kernel function, $R = 3$ (the number of iteration for the robust estimation), $C_v(c_t) = C_v(d_t) = 3, 0$(constant for detections).

In Fig. 2, an example of a sequence of observations is shown. From Fig. 2, we observed that our detection procedure can resist outliers. The sequence of $\{\hat{a}(t)|t = 1, \ldots, n\}$ maintains the underlying pattern of observations and there is no extreme value in the sequence. By applying our robust smoothing to a dataset with some patterns and outliers, the variation patterns become clear. It is important to visualize the variation pattern of the process in order to improve and control the process.

The ratios of detecting the jump within ± 2 points of each true jump point are calculated through the Monte Carlo method with $1,000$ replications, whose results are shown in Figs. 3–5. Figures 3 and 4 show the powers of jump detection in the case of $\delta = 0.2$, when the kernel functions are tricube weights and the Epanechnikov function, respectively. Each bandwidth parameter is optimized by trial and error. The procedure with the Epanechnikov kernel function has higher power than that with the tricube weight function. In addition, the procedure with the Epanechnikov kernel holds a Type I error.

In Fig. 5, the power for $\delta = 0.3$ is shown in the case when the kernel function is the Epanechnikov kernel. At every jump point, the power of detection is more than 0.8.

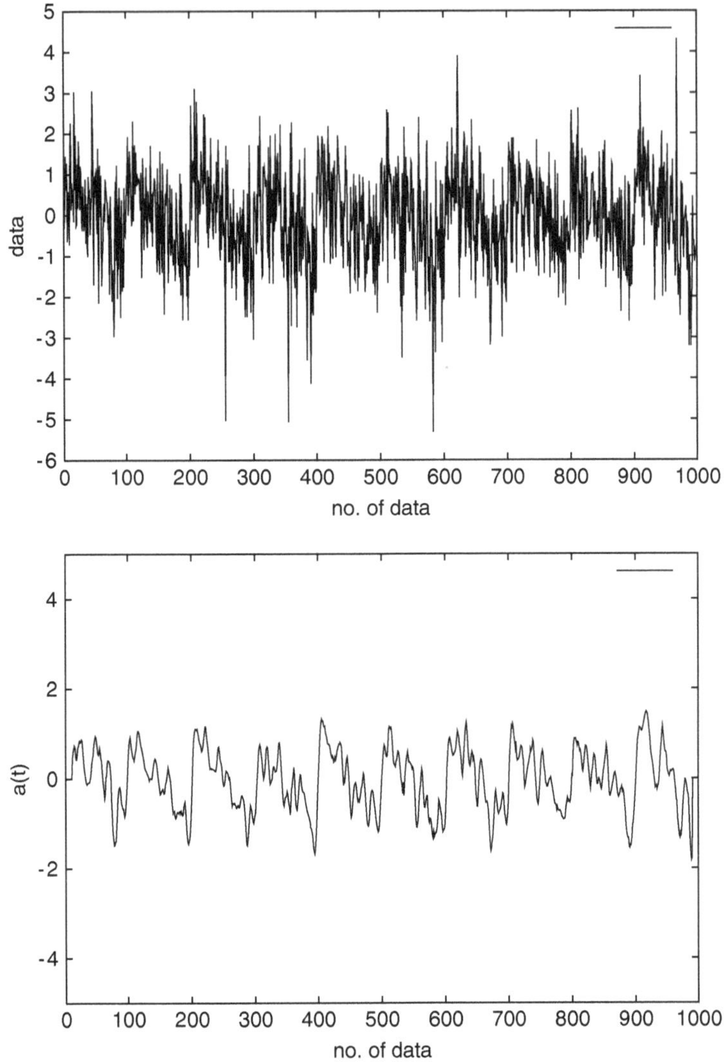

Fig. 2 An example of a sequence of observations for the simulation model and the estimates of $f(t)$ $(\hat{a}(t))$

5 Conclusion

In this study, we proposed a robust procedure for detecting multiple change points, which is able to perform well even when observations have certain systematic patterns such as trends and outliers. From the results of the simulation study, it can be concluded that our procedure is robust. When the Epanechnikov kernel

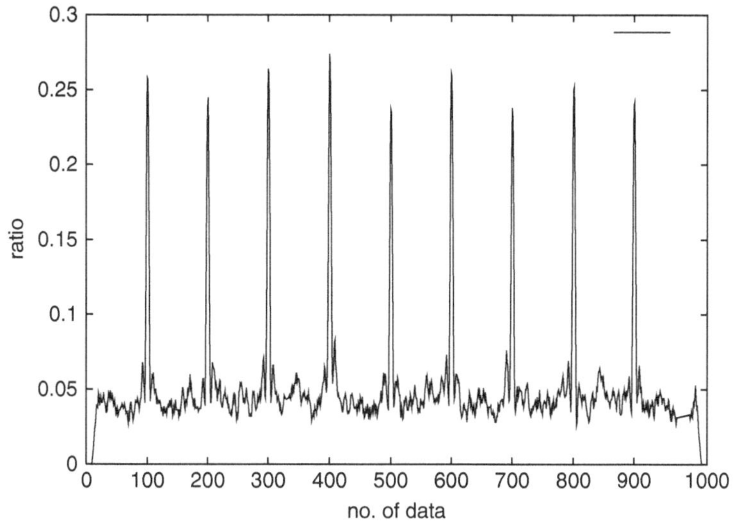

Fig. 3 Ratios of detecting the jump within ±2 points of each true jump point for $\delta = 0.2$ with the tricube weight function

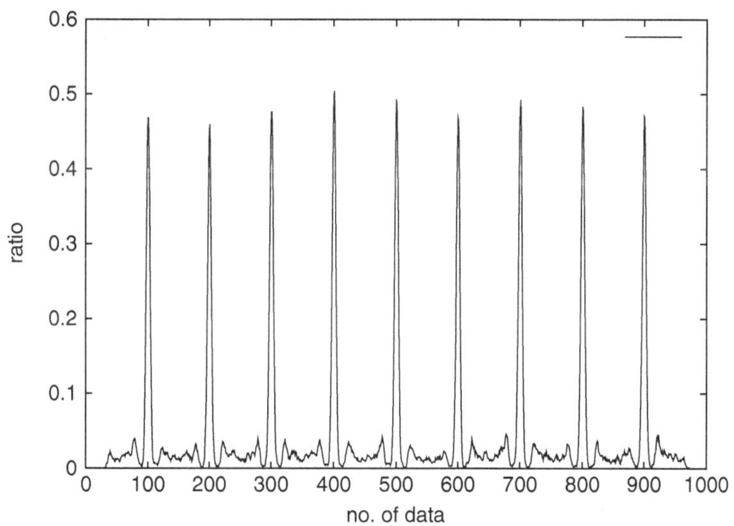

Fig. 4 Ratios of detecting the jump within ±2 points of each true jump point for $\delta = 0.2$ with Epanechnikov kernel function

function is adopted, our procedure shows good performance in terms of the powers of jump detection. The bandwidth is optimized using a trial and error approach. Cleveland (1979) provided an automatic bandwidth selection procedure for robust locally weighted regression. However, this technique requires a large number of

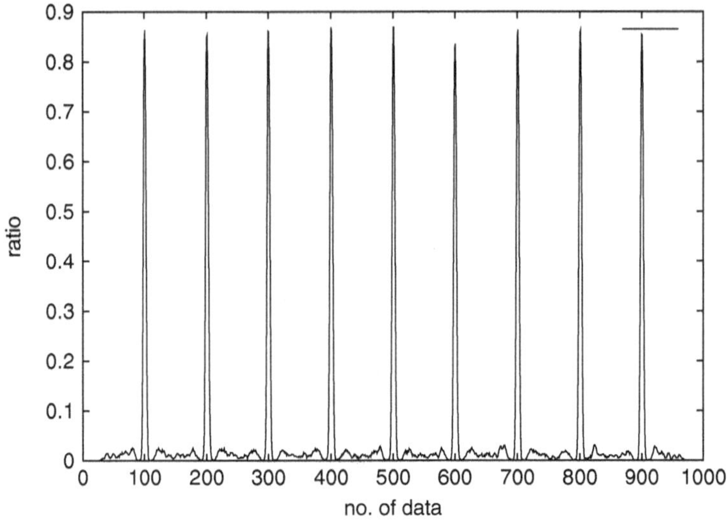

Fig. 5 Ratios of detecting the jump within ±2 points of each true jump point for $\delta = 0.3$ with Epanechnikov kernel function

observations (more than 1,000). Thus, we need to improve this technique for applying our detection procedure to improve and control process in the practical FMS.

References

Cleveland, W. S. (1979). Robust locally weighted regression and smoothing scatterplots. *Journal of the American Statistical Association, 74*, 829–836.

Harnish, P., Nelson, B., & Runger, G. (2009). Process partitions from time-ordered clusters. *Journal of Quality Technology, 41*, 3–17.

Joo, J.-H., & Qiu, P. (2009). Jump detection in a regression curve and its derivative. *Technometrics, 51*, 289–305.

Lenth, R. V. (1989). Quick and easy analysis of unreplicated factorials. *Technometrics, 31*, 469–473.

Qiu, P., & Yandell, B. (1998). A local polynomial jump-detection algorithm in nonparametric regression. *Technometrics, 40*, 141–152.

Sullivan, J. H. (2002). Detection of multiple change points from clustering individual observations. *Journal of Quality Technology, 34*, 371–383.

Risk-Adjusted Cumulative Sum Charting Procedures

Fah F. Gan, Lin Lin, and Chok K. Loke

Abstract Risk-adjusted charting procedures for monitoring the performances of a cardiac surgeon or a group of surgeons have recently gained prominence. Charting procedures developed for manufacturing processes are no longer appropriate because they do not take a patient's risk into account. The first charting procedure, variable life-adjusted display (VLAD) that takes a patient's risk into account was introduced in 1997. The VLAD plots the predicted mortality count minus the observed count cumulatively. The statistic plotted is intuitive and it has gained widespread attention and adoption. However, the run length performance of this chart is still not clearly understood because of the lack of a proper signalling rule. A risk-adjusted cumulative sum (RA-CUSUM) chart based on testing the odds ratio that a patient dies was proposed in 2000. In this paper, we developed and studied a general RA-CUSUM chart of which the RA-CUSUM chart based on odds ratio is a special case. The general RA-CUSUM chart allows testing to be done beyond just testing the odds ratio. One interesting note: although the VLAD and RA-CUSUM chart look seemingly different, we show that the RA-CUSUM chart and the RA-CUSUM chart based on the VLAD's monitoring statistic are in fact the same chart.

Keywords Average run length • Odds ratio • Patient mix • Relative risk • Sensitivity analysis • Sequential probability ratio test Surgical outcomes

1 Introduction

Risk-adjusted charting procedures for monitoring the performances of a surgeon or a group of surgeons have recently gained prominence (see Rogers et al. 2004; Treasure et al. 2004; Sherlaw-Johnson, 2005; Woodall, 2006). Grigg et al. (2003)

F.F. Gan (✉) · L. Lin · C.K. Loke
Department of Statistics and Applied Probability, National University of Singapore,
6 Science Drive 2, Singapore 117546
e-mail: staganff@nus.edu.sg

H.-J. Lenz et al. (eds.), *Frontiers in Statistical Quality Control 10*,
Frontiers in Statistical Quality Control 10, DOI 10.1007/978-3-7908-2846-7_15,

and Grigg and Farewell (2004) provided reviews of the development of risk-adjusted charting procedures. Woodall (2006) provided a comprehensive review of research papers in health-care and public-health surveillance. Other recent papers on risk-adjusted charts include Lovegrove et al. (1999); Gustafson (2000); Treasure et al. (2002); Benneyan and Borgman (2003); Spiegelhalter (2004); Spiegelhalter et al. (2003); Grunkemeier et al. (2003); Sismanidis et al. (2003); Blackstone (2004); Manche and Schembri (2005); Novick et al. (2006); Grigg and Spiegelhalter (2007); Coory et al. (2008); Chang (2008); Biswas and Kalbfleisch (2008); Gan and Tan (2010); Steiner and Jones (2010) and Gandy et al. (2010). Unlike manufacturing processes where the raw material is usually assumed to be of homogeneous quality, patients usually do not have the same surgical risk. Using charting procedures developed for the manufacturing processes amounts to assuming the same risk for all patients. For two surgeons who are equally competent, it is clear that the surgeon who operates on patients with high risks will have a lower success rate than the one who operates on patients with low risks. In other words, it is misleading to use these unadjusted success rates to compare the two surgeons. Steiner et al. (2000) demonstrated that trainee surgeons were performing better than experienced surgeons on the standard cumulative sum (CUSUM) charts while the risk-adjusted cumulative sum (RA-CUSUM) charts showed the opposite.

A patient's preoperative risk can be predicted before a surgery based on the patient's conditions. One commonly used estimate is the euroSCORE which is the predicted mortality $y_i = \exp(\beta_0 + \sum \beta_i u_i)/[1 + \exp(\beta_0 + \sum \beta_i u_i)]$ where u_i's specify the patient's characteristics like age, gender, serum creatinine, systolic pulmonary pressure, neurological dysfunction etc. This model was developed by Roques et al. (1999) by fitting a logistic regression model based on 19030 cardiac operations. An online calculator for the euroSCORE can be found on the website www.euroscore.org. Alternatively, the Parsonnet score (Parsonnet et al. 1989) could be used.

In order to take the risk of a patient into account, Lovegrove et al. (1997; 1999) and Poloniecki et al. (1998) suggested plotting the difference between the preoperative risk y_i and the surgical outcome ($x_i = 1$ if a patient dies or $x_i = 0$ if a patient survives) cumulatively. They termed this chart a variable life-adjusted display (VLAD) and cumulative risk-adjusted mortality (CRAM) chart respectively. This chart is intuitive, easy to understand and it accounts for the risk of a patient by comparing directly the patient's preoperative risk and the outcome of the operation. Treasure et al. (2004) presented some convincing examples using the VLAD that showed both improvement and deterioration in performances of some surgeons. The main criticism of this chart is the lack of a proper signalling rule. A signalling rule provides an objective and quantitative rule for assessing the points plotted and determining appropriate times for taking action. Although Poloniecki et al. (1998) proposed certain control limits for signalling but these limits are not directly interpretable in terms of run length performance. Sherlaw-Johnson (2005) mapped the control limits of the RA-CUSUM chart onto the VLAD but the resulting signalling rule is complicated because the control limits change with inclusion of data from every new surgical operation.

Steiner et al. (2000) proposed a RA-CUSUM chart based on testing the odds ratio that a patient dies, as a way to account for the patient's risk. This chart is derived using the sequential probability ratio test (SPRT) procedure developed by Wald (1947). Such a chart is optimal in terms of run length performance (see Moustakides, 1986). The RA-CUSUM chart and VLAD are developed using two different approaches and are viewed as two different charts (see Rogers et al. 2004; Spiegelhalter 2004; Sherlaw-Johnson 2005; Woodall 2006) but as we shall show in Sect. 3, interestingly the RA-CUSUM chart the CUSUM chart based on the VLAD's monitoring statistic are in fact the same chart.

We will review the VLAD in Sect. 2 and derive the RA-CUSUM chart based on the VLAD's monitoring statistic and the signalling rule using the SPRT approach. In Sect. 3, a general RA-CUSUM chart is developed. The RA-CUSUM chart based on odds ratio developed by Steiner et al. (2000) will be shown to be a special case. In Sect. 4, we will compare the performances of RA-CUSUM charts based on odds ratio and relative risk; assess the sensitivities of RA-CUSUM charts with respect to changes in the underlying risk distribution. This is important because this will allow us to find out how the performance of the RA-CUSUM chart is affected by changes in the distribution of patients' risks. The use of the RA-CUSUM charts based on odds ratio and relative risk in monitoring is illustrated with a real data set in Sect. 5. Finally, conclusions are given in Sect. 6.

2 Risk-Adjusted CUSUM Chart Based on VLAD's Monitoring Statistic

In order to take the risk of a patient into account, Lovegrove et al. (1997; 1999) and Poloniecki et al. (1998) proposed the VLAD chart. Let Y_i denote the predicted preoperative risk of the ith patient under current conditions and let the associated probability density function (pdf) be $f_Y(\cdot\,;\theta)$. Also, let X_i denote the outcome: 1 if the ith patient dies or 0 if the ith patient survives. In this paper, we will use the 30-day post-operative mortality rate for illustration. The VLAD is obtained by plotting $\sum_{i=1}^{n} X_i - \sum_{i=1}^{n} Y_i$ or $\sum_{i=1}^{n} (X_i - Y_i)$ against n, where $X_i - Y_i$ is the VLAD's monitoring statistic. The quantity $\sum_{i=1}^{n} Y_i$ represents the predicted number of deaths up to the nth patient, whereas the quantity $\sum_{i=1}^{n} X_i$ represents the observed number of deaths. The VLAD takes the patients' risks into account by comparing these two quantities directly. This chart is thus intuitive and easy to understand. Except for Sherlaw-Johnson (2005) who mapped a signalling rule from the RA-CUSUM chart onto the VLAD, little progress has been done although the VLAD is gaining wide acceptance among surgeons.

Fig. 1 Plot of penalty-reward w_n (based on testing odds ratio of death) awarded to a surgeon according to a patient's predicted preoperative risk y_n

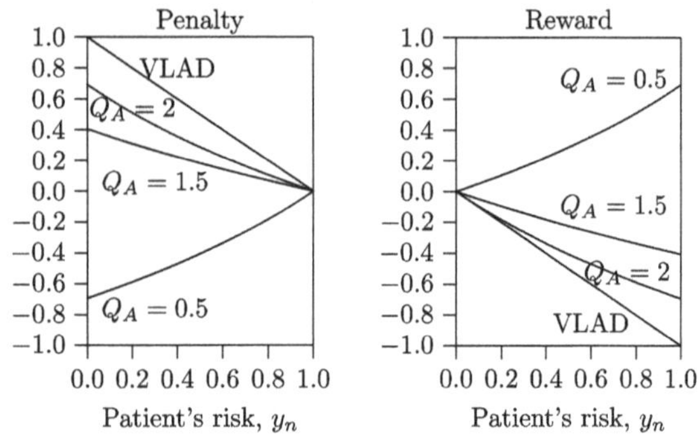

In order to understand the VLAD better, let us define $d(Y_i)$ to be the true probability of the ith patient dying when the patient is operated on by a surgeon. Let the VLAD's monitoring statistic be denoted as

$$Z_i = X_i - Y_i = \begin{cases} 1 - Y_i, & \text{if the } i\text{th patient dies;} \\ -Y_i, & \text{if the } i\text{th patient survives.} \end{cases} \quad (1)$$

Given $Y_i = y_i$, the random variable Z_i is simply a random variable with probabilities $P(Z_i = 1 - y_i) = d(y_i)$ and $P(Z_i = -y_i) = 1 - d(y_i)$. In other words, for the ith patient with a predicted preoperative risk y_i, the chance of dying when operated on by the surgeon is $d(y_i)$. The statistic z_i may be viewed as a penalty-reward score given to the surgeon depending on the patient's preoperative risk and the outcome of the operation. Figure 1 shows the plots of z_i versus preoperative risk y_i. The quantity Q_A is the odds ratio of death defined as $Q_A = \{d(y)/[1-d(y)]\}/\{y/[1-y]\}$. Thus, if a patient dies, the penalty will be heavy if the patient's preoperative risk is low, and the penalty will be small if the patient's preoperative risk is high. Similarly, if a patient survives, the reward will be big if the patient's preoperative risk is high and the reward will be small if the patient's preoperative risk is low. For a chart that does not take the risk into account, Z_i is simply given as

$$Z_i = \begin{cases} 1, & \text{if the } i\text{th patient dies;} \\ 0, & \text{if the } i\text{th patient survives.} \end{cases} \quad (2)$$

Thus, effectively, the VLAD has converted a binary penalty-reward system into a continuous one according to the patient's preoperative risk y_i, and this is precisely how the VLAD takes the patient's risk into account. For the rest of the paper, we assume Z_i follows Eq. 1.

Consider testing $H_0 : d = d_0$ versus $H_A : d = d_A$. If the preoperative risk y remains unchanged for a new operation, we may set $d_0(y) = y$. For the VLAD,

$$E\left\{\sum_{i=1}^{n}(X_i - Y_i)\right\} = nE(X_i - Y_i). \tag{3}$$

Therefore, the cumulative sum has a slope of $E(X_i - Y_i)$ when plotted against n. If the expected chance of survival $E(Y_i)$ is equal to the expected outcome $E(X_i)$, the plot will display no evidence of a sustained slope. On the other hand, if the expected chance of survival differs from the expected outcome, the plot will have a sustained slope of $E(X_i - Y_i)$. This implies that a position on the chart at which the plot has a change in slope indicates the position of a possible change in $E(X_i - Y_i)$. It is thus clear that for the VLAD, it is the slope that is important.

We shall now proceed to derive the RA-CUSUM chart based on the VLAD's monitoring statistic Z_i. We first derive the pdf of Z_i. Let $f_Y(y;\theta)$ be the pdf of Y_i. Using a conditioning result (see Ross, 2006, p. 376 for example), it can be shown that

$$P(Z_i \le z; \theta, d) = \begin{cases} \int_0^1 [1 - d(y)] f_{Y_i}(y;\theta)dy + \int_{1-z}^1 d(y) f_{Y_i}(y;\theta)dy, & z \ge 0; \\ \int_{-z}^1 [1 - d(y)] f_{Y_i}(y;\theta)dy, & z < 0, \end{cases} \tag{4}$$

and thus the pdf of Z_i can be derived as

$$f_Z(z;\theta, d) = \begin{cases} d(1-z) f_Y(1-z;\theta), & z \ge 0; \\ [1 - d(-z)] f_Y(-z;\theta), & z < 0. \end{cases} \tag{5}$$

Now consider testing $H_0 : d = d_0$ versus $H_A : d = d_A$, using the SPRT approach, a RA-CUSUM chart can be obtained by plotting

$$S_n = \max\{0, S_{n-1} + W_n\}, \tag{6}$$

against n and a signal is issued if $S_n > h$ for some chart limit $h > 0$, where $S_0 = 0$ and W_n is the log of the likelihood ratio of densities,

$$W_n = \log\left\{\frac{f_Z(Z_n;\theta, d_A)}{f_Z(Z_n;\theta, d_0)}\right\}. \tag{7}$$

Using Eq. 5, it can be shown that

$$W_n = \begin{cases} \log\{d_A(1 - Z_n)/d_0(1 - Z_n)\}, & Z_n \ge 0; \\ \log\{[1 - d_A(-Z_n)]/[1 - d_0(-Z_n)]\}, & Z_n < 0. \end{cases} \tag{8}$$

This completes the derivation of a RA-CUSUM based on the VLAD's monitoring statistic Z_n. The choice of the functions d_0 and d_A will be explained in details in the next section.

Steiner et al. (2000) considered testing the odds ratio of death under H_0 : Odds Ratio $= Q_0$ versus H_A : Odds Ratio $= Q_A$. For testing these hypotheses, $d_0(y)$ and $d_A(y)$ would take the form $d_0(y) = Q_0 y/(1 - y + Q_0 y)$ and $d_A(y) = Q_A y/(1 - y + Q_A y)$. The statistic W_n can then be obtained as

$$W_n = \begin{cases} \log\left\{\dfrac{(Z_n + Q_0 - Q_0 Z_n)Q_A}{(Z_n + Q_A - Q_A Z_n)Q_0}\right\}, & Z_n \geq 0; \\ \log\left\{\dfrac{1 + Z_n - Z_n Q_0}{1 + Z_n - Z_n Q_A}\right\}, & Z_n < 0. \end{cases} \quad (9)$$

If we choose $Q_0 = 1$ and $Q_A = 2$ as in Steiner et al. (2000), the RA-CUSUM chart based on the VLAD's monitoring statistic Z_n can then be obtained by plotting

$$S_n = \max\left\{0, S_{n-1} + \log\left(\frac{2}{2 - Z_n}\right)1_{\{Z_n \geq 0\}} + \log\left(\frac{1}{1 - Z_n}\right)1_{\{Z_n < 0\}}\right\}, \quad (10)$$

where $1_{\{\cdot\}}$ is the indicator function. Steiner et al. (2000) compared their RA-CUSUM chart with a 'CUSUM' chart based on the VLAD's monitoring statistic $Z_n = X_n - Y_n$ directly and obtained by plotting

$$S_n = \max\{0, S_{n-1} + Z_n\}, \quad (11)$$

against n, where $S_0 = 0$ and signal when $S_n > h$. Note that this chart does not depend on any alternative hypothesis. It is a cumulative sum chart of some sort but it is not Page's (1954) CUSUM chart because Eq. 11 is not based on the SPRT procedure. Thus, Moustakides's (1986) optimality result for a CUSUM chart does not hold for this chart.

3 A General Risk-Adjusted Cumulative Sum Chart

In this section, we will first describe the standard (non-risk-adjusted) CUSUM chart for monitoring the binary outcomes. We then describe Steiner's RA-CUSUM chart based on testing the odds ratio. We proceed to derive a general RA-CUSUM chart and show that it is mathematically identical to the RA-CUSUM chart based on VLAD's monitoring statistic derived in the previous section. Steiner's RA-CUSUM chart will also be shown to be a special case of the general RA-CUSUM chart.

Let X_i denote the outcome of the ith patient as defined in the previous section. The probability function of X_i is given as $P(X_i = x_i) = p^{x_i}(1 - p)^{1-x_i}$ where $x_i = 0$ or 1. Consider testing $H_0 : p = p_0$ versus $H_A : p = p_A$. Based on the

SPRT approach, the standard CUSUM chart (Reynolds and Stoumbos, 1999) for testing these hypotheses is obtained by plotting

$$S_n = \max\{0, S_{n-1} + W_n\}, \tag{12}$$

against n where

$$W_n = \begin{cases} \log\{p_A/p_0\}, & \text{if the } n\text{th patient dies;} \\ \log\{(1-p_A)/(1-p_0)\}, & \text{if the } n\text{th patient survives,} \end{cases} \tag{13}$$

and a signal is issued when $S_n > h$. The statistic W_n may be interpreted as the penalty-reward given to a surgeon based on the outcome of the nth operation. Note that the penalty and reward are constants and they do not depend on the preoperative risk of the patient Y_n, thus the standard CUSUM chart does not take the patient's risk into account.

If y is the predicted probability that a patient dies, then the odds that the patient dies is $y/(1-y)$. Steiner et al. (2000) considered testing the odds ratio under H_0 : $p_0/(1-p_0) = Q_0 y/(1-y)$ versus $H_A : p_A/(1-p_A) = Q_A y/(1-y)$ for all y and based on Eq. 13, they showed that W_n was given as

$$W_n = \begin{cases} \log\left\{\dfrac{(1-Y_n+Q_0Y_n)Q_A}{(1-Y_n+Q_AY_n)Q_0}\right\}, & \text{if the } n\text{th patient dies;} \\ \log\left\{\dfrac{1-Y_n+Q_0Y_n}{1-Y_n+Q_AY_n}\right\}, & \text{if the } n\text{th patient survives.} \end{cases} \tag{14}$$

To derive a general RA-CUSUM chart, consider testing the risk under $H_0 : d = d_0$ versus $H_A : d = d_A$. Under these hypotheses, $p_0 = d_0(y_n)$ and $p_A = d_A(y_n)$. Based on Eq. 13, the statistic W_n becomes

$$W_n = \begin{cases} \log\{d_A(Y_n)/d_0(Y_n)\}, & \text{if the } n\text{th patient dies;} \\ \log\{[1-d_A(Y_n)]/[1-d_0(Y_n)]\}, & \text{if the } n\text{th patient survives,} \end{cases} \tag{15}$$

or in terms of Z_n or $X_n - Y_n$:

$$W_n - \begin{cases} \log\{d_A(1-Z_n)/d_0(1-Z_n)\}, & Z_n \geq 0; \\ \log\{[1-d_A(-Z_n)]/[1-d_0(-Z_n)]\}, & Z_n < 0. \end{cases} \tag{16}$$

If we compare Eqs. 8 and 16, we find that they are identical and thus we have shown that the general RA-CUSUM chart and the RA-CUSUM chart based on the VLAD's monitoring statistic are identical.

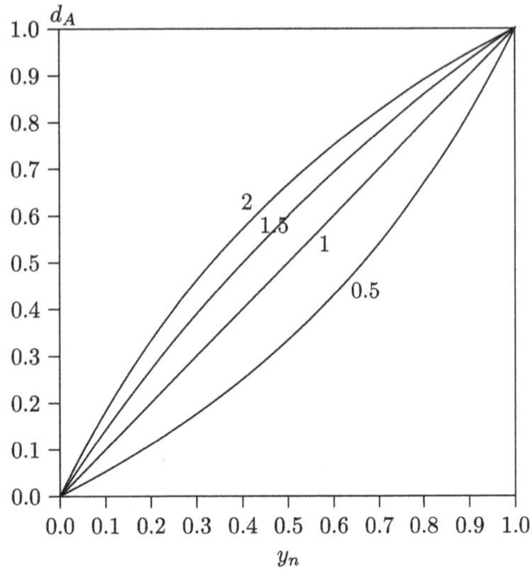

Fig. 2 Plots of d_A against the patient's predicted preoperative risk for $Q_A = 0.5, 1, 1.5, 2$ based on testing odds ratio of death

By testing odds ratio, $H_0 : d_0(y)/[1 - d_0(y)] = Q_0 y/(1 - y)$ versus $H_A : d_A(y)/[1-d_A(y)] = Q_A y/(1-y)$, we will obtain the functions $d_0(y)$ and $d_A(y)$ as

$$d_0(y) = Q_0 y/(1 - y + Q_0 y), \tag{17}$$

$$d_A(y) = Q_A y/(1 - y + Q_A y). \tag{18}$$

Substituting Eqs. 17 and 18 into Eq. 15, we will obtain Eq. 14 which is monitoring statistic of the RA-CUSUM chart developed by Steiner et al. (2000). In other words, every set of unique functions $d_0(y)$ and $d_A(y)$ defines a unique RA-CUSUM chart.

Before we proceed to derive the RA-CUSUM chart based on testing the relative risk, we will examine the penalty-reward system defined by testing the odds ratio. Figure 1 shows the plot of w_n versus the patient's risk y_n for $Q_0 = 1$ and $Q_A = 0.5, 1.5$ and 2. The plot of z_n for the original VLAD remains the same irrespective of any alternative hypothesis. In contrast, the plot of w_n for the RA-CUSUM chart changes with the value of Q_A. Note that for $Q_A = 0.5$, the penalty is negative when a patient dies and the reward is positive when a patient survives. This means that the RA-CUSUM chart for $Q_A = 0.5$ signals for improvement. Figure 2 contains plots of d_A against the patient's risk for various values of Q_A. As expected, for a patient with preoperative risk y_n, the probability that a patient dies after operated on by the surgeon increases as Q_A increases and decreases as Q_A decreases.

Equation 15 is general and it defines the SPRT statistic according to the functions d_0 and d_A. We now consider testing relative risk R instead of odds ratio under $H_0 : d_0(y) = R_0 y$ versus $H_A : d_A(y) = R_A y$ for all y. Again if the preoperative risk y is based on current conditions, we may set $R_0 = 1$. The function d_A is defined as

$$d_A(y) = \begin{cases} 1, & \text{if } y = 1 \text{ and } R_A \le 1; \\ R_A y, & \text{if } y < 1 \text{ and } R_A \le 1; \\ R_A y, & \text{if } y < 1/R_A \text{ and } R_A > 1; \\ 1, & \text{if } y \ge 1/R_A \text{ and } R_A > 1. \end{cases} \quad (19)$$

Figure 3 contains the plots of d_A versus y for various values of R_A. Take $R_A = 1.47$ for example, the probability that a patient dies under new conditions is 47% higher if the patient's preoperative risk is less than $1/1.47$. For a patient with a preoperative risk greater than $1/1.47$, the probability that the patient dies is 1.

The statistic W_n can be derived using Eqs. 15 and 19 as

$$W_n = \begin{cases} \log(R_A), & \text{if } Y_n < 1/R_A \text{ and the } n\text{th patient dies;} \\ \log(1/Y_n), & \text{if } Y_n \ge 1/R_A \text{ and the } n\text{th patient dies;} \\ \log\left(\frac{1-R_A Y_n}{1-Y_n}\right), & \text{if } Y_n < 1/R_A \text{ and the } n\text{th patient survives;} \\ -\infty, & \text{if } Y_n \ge 1/R_A \text{ and the } n\text{th patient survives.} \end{cases} \quad (20)$$

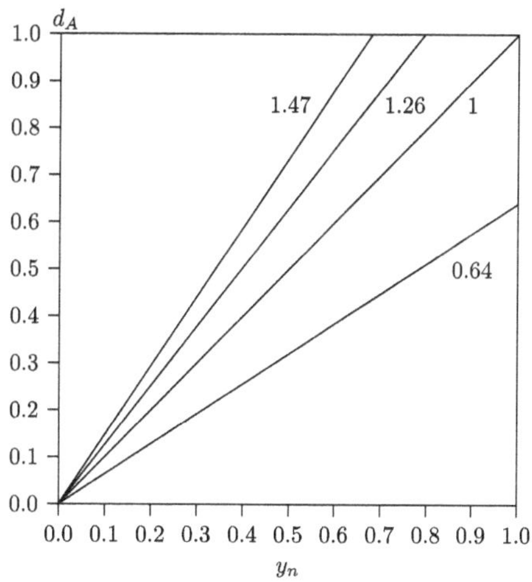

Fig. 3 Plot of d_A against the patient's predicted preoperative risk for $R_A = 0.64, 1, 1.26$ and 1.47 based on testing relative risk

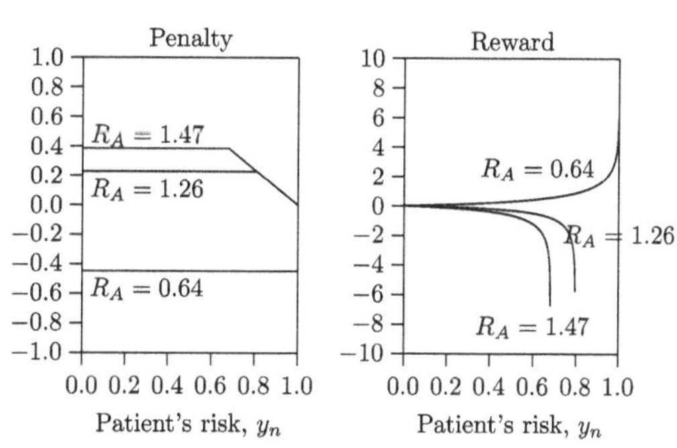

Fig. 4 Plot of penalty-reward w_n (based on testing relative risk) awarded to a surgeon according to a patient's predicted preoperative risk y_n

Figure 4 shows the plot of w_n versus the patient's preoperative risk y_n for $R_0 = 1$ and $R_A = 0.64, 1.26$ and 1.47. As a point of theoretical interest, for $R_A \geq 1$ and $Y_i \geq 1/R_A$, the reward W_n is negative infinity. This is because under the alternative hypothesis, the patient will die with probability one, but however, if the patient survives an operation, then an infinite reward will be given. This will cause the RA-CUSUM chart to be reset to zero.

4 Sensitivity Analysis

In this section, we investigate the sensitivity of the RA-CUSUM charts based on the average run length (ARL). In this context, the run length RL is defined to be the number of patients operated on until a signal is issued and the ARL is given as $E(RL)$. The collocation procedure developed by Knoth (2005, 2007) and Hackbusch (1995) is adapted here to compute the ARL accurately using the distribution function of W_n given in the Appendix.

From analysis of real data sets, the patient's risk is found to have a distribution similar to beta$(1, 3)$. In our investigation, we thus assume the patient's risk follows a beta$(1, 3)$ distribution. For detecting a deterioration in a surgeon's performance, we consider RA-CUSUM charts optimal in detecting $Q_A = 1.1$, 1.2, 1.3, 1.4, 1.5, 2.0 and 3.0. For detecting an improvement in a surgeon's performance, we consider RA-CUSUM charts optimal in detecting $Q_A = 0.9$, 0.8, 0.7, 0.6, 0.5, 0.2 and 0.1. We have chosen Q_A to be as small as 1.1 and this corresponds to a 10% increase in a patient's odds of dying. The in-control ARL is set at 100 here for illustration. Tables for other values of in-control ARL show similar results.

Some researchers (see Rogers et al. 2004 for example) have voiced their concerns about the effect of changes in the underlying risk distribution on the performance of the RA-CUSUM chart. This is a valid concern and in our first investigation, we consider a shift of the underlying risk distribution from beta $(1, 3)$ to the following distributions: beta $(1, 2)$, beta $(1, 2.5)$, beta$(1, 4)$ and beta $(1, 5)$, and then examine the effect on the in-control ARL. The resulting ARL's are displayed in Table 1. First of all, note that if the underlying distribution is beta $(1, 3)$, the in-control ARL of these charts is given by 100. If the risk distribution shifts to beta $(1, 2.5)$ or beta $(1, 2)$ which is more skewed to the right which means that there are more high-risk patients, the in-control ARL decreases by about 3–13%. If the risk distribution changes to beta$(1, 4)$ or beta $(1, 5)$ which means that there are more low-risk patients, the in-control ARL increases by about 12–31%. This table shows clearly that the performance of a RA-CUSUM chart is affected by changes in the underlying risk distribution. It is thus important to monitor the underlying risk distribution simultaneously (see Loke and Gan 2012 for further details) so that any conclusions drawn from a RA-CUSUM chart should be treated with caution when the underlying risk distribution has shifted.

Table 2 contains the ARL profiles of the various RA-CUSUM charts for detecting changes in the odds ratio. As expected, the ARL of these charts decreases as the

Table 1 In-control average run lengths of RA-CUSUM charts based on testing odds ratio corresponding to various underlying risk distributions

Risk	Q_A	1.1	1.2	1.3	1.4	1.5	2.0	3.0
Distribution	h	0.308	0.558	0.765	0.940	1.09	1.607	2.125
beta(1,2)		91	91	92	92	92	94	95
beta(1,2.5)		95	95	95	96	96	96	97
beta(1,3)		100	100	100	100	100	100	100
beta(1,4)		112	111	111	110	110	109	107
beta(1,5)		124	123	123	122	121	119	116
Risk	Q_A	0.9	0.8	0.7	0.6	0.5	0.2	0.1
Distribution	h	0.335	0.652	0.954	1.242	1.521	2.330	2.616
beta(1,2)		90	90	90	89	89	87	87
beta(1,2.5)		95	95	94	94	94	93	93
beta(1,3)		100	100	100	100	100	100	100
beta(1,4)		112	112	113	113	113	115	115
beta(1,5)		125	126	126	127	127	130	131

actual odds ratio deviates from $Q = 1$. Table 2 also shows that these charts are in fact optimal in detecting the intended odds ratios and this is consistent with the optimality result obtained by Moustakides (1986).

Let $d_{A,Q}(y)$ and $d_{A,R}(y)$ be the functions defined in Eqs. 18 and 19 respectively. In order to compare RA-CUSUM charts based on odds ratio and relative risk, we consider matching the average weighted risk

$$\int_0^1 d_{A,Q}(y) f_Y(y;\theta)dy = \int_0^1 d_{A,R}(y) f_Y(y;\theta)dy, \tag{21}$$

to obtain $d_{A,Q}(y)$ and $d_{A,R}(y)$ that are comparable. Table 3 contains the in-control ARL's of RA-CUSUM charts based on relative risk, corresponding to various risk distributions. Table 4 contains the ARL profiles of these charts with respect to changes in relative risk. As expected, the ARL of the charts decreases as the actual relative risk deviates from $R = 1$. Table 4 also shows that these charts are in fact optimal in detecting the intended relative risk and this is again consistent with the optimality result obtained by Moustakides (1986).

The optimal result obtained by Moustakides (1986) ensures that the RA-CUSUM chart based on odds ratio is optimal among all charts based on odds ratio. The same result also ensures that the RA-CUSUM chart based on relative risk is optimal among all charts based on relative risk. However, the result does not tell us which is optimal when the two RA-CUSUM charts are compared. The run length profiles in Table 2 are very similar to those in Table 4, with the RA-CUSUM chart based on relative risk being slightly more sensitive at the intended shift, except for RA-CUSUM charts for detecting $R_A = 1.06$ and $R_A = 1.11$. If we compare the run length profiles in Table 1 with those in Table 3, we will find that RA-CUSUM charts based on odds ratio are less sensitive to changes in the risk distribution

Table 2 Sensitivity analysis: ARL profiles of RA-CUSUM charts based on testing odds ratio optimal in detecting Q_A

	Q_A	1.1	1.2	1.3	1.4	1.5	2.0	3.0
Q	h	0.308	0.558	0.765	0.940	1.09	1.607	2.125
1.0		100.0	100.0	100.0	100.0	100.0	100.0	100.0
1.1		78.5	78.5	78.5	78.6	78.7	79.3	80.2
1.2		64.2	64.2	64.2	64.3	64.4	65.0	66.3
1.3		54.3	54.2	54.2	54.2	54.3	54.8	56.1
1.4		47.1	46.9	46.8	46.8	46.8	47.3	48.5
1.5		41.7	41.5	41.3	41.2	41.2	41.5	42.6
2.0		27.4	27.0	26.7	26.5	26.3	26.1	26.4
2.5		21.2	20.8	20.5	20.2	20.1	19.6	19.5
3.0		17.9	17.5	17.1	16.9	16.7	16.1	15.8
4.0		14.3	13.9	13.5	13.3	13.1	12.4	12.0
	Q_A	0.9	0.8	0.7	0.6	0.5	0.2	0.1
Q	h	0.335	0.652	0.954	1.242	1.521	2.330	2.616
1.0		100.0	100.0	100.0	100.0	100.0	100.0	100.0
0.9		78.5	78.6	78.7	78.9	79.3	80.8	81.6
0.8		62.3	62.3	62.3	62.6	62.9	65.1	66.2
0.7		50.1	49.9	49.8	49.9	50.1	52.2	53.5
0.6		40.8	40.4	40.1	40.0	40.1	41.8	42.9
0.5		33.6	33.0	32.6	32.4	32.3	33.3	34.3
0.4		27.9	27.3	26.8	26.4	26.1	26.5	27.2
0.3		23.4	22.7	22.2	21.7	21.3	21.1	21.5
0.2		19.7	19.0	18.4	17.9	17.5	16.7	16.9
0.1		16.6	16.0	15.4	14.8	14.3	13.2	13.1

Table 3 In-control average run lengths of RA-CUSUM charts based on testing relative risk corresponding to various underlying risk distributions

Risk	R_A	1.06	1.11	1.16	1.21	1.26	1.47	1.83
Distribution	h	0.313	0.579	0.808	0.998	1.165	1.742	2.303
beta(1,2)		62	62	63	64	65	70	78
beta(1,2.5)		80	80	81	81	82	83	88
beta(1,3)		100	100	100	100	100	100	100
beta(1,4)		142	141	141	141	140	136	129
beta(1,5)		184	183	183	183	182	177	163
Risk	R_A	0.94	0.88	0.80	0.73	0.64	0.32	0.18
Distribution	h	0.326	0.641	0.939	1.220	1.491	2.288	2.583
beta(1,2)		68	70	71	73	75	80	82
beta(1,2.5)		84	84	85	86	87	89	91
beta(1,3)		100	100	100	100	100	100	100
beta(1,4)		135	133	131	129	127	121	120
beta(1,5)		171	167	163	159	155	144	140

Table 4 Sensitivity analysis: ARL profiles of RA-CUSUM charts based on testing relative risk optimal in detecting R_A

	R_A	1.06	1.11	1.16	1.21	1.26	1.47	1.83
R	h	0.313	0.579	0.808	0.998	1.165	1.742	2.303
1.00		99.9	100.0	100.0	100.0	100.0	99.9	100.1
1.06		79.1	79.1	79.1	79.2	79.4	80.0	81.4
1.11		64.5	64.4	64.4	64.6	64.7	65.5	67.4
1.16		54.2	54.1	54.0	54.1	54.2	54.9	57.0
1.21		46.9	46.7	46.6	46.6	46.6	47.2	49.1
1.26		41.5	41.2	41.0	40.9	40.9	41.3	43.1
1.47		27.3	26.8	26.4	26.2	26.0	25.6	25.3
1.66		20.8	20.7	20.3	19.9	19.9	19.2	19.1
1.83		18.4	17.9	17.5	17.2	16.9	15.9	15.5
2.13		14.4	14.3	14.0	14.0	13.6	12.6	11.8
	R_A	0.94	0.88	0.80	0.73	0.64	0.32	0.18
R	h	0.326	0.641	0.939	1.220	1.491	2.288	2.583
1.00		100.0	100.0	100.0	100.0	100.0	100.0	100.0
0.94		77.1	77.1	77.2	77.4	77.7	79.1	79.8
0.88		60.1	60.1	60.1	60.3	60.6	62.5	63.4
0.80		47.8	47.6	47.5	47.6	47.8	49.5	50.5
0.73		38.7	38.3	38.1	38.0	38.1	39.4	40.3
0.64		31.7	31.3	31.0	30.7	30.7	31.5	32.2
0.54		26.8	25.9	25.5	25.1	24.9	25.3	25.8
0.44		22.4	21.8	20.8	20.5	20.3	20.7	21.1
0.32		18.6	18.1	17.6	17.2	16.8	16.3	16.4
0.18		15.7	15.2	14.7	14.3	13.9	13.0	12.9

than those based on relative risk. This suggests that RA-CUSUM charts based on odds ratio will be more reliable if changes in the underlying risk distribution are a concern.

5 Application to Acute Myocardial Infarction Data

The data set contains outcomes of patients with an acute myocardial infarction (more commonly known as heart attack) who are admitted to an anonymous hospital, collected as part of the NHS Research and Development funded EMMACE-1 (Evaluation of Methods and Management of Acute Coronary Events) Study (Dorsch et al., 2001). The post-operative outcomes after 30 days were collected for these patients admitted over a 3-month period. The mortality risk for each patient was both calculated and authenticated locally at the hospital. A total of 123 patients were observed and a cognizance of 27 deaths resulted in a mortality rate of 21.95%. Probability plot was used to identify the distribution and the data was found to be well fitted by a beta distribution with shape parameters $\alpha = 1$ and $\beta = 3$.

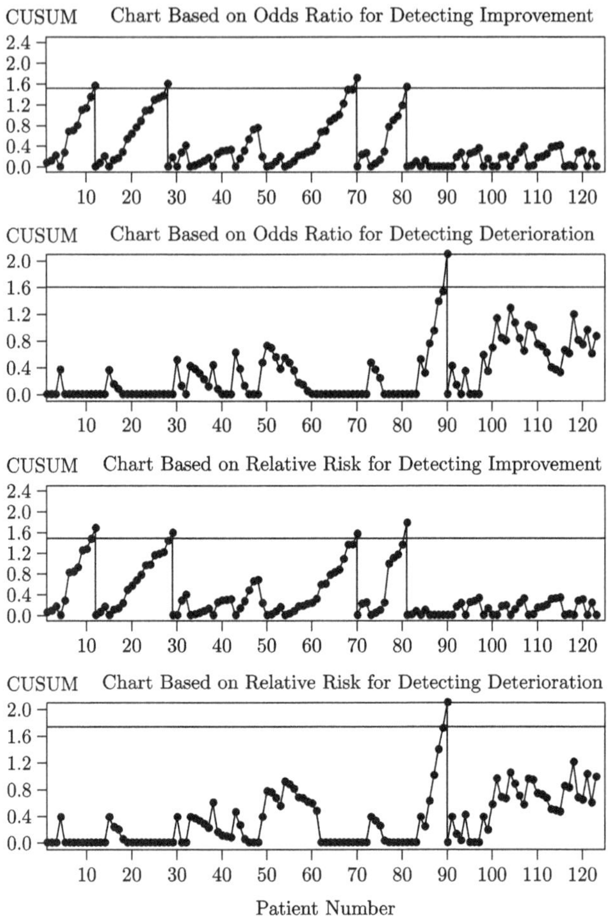

Fig. 5 Risk-adjusted CUSUM charts for monitoring acute myocardial infarction data collected as part of the NHS Research and Development funded EMMACE-1 (Evaluation of Methods and Management of Acute Coronary Events) Study (Dorsch et al., 2001). Both schemes show possible improvement with four signals issued, follow by a possible deterioration in performance with a signal issued at the 89th patient

Assume that we are interested in designing a RA-CUSUM chart which is optimal in detecting a deterioration in the performance with the odds ratio of mortality risk increased to two times the in-control odds ratio, that is $Q_A = 2$. This will correspond to a relative risk of $R_A = 1.47$ by matching the average weighted risk in Eq. 21. Also, assume that a RA-CUSUM chart which is optimal in detecting an improvement with the odds ratio of mortality risk decreased to half of the in-control odds ratio, that is $Q_A = 0.5$. This will correspond to a relative risk of $R_A = 0.64$ using Eq. 21. Further assume that in-control ARL is set at 100. Using the results from the previous section, the chart limit h for the detection of a deterioration will be set as 1.607 and 1.742 for the RA-CUSUM charts based on odds ratio and relative risk respectively. Likewise, the chart limit h for the detection of an improvement will be set as 1.521 and 1.491 for the RA-CUSUM charts based on odds ratio and relative risk respectively. The RA-CUSUM charts constructed are displayed in Fig. 5. These figures show that the RA-CUSUM charts based on odds ratio and relative risk are similar. Both schemes show possible improvement with four signals issued, follow by a possible deterioration in performance with a signal issued at the 89th patient.

6 Conclusions

To monitor a surgeon's performance, especially to detect a deterioration is crucial in saving patients' lives. The VLAD is now popularly used in hospitals for monitoring, but its lack of a proper signalling rule makes it less effective in monitoring a surgeon's performance. In this paper, we have used the SPRT approach to derive a RA-CUSUM chart based on the VLAD's monitoring statistic (see Eq. 10). Such a chart not only has a well defined chart limit and it is also optimal. The main contribution of this paper is the development of a general RA-CUSUM chart in which the RA-CUSUM chart based on odds ratio developed by Steiner et al. (2000) is shown to be a special case. With this general RA-CUSUM chart, we are now able to test other forms of risks as specified in the null and alternative hypotheses. Thus, in addition to testing odds ratio, we also consider testing relative risk. The main advantage of using relative risk is that it is easier to understand and interpret for surgeons and patients alike. Our sensitivity analysis also shows that both types of RA-CUSUM charts are sensitive to changes in the underlying distribution. Thus, this suggests the need to monitor the risk distribution so that meaningful conclusions drawn from the RA-CUSUM charts can be done taking the underlying risk distribution into account. Between the two types of RA-CUSUM charts, our sensitivity analysis shows that the one based on odds ratio is less sensitive to changes in the underlying risk distribution, thus this chart is preferred if changes in the underlying risk distribution are a major concern. Finally, although we have illustrated implementation of the RA-CUSUM charts based on acute myocardial infarction data, these charts can be applied equally well to various types of surgical operations as long as the risk can be predicted.

Acknowledgements The first and third authors are supported by the Academic Research Fund Tier 1 (R-155-000-092-112), Ministry of Education, Singapore. We wish to thank Dr Alistair Hall for providing the data from the EMMACE-1 Study and the permission to use it here.

Appendix

(A) Probability Distribution of W for Testing Odds Ratio

For odds ratio,

$$W_i = \begin{cases} \log\left[\frac{Q_A}{1-Y_i+Q_AY_i}\right], & \text{if the } i\text{th patient dies;} \\ \log\left[\frac{1}{1-Y_i+Q_AY_i}\right], & \text{if the } i\text{th patient survives.} \end{cases}$$

The pdf of W can be obtained using a conditioning approach (see Ross, 2006, p. 376 for example) as

$$f_W(w;\theta,d) = \begin{cases} \frac{Q(e^w-Q_A)}{Q_A-e^wQ_A+Q(e^w-Q_A)}\frac{Q_A}{e^w(Q_A-1)}f_Y(\frac{e^w-Q_A}{e^w(1-Q_A)};\theta), & w \geq 0; \\ \frac{1-Q_Ae^w}{1-Q_Ae^w+Q(e^w-1)}\frac{1}{e^w(1-Q_A)}f_Y(\frac{e^w-1}{e^w(1-Q_A)};\theta), & w < 0, \end{cases}$$

for $Q_A > 1$ and

$$f_W(w;\theta,d) = \begin{cases} \frac{1-e^wQ_A}{1-e^wQ_A+Q(e^w-1)}\frac{1}{e^w(1-Q_A)}f_Y(\frac{e^w-1}{e^w(1-Q_A)};\theta), & w \geq 0; \\ \frac{Q(e^w-Q_A)}{Q_A-e^wQ_A+R(e^w-Q_A)}\frac{Q_A}{e^w(Q_A-1)}f_Y(\frac{e^w-Q_A}{e^w(1-Q_A)};\theta), & w < 0, \end{cases}$$

for $Q_A < 1$.

(B) Probability Distribution of W for Testing Relative Risk

If $R_A \geq 1$

$$W_i = \begin{cases} \log(R_A), & \text{if the } i\text{th patient dies and } Y_i < 1/R_A; \\ \log(1/Y_i), & \text{if the } i\text{th patient dies and } Y_i \geq 1/R_A; \\ \log\left(\frac{1-R_AY_i}{1-Y_i}\right), & \text{if the } i\text{th patient survives and } Y_i < 1/R_A; \\ -\infty, & \text{if the } i\text{th patient survives and } Y_i \geq 1/R_A. \end{cases}$$

For $R \geq 1$, the probability distribution of W can be derived as

$$f_W(w;\theta,d) = \begin{cases} 0, & \text{if } w < 0 \text{ and } w \leq \log\left[\frac{R-R_A}{R-1}\right]; \\ \left[1 - R(\frac{1-e^w}{R_A-e^w})\right] \\ \frac{e^w(R_A-1)}{(R_A-e^w)^2}f_{Y_i}(\frac{1-e^w}{R_A-e^w};\theta), & \text{if } w < 0 \text{ and } w > \log\left[\frac{R-R_A}{R-1}\right]; \\ \frac{1}{e^w}f_{Y_i}(\frac{1}{e^w};\theta), & \text{if } 0 \leq w < \log(R_A) \text{ and} \\ & w \leq \log(R); \\ R(\frac{1}{e^w})^2 f_{Y_i}(\frac{1}{e^w};\theta), & \text{if } 0 \leq w < \log(R_A) \text{ and,} \\ & w > \log(R), \end{cases}$$

$$P(W_i = \log(R_A);\theta,d) = \begin{cases} \int_0^{\frac{1}{R_A}} Ryf_{Y_i}(y;\theta)dy, & \text{if } R_A > R; \\ \int_0^{\frac{1}{R}} Ryf_{Y_i}(y;\theta)dy \\ + \int_{\frac{1}{R}}^{\frac{1}{R_A}} f_{Y_i}(y;\theta)dy, & \text{if } R_A \le R, \end{cases}$$

$$P(W_i = -\infty;\theta,d) = \begin{cases} 0, & \text{if } R_A \le R; \\ \int_{\frac{1}{R_A}}^{\frac{1}{R}} (1-Ry)f_{Y_i}(y;\theta)dy, & \text{if } R_A > R. \end{cases}$$

For $R < 1$, the pdf of W can be derived as

$$f_W(w;\theta,d) = \begin{cases} \left[(1 - R(\frac{1-e^w}{R_A-e^w})\right] \\ f_{Y_i}(\frac{1-e^w}{R_A-e^w};\theta)\frac{e^w(R_A-1)}{(R_A-e^w)^2}, & \text{if } w < 0; \\ R(\frac{1}{e^w})^2 f_{Y_i}(\frac{1}{e^w};\theta), & \text{if } 0 \le w < \log(R_A); \\ 0, & \text{if } w \ge \log(R_A), \end{cases}$$

$$P(W_i = \log R_A;\theta,d) = \int_0^{\frac{1}{R_A}} Ryf_{Y_i}(y;\theta)dy,$$

$$P(W_i = -\infty;\theta,d) = \int_{\frac{1}{R_A}}^{1} (1-Ry)f_{Y_i}(y;\theta)dy.$$

If $R_A < 1$,

$$W_i = \begin{cases} \log(R_A), & \text{if the } i\text{th patient dies;} \\ \log\left(\frac{1-R_AY_i}{1-Y_i}\right), & \text{if the } i\text{th patient survives.} \end{cases}$$

For $R \le 1$,

$$f_W(w;\theta,d) = \begin{cases} (1 - R\frac{e^w-1}{e^w-R_A})\frac{e^w(1-R_A)}{(e^w-R_A)^2} f_{Y_i}(\frac{e^w-1}{e^w-R_A};\theta)dy, & \text{if } w \ge 0; \\ 0, & \text{if } w < 0, \end{cases}$$

$$P(W_i = \log R_A;\theta,d) = \int_0^1 Ryf_{Y_i}(y;\theta)dy.$$

For $R > 1$,

$$f_W(w;\theta,d) = \begin{cases} (1 - R\frac{e^w-1}{e^w-R_A})\frac{e^w(1-R_A)}{(e^w-R_A)^2} \\ \qquad f_{Y_i}(\frac{e^w-1}{e^w-R_A};\theta)dy, & \text{if } 0 \le w < \log[\frac{(R-R_A)}{(R-1)}]\,; \\ 0, & \text{else,} \end{cases}$$

$$P(W_i = \log(R_A);\theta,d) = \int_0^{\frac{1}{R}} Ryf_{Y_i}(y;\theta)dy + \int_{\frac{1}{R}}^1 f_{Y_i}(y;\theta)dy.$$

References

Benneyan, J. C., & Borgman, A. D. (2003). Risk-adjusted sequential probability ratio tests and longitudinal surveillance methods. *International Journal for Quality in Health Care, 15*, 5–6.

Biswas, P., & Kalbfleisch, J. D. (2008). A Risk-adjusted CUSUM in continuous time based on the cox model. *Statistics in Medicine, 27*, 3382–3406.

Blackstone, E. H. (2004). Monitoring surgical performance. *The Journal of Thoracic and Cardiovascular Surgery, 128*, 807–810.

Chang, T. C. (2008). Cumulative sum schemes for surgical performance monitoring. *Journal of the Royal Statistical Society Series A, 171*, 407–432.

Coory, M., Duckett, S., & Sketcher-Baker K. (2008). Using control charts to monitor quality of hospital care with administrative data. *International Journal for Quality in Health Care, 20*, 31–39.

Dorsch, M. F., Lawrance, R. A., Sapsford, R. J., Oldham, J., Greenwood, D. C., Jackson, B. M., Morrell, C., Ball, S. G., Robinson, M. B., Hall, A. S., & The EMMACE Study Group. (2001). A simple benchmark for evaluating quality of care of patients following acute myocardial infarction. *Heart, 86*, 150–154.

Gan, F. F., & Tan, T. (2010) Risk-adjusted number-between failures charting procedures for monitoring a patient care process for acute myocardial infarctions. *Health Care Management Science, 13*, 222–233.

Gandy, A., Kvaloy, J. T., Bottle, A., & Zhou, F. (2010) Risk-adjusted monitoring of time to event. *Biometrika, 97*, 375–388.

Grigg, O., & Farewell, V. (2004). An overview of risk-adjusted charts. *Journal of the Royal Statistical Society, Series A, 167*, 523–539.

Grigg, O., & Spiegelhalter, D. J. (2007). A simple risk-adjusted exponentially weighted moving average. *Journal of the American Statistical Association, 102*, 140–152.

Grigg, O., Farewell, V., & Spiegelhalter, D. J. (2003). Use of risk-adjusted CUSUM and RSPRT charts for monitoring in medical contexts. *Statistical Methods in Medical Research, 12*, 147–170.

Grunkemeier, G. L., Wu, Y. X., & Furnary, A.P. (2003). Cumulative sum techniques for assessing surgical results. *Annals of Thoracic Surgery, 76*, 663–667.

Gustafson, T. L. (2000). Practical risk-adjusted quality control charts for infection control. *American Journal of Infection Control, 28*, 406–414.

Hackbusch, W. (1995). *Integral equations: Theory and numerical treatment* (International series of numerical mathematics, Vol. 120). Basel/Boston/Berlin: Birkhäuser.

Knoth, S. (2005). Accurate ARL computation for EWMA-S^2 control charts. *Statistics and Computing, 15*, 341–352.

Knoth, S. (2007). Accurate ARL computation for EWMA control charts monitoring normal mean and variance simultaneously. *Sequential Analysis, 26*, 251–263.

Loke, C. K., & Gan, F. F. (2012) Joint monitoring scheme for clinical failures and predisposed risks. *Quality Technology and Quantitative Management, 9*, 3–21.

Lovegrove, J., Valencia, O., Treasure, T., Sherlaw-Johnson, C., & Gallivan, S. (1997). Monitoring the results of cardiac surgery by variable-life-adjusted display. *The Lancet, 350*, 1128–1130.

Lovegrove, J., Sherlaw-Johnson, C., Valencia, O., & Gallivan, S. (1999). Monitoring the performance of cardiac surgeons. *Journal of the Operational Research Society, 50*, 685–689.

Manche, A., & Schembri, K. (2005). Evaluating a trainee's progress in surgical dexterity. *Malta Medical Journal, 17*, 49–53.

Moustakides, G. V. (1986). Optimal stopping times for detecting changes in distributions. *The Annals of Statistics, 14*, 1379–1387.

Novick, R. J., Fox, S. A., Stitt, L. W., & Forbes, T. L. (2006). Direct comparison of risk-adjusted and non-risk-adjusted CUSUM analysis of coronary artery bypass surgery outcomes. *The Journal of Thoracic and Cardiovascular Surgery, 132*, 386–391.

Page, E. S. (1954). Continuous inspection schemes. *Biometrika, 41*, 100–115.

Parsonnet, V., Dean, D., & Bernstein, A. D. (1989). A method of uniform stratification of risks for evaluating the results of surgery in acquired adult heart disease. *Circulation, 779* (Suppl 1), 1–12.

Poloniecki, J., Valencia, O., & Littlejohns, P. (1998). Cumulative risk adjusted mortality chart for detecting changes in death rate: Observational study of heart surgery. *British Medical Journal, 316*, 1697–1700.

Reynolds, M. R., Jr., & Stoumbos, Z. G. (1999). A CUSUM chart for monitoring a proportion when inspecting continuously. *Journal of Quality Technology, 31*, 87–108.

Rogers, C. A., Reeves, B. C., Caputo, M., Ganesh, J. S., Bosner, R. S., & Angelini, G. D. (2004). Control chart methods for monitoring cardiac surgical performance and their interpretation. *The Journal of Thoracic and Cardiovascular Surgery, 128*, 811–819.

Roques F., Nashef S. A., Michel P., Gauducheau E., de Vincentiis C., Baudet E., Cortina J., David M., Faichney A., Gabrielle F., Gams E., Harjula A., Jones M. T., Pintor P. P., Salamon R., & Thulin L. (1999). Risk factors and outcome in European cardiac surgery: Analysis of the EuroSCORE multinational database of 19030 patients. *European Journal of Cardio-Thoracic Surgery, 15*(6), 816–22; discussion 822–823.

Ross, S. (2006). *A first course in probability* (7th ed.). New Jersey: Pearson Prentice Hall.

Sherlaw-Johnson, C. (2005). A method for detecting runs of good and bad clinical outcomes on variable life-adjusted displays (VLAD) charts. *Health Care Management Science, 8*, 61–65.

Sismanidis, C., Bland, M., & Poloniecki, J. (2003). Properties of the cumulative risk-adjusted mortality (CRAM) chart, including the number of deaths before a doubling of the death rate is detected. *Medical Decision Making, 23*, 242–251.

Spiegelhalter, D. J. (2004). Monitoring clinical performance: A commentary. *The Journal of Thoracic and Cardiovascular Surgery, 128*, 820–822.

Spiegelhalter, D., Grigg, O., Kinsman, R., & Treasure, T. (2003). Risk-adjusted sequential probability ratio tests: Applications to Bristol, shipman and adult cardiac surgery. *International Journal for Quality in Health Care, 15*, 7–13.

Steiner, S. H., & Jones, M. (2010). Risk-adjusted survival time monitoring with an updating exponentially weighted moving average (EWMA) control chart. *Statistics in Medicine, 29*, 444–454.

Steiner, S. H., Cook, R. J., Farewell, V. T., & Treasure, T. (2000). Monitoring surgical performances using risk-adjusted cumulative sum charts. *Biostatistics, 1*, 441–452.

Treasure, T., Valencia, O., Sherlaw-Johnson, C., & Gallivan, S. (2002). Surgical performance measurement. *Health Care Management Science, 5*, 243–248.

Treasure, T., Gallivan, S., & Sherlaw-Johnson, C. (2004). Monitoring cardiac surgical performance: A commentary. *The Journal of Thoracic and Cardiovascular Surgery, 128*, 823–825.

Wald, A. (1947). *Sequential analysis*. New York: John Wiley.

Woodall, W. H. (2006). The use of control charts in health-care and public-health surveillance (with discussion). *Journal of Quality Technology, 38*, 89–134.

Bayesian Sampling Plans for Inspection by Variables

Peter-Th. Wilrich

Abstract We select an appropriate sampling plan from a sampling inspection system for inspection by variables, i.e. ISO 3951, and base the lot acceptance decision and the adaptation of the sampling plan directly on the a posteriori distribution of the fraction of nonconforming items in the lots and especially the a posteriori estimate of the probability of the fraction of nonconforming items in the lot being larger than the acceptance quality limit AQL. We do not assume a prior distribution of the fraction nonconforming in the lots because the production process does not directly generate fractions of nonconforming items but items with a quantitative characteristic assumed to be normally distributed with parameters μ (lot mean) and σ^2 (within-lot variance) varying in time, respectively from lot to lot. The process curve, i.e. the two-dimensional distribution of the lot means and the within-lots variances is assumed to be Normal-scaled-inverse-chi-squared. In our hierarchical Bayes model we estimate the parameters of the process curve directly by exponentially weighted means and variances of the sample averages and the sample variances of the already inspected lots. Switching between tightened, normal and reduced inspection turns out to be more straightforward than with the switching rules of ISO 3951. Furthermore, if the process curve of the variance is stable, it is possible to switch from sampling plans with unknown variance to sampling plans with known variance. We apply the Bartlett test for switching to σ-plans and a CUSUM-s^2-chart for switching back to s-plans.

Keywords Bayesian sampling plans • Inspection by variables • Sampling plans • Variables sampling

---

P.-T. Wilrich (✉)
Institute for Statistics and Econometrics, Freie Universität Berlin, Garystrasse 21, D-14195 Berlin, Germany
e-mail: wilrich@wiwiss.fu-berlin.de

H.-J. Lenz et al. (eds.), *Frontiers in Statistical Quality Control 10*,
Frontiers in Statistical Quality Control 10, DOI 10.1007/978-3-7908-2846-7_16,

1 Introduction

ISO 2859–1 (1999), ISO 3951–1 (2005) and ISO 3951–2 (2005) specify acceptance sampling systems for inspection by attributes (ISO 2859–1 1999) or for inspection by variables (ISO 3951–1 2005, ISO 3951–2 2005) to be used for a continuing series of lots stemming from one and the same production process. Their purpose "is

(1) To induce a supplier through the economic and psychological pressure of lot non-acceptance to maintain a process average at least as good as the specified acceptance quality limit, while at the same time
(2) Providing an upper limit for the risk of the consumer of accepting the occasional poor lot" (ISO 2859–1 1999, p.1).

In order to achieve purpose (2), sampling plans $(n,\ p^\star)$ are provided: an incoming lot is accepted if the estimate of the fraction of nonconforming units in the lot derived from a sample of size n chosen randomly from the lot is not larger than the acceptability constant $p^\star$, otherwise it is rejected.

In order to achieve purpose (1) rules for switching between three inspection severities (normal inspection, reduced inspection, tightened inspection) with different sampling plans $(n,\ p^\star)$ and a fourth inspection stage "discontinuation of inspection" have to be applied.

Users of these ISO sampling systems often complain about two deficiencies:

1. Regardless of whether a lot is rejected or accepted they would like to get a probability of the lot being actually unacceptable, i.e. having a fraction of nonconforming units larger than the acceptance quality limit p_{AQL}. Instead, they only know the probability of the lot being accepted or rejected in relation to its fraction of nonconforming units, i.e. they get probabilities describing the quality of the acceptance sampling procedure and not the quality of the inspected lot.
2. They are aware of switching between the inspection severities but they would like to know what that means in terms of the distribution of the fraction of nonconforming units in the lots. Especially they are interested in estimates of the expectation and of the standard deviation of the distribution of the lot means and the within-lot standard deviations, i.e. the process curve, and in estimates of the probability distribution of the fraction of nonconforming units in the lot.

Wilrich (2010) has proposed a Bayesian approach that answers both questions for the case of sampling by attributes. In this paper we transfer this approach from sampling by attributes to sampling by variables. The lot acceptance decision is directly based on the a posteriori distribution of the fraction of nonconforming units in the lot and especially the a posteriori estimate of the probability of the fraction of nonconforming units in the lot being larger than the (specified) acceptance quality limit p_{AQL}.

2 The Model

Let X be a quality characteristic for which a lower specification limit L and an upper specification limit U are specified. A unit is conforming if its value x of X is not smaller than L and not larger than U, $L \leq X \leq U$; otherwise it is nonconforming. A lot $t; t = 1, 2, \ldots$ represents the units produced by the production process at time t. We assume that the distribution of X in lot t is (approximately) normal with mean μ_t and within-lot standard deviation σ_t. We further assume that X can be measured with negligible error, i.e. with standard deviation no more than 10% of the within-lot standard deviation. Under these assumptions the fraction of nonconforming units in lot t is

$$p_t = \Phi\left(\frac{\mu_t - U}{\sigma_t}\right) + \Phi\left(\frac{L - \mu_t}{\sigma_t}\right) \tag{1}$$

where $\Phi(\cdot)$ denotes the cumulative distribution function of the standardized normal distribution.

We use a hierarchical Bayes model with two levels of random variation (Stange 1977); at the first level we have randomly distributed lot means μ_t and within-lot standard deviations σ_t and hence, fractions p_t of nonconforming units in the lots, and at the second level we have the randomly distributed quality characteristic X of the units of the sample drawn randomly from the lot.

Level 1: We assume that the within-lot variance σ_t^2 is the value of the random variable σ^2 that follows a scaled inverse χ^2–distribution with parameters $\nu_{0,t}$ and $\sigma_{0,t}^2$, i.e. $(\nu_{0,t} - 2)\sigma_{0,t}^2/\sigma^2$ is $\chi^2_{\nu_{0,t}}$-distributed with $\nu_{0,t}$ degrees of freedom. The probability density of σ^2 is

$$f(\sigma^2) = \frac{((\nu_{0,t} - 2)/2)^{\nu_{0,t}/2}}{\Gamma(\nu_{0,t}/2)} \sigma_{0,t}^{\nu_{0,t}} \sigma^{-(\nu_{0,t}+2)} \exp\left(-\frac{\nu_{0,t} - 2}{2}\sigma^2/\sigma_{0,t}^2\right). \tag{2}$$

σ^2 has expectation $E(\sigma^2) = \sigma_{0,t}^2$ and variance $V(\sigma^2) = 2\sigma_{0,t}^4/(\nu_{0,t} - 4)$ for $\nu_{0,t} \geq 5$.

The lot mean μ_t is the value of the random variable μ. The conditional distribution of μ for the given value σ^2 is a normal distribution with expectation $\mu_{0,t}$ and variance $\sigma^2/n_{0,t}$. It has the probability density

$$f(\mu|\sigma^2) = \frac{1}{\sqrt{2\pi}\sigma/\sqrt{n_{0,t}}} \exp\left(-\frac{n_{0,t}}{2}\left(\frac{\mu - \mu_{0,t}}{\sigma}\right)^2\right). \tag{3}$$

The joint distribution of (μ, σ^2), the process curve of (μ, σ^2), has the probability density

$$f(\mu,\sigma^2) = f(\sigma^2)f(\mu|\sigma^2) = \frac{((\nu_{0,t}-2)/2)^{\nu_{0,t}/2}\sqrt{n_{0,t}}}{\sqrt{2\pi}\,\Gamma(\nu_{0,t}/2)}\sigma_{0,t}^{\nu_{0,t}}\sigma^{-(\nu_{0,t}+3)} \times$$

$$\times \exp\left(-\frac{1}{2\sigma^2}\left((\nu_{0,t}-2)\sigma_{0,t}^2 + n_{0,t}(\mu-\mu_{0,t})^2\right)\right). \quad (4)$$

We denote this process curve of (μ,σ^2) as Normal-scaled-inverse-chisqared distribution. In this distribution μ has expectation $E(\mu) = \mu_{0,t}$ and variance $V(\mu) = \sigma_{0,t}^2/n_{0,t}$.

The marginal distribution of μ has the probability density

$$f(\mu) = \frac{((\nu_{0,t}-2)/2)^{\nu_{0,t}/2}\sqrt{n_{0,t}}}{\sqrt{2\pi}\,\Gamma(\nu_{0,t}/2)}\sigma_{0,t}^{\nu_{0,t}}\frac{\Gamma((\nu_{0,t}+1)/2)2^{(\nu_{0,t}+1)/2}}{\left((\nu_{0,t}-2)\sigma_{0,t}^2 + n_0(\mu-\mu_{0,t})^2\right)^{(\nu_{0,t}+1)/2}}. \quad (5)$$

This is a scaled t-distribution because the random variable

$$t_{\nu_{0,t}} = \sqrt{n_{0,t}}\frac{\mu-\mu_{0,t}}{\sigma_{0,t}}\sqrt{\nu_{0,t}/(\nu_{0,t}-2)} \quad (6)$$

follows the $t_{\nu_{0,t}}$-distribution with $\nu_{0,t}$ degrees of freedom.

The joint distribution of (μ,σ^2) has four (unknown) parameters: $\mu_{0,t}$ is the process average, i.e. the expectation of the lot mean μ at time t, $n_{0,t}$ determines the variance of μ at time t, $\sigma_{0,t}^2$ is the expectation of the within-lot variance σ^2 at time t, and $\nu_{0,t}$ determines the variance of σ^2 at time t.

Lot t has the unknown lot mean μ_t and the unknown within-lot variance σ_t^2 and hence, the unknown fraction p_t of nonconforming units according to Eq. (1). The expected fraction of nonconforming units in lot t is

$$p_{expected,t} = \Phi\left(\frac{\mu_{0,t}-U}{\sigma_{0,t}}\right) + \Phi\left(\frac{L-\mu_{0,t}}{\sigma_{0,t}}\right) \quad (7)$$

Given p_{AQL} as the borderline between acceptable and unacceptable lots the probability of a lot being acceptable is

$$P(p_{expected,t} \le p_{AQL}) = \int_{p(\mu,\sigma^2)\le p_{AQL}} f(\mu,\sigma^2)\,d\mu\,d\sigma^2 \quad (8)$$

that is unknown because $f(\mu,\sigma^2)$ according to Eq. (4) is unknown.

The reason for the choice of the Normal-scaled-inverse-chisqared distribution as process curve of (μ,σ^2) will be discussed in the following section. It should be noted that in this distribution μ and σ^2 are uncorrelated but not independent. The conditional distribution of μ given σ^2 has an expectation that is not related to the given σ^2 whereas the variance of μ is related to the given σ^2 by $V(\mu) = \sigma^2/n_{0,t}$. One could argue that this is unrealistic, however, it makes sense to assume that in case the within-lot variance σ_t^2 is small the variance $V(\mu)$ of the lot mean is also small, and vice versa.

Level 2: We assume that the quality characteristic X of a unit of lot t follows the normal distribution with expectation μ_t and variance σ_t^2. X has the probability density

$$f(x) = \frac{1}{\sqrt{2\pi}\sigma_t} \exp\left(-\frac{1}{2}\left(\frac{x-\mu_t}{\sigma_t}\right)^2\right). \tag{9}$$

Level 1 and level 2 combined: Under the distributional assumptions for level 1 and level 2 the joint distribution of the sample mean $\bar{x}_t$ and the sample variance s_t^2 in a sample of size n, randomly chosen from lot t that has been randomly chosen from the Normal-scaled-inverse-chisqared process curve, has the probability density

$$f(\bar{x}_t, s_t^2) = \frac{1}{\sqrt{\pi}} \sqrt{\frac{n_{0,t}n}{n_{0,t}+n}} \frac{\Gamma((n_{0,t}+n)/2)}{\Gamma(\nu_{0,t}/2)\Gamma(\nu/2)} \nu^{\nu/2} \left((\nu_{0,t}-2)\sigma_{0,t}^2\right)^{\nu_{0,t}/2} \times$$

$$\times \frac{(s_t^2)^{\nu/2-1}}{\left((\nu_{0,t}-2)\sigma_{0,t}^2 + \nu s_t^2 + n_{0,t}n(\bar{x}_t - \mu_{0,t})^2/(n_{0,t}+n)\right)^{(\nu_{0,t}+n)/2}} \tag{10}$$

with $\nu = n - 1$. In this distribution expectation and variance of $\bar{x}_t$ and s_t^2 are

$$E(\bar{x}_t) = \mu_{0,t} \qquad E(s_t^2) = \sigma_{0,t}^2$$
$$V(\bar{x}_t) = \left(\frac{1}{n_{0,t}} + \frac{1}{n}\right)\sigma_{0,t}^2 \qquad V(s_t^2) = \frac{2(\nu+\nu_{0,t}-2)}{\nu(\nu_{0,t}-4)}\sigma_{0,t}^4 \tag{11}$$

3 The Estimation of the Parameters of the Process Curve and the Lot Acceptance Decision

Having reached lot k we estimate the parameters of the process curve directly (using an empirical Bayes approach) from the sample means $\bar{x}_t$ and the sample variances s_t^2 of the samples taken from the lots $t = 1, 2, \ldots, k$.

Under the assumption that the parameters of the process curve do not change in time ($\mu_{0,t} = \mu_0, n_{0,t} = n_0, \sigma_{0,t} = \sigma_0, \nu_{0,t} = \nu_0$ for $t - 1, 2, \ldots, k$), means and variances of the k sample means $\bar{x}_t$ and the k sample variances s_t^2 could be used as moment estimators of the respective theoretical parameters and hence, estimators of the parameters of the process curve could be derived. However, the assumption of a constant process curve is contradictory to the purpose of the sampling procedure to detect a possible alteration of the process curve. Hence, we have to estimate the parameters of the process curve more based on the most recent observations (by giving them a larger weight) than on the more elderly observations. To do this, we use exponentially weighted moving averages of the sample means $\bar{x}_t$ and the sample variances s_t^2,

$$\begin{aligned} \hat{\mu}_k &= (1-\gamma)\hat{\mu}_{k-1} + \gamma\hat{x}_k \\ \hat{\sigma}_k^2 &= (1-\gamma)\hat{\sigma}_{k-1}^2 + \gamma s_k^2 \end{aligned} \qquad k = 2, 3, \ldots \quad (12)$$

and start the estimation with

$$\hat{\mu}_1 = \bar{x}_1; \qquad \hat{\sigma}_1^2 = s_1^2. \qquad (13)$$

In addition, we need the exponentially weighted moving averages of the variances of the sample means and of the variances of the sample variances,

$$\begin{aligned} \hat{\sigma}_{\bar{x},k}^2 &= (1-\gamma)\hat{\sigma}_{\bar{x},k-1}^2 + \gamma(\bar{x}_k - \hat{\mu}_k)^2 \\ \hat{\sigma}_{s^2,k}^2 &= (1-\gamma)\hat{\sigma}_{s^2,k-1}^2 + \gamma(s_k^2 - \hat{\sigma}_k^2)^2 \end{aligned} \qquad k = 3, 4, \ldots \quad (14)$$

and start the estimation with

$$\hat{\sigma}_{\bar{x},2}^2 = (\bar{x}_2 - \bar{x}_1)^2/2; \qquad \hat{\sigma}_{s^2,2}^2 = (s_2^2 - s_1^2)^2/2. \qquad (15)$$

γ can be chosen arbitrarily between 0 and 1. For $\gamma = 1$ the estimates are equal to the respective values of the present sample. For smaller γ, the exponentially weighted moving averages are based more on past sample results. We choose $\gamma = \max(1/i, 0.1)$. This choice assures that in the beginning, for lots $i = 1, 2, \ldots, 1/\gamma$, averages are calculated instead of moving averages in order not to overweight past sample results.

An assumption underlying the application of sampling systems is a production process with an acceptable process curve, i.e. one that causes very few lot rejections. Hence, we base the updating procedure for the estimation of the parameters of the process curve only on accepted lots. Lots being rejected are interpreted as outlier lots with an abnormal large fraction of nonconforming units that are not generated by the process curve but by an irregular situation. On the other hand, frequently occurring rejections of lots might signalize a deterioration of the process curve. Hence, if more than two consecutive lots are rejected, the updating procedure starts to include all consecutively following rejected lots. However, in practice one would rather stop the application of the sampling system in such a situation and investigate the reason for the deterioration of the process curve.

The updated exponentially moving averages $\hat{\mu}_k, \hat{\sigma}_k^2, \hat{\sigma}_{\bar{x},k}^2, \hat{\sigma}_{s^2,k}^2$ are estimators of the parameters $\mu_{0,k}, \sigma_{0,k}^2, V(\bar{x}_k), V(s_k^2)$ according to Eq. (11). If these estimators are plugged into formulae Eq. (11) estimators

$$\begin{aligned} \hat{n}_{0,k} &= \max\left(1, \frac{1}{\hat{\sigma}_{\bar{x},k}^2/\hat{\sigma}_k^2 - 1/n}\right) \\ \hat{\nu}_{0,k} &= \frac{4\nu K + 2\nu - 4}{\nu K - 2} \end{aligned} \qquad (16)$$

with $K = \hat{\sigma}_{s^2,k}^2/(\hat{\sigma}_k^2)^2$ are derived.

The Normal-scaled-inverse-chi-squared distribution Eq. (4) with parameters equal to the estimates Eqs. (12) and (16) obtained from lots $t = 1, 2, \ldots, k$, is the estimate of the process curve at lot k and is used as the a priori distribution

$$f_{prior}(\mu_{k+1}, \sigma_{k+1}^2; \hat{\mu}_k, \hat{\sigma}_k^2, \hat{n}_{0,k}, \hat{\nu}_{0,k}) =$$

$$= \frac{((\hat{\nu}_{0,k} - 2)/2)^{\hat{\nu}_{0,k}/2} \sqrt{\hat{n}_{0,k}}}{\sqrt{2\pi} \Gamma(\hat{\nu}_{0,k}/2)} \hat{\sigma}_k^{\hat{\nu}_{0,k}} \sigma_{k+1}^{-(\hat{\nu}_{0,k}+3)} \times$$

$$\times \exp\left(-\frac{1}{2\sigma_{k+1}^2} \left((\hat{\nu}_{0,k} - 2)\hat{\sigma}_k^2 + \hat{n}_{0,k}(\mu_{k+1} - \hat{\mu}_k)^2 \right) \right) \tag{17}$$

for the Bayesian analysis of sample $k + 1$ from lot $k + 1$. The likelihood function of $(\mu_{k+1}, \sigma_{k+1}^2)$ under the normal distribution is

$$l(\mu_{k+1}, \sigma_{k+1}^2; \bar{x}_{k+1}, s_{k+1}^2) =$$

$$= \frac{1}{(2\pi\sigma_{k+1}^2)^{n/2}} \exp\left(-\frac{1}{2\sigma_{k+1}^2} \left((n-1)s_{k+1}^2 - (\bar{x}_{k+1} - \mu_{k+1})^2 \right) \right). \tag{18}$$

The probability density of the a posteriori distribution of $(\mu_{k+1}, \sigma_{k+1}^2)$ in lot $k + 1$ is

$$f_{post}(\mu_{k+1}, \sigma_{k+1}^2; \hat{\mu}_k, \hat{\sigma}_k^2, \hat{n}_{0,k}, \hat{\nu}_{0,k}) \sim$$

$$\sim l(\mu_{k+1}, \sigma_{k+1}^2; \bar{x}_{k+1}, s_{k+1}^2) \cdot f_{prior}(\mu_{k+1}, \sigma_{k+1}^2; \hat{\mu}_k, \hat{\sigma}_k^2, \hat{n}_{0,k}, \hat{\nu}_{0,k})$$

$$\sim \sigma_{k+1}^{-(\hat{\nu}_{0,k}+3+n)} \exp\left(-\frac{C}{2\sigma_{k+1}^2} \right) \tag{19}$$

where

$$\begin{aligned}
C &= (\hat{\nu}_{0,k} - 2)\hat{\sigma}_k^2 + (n-1)s_{k+1}^2 + \hat{n}_{0,k}(\mu_{k+1} - \hat{\mu}_k)^2 + (\bar{x}_{k+1} - \mu_{k+1})^2 \\
&= (\hat{\nu}_{0,k} - 2)\hat{\sigma}_k^2 + (n-1)s_{k+1}^2 + \frac{(\bar{x}_{k+1} - \hat{\mu}_k)^2}{1/\hat{n}_{0,k} + 1/n} \\
&\quad + (\hat{n}_{0,k} + n)(\mu_{k+1} - \mu_{post,k+1})^2 \\
&= (\hat{\nu}_{0,k} + n - 2)\hat{\sigma}_{post,k+1}^2 + (\hat{n}_{0,k} + n)(\mu_{k+1} - \mu_{post,k+1})^2 \\
&= (\nu_{0,post,k+1} - 2)\sigma_{post,k+1}^2 + n_{0,post,k+1}(\mu_{k+1} - \mu_{post,k+1})^2
\end{aligned} \tag{20}$$

with

$$\begin{aligned}
\mu_{post,k+1} &= \frac{\hat{n}_{0,k}\hat{\mu}_k + n\bar{x}_{k+1}}{\hat{n}_{0,k} + n} \\
n_{0,post,k+1} &= \hat{n}_{0,k} + n
\end{aligned}$$

$$\sigma^2_{post,k+1} = \frac{(\hat{\nu}_{0,k} - 2)\hat{\sigma}_k^2 + (n-1)s_{k+1}^2 + \frac{(\bar{x}_{k+1}-\hat{\mu}_k)^2}{1/\hat{n}_{0,k}+1/n}}{\hat{\nu}_{0,k} - 2 + n}$$

$$\nu_{0,post,k+1} = \hat{\nu}_{0,k} + n \tag{21}$$

Apparently, the probability density of the a posteriori distribution according to Eqs. (19), (20) and (21) is of the same type as the probability density of the a priori distribution according to Eq. (17): Eq. (19) is derived from Eq. (17) if $\hat{\mu}_k, \hat{n}_{0,k}, \hat{\sigma}_k^2, \hat{\nu}_{0,k}$ are substituted by $\mu_{post,k+1}, n_{0,post,k+1}, \sigma^2_{post,k+1}, \nu_{0,post,k+1}$, respectively. This follows from the fact that the Normal-scaled-inverse-chisqared distribution of (μ, σ^2) and the Normal-chi-squared distribution of $(\bar{x}, s^2)$ are conjugated – and this was the reason for the choice of the model. Since we know the type of the a posteriori distribution, we only need to determine its parameters according to Eq. (21).

The a priori estimate of the fraction p_{k+1} of nonconforming units in lot $k+1$ is

$$\hat{p}_{k+1,prior} = \Phi\left(\frac{\hat{\mu}_k - U}{\hat{\sigma}_k}\right) + \Phi\left(\frac{L - \hat{\mu}_k}{\hat{\sigma}_k}\right) \tag{22}$$

and the a posteriori estimate of the fraction p_{k+1} of nonconforming units in lot $k+1$ is

$$\hat{p}_{k+1,post} = \Phi\left(\frac{\mu_{post,k+1} - U}{\sigma_{post,k+1}}\right) + \Phi\left(\frac{L - \mu_{post,k+1}}{\sigma_{post,k+1}}\right) \tag{23}$$

The probability of lot $k+1$ having a fraction p_{k+1} of nonconforming units smaller than p_{AQL} is a priori

$$P_{prior}(p_{k+1} \le p_{AQL}) = \int\limits_{\substack{p(\mu_{k+1}, \sigma^2_{k+1}) \\ \le p_{AQL}}} f_{prior}(\mu_{k+1}, \sigma^2_{k+1})\, d\mu_{k+1}\, d\sigma^2_{k+1} \tag{24}$$

and a posteriori

$$P_{post}(p_{k+1} \le p_{AQL}) = \int\limits_{\substack{p(\mu_{k+1}, \sigma^2_{k+1}) \\ \le p_{AQL}}} f_{post}(\mu_{k+1}, \sigma^2_{k+1})\, d\mu_{k+1}\, d\sigma^2_{k+1} \tag{25}$$

The latter one is used for the acceptance decision. In the classical approach to sampling inspection we accept lot $k+1$ if the MVUE estimator $\hat{p}_{k+1}$ of the fraction of nonconforming units in the lot (Bowker et al. 1952) is not larger than the acceptability constant $p^\star$ of the sampling plan. In our Bayes approach we accept lot $k+1$ if the a posteriori probability $P_{post}(p_{k+1} \le p_{AQL})$ for lot $k+1$ having a fraction p_{k+1} of nonconforming units smaller than p_{AQL}, is not smaller than a predetermined probability limit A, $P_{post}(p_{k+1} \le p_{AQL}) \ge A$, or equivalently, if $P_{post}(p_{k+1} > p_{AQL}) = 1 - P_{post}(p_{k+1} \le p_{AQL}) \le 1 - A = R$. We reject it otherwise. The sampling plan now consists of the sample size n and the probability limit R. A larger value of R implies a larger probability of accepting a lot, and vice versa. Since we assume that the sampling procedure is intended to be applied to a production process of lots having a process average not larger than p_{AQL} (and in this case all lots should be accepted) we choose $R = 0.8$.

This decision procedure works for all lots except for the first and the second one because in the beginning we do not have estimates of the parameters of the process curve and hence no a priori distributions for these lots. Therefore, we start with a "non-informative" a priori distribution and choose the improper uniform distribution in the range between $-\infty$ and ∞ as "non-informative" for μ_0 and the improper probability density $1/\sigma_0^2$ as "non-informative" for σ_0^2, i.e. $f_{prior}(\mu_0, \sigma_0^2) = 1/\sigma_0^2$.

The acceptance decision rule $P_{post}(p_{k+1} > p_{AQL}) \le R$ includes an intrinsic adaptive procedure: if the a posteriori estimate of the fraction of nonconforming units in lot k is large, the a priori estimate of the fraction of nonconforming units in lot $k+1$ increases and hence, the probability $P_{post}(p_{k+1} > p_{AQL})$ of the fraction of nonconforming units above p_{AQL} also increases so that rejection of lot $k+1$ becomes more likely, and vice versa. In the classical sampling inspection system of ISO 3951 a similar effect is achieved by switching to tightened inspection under which the sample size remains unchanged but the acceptability constant $p^\star$ becomes smaller. This switch is unnecessary in the Bayesian approach.

4 Some Simulated Scenarios

As the basis of an illustration of the performance of the Bayesian sampling plans we adapt the example in Sect. 15.3.2.4 of ISO 3951–2 (2005): The minimum temperature of operation for a certain device is specified as $L = 60$°C and the maximum temperature as $U = 70$°C. The quality requirement for a series of lots of equal size $N = 96$ is expressed by an acceptance quality limit $p_{AQL} = 4.0\%$, i.e. lots are acceptable if they come from a process with an expected fraction of nonconforming units in the lots not larger than 4.0%. For $N = 96$ and $p_{AQL} = 4.0\%$, ISO 3951 – 2, Tables A.2 and G.1, give the sample size $n = 13$ (for normal and tightened inspection), the acceptability constant $p_1^\star = 0.1154$ for normal inspection and $p_0^\star = 0.07537$ for tightened inspection. The switching rules of the Standard are applied, except that switching to reduced inspection is not installed. The Bayesian approach is used in parallel, with $n = 13$, $p_{AQL} = 4.0\%$ and $A = 0.2$ or $R = 0.8$.

We present five different scenarios concerning the production process. $M = 100$ lot means μ_t are generated randomly by drawing random numbers from a normal distribution with process average (expectation of the lot mean) μ_{μ_t} and variance

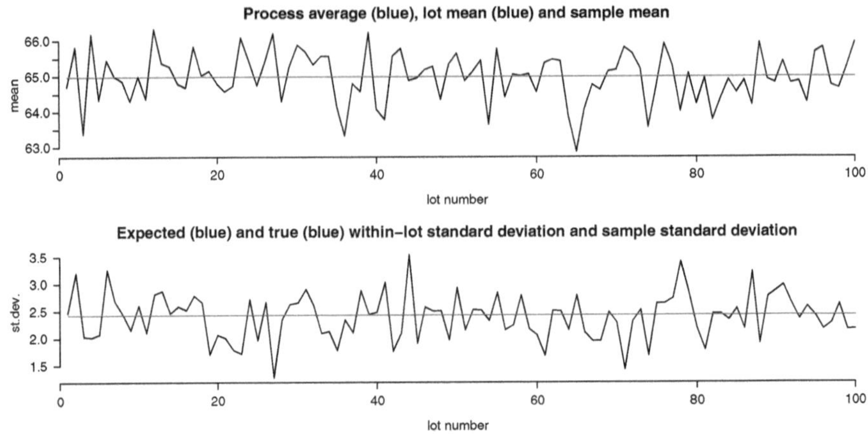

Fig. 1 Lot means $\mu_t = \mu_{\mu_t} = 65$ (*blue*) and sample means $\bar{x}_t$ (*upper graph*); within-lot standard deviations $\sigma_t = \mu_{\sigma_t} = 2.435$ (*blue*) and sample standard deviations s_t (*lower graph*) for scenario 1

$\sigma^2_{\mu_t}$ of the lot means. Within-lot variances σ_t^2 are generated randomly by drawing random numbers from a chi-squared distribution with expectation $\mu_{\sigma_t^2}$ and variance $\sigma^2_{\sigma_t^2}$. A sample of size $n = 13$ is drawn randomly from each lot, and the sample mean $\bar{x}_t$ and the sample variance s_t^2 are determined.

4.1 Scenario 1: Constant Lot Means and Within-Lot Standard Deviations

We set $\mu_{\mu_t} = 65$, $\sigma^2_{\mu_t} = 0$, $\mu_{\sigma_t^2} = 2.435^2$, $\sigma^2_{\sigma_t^2} = 0$ for all $t = 1, 2, \ldots$: The constant process average lies in the middle of the specification interval (L, U) and the expectation of the within-lot variance is chosen so that the expected fraction of nonconforming units in a lot is $0.04 = p_{AQL}$. Since the variance of the lot mean and the variance of the within-lot variance are 0, we have the unrealistic case of all lots having identical means and within-lot variances.

Figure 1 shows the lot means μ_t, all equal to the process average $\mu_{\mu_t} = 65$ (blue), and the sample means $\bar{x}_t$ in the upper graph and the within-lot standard deviations σ_t, all equal to the expected within-lot standard deviation $\mu_{\sigma_t} = 2.435$, and the sample standard deviations s_t in the lower graph. Figure 2 shows in the upper graph the fractions of nonconforming units in the lots (blue), all equal to the expected fraction $p_{AQL} = 0.04$ of nonconforming units in the lots, their MVUE estimates (black), a priori (green) and a posteriori (red) estimates; red circles indicate rejection by ISO 3951, symbol "N" the start of normal inspection and symbol "T" the start of tightened inspection. In the lower graph the expected (blue), a priori (green) and a posteriori (red) probabilities of the fractions of nonconforming

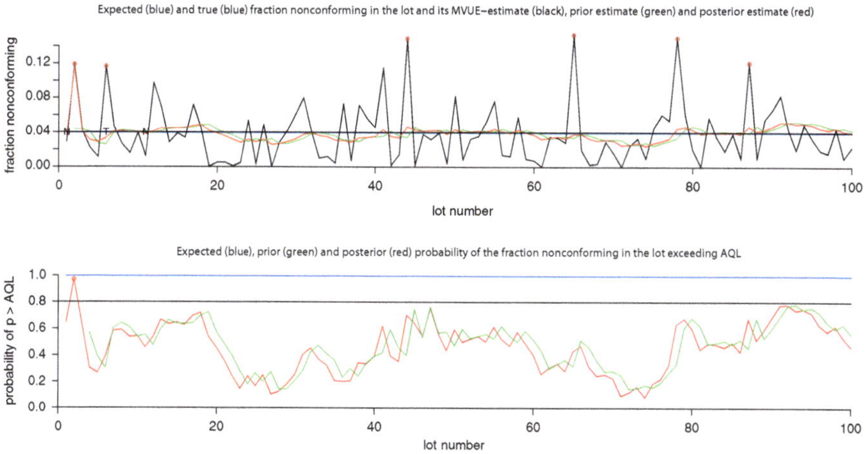

Fig. 2 *Upper graph*: Fractions $p_t = p_{AQL} = 0.04$ of nonconforming units in the lots (*blue*) and their MVUE estimates (*black*), a priori (*green*) and a posteriori (*red*) estimates for scenario 1. *Red circles* indicate rejection by ISO 3951, N = start of normal inspection, T = start of tightened inspection. *Lower graph*: Expected, a priori and a posteriori probabilities of the fractions of nonconforming units in the lots exceeding p_{AQL} for scenario 1; *red circles* indicate rejections by the Bayes method

units in the lots exceeding p_{AQL} are shown; red circles indicate rejections by the Bayes method. Since the process average, expressed as the expected fraction 0.04 of nonconforming units in the lots, is equal to $p_{AQL} = 0.04$, we have the limiting case where all lots should be accepted. In our simulation run, 94% of the 100 lots have been accepted by ISO and 99% by the Bayes method.

4.2 *Scenario 2: Process Curve Constant in Time*

We set $\mu_{\mu_t} = 65, \sigma^2_{\mu_t} = 0.5^2, \mu_{\sigma^2_t} = 2.435^2, \sigma^2_{\sigma^2_t} = 1.0$ for all $t = 1, 2, \ldots$: The constant process average lies in the middle of the specification interval (L, U) and the expectation of the within-lot variance is chosen so that the expected fraction of nonconforming units in a lot is $0.04 = p_{AQL}$. The lot means vary with a standard deviation equal to 0.5 and the within-lot variances vary with a standard deviation equal to 1.0.

Figure 3 corresponds to Fig. 1, however, now the lot means μ_t vary between 63.9 and 66.5 and the within-lot standard deviations σ_t vary between 1.9 and 3.0. Figure 4 corresponds to Fig. 2: the fractions p_t of nonconforming units in the lots vary between 0.014 and 0.113. Fifty-six percent of the lots have a fraction of nonconforming units larger than p_{AQL}. However, since the process average, expressed as the expected fraction 0.04 of nonconforming units in the lots, is equal to $p_{AQL} = 0.04$, we have the limiting case where all lots should be accepted.

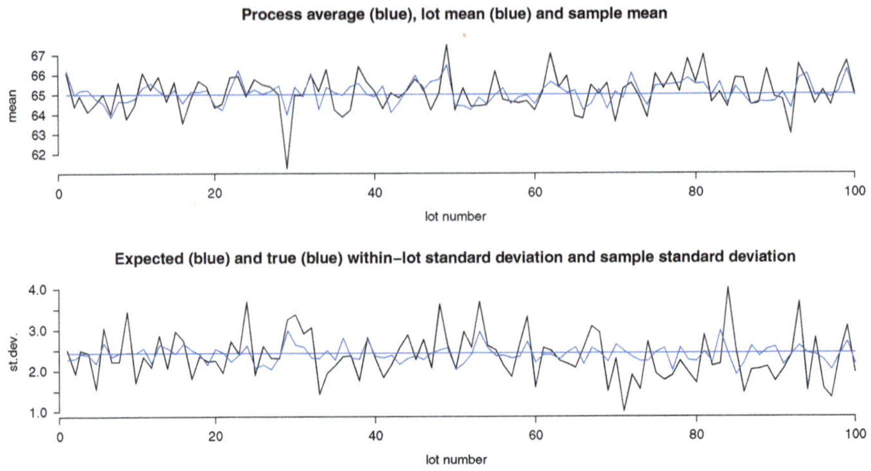

Fig. 3 Process average μ_{μ_t} (*blue*), lot means μ_t (*blue*) and sample means $\bar{x}_t$ (*upper graph*); expected within-lot standard deviation μ_{σ_t} (*blue*), within-lot standard deviations σ_t (*blue*) and sample standard deviations s_t (*lower graph*) for scenario 2

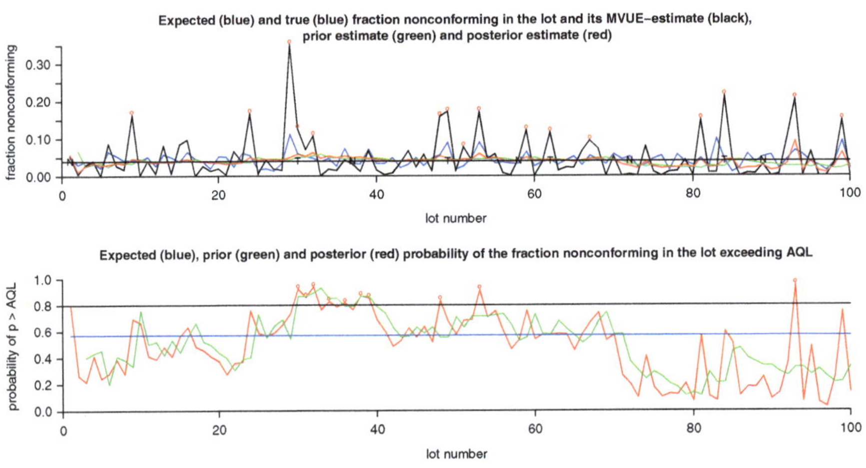

Fig. 4 *Upper graph*: Expected (*blue*) and true (*blue*) fractions of nonconforming units in the lots and their MVUE estimates (*black*), a priori (*green*) and a posteriori (*red*) estimates for scenario 2. *Red circles* indicate rejection by ISO 3951, N = start of normal inspection, T = start of tightened inspection. *Lower graph*: Expected, a priori (*green*) and a posteriori (*red*) probabilities of the fractions of nonconforming units in the lots exceeding p_{AQL} for scenario 2; *red circles* indicate rejections by the Bayes method

In our simulation run, 84% of the 100 lots have been accepted by ISO and 90% by the Bayes method. Figure 5 shows the true parameters $\mu_{\mu_t} = 65, \sigma^2_{\mu_t} = 0.5^2, \mu_{\sigma^2_t} = 2.435^2, \sigma^2_{\sigma^2_t} = 1.0^2$ of the production process and their estimates being updated from lot to lot. The Bayesian analysis for lot 50 is depicted in Fig. 6. The graph in the upper left shows a contour plot of the process curve (blue) and a contour plot of the current a priori distribution (green), in the upper right the likelihood function (black) based on the results of the sample of size $n = 13$, and in the lower left the current a posteriori distribution (red) of (μ, σ). Iso-p curves are given in black. In the lower right the distribution functions of the fraction of nonconforming units in the lots, expected (blue), a priori (green) and a posteriori (red), are displayed. The acceptance quality limit is indicated as a blue vertical line. The intersections of this line with the probability distribution functions give the probabilities $P_{expected}(p_{k+1} \leq p_{AQL})$, $P_{prior}(p_{k+1} \leq p_{AQL})$ and $P_{post}(p_{k+1} \leq p_{AQL})$. Lot 50 is accepted because $P_{post}(p_{k+1} \leq p_{AQL}) = 0.423$ is larger than $A = 0.2$.

4.3 Scenario 3: Sudden Shifts of the Process Average

Scenario 3 starts with $\mu_{\mu_t} = 65, \sigma^2_{\mu_t} = 0.5^2, \mu_{\sigma^2_t} = 2.304^2, \sigma^2_{\sigma^2_t} = 1.0^2$. The expected fraction of nonconforming units in the lots is $p_{expected} = 0.03$. Beginning with lot 34 the process average μ_{μ_t} is shifted to 66 and the expectation of the within-lot variance is set to $\mu_{\sigma^2_t} = 2.354^2$ so that the expected fraction of nonconforming units in the lots is $p_{expected} = 0.05$. Beginning with lot 67 the process average μ_{μ_t} is shifted to 67 and the expectation of the within-lot variance is set to $\mu_{\sigma^2_t} = 2.030^2$ so that the expected fraction of nonconforming units in the lots is now $p_{expected} = 0.07$.

Figure 7 shows in the upper graph in blue color the expected fractions of nonconforming units in the lots ($p_{expected} = 0.03$ for lot 1–33, $p_{expected} = 0.05$ for lot 34–66, $p_{expected} = 0.07$ for lot 67–100), the true fractions of nonconforming units in the lots (blue), their MVUE estimates (black), a priori (green) and a posteriori (red) estimates; red circles indicate rejection by ISO 3951, symbol "N" the start of normal inspection, symbol "T" the start of tightened inspection and symbol "S" inspection stop; however, their simulation is continued at tightened inspection. In the lower graph the expected (blue), a priori (green) and a posteriori (red) probabilities of the fractions of nonconforming units in the lots exceeding p_{AQL} are shown; red circles indicate rejections by the Bayes method. In the first period (lots 1–33) the process average, 0.03, is smaller than p_{AQL}, and hence, all lots should be accepted; in period 2 (lots 34–66) it is 0.05 and in period 3 (lots 67–100) it is 0.07, hence larger than p_{AQL}, and all lots should be rejected. In our simulation run, 1, 2 and 19 lots have been rejected by ISO and 0, 1 and 22 by the Bayes method in periods 1, 2 and 3, respectively. There is no essential difference between the performance of the ISO 3951 and the Bayes method. Figure 8 shows the true parameters $\mu_{\mu_t} = 65, 66, 67, \sigma^2_{\mu_t} = 0.5^2, \mu_{\sigma^2_t} = 2.304^2, 2.354^2, 2.030^2, \sigma^2_{\sigma^2_t} = 1.0^2$ of the production process and their estimates being updated from lot to lot.

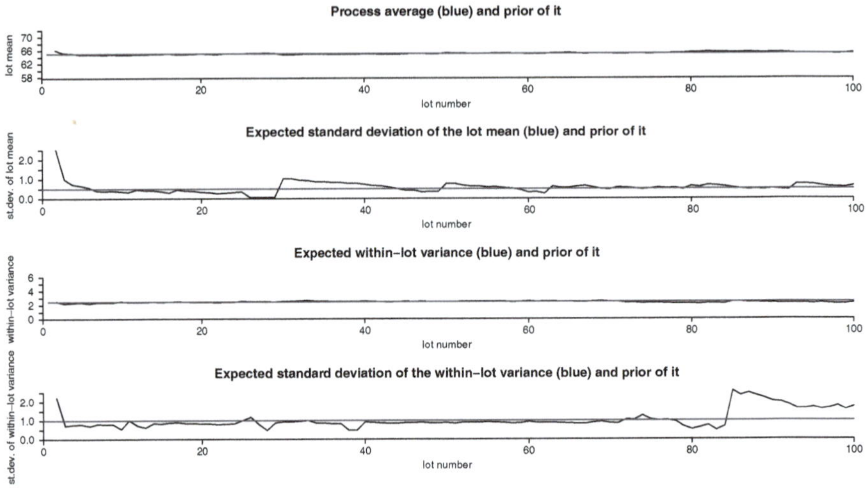

Fig. 5 True parameters $\mu_{\mu_t} = 65, \sigma^2_{\mu_t} = 0.5^2, \mu_{\sigma^2_t} = 2.0^2, \sigma^2_{\sigma^2_t} = 1.0^2$ of the production process and their estimates for scenario 2

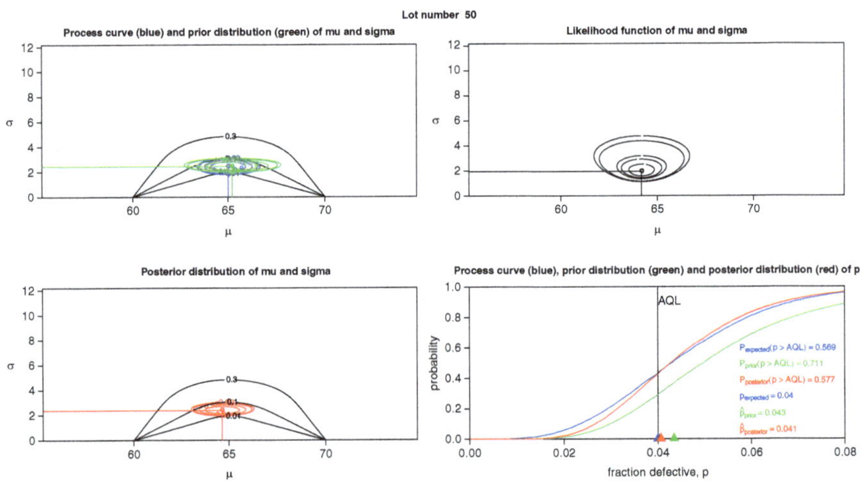

Fig. 6 Bayesian analysis for lot 50 of scenario 2. The graph in the *lower left* shows a contour plot of the process curve (*blue*) and of its (updated) a priori estimate (*green*). The graph in the *upper left* shows a contour plot of the likelihood function. The graph in the *lower right* shows a contour plot of the a posteriori estimate of the process curve (*red*). The graph in the *upper right* shows the process curve, the a priori and the a posteriori distribution function of the fraction of nonconforming units in lot 50. Since $P_{post}(p_{50} \leq p_{AQL}) = 0.423 > A = 0.2$, lot 50 is accepted. The *black contours* are Iso-p lines

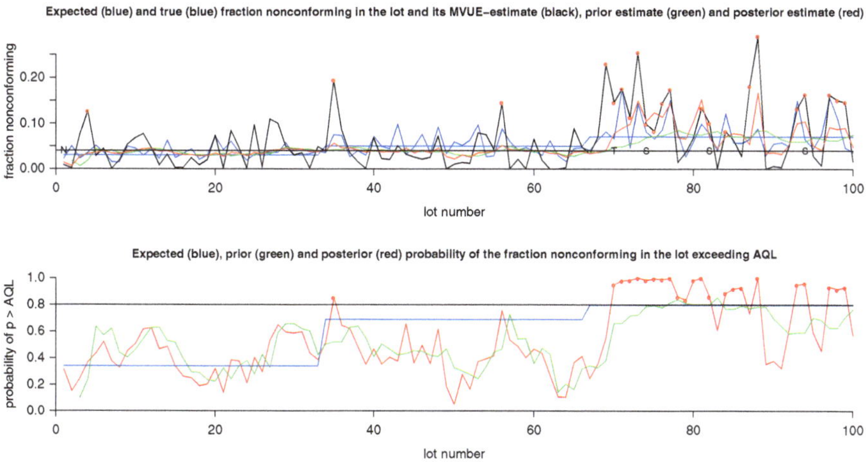

Fig. 7 *Upper graph*: Expected (*blue*) and true (*blue*) fractions of nonconforming units in the lots and their MVUE estimates (*black*), a priori (*green*) and a posteriori (*red*) estimates for scenario 3. *Red circles* indicate rejection by ISO 3951, $N =$ start of normal inspection, $T =$ start of tightened inspection, $S =$ inspection stop. *Lower graph*: Expected (*blue*), a priori (*green*) and a posteriori (*red*) probabilities of the fractions of nonconforming units in the lots exceeding p_{AQL} for scenario 3; *red circles* indicate rejections by the Bayes method

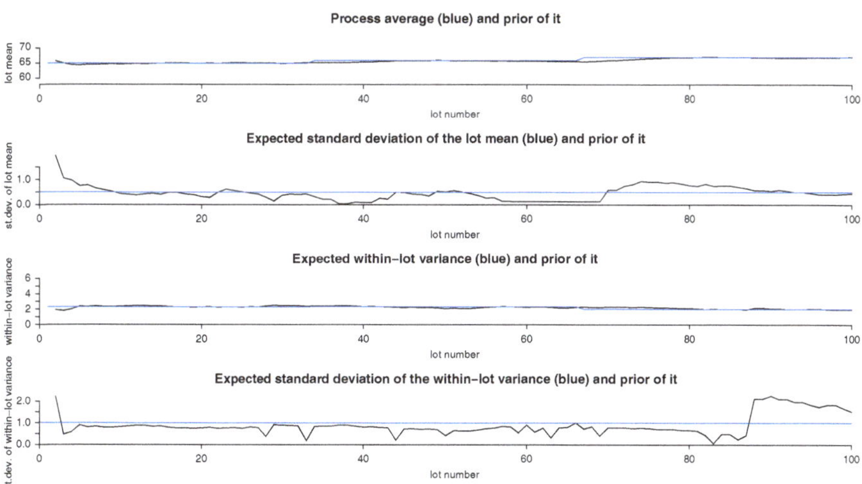

Fig. 8 True parameters $\mu_{\mu_t} = 65, 66, 67$, $\sigma^2_{\mu_t} = 0.5^2$, $\mu_{\sigma^2_t} = 2.304^2, 2.354^2, 2.030^2$, $\sigma^2_{\sigma^2_t} = 1.0^2$ of the production process and their estimates for scenario 3

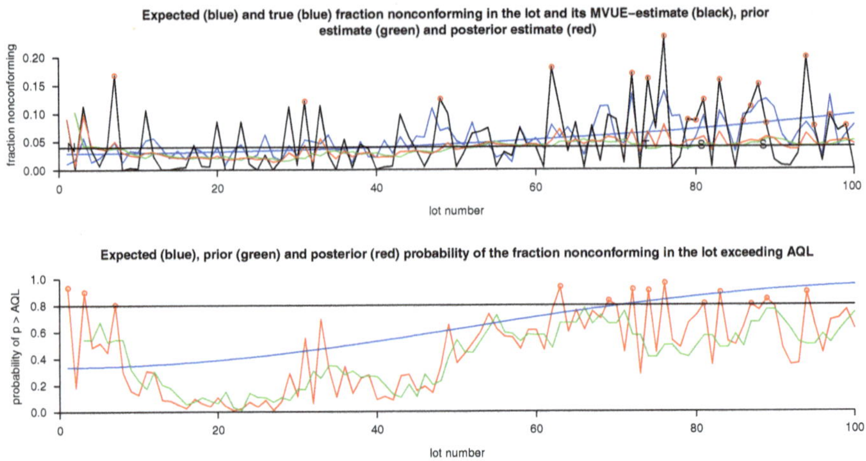

Fig. 9 *Upper graph*: Expected (*blue*) and true (*blue*) fractions of nonconforming units in the lots and their MVUE estimates (*black*), a priori (*green*) and a posteriori (*red*) estimates for scenario 4. *Red circles* indicate rejection by ISO 3951, $N =$ start of normal inspection, $T =$ start of tightened inspection, $S =$ inspection stop. *Lower graph*: Expected (*blue*), a priori (*green*) and a posteriori (*red*) probabilities of the fractions of nonconforming units in the lots exceeding p_{AQL} for scenario 4; *red circles* indicate rejections by the Bayes method

4.4 Scenario 4: Drift of the Process Curve

Scenario 4 starts with the same parameters as scenario 3: $\mu_{\mu_t} = 65, \sigma^2_{\mu_t} = 0.5^2, \mu_{\sigma^2_t} = 2.304^2, \sigma^2_{\sigma^2_t} = 1.0$. Then the process average μ_{μ_t} increases linearly from 65 at lot 1 to 67 at lot 100. This linearly increasing process average produces an expected fraction of nonconforming units in the lots increasing nonlinearly from 0.03 at lot 1 to 0.098 at lot 100; up to lot 40 the process average is smaller than p_{AQL}, from lot 41 to 100 it is larger.

Figure 9 shows in the upper graph in blue color the expected fractions of nonconforming units in the lots ($p_{expected} = 0.03$ for lot 1 to $p_{expected} = 0.098$ at lot 100), the true fractions of nonconforming units in the lots (blue), their MVUE estimates (black), a priori (green) and a posteriori (red) estimates; red circles indicate rejection by ISO 3951, symbol “N” the start of normal inspection, symbol “T” the start of tightened inspection and symbol “S” inspection stop; however, the simulation is continued at tightened inspection. In the lower graph the expected (blue), a priori (green) and a posteriori (red) probabilities of the fractions of nonconforming units in the lots exceeding p_{AQL} are shown; red circles indicate rejections by the Bayes method. Eighty-one percent of the lots have been accepted by ISO and 87% by the Bayes method. Up to lot 40 where all lots should be accepted, both methods reject only very few lots. Again there is no essential difference between the performance of the ISO 3951 and the Bayes method.

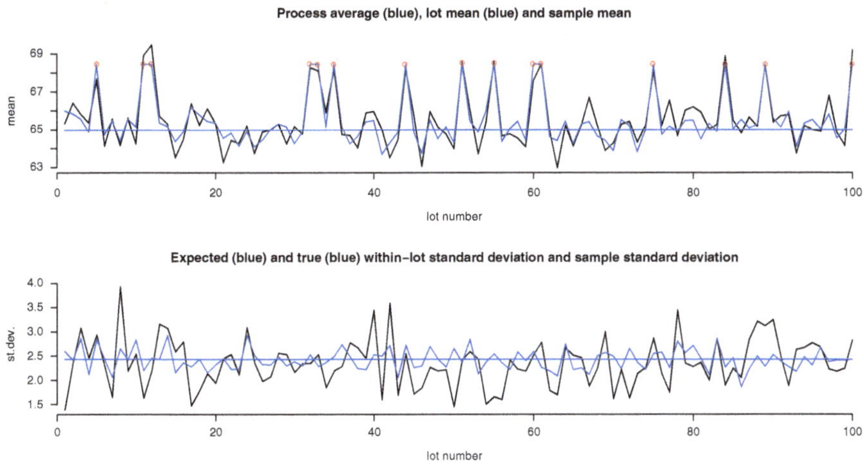

Fig. 10 Process average μ_{μ_t} (*blue*), lot means μ_t (blue) and sample means $\bar{x}_t$ (*upper graph*); expected within-lot standard deviation μ_{σ_t} (*blue*), within-lot standard deviations σ_t (*blue*) and sample standard deviations s_t (*lower graph*) for scenario 5; outlying lot means are indicated by a *red point*

4.5 Scenario 5: A Process with Outlier Lots

The process curve is identical to that of scenario 2: $\mu_{\mu_t} = 65, \sigma^2_{\mu_t} = 0.5^2, \mu_{\sigma_t^2} = 2.435^2, \sigma^2_{\sigma_t^2} = 1.0$ for all $t = 1, 2, \ldots$. However, there is a probability of 0.2 that the lot will be an outlier lot with a lot mean μ_t chosen from a uniform distribution between 67 and 68.

Figure 10 shows the process average $\mu_{\mu_t} = 65$ (blue), the lot means μ_t (blue) and the and sample means $\bar{x}_t$ in the upper graph. There are 15 outlying lot means and they are designated by a red point. The lower graph shows the within-lot standard deviations σ_t (blue) and the sample standard deviations s_t. Figure 11 shows in the upper graph the fractions of nonconforming units in the lots (blue), their MVUE estimates (black), a priori (green) and a posteriori (red) estimates; red circles indicate rejection by ISO 3951, symbol "N" the start of normal inspection, symbol "T" the start of tightened inspection and symbol "S" the discontinuation of inspection. ISO switches frequently between normal and tightened inspection and would discontinue the inspection procedure at lots 14 and 44. In the lower graph the expected (blue), a priori (green)and a posteriori (red) probabilities of the fractions of nonconforming units in the lots exceeding p_{AQL} are shown; red circles indicate rejections by the Bayes method. Altogether, 78% of the 100 lots have been accepted by ISO and 80% by the Bayes method. Bayes rejects 14 of the 15 outlier lots, ISO all 15. Figure 12 shows that, due to some of the outlier lots included in the update of the estimate of the process curve, the estimated standard deviations vary more than in Fig. 5. However, the estimates of the process average and of the expectation of the within-lot standard deviation are (almost) unbiased.

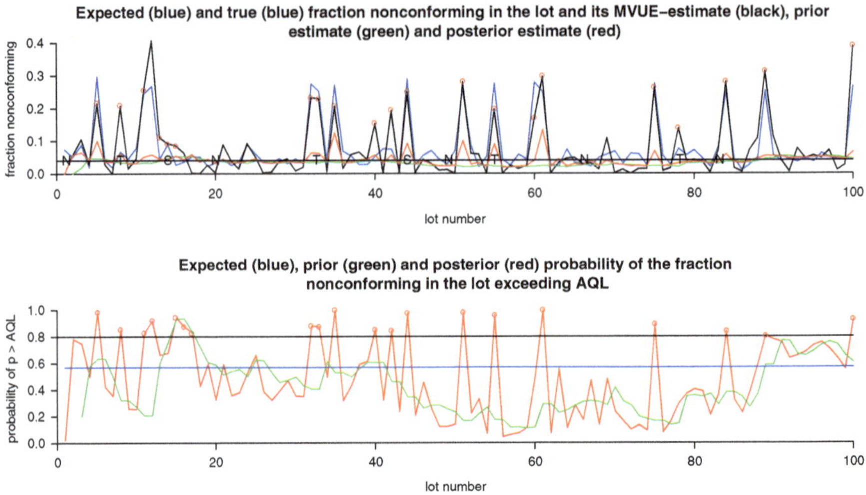

Fig. 11 *Upper graph*: Expected (*blue*) and true (*blue*) fractions of nonconforming units in the lots and their MVUE estimates (*black*), a priori (*green*) and a posteriori (*red*) estimates for scenario 5. *Red circles* indicate rejection by ISO 3951, $N =$ start of normal inspection, $T =$ start of tightened inspection, $S =$ inspection stop. *Lower graph*: Expected (*blue*), a priori (*green*) and a posteriori (*red*) probabilities of the fractions of nonconforming units in the lots exceeding p_{AQL} for scenario 5; *red circles* indicate rejections by the Bayes method

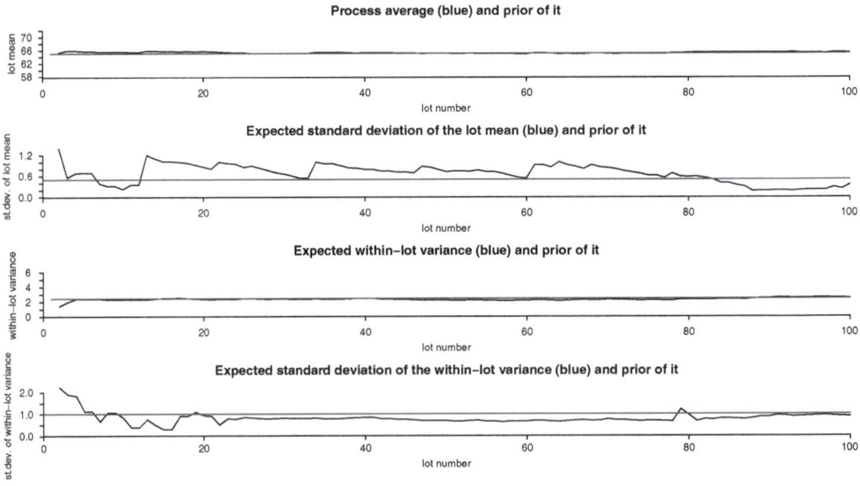

Fig. 12 True parameters $\mu_{\mu_t} = 65, \sigma^2_{\mu_t} = 0.5^2, \mu_{\sigma^2_t} = 2.435^2, \sigma^2_{\sigma^2_t} = 1.0^2$ of the production process and their estimates for scenario 5

5 Switch Between s-Plans and σ-Plans

It is often observed that the within-lot standard deviations σ_t vary less than the lot means μ_t. Eventually the updated estimates of the standard deviation may indicate a constant standard deviation σ_t with a value that is sufficiently well known. This allows switching from sampling plans for unknown σ (s-plans) to sampling plans for known σ (σ-plans). Their advantage is the smaller sample size at almost unaltered risk.

ISO 3951, Part 1, paragraph 22 "Switching between 's' and 'σ' methods" recommends: "If it appears that the value of s has been in control for at least 20 consecutive lots, the (weighted) root mean square value of s may be presumed to be σ, the 'known' standard deviation of the process, and the 'σ' method may be adopted. In order to verify that the variability remains under control, the value of s should still be calculated and plotted on a control chart."
In ISO 3951 the corresponding sample sizes n_σ and n_s are

n_σ	3	4	6	9	13	18	25	35	50	70	95	125	160	200	250
n_s	2	3	4	6	8	10	12	15	18	21	25	32	40	50	65

This correspondence can roughly be expressed by the formula $n_\sigma = n_s^{3/4}$, rounded to integers.

Since, in the beginning of the application of the sampling procedure or after a switch from the σ-plan to the s-plan, we do not have a reliable estimate of the standard deviation σ_t, we cannot run a s^2-chart. Therefore, in order to follow the rule of ISO 3951, we apply the Bartlett test for variance homogeneity to the last 20 observed variances s^2. If the null hypothesis of variance homogeneity is not rejected at the 1% significance level we switch to the corresponding σ-plan and use the mean of the last 20 variances s^2 as 'known' variance σ_0^2. Further, we start running a CUSUM-s^2-chart. We switch back to the s-plan if the CUSUM-s^2-chart signalizes that σ is no longer equal to the 'known' standard deviation σ_0.

If the switch from the s-plan to the σ-plan is signalized at lot $k = k_0$, the updated variances according to Eq. (14) have to be adjusted from the sample size n_s to the sample size n_σ:

$$\hat{\sigma}_{\bar{x},k_0,new} = \frac{n_s}{n_\sigma}\hat{\sigma}_{\bar{x},k_0}$$

$$\hat{\sigma}_{s^2,k_0,new} = \frac{n_s - 1}{n_\sigma - 1}\hat{\sigma}_{s^2,k_0}; \qquad (26)$$

$\hat{\mu}_{k_0}$ and $\hat{\sigma}_{k_0}^2$ remain unchanged. The estimation of the parameters of the process curve has to be altered as follows: From Eq. (11) and $V(\mu_t) = \sigma_{0,t}^2 / n_{0,t}$ we get

$$V(\mu_t) = V(\bar{x}_t) - \frac{\sigma_{0,t}^2}{n} \qquad (27)$$

and estimate $V(\mu_t)$ at lot k by

$$\hat{V}(\mu_k) = \hat{\sigma}^2_{\bar{x},k} - \frac{\sigma_0^2}{n} := \hat{\tau}_k^2 \tag{28}$$

with $n = n_\sigma$. The a priori distribution and the a posteriori distribution become

$$f_{prior}(\mu_{k+1}; \hat{\mu}_k, \hat{\tau}_k^2) = \frac{1}{\sqrt{2\pi}\,\hat{\tau}_k} \exp\left(-\frac{1}{2\hat{\tau}_k^2}(\mu_{k+1} - \hat{\mu}_k)^2\right) \tag{29}$$

and

$$f_{post}(\mu_{k+1}; \mu_{post,k+1}, \sigma^2_{post,k+1}) = \frac{1}{\sqrt{2\pi}\sigma_{post,k+1}} \exp\left(-\frac{1}{2\sigma^2_{post,k+1}}(\mu_{k+1} - \mu_{post,k+1})^2\right), \tag{30}$$

respectively, with

$$\mu_{post,k+1} = \frac{(1/\hat{\tau}_k^2)\hat{\mu}_{0,k} + (n/\sigma_0^2)\bar{x}_{k+1}}{(1/\hat{\tau}_k^2) + (n/\sigma_0^2)}$$
$$\sigma^2_{post,k+1} = \frac{1}{(1/\hat{\tau}_k^2) + (n/\sigma_0^2)}. \tag{31}$$

The a priori estimate of the fraction p_{k+1} of nonconforming units in lot $k+1$ is now

$$\hat{p}_{k+1,prior} = \Phi\left(\frac{\hat{\mu}_k - U}{\sigma_0}\right) + \Phi\left(\frac{L - \hat{\mu}_k}{\sigma_0}\right) \tag{32}$$

and the a posteriori estimate of the fraction p_{k+1} of nonconforming units in lot $k+1$ is

$$\hat{p}_{k+1,post} = \Phi\left(\frac{\mu_{post,k+1} - U}{\sigma_0}\right) + \Phi\left(\frac{L - \mu_{post,k+1}}{\sigma_0}\right). \tag{33}$$

All other equations remain unchanged.

When switching back from the σ-plan to the s-plan at lot $k = k_1$, the updated variances according to Eq. (14) have to be adjusted from the sample size n_σ to the sample size n_s:

$$\hat{\sigma}_{\bar{x},k_1,new} = \frac{n_\sigma}{n_s}\hat{\sigma}_{\bar{x},k_1}$$
$$\hat{\sigma}_{s^2,k_1,new} = \frac{n_\sigma - 1}{n_s - 1}\hat{\sigma}_{s^2,k_1}. \tag{34}$$

We apply the switching rules to the lots of Scenario 1 where the within-lot standard deviation is constant, $\sigma_t = 2.435$ for $t = 1, \ldots, N$. Figure 13 presents the

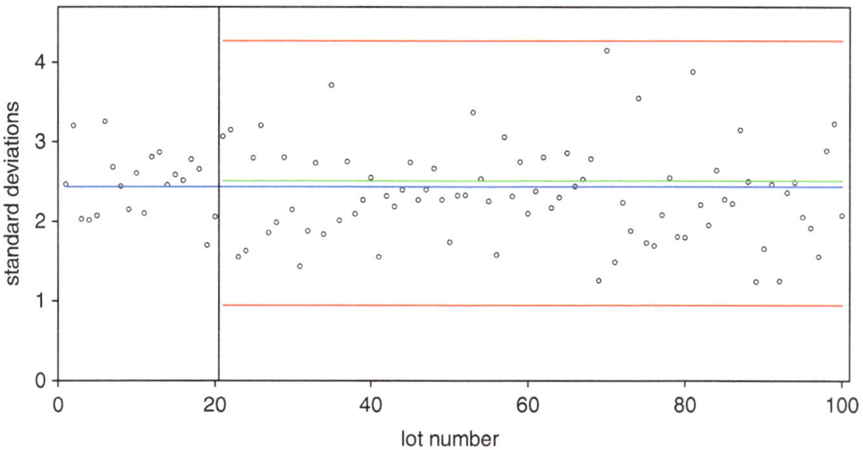

Fig. 13 Standard deviations s_t and the constant σ_t (*blue*) of scenario 1. From lot 20 the σ-plan with σ according to the *green line* is used. The *red lines* are the 99% action limits of the Shewhart-s-chart

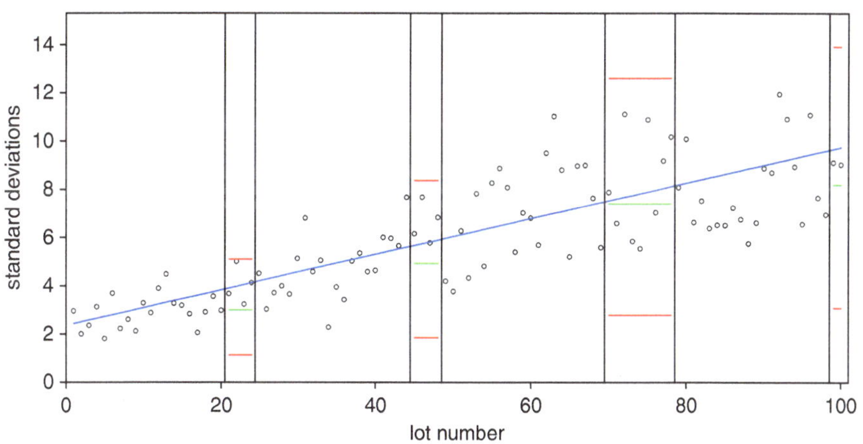

Fig. 14 Standard deviations s_t and the linearly increasing σ_t (*blue*). Switches are indicated by *vertical lines*. In the intervals where a *green line* is shown the σ-plan is used with the 'known' σ indicated by the *green line*

standard deviations s_t as points and the constant σ_t as blue horizontal line. At $t = 20$ the Bartlett test is applied and indicates variance homogeneity and hence, the σ-plan with $n_\sigma = 8$ corresponding to the s-plan with $n_s = 13$ is applied. The square root of the mean of the first 20 variances s_t^2, $\sigma_0 = 2.51$, is used as 'known' standard deviation. In the figure σ_0 is indicated as a horizontal green line. The horizontal red lines are the 99% action limits of the Shewhart-s-chart, however, this chart is not used. Instead a CUSUM-s^2-chart is started which does not indicate an out-of-control situation so that the σ-plan is used until lot 100.

In Fig. 14 the within-lot standard deviation increases from $\sigma_1 = 2.435$ to $\sigma_{100} = 9.74$ (blue). At $t = 20$ the Bartlett test is applied and indicates variance homogeneity and hence, the σ-plan with $n_\sigma = 8$ corresponding to the s-plan with $n_s = 13$ is applied. The square root of the mean of the first 20 variances s_t^2, $\sigma_0 = 3.1$, is used as 'known' standard deviation. At lot 24 the CUSUM-s^2-chart signals an increasing σ_t and sampling switches back to the s-plan until lot 44 with a switch to the σ-plan, ... We observe long periods where the s-plan is used and short periods with the use of the σ-plan.

6 Conclusions

The performance of the ISO 3951 procedure (with the switching rules) and the Bayes method do not show essential differences. Compared with ISO 3951 the computational effort of the Bayesian sampling inspection is larger, however, it has some advantages:

- The acceptance decision does not end only with a decision "lot accepted" or "lot rejected" but in addition with the estimated probability of the fraction of nonconforming units in the lot being larger than p_{AQL} (or another established value). Hence, the user gets an idea of the reliability of the decision. In addition, he can establish other border lines for this estimated probability, in order to sort lots into two or more groups with higher or lower probability of fractions nonconforming exceeding p_{AQL} being used as different grades of product.
- After having inspected only a few lots, the user will have a reliable knowledge of the process curve, especially its process average. This quality history facilitates steps towards quality improvements. In addition, it can be used for switches to reduced inspection and to σ-plans with smaller sample size.

References

Bowker, A. H., & Goode, H. P. (1952) *Sampling inspection by variables* (Chap. 11). New York/Toronto/London: McGraw-Hill.

ISO 2859–1 (1999) *Sampling procedures for inspection by attributes – Part 1: Sampling schemes indexed by acceptance quality limit (AQL) for lot-by-lot inspection*. Geneva: International Standardisation Organisation.

ISO 3951–1 (2005) *Sampling procedures for inspection by variables – Part 1: Specification for single sampling plans indexed by acceptance quality limit (AQL) for lot-by-lot inspection – Single quality characteristic and single AQL*. Geneva: International Standardisation Organisation.

ISO 3951–2 (2005) *Sampling procedures for inspection by variables – Part 2: General specification for single sampling plans indexed by acceptance quality limit (AQL) for lot-by-lot*

inspection of independent quality characteristics. Geneva: International Standardisation Organisation.
Stange, K. (1977) *Bayes-Verfahren. Schätz- und Testverfahren bei Berücksichtigung von Vorinformationen*. Berlin/Heidelberg/New York: Springer.
Wilrich, P.-Th. (2010) A new approach to Bayesian sampling plans. In H.-J. Lenz, P.-Th. Wilrich, & W. Schmid (Eds.), *Frontiers in statistical quality control 9* (pp. 191–206). Berlin/Heidelberg: Physica

Quality Assessment in the Presence of Additional Data in Photovoltaics

Sabine Meisen, Andrey Pepelyshev, and Ansgar Steland

Abstract Acceptance sampling represents an important tool for quality control. The practical methods of choice for non-normal variables are attribute sampling and variables sampling assuming normality applied to averages instead of single observations. Both methods usually lead to very large sample sizes and are therefore infeasible in practice if observations are expensive. We discuss and extend recent results developed for the photovoltaic industry and actively used there. Here – and presumably in other industries as well – additional data are available which can be used to construct valid and asymptotically optimal sampling plans for non-normal measurements. Consistency and asymptotic optimality of the sampling plans, which are random in our setup, as well as asymptotic normality of the required sample size are established under weak assumptions. We also provide sensitivity studies dealing with the effects of a systematic bias (shift) between the additional data and the lot (shipment), which may matter in practice. The new plans are investigated by simulations to some extent.

Keywords Acceptance sampling • Limit theorems • Quality control • Renewable energies • Sensitivity • Thin-film technology

S. Meisen (✉)
Institute for Medical Statistics, RWTH Aachen University, Aachen, Germany
e-mail: smeisen@ukaachen.de

A. Pepelyshev · A. Steland
Institute of Statistics, RWTH Aachen University, Aachen, Germany
e-mail: pepelyshev@stochastik.rwth-aachen.de; steland@stochastik.rwth-aachen.de

H.-J. Lenz et al. (eds.), *Frontiers in Statistical Quality Control 10*,
Frontiers in Statistical Quality Control 10, DOI 10.1007/978-3-7908-2846-7_17,

1 Introduction

In industry, quality control of lots or shipments of produced items is an important practical problem, particularly when high-quality leadership is a strategic goal. Delivering shipments of bad quality to customers may result in expensive law suits. In the photovoltaic industry, the expectations of customers in terms of quality of delivered photovoltaic modules (PV modules) are very high, presumably due to the fact that state of the art semi-conductor technologies are at the core of the business, thus being associated with digital precision. Acceptance sampling, which deals with the problem of determining the minimal sample size necessary to control the producer's as well as the consumer's risk, is therefore an important practical approach to the problem, cf. the recent monograph (Schilling et al. 2009). The customer is interested in the quality of his or her shipment of modules and not in the average outgoing quality. Thus, manufacturers interested in customer satisfaction should control production on the basis of outgoing shipments.

In this article, we study the classic acceptance sampling problem under the general assumption that the measurements (the power output in photovoltaics) may follow an arbitrary continuous distribution function (d.f.). To handle this case, we use additional data in the form of a historic data set, a situation which is typically present in photovoltaics. We derive sampling plans assuming a less general distributional model as in our previous work (Steland et al. 2009), but the results of the present article are valid under less restrictive assumptions. Further, the model of the present article nicely allows us to investigate sensitivity and robustness issues, respectively. We study the effect on the sampling plans, when the distributions of the shipment and the lab samples differ, which is of substantial interest for applications, where a systematic bias is of primary concern.

Our setup is as follows: We consider a shipment $X_1, \dots, X_N \sim F$ with a common distribution function F which is assumed to be continuous and strictly increasing with a finite fourth moment. Here and in what follows, $\sim$ always indicates that the random variables are independent and identically distributed. However, the stochastic relationship *between* samples may be arbitrary, i.e. they are not required to be independent. Let $\mu = E(X_1)$ and $\sigma^2 = E(X_1 - \mu)^2 \in (0, \infty)$. In the photovoltaic problem motivating our work, X_i represents the true but random power output of the ith module, which is classified as non-conforming, i.e. being of low quality, if $X_i \leq \tau$ for some constant τ. In practice, one puts $\tau = \mu(1 - \varepsilon)$ where ε is the tolerance. The additional data are provided in the form of a historic sample $Y_1, \dots, Y_m \sim F$ of size m. Conditions on m will be given in Sect. 3. Clearly, the expected fraction of non-conforming modules in the shipment is given by

$$p = P(X_1 \leq \tau) = F(\tau).$$

Suppose we have fixed two numbers $0 < AQL < RQL < 1$, namely the *acceptable quality level* (AQL) and the *rejectable quality level* (RQL), such that the lot should be accepted if $p \leq AQL$, whereas it should be rejected when $p \geq RQL$.

Since p is unknown to us and checking all modules is infeasible, our aim is to decide on the basis of a control sample $X_1', \ldots, X_n'$ of n measurements, n as small as possible, whether or not the shipment has to be accepted with controlled error probabilities of false decisions. Suppose that the decision is based on a statistic $T_n' = T_n'(X_1', \ldots, X_n')$ using the decision rule to accept the shipment if and only if $T_n' > c$. A natural choice is to use a standardized sum statistic, i.e. to accept if

$$T_n' = \sqrt{n}\frac{\overline{X}_n' - \tau}{\sigma} > c.$$

Then a solution (n, c) to the above problem, defined rigorously in Sect. 3, is called a *sampling plan*.

In the case of normally distributed items, the optimal solution is well known and indeed based on the statistic T_n'. The resulting procedure is called *variables sampling*. However, in photovoltaics the power output measurements of PV modules are usually non-normal, such that variables sampling yields invalid sampling plans and is therefore not applicable. Indeed, all kinds of distributional shapes appear in practice. We discuss some of the factors leading to that unpleasant empirical fact in the next section. It is well known, and we shall provide the relevant arguments in Sect. 3 when deriving the plans, that the optimal sampling plan for an arbitrary d.f. F depends on that unknown F.

In photovoltaics, our key application for the methodology discussed in the present article, additional data from the production line, the so-called *flasher report tables*, are available and can be used in the construction of sampling plans. This key idea has been used in Herrmann et al. (2416) to develop a photovoltaic-specific two-stage decision procedure using those flash data in order to construct valid sampling plans for normal as well as non-normal data. The procedure has also been implemented in a software tool which is used in the photovoltaic industry, cf. Herrmann et al. (2010). In Steland et al. (2009), we elaborated on the new procedure addressing the non-normal case by establishing its asymptotic optimality as well as its asymptotic distribution assuming a general location scale model for the additional data. The mathematical proofs required advanced tools from probability theory such as empirical process theory and the functional delta method in metric spaces. Unfortunately, the formula for the asymptotic variance turned out to be rather complex and first simulations indicated that the variance of the estimated sample size is rather high even when hundreds of additional measurements are available. Thus, a natural question is whether it is possible to construct similar sampling plans leading to simpler formulas, simpler derivations and better accuracy in practice. For another recent proposal we refer to Herrmann et al. (2010). We show that indeed concise proofs can be given for the approximation theorem behind the construction of the sampling plans as well as for the central limit theorem for the estimated sample size. Here the classic Bahadur-Kiefer representation of sample quantiles plays a key role in the derivations. Further, we provide a new result on the strong consistency of the estimated sample size, thus strengthening the weak consistency implied by the central limit theorem, and the weak consistency of the estimated critical value under weak assumptions.

A further major goal of the present work is to study the effect of a systematic bias affecting both the lot and lab measurements and the historic data, respectively, in order to get a better understanding of what happens in this case. Indeed, such a sensitivity analysis with respect to the sampling distribution – depending on the viewpoint it can be regarded as a robustness study as well – can provide valuable insights into the stability of a procedure. We will study both a model where a constant bias is present and a model where an asymptotically vanishing bias is assumed. The latter approach has a nice interpretation in terms of an *asymptotic learning effect* and shows that a local bias which is sufficiently small in larger control samples has no effect on the sampling plan. Although it is not surprising that a constant bias has such an effect, our findings show that the bias does not affect the required sample size, which greatly simplifies its treatment in practice.

The organization of the article is as follows. In Sect. 2, we provide some information on the photovoltaic background of the problem. Section 3 reviews the derivation of acceptance sampling plans and introduces the sampling plans proposed for the above setting including results on their asymptotic optimality. The study on the effects of a systematic bias and asymptotic learning is presented in Sect. 4. The result on the asymptotic normality is given in Sect. 5. Finally, Sect. 6 provides numerical results from a Monte Carlo study to assess the accuracy of the proposed sampling plans. We reveal an interesting and surprisingly strong effect of the algorithm used to calculate a sample quantile. Proofs of the results are postponed to an appendix.

2 Background on and Application to Photovoltaics

In the present section, we give a brief account of the photovoltaic background which motivated our research and the way how we approached the problem.

Photovoltaics represents one of the key technologies having the potential to provide a substantial contribution to the world's energy problem. Presumably, the main reason why the market share of solar energy is still relatively small compared to its potential and benefits is the fact that the costs per watt are still rather high. Although the costs have been substantially decreased in recent years, research still focuses on further reductions in costs, either by increasing the efficiency of a given solar cell technology or by developing new technologies, e.g. by employing cheaper materials and chemicals.

The economic life time of a photovoltaic system (PV system) ranges between 20 and 30 years. Thus, the quality in terms of the power output of the modules at delivery is a crucial parameter for the profitability of such an investment. Even small departures from the nominal power output accumulate to considerable losses over the years. Assessing the quality of PV modules, which is done under standard conditions (STC) in a lab, is therefore an important issue for quality control.

PV modules are an interconnected assembly of solar cells. To protect the cells from damage during manufacturing, delivery and usage, they are embedded between a tedlar plate on the bottom, a tempered glass on the top, and framed, usually with an aluminium frame. Since a single module can produce only a limited amount of electricity, around 200 W under STC, a PV system consists of many connected PV modules.

There are two common technologies to manufacture PV modules: Crystalline modules use silicon solar cells produced from solid Si wafers, whereas the CIS thin-film technology applies copper (C), indium (I) and selenium (S) in a layer construction of around 2 μm onto a substrate. The electrical properties such as the spectrum of the sun light transformed into electricity, the loss of efficiency when exposed to heat, a serious issue for systems installed in Southern Europe or Africa, or the efficiency when there are clouds as it is often the case in Northern Europe, heavily depend on the technology and various other physical parameters of the chosen module type.

Calibrating a PV module in a testing laboratory is also a different problem, since a couple of factors may complicate collecting measurements and also may lead to considerable difference of indoor and outdoor measurements. As reviewed and experimentally analyzed in Virtuani et al. (2010), the following effects matter in practice:

(i) Measurement related sweep-time effects refereing to the influence of the duration to complete an IV scan. Depending on the selected flash tester, which generates a pulse of calibrated light, such a scan can be based on up to 100 flashes. The duration of each flash is typically 100 ms. For details on accurate testing of PV panels we refer to Roy et al. (2010).

(ii) Spectral mismatch arising when using a reference cell with a spectral response different from that of the device under test; its size depends on the spectral irradiance distribution of the spectral simulator with respect to the reference spectrum AM 1.5G.

(iii) Finally, thin-film modules are affected by the effect of light soaking, since the performance (even under standard test conditions) depends on the history of module (exposure to light or storage in the dark). This phenomenon called *light soaking* is in effect at the time of delivery but disappears when the modules are exposed to sun light for several days. The light soaking effect were first reported in Ruberto et al. (1987) and is addressed to the tunneling of electrons trapped in deep states of CdS to holes in the CIS layer valence band under illumination, resulting in an increase in the open-circuit voltage and fill factor. Rau et al. (1989) found such effects in thin films; here annealing at 80°C decreases the dark conductivity and illuminating the annealed solar cell by white light reestablished the previous state. During that light soaking period the performance can increase by 2–5%, cf. Kuurne et al. (2008). In industrial practice, it can even be larger.

Finally, it is common practice in industry to classify the produced PV modules in classes. As a consequence of the above discussion, when analyzing *comparable* modules, i.e. modules of the same technology and power rating satisfying additional criteria for inclusion or exclusion in a study, the true distribution of measurements may have any form. Especially, measurements are typically non-normal, thus violating the classic assumptions in statistical acceptance sampling by variables.

Relying on ad hoc proposals such as forming subgroups and then applying variables sampling to the subgroup means to ensure approximatively the normal assumption, is not feasible due to the high costs of taking control measurements, since this procedure leads to enormous sample sizes. Moreover, taking the control measurements is very expensive. For the same reason, applying attribute sampling is no reasonable solution.

Solar cells are manufactured in a production line. The performance of each module is measured in a sun simulator using short flashes. These measurements are therefore called flash measurements and form the flash data tables. In present days, they are routinely collected by manufacturers, thus often large samples are available. However, these cheap measurements may differ from the measurements taken in a photovoltaic laboratory. One should check carefully, whether a given flasher report table follows the same distribution as the shipment before applying the methods discussed in the present article. Some standard tests and their application to real photovoltaic data are described in Herrmann et al. (2010). For a new approach to the problem we refer to the recent work Steland et al. (2011).

3 The Acceptance Sampling Problem

Acceptance sampling is a well established field of statistics and quality control – at least for the classic distributional assumptions. For basic notions we therefore refer to Schilling et al. (2009). Our goal is to find an acceptance sampling plan (n, c), i.e. a sample size n for a control sample and a critical value c, such that

$$P(T_n' > c) \geq 1 - \alpha, \qquad p \leq AQL, \tag{1}$$

and

$$P(T_n' > c) \leq \beta, \qquad p \geq RQL. \tag{2}$$

Here α is an upper bound on the probability that the shipment is rejected when it is of high quality, thus controlling the producer's risk, whereas β is the consumer's risk that the shipment is accepted although it is of low quality. Our derivations below will show, as a side-product, that $P(T_n' > c)$, the operating characteristic (OC), is a function of the quality level p. Approximations based on large sample theory will then allow us to solve the problem to construct appropriate sampling plans. We first discuss the unrealistic case that the underlying distribution is known and then proceed to a general solution for an arbitrary unknown distribution of the measurements.

3.1 The Case of a Known Distribution

Let us assume that F and therefore $\mu = \int x\,dF(x)$ as well as $\sigma^2 = \int x^2\,dF(x) - \mu^2$ are known. Using the crucial relationship $F(\tau) = p \Leftrightarrow \tau = F^{-1}(p)$, we obtain

$$T_n' > c \quad \Leftrightarrow \quad \sqrt{n}\frac{\overline{X}_n' - \mu}{\sigma} > c + \frac{\sqrt{n}(F^{-1}(p) - \mu)}{\sigma}.$$

By virtue of the central limit theorem, we get the approximation

$$P\left(\sqrt{n}\,\frac{\overline{X}_n' - \tau}{\sigma} > c\right) \approx 1 - \Phi\left(c + \frac{\sqrt{n}(F^{-1}(p) - \mu)}{\sigma}\right). \tag{3}$$

Notice that this approximation requires n to be large. However, since statistical inference should never be based on too few observations, assuming that the central limit theorem provides a sufficiently accurate approximation should not be too restrictive for many distributions F. Further, one may check the accuracy of the above approximation after calculating the sampling plan, such that c is fixed, using historic data which is available by assumption. This could be done, for instance, by estimating the Berry-Esséen upper bound or by means of a simulation study, the latter approach being preferable.

Thus, Eqs. 1 and 2 are approximately satisfied, if we select (n, c) such that

$$1 - \Phi\left(c + \frac{\sqrt{n}(F^{-1}(AQL) - \mu)}{\sigma}\right) \geq 1 - \alpha \tag{4}$$

and

$$1 - \Phi\left(c + \frac{\sqrt{n}(F^{-1}(RQL) - \mu)}{\sigma}\right) \leq \beta \tag{5}$$

hold true. Since Φ is strictly increasing, these inequalities are equivalent to

$$\Phi^{-1}(\alpha) - \frac{\sqrt{n}(F^{-1}(AQL) - \mu)}{\sigma} \geq c \geq \Phi^{-1}(1 - \beta) - \frac{\sqrt{n}(F^{-1}(RQL) - \mu)}{\sigma}.$$

Consider the left and right sides of the above chain of inequalities as functions in the real variable n. Then we arrive at the following proposition.

Proposition 1. *The optimal sampling plan (n, c) is obtained as the intersection of the mappings*

$$n \longmapsto \Phi^{-1}(\alpha) - \frac{\sqrt{n}(F^{-1}(AQL) - \mu)}{\sigma}$$

and

$$n \longmapsto \Phi^{-1}(1 - \beta) - \frac{\sqrt{n}(F^{-1}(RQL) - \mu)}{\sigma}.$$

Equating the above mappings leads to

$$\Phi^{-1}(\alpha) - \Phi^{-1}(1-\beta) = \frac{\sqrt{n}}{\sigma}(F^{-1}(AQL) - F^{-1}(RQL)).$$

By assumption, F is strictly increasing, which allows us to solve the equation for n and c leading the following result.

Theorem 1. *For known distributional parameters μ and σ, the optimal sampling plan is given by*

$$n_\infty(\mu,\sigma) = \frac{\sigma^2\left(\Phi^{-1}(\alpha) - \Phi^{-1}(1-\beta)\right)^2}{(F^{-1}(AQL) - F^{-1}(RQL))^2}, \tag{6}$$

$$c(\mu,\sigma) = \Phi^{-1}(\alpha) - \frac{\sqrt{n}(F^{-1}(AQL) - \mu)}{\sigma}. \tag{7}$$

In particular, the optimal sampling plan depends on the unknown d.f. F of the measurements.

Theorem 1 shows that the asymptotically optimal sampling plan requires knowledge of the location μ and dispersion σ as well as knowledge of the distribution of the quality measurements. But as those quantities are unknown to us, the sampling plan can not be applied in practice.

3.2 The Case of an Unknown Distribution

Clearly, the case that F is known to us is only of theoretical interest. Thus, we shall now assume that F is an arbitrary d.f. such that the fourth moment is finite. The basic idea is now to estimate the unknown quantities in the formulas derived in the previous subsection. However, various mathematical problems now arise. One has to establish an approximation of the OC curve leading to the same sampling plans one obtains when estimating unknown quantities in formulas (Eq. 6) and (Eq. 7). It turns out that within the framework of the present article, a transparent proof of such an approximation can be given without relying on empirical process theory as in Steland et al. (2009). Further, we present a new result establishing strong consistency of the estimated sample size and weak consistency of the estimated critical value under fairly weak assumptions.

We shall now derive the sampling plan in the case of unknown parameters μ and σ when additional data $Y_1, \ldots, Y_m$ are given. Statistical theory suggests using the modified decision rule

$$T_n := \sqrt{n}\,\frac{\overline{X'_n} - \tau}{S_m} > c,$$

where $S_m^2 = \frac{1}{m-1}\sum_{i=1}^m (Y_i - \overline{Y}_m)^2$ with $\overline{Y}_m = \frac{1}{m}\sum_{i=1}^m Y_i$. Let us introduce the empirical distribution function of the historic sample defined by

$$F_m(y) = \frac{1}{m}\sum_{i=1}^m 1(Y_i \le y), \qquad y \in \mathbb{R}.$$

Here $1(A) = 1$, if the expression A (defining an event) is true, and $1(A) = 0$, if A is false. As usual,

$$F_m^{-1}(p) = \inf\{t \in \mathbb{R} : F_m(t) \ge p\}, \qquad p \in (0,1), \tag{8}$$

denotes the left continuous empirical quantile function, i.e. $F_m^{-1}(p) = Y_{(\lfloor np \rfloor + 1)}$, where $Y_{(1)} \le \cdots \le Y_{(m)}$ denotes the order statistic.

In the sequel, we need the following regularity assumption on the sample sizes n and m.

Assumption (A): $\frac{n}{m} \to 0$, as $n, m \to \infty$.

The reasoning behind that assumption is the following: The construction of the asymptotically optimal procedure requires certain approximations as in Eq. 3. Now both n and m have to be large, but, in addition, n/m has to be small. However, that condition is not restrictive in practice, as long as the historic data set is large enough. This is the typical case in photovoltaics and presumably in other areas of application as well.

The approximation of the operating characteristic $P(T_n > c)$ is now more involved and given in the following theorem, which is an analogue of Steland et al. (2009, Theorem 3.1).

Theorem 2. *Suppose F is a continuous and strictly increasing d.f. with a finite second moment. If $X_1', \ldots, X_n'$, the control sample, and $Y_1, \ldots, Y_m$, the historic data set, are random samples satisfying Assumptions (A), then there exists a sequence $\delta_n(p)$, $n \in \mathbb{N}$, of random variables with $\delta_n(o) = o_P(1)$, such that for all $c \in \mathbb{R}$*

$$P(T_n > c) = P\left(\sqrt{n}\frac{\overline{X}_n' - \mu}{\sigma} + \delta_n(p) > c + \frac{\sqrt{n}(F_m^{-1}(p) - \overline{Y}_m)}{S_m}\right)$$

leading to the approximation

$$P(T_n > c) \approx 1 - \Phi\left(c + \frac{\sqrt{n}(F_m^{-1}(p) - \overline{Y}_m)}{S_m}\right).$$

A proof of this result is given in the appendix. Notice that the result holds true under the weak assumption of a finite second moment of the underlying distribution.

Repeating the derivations of the previous subsection, we obtain the following analogue of Proposition 1.

Proposition 2. *The optimal sampling plan* (n, c) *is obtained as the intersection of the mappings*

$$n \longmapsto \Phi^{-1}(\alpha) - \frac{\sqrt{n}(F_m^{-1}(AQL) - \overline{Y}_m)}{S_m}$$

and

$$n \longmapsto \Phi^{-1}(1-\beta) - \frac{\sqrt{n}(F_m^{-1}(RQL) - \overline{Y}_m)}{S_m}.$$

One has to take into account that F_m (and therefore F_m^{-1}) is not strictly increasing and $F_m^{-1}(AQL) < F_m^{-1}(RQL)$ may not hold. However, the latter condition holds true if m is large enough, which allows us to solve algebraically the equation as in the previous subsection leading to formulas for n and c depending on F_m^{-1}. Further, the empirical quantiles $F_m^{-1}(p)$ converge to $F^{-1}(p)$ with probability 1, since that property is well known to be equivalent to $F_m(y) \to F(y)$ with probability 1, as $m \to \infty$, which of course holds true, and $\overline{Y}_m$ to μ as well as S_m to σ, as $m \to \infty$, with probability 1, as long as the underlying d.f. F has a finite second moment. A more refined argument yields the following result, which is proved in the appendix.

Theorem 3. *Suppose* F *is a d.f. with a finite fourth moment. For unknown distributional parameters, the estimated sampling plan is given by*

$$n_m = \frac{S_m^2 \left(\Phi^{-1}(\alpha) - \Phi^{-1}(1-\beta)\right)^2}{\left(F_m^{-1}(AQL) - F_m^{-1}(RQL)\right)^2},$$

$$c_m = \Phi^{-1}(\alpha) - \frac{\sqrt{n}(F_m^{-1}(AQL) - \overline{Y}_m)}{S_m}.$$

It converges to the optimal sampling plan in the sense that

$$n_m/n_\infty(\mu,\sigma) \overset{a.s.}{\to} 1 \quad \text{and} \quad c_m - c(\mu,\sigma) \overset{P}{\to} 0,$$

as $m \to \infty$, *provided Assumption (A) holds true as well as* $F'(F^{-1}(AQL)) > 0$ *and* $F'(F^{-1}(RQL)) > 0$.

It is worth mentioning that the asymptotic optimality holds true under the weak regularity assumption of a finite fourth moment, the minimal assumption under which the statistic S_m^2 is meaningful in the sense of strong consistency and asymptotic normality. The proof given in the appendix shows that the strong consistency of n_m even holds under the weaker condition of a finite second moment. However, although the assumption of finite higher moments is not regarded as an issue in photovoltaics, since measurements are often even bounded by definition of the sampling process, it may be in other areas of applications.

4 The Effect of a Bias and Asymptotic Learning

In practice, it may happen that the distribution of the historic data set, $Y_1, \ldots, Y_m$, and the distribution of control measurements $X_1', \ldots, X_n'$, which are made in a laboratory, do not coincide. Therefore the present section is devoted to a study of the effect of departures from the assumption $X_i \sim F$. We are interested in the effect of a systematic bias, as it may happen when using a differently calibrated measurement system to measure modules of the lot and of the control sample. We will see that in this setting the optimal sampling plan depends on the bias, even asymptotically. That result is interesting in its own right, but can be used in practice as well, in order to correct for such a systematic bias, provided an estimate of the bias is available.

The next step of our analysis is then to model the bias as a function of the sample size n which tends to 0, as n approaches ∞, i.e. to consider local alternatives, and to ask under which conditions on n and m there is no asymptotic effect of that disturbance. Considering a sequence of distributional models indexed by the optimal sample size n has two interpretations, both of which are meaningful. Firstly, one may conduct a sequence of experiments where the risk probabilities α and β are decreased from experiment to experiment leading to larger sample sizes n. Then a bias of the order $o(1)$ can be interpreted as a *learning effect* when conducting more and more experiments. Secondly, when conducting one experiment, a model with a bias of order $o(1)$ captures the fact that presumably more efforts are spent on obtaining better measurements with smaller bias, when analyzing large samples to get very precise results, which are more expensive than conducting small experiments.

4.1 The Effect of a Fixed Systematic Bias

Let us assume that the random variables $X_1, \ldots, X_N$ representing the shipment (lot) satisfy

$$X_1 + \gamma, \ldots, X_N + \gamma \sim F, \qquad N \in \mathbb{N},$$

for a given constant shift γ, whereas the historic sample is not affected by the shift, i.e. $Y_1, \ldots, Y_m \sim F$. This means that the distribution of the shipment measurements is equal in distribution to the historical measurements after adding γ. Since the control measurements are selected from the shipment, we have $X_1' + \gamma, \ldots, X_n' + \gamma \sim F$ as well. Equivalently, we could assume that $X_1, \ldots, X_N \sim F$ and $Y_1 - \gamma, \ldots, Y_m - \gamma \sim F$, but the above formulation simplifies the derivations.

Now the fraction of non-conforming modules satisfies the equation

$$p = E\left(\frac{1}{N}\sum_{i=1}^{N} 1(X_i \leq \tau)\right) = F(\tau + \gamma)$$

or, equivalently, $\tau = F^{-1}(p) - \gamma$. It turns out that when going through all derivations given in the previous section and the proof of Theorem 2 in the appendix, we arrive at the following approximation of the operating characteristic

$$P(T_n' > c) \approx 1 - \Phi\left(c + \frac{\sqrt{n}(F_m^{-1}(p) - \overline{Y}_m)}{S_m} - \frac{\sqrt{n}\gamma}{S_m}\right)$$

leading to the optimal sampling plan

$$n_\gamma = \frac{S_m^2(\Phi^{-1}(\alpha) - \Phi^{-1}(1-\beta))^2}{(F_m^{-1}(AQL) - F_m^{-1}(RQL))^2}, \tag{9}$$

$$c_\gamma = \Phi^{-1}(\alpha) - \frac{\sqrt{n}(F_m^{-1}(AQL) - \overline{Y}_m)}{S_m} + \frac{\sqrt{n}\gamma}{S_m}. \tag{10}$$

The interpretation of these results is as follows: The bias (location shift) γ has only an effect on the critical value, but the optimal sample size remains the same. As a consequence, we may formulate the following rule of thumb: If the measurements of the shipment are shifted by γ compared to the historic sample measurements, one can apply the optimal sampling plan derived in the previous section when correcting the critical value by the additive term $\sqrt{n}\gamma / S_m$.

4.2 The Effect of Asymptotic Learning

Suppose now that

$$X_1 + \gamma_n, \dots, X_N + \gamma_n \sim F, \qquad N \in \mathbb{N}, \tag{11}$$

for some sequence $\{\gamma_n\}$ of real numbers converging to 0 as $n \to \infty$.

Assumption (B): Suppose the sample sizes n, m and the sequence $\{\gamma_n\}$ are selected such that

$$\frac{n}{m} = o(1) \quad \text{and} \quad \sqrt{m}\gamma_n = o(1),$$

as $n, m \to \infty$.

Suppose γ_n is chosen as

$$\gamma_n = \Gamma n^{-\xi}, \qquad n \geq 1,$$

for positive constants Γ and ξ. Then it is easily seen that Assumption (B) is satisfied if in addition to $n/m = o(1)$ the condition

$$\frac{m}{n^{2\xi}} = o(1)$$

holds true. For brevity of presentation, we omit the proof of the following result and refer to Meisen et al. (2010) for details.

Theorem 4. *Under the sampling model Eq. 11 and Assumption (B), the approximation of the operating characteristic obtained in Theorem 2 still holds true, such that the sampling plan (n_m, c_m) given in Theorem 3 is asymptotically optimal in this case.*

5 Asymptotic Normality

The present section is devoted to a study of the asymptotic distribution of the optimal sample size, as the sample size m of the historic data set tends to ∞. It turns out that in the present setting a proof of the asymptotic normality can be based on the Bahadur-Kiefer representation of sample quantiles and the delta method in $\mathbb{R}^3$ combined with the multivariate central limit theorem.

For simplicity of presentation, we use the abbreviations

$$p_\alpha = AQL \qquad \text{and} \qquad p_\beta = RQL$$

in what follows.

Theorem 5. *Suppose the historic data set as well as the lot and control measurements are distributed according to a strictly increasing and continuous d.f. F such that $\int x^4 dF(x) < \infty$ and*

$$F'(F^{-1}(p_\alpha)) > 0 \quad \text{as well as} \quad F'(F^{-1}(p_\beta)) > 0.$$

Let $n_\infty = n_\infty(\mu, \sigma)$ and n_m be as in Theorem 1 and Theorem 3. If Assumption (A) holds true, n_m is asymptotically normal,

$$\sqrt{m}(n_m - n_\infty) \xrightarrow{d} N(0, \eta^2), \qquad m \to \infty,$$

for $\eta^2 = \mathbf{g}\Sigma\mathbf{g}'$, where

$$\mathbf{g} = \frac{C(\alpha, \beta)}{(F^{-1}(p_\alpha) - F^{-1}(p_\beta))^3} \cdot (-2\sigma^2, 2\sigma^2, (F^{-1}(p_\alpha) - F^{-1}(p_\beta))'$$

and

$$\Sigma = \begin{pmatrix} \Sigma_{\alpha\alpha} & \Sigma_{\alpha\beta} & \zeta_{Y\alpha} \\ \Sigma_{\alpha\beta} & \Sigma_{\beta\beta} & \zeta_{Y\beta} \\ \zeta_{Y\alpha} & \zeta_{Y\beta} & \zeta_{YY} \end{pmatrix}$$

with entries

$$\Sigma_{\alpha\alpha} = \frac{p_\alpha(1-p_\alpha)}{(F'(F^{-1}(p_\alpha)))^2}, \qquad \Sigma_{\beta\beta} = \frac{p_\beta(1-p_\beta)}{(F'(F^{-1}(p_\beta)))^2},$$

$$\Sigma_{\alpha\beta} = \frac{p_\alpha(1-p_\beta)}{(F'(F^{-1}(p_\alpha)))\,(F'(F^{-1}(p_\beta)))}$$

$$\zeta_{Y\alpha} = \frac{E((Y_1-\mu)^2-\sigma^2)(p_\alpha - 1(Y_1 \le F^{-1}(p_\alpha)))}{F'(F^{-1}(p_\alpha))},$$

$$\zeta_{Y\beta} = \frac{E((Y_1-\mu)^2-\sigma^2)(p_\beta - 1(Y_1 \le F^{-1}(p_\beta)))}{F'(F^{-1}(p_\beta))},$$

$$\zeta_{YY} = Var\,((Y_1-\mu)^2),$$

leading to

$$\eta^2 = \frac{4\sigma^4(\Phi^{-1}(\alpha) - \Phi^{-1}(1-\beta))^4}{(p_\alpha - p_\beta)^6}\big[4\sigma^4(\Sigma_{\alpha\alpha} - 2\Sigma_{\alpha\beta} + \Sigma_{\beta\beta})$$
$$+ 4\sigma^2(F^{-1}(p_\alpha) - F^{-1}(p_\beta))(\zeta_{Y\beta} - \zeta_{Y\alpha}) + \zeta_{YY}(F^{-1}(p_\alpha) - F^{-1}(p_\beta))^2\big].$$

It is worth mentioning that the above formulas are more transparent than those obtained in Steland et al. (2009). Although it is interesting that the estimated sample size is asymptotically normal, the result is of limited value for practical purposes, since simulations have shown that the convergence is rather slow. Consequently, it is not clear whether the construction of asymptotic confidence intervals based on the above result would yield intervals with accurate coverage probabilities.

6 Simulations

We conducted small-scale simulations in order to investigate to some extent the accuracy of the new sampling plans under some models. We were also interested in the effect of the method used to calculate a sample quantile, as standard statistical software usually offers several methods.

As a kind of benchmark model, we selected the normal distribution with mean 220 and variance 4. The reason is that photovoltaic modules are often traded with a nominal power output of 220 W. The variance, 4, captures to some extent the variability observed in practice, although that varies with technology. Two kinds of departures from the normality assumption were studied: One-sided contaminations inducing skewness in the data samples and symmetric contaminations inducing, e.g. a different kurtosis. The amount of contamination to induce these effects was chosen as 20%, and the mean and the variance of the contaminating subpopulations was chosen between 210–240 and 4–8, respectively.

Table 1 Characteristics of the distribution of n_m and c_m for Model 1

m	$E(n_m)$	$sd(n_m)$	$q_{0.25}$	$q_{0.50}$	$q_{0.75}$	$E(c_m)$	$sd(c_m)$
100	195.4	1144.0	30	61	143	19.4	16.9
250	79.1	77.5	37	58	94	15.3	5.6
500	74.9	44.5	46	64	91	15.5	3.9
5,000	65.6	10.5	58	65	72	14.9	1.1
50,000	64.8	3.2	63	65	67	14.9	0.3

Table 2 Characteristics of the distribution of n_m and c_m for Model 2

m	$E(n_m)$	$sd(n_m)$	$q_{0.25}$	$q_{0.50}$	$q_{0.75}$	$E(c_m)$	$sd(c_m)$
100	317.6	2568.7	44	98	231	38.2	34.3
250	123.7	133.9	54	89	148	30.6	11.8
500	118.6	72.7	70	101	145	31.4	8.1
5,000	103.6	17.5	91	102	114	30.4	2.3
50,000	102.5	5.5	99	102	106	30.4	0.7

Table 3 Characteristics of the distribution of n_m and c_m for Model 3

m	$E(n_m)$	$sd(n_m)$	$q_{0.25}$	$q_{0.50}$	$q_{0.75}$	$E(c_m)$	$sd(c_m)$
100	679.1	5724.1	92	193	455	23.6	22.2
250	256.2	253.1	117	186	303	18.7	7.0
500	241.7	143.6	148	206	293	19.0	4.8
5,000	211.3	33.7	188	208	232	18.2	1.4
50,000	209.2	10.8	202	209	216	18.2	0.4

For each parameter combination given by the sample size m of the historic data set and the parameters of the above mixture model, we calculated Monte Carlo estimates for the expected required sample size, $E(n_m)$, and the associated standard deviation of n_m based on 50,000 replications. In addition, the quartiles $q_{0.25}, q_{0.5}$ and $q_{0.75}$ of the distribution of n_m are reported, which enables us to judge the skewness of the distribution of n_m. Finally, the expected critical value, $E(c_m)$, and its standard deviation are provided.

6.1 *One-Sided Contaminations*

Data sets according to the following models were simulated:

$$\begin{aligned}
&\text{Model 1:} \quad X_i \sim F_1 = N(220, 4), \\
&\text{Model 2:} \quad X_i \sim F_2 = 0.1N(210, 6) + 0.9N(230, 4), \\
&\text{Model 3:} \quad X_i \sim F_3 = 0.9N(220, 4) + 0.1N(230, 8).
\end{aligned} \tag{12}$$

The parameters were specified as $AQL = 2\%$, $RQL = 5\%$ and $\alpha = \beta = 5\%$. Tables 1–3 show the simulation results for these models.

Table 4 Characteristics of the distribution of n_m and c_m for Model 4

m	$E(n_m)$	$sd(n_m)$	$q_{0.25}$	$q_{0.50}$	$q_{0.75}$	$E(c_m)$	$sd(c_m)$
100	624.8	23533.2	79	161	364	31.7	34.2
250	203.9	191.6	96	150	242	25.2	9.3
500	193.7	111.4	121	167	234	25.5	6.4
5,000	169.2	26.3	150	167	185	24.5	1.8
50,000	167.8	8.2	162	168	173	24.5	0.6

Table 5 Characteristics of the distribution of n_m and c_m for Model 5

m	$E(n_m)$	$sd(n_m)$	$q_{0.25}$	$q_{0.50}$	$q_{0.75}$	$E(c_m)$	$sd(c_m)$
100	1951.6	22814.7	284	578	1319	55.9	52.9
250	727.9	677.7	347	539	872	44.6	16.3
500	705.7	405.9	438	606	849	45.5	11.5
5,000	616.9	96.3	550	607	676	43.8	3.3
50,000	608.6	29.6	588	608	629	43.6	1.0

Table 6 Characteristics of the distribution of n_m and c_m for Model 6

m	$E(n_m)$	$sd(n_m)$	$q_{0.25}$	$q_{0.50}$	$q_{0.75}$	$E(c_m)$	$sd(c_m)$
100	1023.9	19714.2	152	309	709	41.6	39.2
250	392.8	390.0	186	287	461	33.3	12.5
500	374.3	215.0	233	321	453	33.8	8.5
5,000	327.3	51.6	291	322	358	32.5	2.4
50,000	323.5	15.8	312	323	334	32.4	0.8

6.2 Symmetric Contaminations

Let us now study symmetric contaminations according to the models

$$\begin{aligned} &\text{Model 4:} \quad X_i \sim F_4 = 0.2N(210, 8) + 0.6(220, 4) + 0.2N(230, 8), \\ &\text{Model 5:} \quad X_i \sim F_5 = 0.2N(200, 8) + 0.6(220, 4) + 0.2N(240, 8), \\ &\text{Model 6:} \quad X_i \sim F_6 = 0.2N(210, 4) + 0.6(220, 4) + 0.2N(230, 4), \\ &\text{Model 7:} \quad X_i \sim F_7 = 0.2N(200, 4) + 0.6(220, 4) + 0.2N(240, 4). \end{aligned} \tag{13}$$

Notice that in all models the observations have a mean of 220. In Model 4, we consider the case that one fifth of the contaminated data have the mean 210, whereas another fifth scatters around 230. The variance for these two subpopulations is 8. Model 6 is similar to Model 4 except that the variance is fixed at 4. In Models 5 and 7 the contaminating subpopulations have means 200 and 240. The empirical results for these models are provided in Tables 4–7.

It is interesting to note that $q_{0.5}$ for $m = 250$ is typically smaller then $q_{0.5}$ for $m = 100$ since the use of the empirical distribution leads to rounding (discretion). For example, for a sample of size 250, the 0.02-quantile is the fifth order statistics of the sample ($250 \times 0.02 = 5$) and the 0.05-quantile is the thirteenth-order statistics of the sample ($250 \times 0.05 = 12.5 \neq 13$)

Table 7 Characteristics of the distribution of n_m and c_m for Model 7

m	$E(n_m)$	$sd(n_m)$	$q_{0.25}$	$q_{0.50}$	$q_{0.75}$	$E(c_m)$	$sd(c_m)$
100	3897.3	49612.1	572	1,160	2,643	76.7	72.8
250	1453.1	1435.4	686	1,059	1,718	60.8	22.8
500	1406.5	817.9	872	1,201	1,698	62.2	15.9
5,000	1221.0	190.6	1,085	1,202	1,337	59.7	4.5
50,000	1204.8	58.5	1,164	1,204	1,244	59.5	1.4

6.3 *Effect of Quantile Algorithms*

Statistical software such as R or SAS implements various standard procedures to calculate sample quantiles. R implements nine algorithms discussed in Hyndman et al. (1996). Algorithm 1 corresponds to the left inverse of the empirical distribution function, the definition we use in this article, see Eq. 8. However, R's default algorithm is Algorithm 7. SAS's PROC UNIVARIATE also provides several methods; the default is to use the average of the npth and $(np + 1)$-th order statistic, if np is an integer, and the $(\lfloor np \rfloor + 1)$th order statistic, otherwise, corresponding to Algorithm 2.

We used the parameters $\alpha = \beta = 0.05$, $AQL = 0.02$ and $RQL = 0.05$ and simulated data according to the model

$$X_i \sim \mathrm{N}(220, 4), \quad i = 1, \ldots, m.$$

For each algorithm and sample size $m = 100$, 250, 500, 5,000, the same statistical quantities as above were estimated using 50,000 simulation runs. For better comparison, for each algorithm the same random numbers were used by initializing the random number generator using the statement set.seed(17). The results are presented in Tables 8 and 9.

For Algorithm 1, we can see that the median $q_{0.5}$ is close to 65 for small m but the standard deviation of n_m is very large. In general, results for Algorithms 2–6 and 8–9 are rather close to the results for Algorithm 1; in some cases even identical to the results for Algorithm 1. However, the results for Algorithm 7, which is used in R by default, are worse than the results when using Algorithm 1. In particular, the median $q_{0.5}$ is substantially larger than 65. For $m \geq 5,000$, there are no notable differences.

7 Discussion

Sampling plans have been proposed for variables sampling when the true but unknown distribution is completely unknown. These plans require additional historic sampling information, which is usually available in photovoltaics, the key

Table 8 Results for Algorithms 1–5

m	$E(n_m)$	$sd(n_m)$	$q_{0.25}$	$q_{0.50}$	$q_{0.75}$	$E(c_m)$	$sd(c_m)$
			Algorithm 1				
100	195.4	1144.0	30	61	143	19.4	16.9
250	79.1	77.5	37	58	94	15.3	5.6
500	74.9	44.5	46	64	91	15.5	3.9
5,000	65.6	10.5	58	65	72	14.9	1.1
			Algorithm 2				
100	162.8	424.6	40	75	156	19.2	12.3
250	95.1	93.3	45	70	112	16.7	6.2
500	78.1	44.8	49	67	95	15.8	3.9
5,000	65.9	10.5	59	65	72	15	1.1
			Algorithm 3				
100	195.4	1144.0	30	61	143	19.4	16.9
250	101.6	114.4	44	70	118	17.3	7.0
500	74.9	44.5	46	64	91	15.5	3.9
5,000	65.6	10.5	58	65	72	14.9	1.1
			Algorithm 4				
100	195.4	1144.0	30	61	143	19.4	16.9
250	88.2	88.9	40	64	104	16.2	6.1
500	74.9	44.5	46	64	91	15.5	3.9
5,000	65.6	10.5	58	65	72	14.9	1.1
			Algorithm 5				
100	162.8	424.6	40	75	156	19.2	12.3
250	95.1	93.3	45	70	112	16.7	6.2
500	78.1	44.8	49	67	95	15.8	3.9
5,000	65.9	10.5	59	65	72	15	1.1

area of application we have in mind where the methods are already in active use. Our theoretical results show that the proposed sampling plans are consistent and asymptotically optimal under very weak assumptions. Moreover, the estimated sample size satisfies a central limit theorem. Whether or not those results remain valid when the measurements are dependent is still an open issue. However, our results do not require independence of the historic sample, the control measurements and the shipment. Indeed, *any* kind of dependence between the samples is allowed. Hence our results are valid both when drawing randomly the control measurements from the shipment and when using observations independent from the shipment, which is, e.g. the case when the modules used for the control measurements are not released to the customer.

Our simulations indicate, firstly, that the presented methodology provides accurate sampling plans for a wide range of distributions, provided the size of the historic data set is sufficiently large. However, the estimated sample size is affected by a substantially large variance, which hinders practical application. Here further research is in order to develop procedures with reduced variability, which would lead to improved sampling plans which can be used for smaller historic data sets.

Table 9 Results for Algorithms 6–9

m	$E(n_m)$	$sd(n_m)$	$q_{0.25}$	$q_{0.50}$	$q_{0.75}$	$E(c_m)$	$sd(c_m)$
			Algorithm 6				
100	174	689.3	30	61	140	19.0	15.0
250	87.7	87.9	40	64	103	16.2	6.1
500	74.7	44.2	46	64	91	15.5	3.9
5,000	65.5	10.5	58	65	72	14.9	1.1
			Algorithm 7				
100	301.6	1405.4	51	104	239	23.1	19.1
250	108.5	107	50	79	129	17.5	6.6
500	83.2	49.3	51	71	101	16.1	4.1
5,000	66.2	10.6	59	65	73	15.0	1.1
			Algorithm 8				
100	157.1	437.8	36	69	147	18.9	12.5
250	92.0	90.1	43	68	109	16.5	6.1
500	76.8	44.3	48	66	93	15.7	3.8
5,000	65.8	10.5	58	65	72	15.0	1.1
			Algorithm 9				
100	157.8	431.0	37	71	149	18.9	12.4
250	92.7	90.7	44	68	110	16.5	6.1
500	77.1	44.4	48	66	94	15.7	3.8
5,000	65.8	10.5	58	65	72	15.0	1.1

The simulation study of the effect of the algorithm used to calculate sample quantiles reveals a striking and unexpected strong effect for small sample sizes. All algorithms estimate quantiles by calculating a function of at most two successive order statistics, i.e. they perform a smoothing operation in some cases. The effect on the results is surprisingly strong and points to the conjecture that, in general, improved smoothing may lead to better sampling plans. Again, future research is in order to reveal to the extent to which improved sampling plans can be constructed by using refined quantile estimation algorithms based on smoothing techniques.

Acknowledgements The authors acknowledge financial support from the German Federal Ministry of the Environment, Nature Conservation and Nuclear Safety (grant no. 0325226).

Appendix: Proofs

In this mathematical appendix, we use the following notation. For a function f defined on some subset of the p-dimensional Euclidean space $\mathbb{R}^p$, $\dot{f}$ denotes the gradient of f. When there is no danger of confusion, we identify column and row vectors. The symbol $\stackrel{P}{\to}$ denotes convergence in probability and $\stackrel{d}{\to}$ the convergence in distribution of a sequence of random variables and random vectors. For brevity of presentation, we use the stochastic o and O symbols: $o_P(1)$ stands for a random sequence R_n with $R_n \xrightarrow{P} 0$, as $n \to \infty$, whereas $O_P(1)$ denotes a sequence R_n such that for any $\varepsilon > 0$ one may find a constant C with $P(|R_n| > C) < \varepsilon$ for all n. As is well known, $R_n = O_P(1)$ holds true, if R_n converges in distribution. We also frequently use rules such as $O_P(1) \cdot o(1) = o_P(1)$.

We need the following auxiliary results on the Bahadur representation of sample quantiles, a classical result dating back to Bahadur et al. (1966) and Kiefer et al. (1967), which in turn implies their joint asymptotic normality. For the reader's convenience, we provide those results in some detail.

Theorem 6. *(i) Let $p \in (0,1)$ and suppose $F'(F^{-1}(p)) > 0$. Then*

$$F_m^{-1}(p) = F^{-1}(p) + \frac{F(F^{-1}(p)) - F_m(F^{-1}(p))}{F'(F^{-1}(p))} + o_P(m^{-1/2}).$$

(ii) Suppose $0 < p_1 < \cdots < p_k < 1$ and $F'(F^{-1}(p_i)) > 0$ for $i = 1, \ldots, k$. Then

$$\sqrt{m}(F_m^{-1}(p_1) - F^{-1}(p_1), \ldots, F_m^{-1}(p_k) - F^{-1}(p_k)) \xrightarrow{d} N(0, S_k),$$

where $S_k = (s_{ij})$ is a symmetric $k \times k$ matrix with entries

$$s_{ij} = \frac{p_i(1-p_i)}{F'(F^{-1}(p_i))},$$

for $1 \le i, j \le k$.

□

Proof. (Theorem 2)
To prove the assertion, we follow arguments used in Steland et al. (2009). Straightforward algebra leads to

$$T_n > c \Leftrightarrow \sqrt{n}\left(\frac{\overline{X_n'} - \mu}{\sigma} + T_1(n) + T_2(n) - T_3(n)\right) > c + \frac{\sqrt{n}(F_m^{-1}(p) - \overline{Y}_m)}{S_m}$$

where

$$T_1(n) = \frac{\overline{X'}_n - \mu}{\sigma} \cdot \frac{\sigma - S_m}{S_m},$$

$$T_2(n) = \frac{F_m^{-1}(p) - F^{-1}(p)}{S_m},$$

$$T_3(n) = \frac{\overline{Y}_m - \mu}{S_m}.$$

The assertion follows, if we show that

$$\delta_n(p) = \sqrt{n}(T_1(n) + T_2(n) - T_3(n)) = o_P(1),$$

as $n \to \infty$. Clearly, $\sqrt{n}T_1(n) = o_P(1)$ by Slutzky's lemma, since $\sqrt{n}(\overline{X}_n' - \mu)/\sigma \xrightarrow{d} N(0,1)$, as $m \to \infty$, and $(\sigma - S_m)/\sigma \xrightarrow{P} 0$, as $n \to \infty$. Next notice that

$$\sqrt{n}T_2(n) = \frac{1}{S_m}\frac{\sqrt{n}}{\sqrt{m}}\sqrt{m}(F_m^{-1}(p) - F^{-1}(p)).$$

The first factor converges to $1/\sigma$, in probability, the second one is $o(1)$ by Assumption (A) and the third factor is asymptotically normal by Theorem 6, as $m \to \infty$. Thus, $T_2(n) = o_P(1)$. Similarly, we have $\sqrt{n}T_3(n) = \frac{\sqrt{n}}{\sqrt{m}}\frac{\sigma}{S_m}\sqrt{m}\frac{\overline{Y}_m - \mu}{\sigma} = o_P(1)$ by Assumption (A). Let us now check the approximation for $P(T_n > c)$. Clearly,

$$U_n = \sqrt{n}(\overline{X}_n' - \mu)/\sigma + \delta_n(p) \xrightarrow{d} N(0,1),$$

as $n \to \infty$. This implies that $\sup_{z \in \mathbb{R}} |P(U_n \le z) - \Phi(z)| = o(1)$, by virtue of the Glivenko-Cantelli theorem, which completes the proof, since

$$\begin{aligned} |P(T_n > c) - & [1 - \Phi(c + \sqrt{n}(F_m^{-1}(p) - \overline{Y}_m))]| \\ &= |P(U_n > c) - [1 - \Phi(c + \sqrt{n}(F_m^{-1}(p) - \overline{Y}_m))]| \\ &\le \sup_{z \in \mathbb{R}} |P(U_n > z) - [1 - \Phi(z)]| = o(1). \end{aligned}$$

□

Proof. (*Theorem 3*)
Obviously,

$$\frac{n_m}{n_\infty(\mu,\sigma)} = \frac{S_m^2}{\sigma^2}\frac{(F^{-1}(AQL) - F^{-1}(RQL))^2}{(F_m^{-1}(AQL) - F_m^{-1}(RQL))^2}.$$

The first factor on the right-hand side converges to 1, almost surely, as $m \to \infty$, as well as the second one, since $F_m^{-1}(AQL) - F_m^{-1}(RQL) \to F^{-1}(AQL) - F^{-1}(RQL)$, almost surely, as $m \to \infty$. To show the second assertion, recall that

$$c_m = \Phi^{-1}(\alpha) - \frac{\sqrt{n}(F_m^{-1}(AQL) - \overline{Y}_m)}{S_m},$$

$$c(\mu,\sigma) = \Phi^{-1}(\alpha) - \frac{\sqrt{n}(F^{-1}(AQL) - \mu)}{\sigma},$$

and notice that

$$c_m - c(\mu,\sigma) = \frac{\sigma}{S_m}W_{n,m},$$

where

$$W_{n,m} = \left[-\frac{\sqrt{n}(F_m^{-1}(AQL) - \overline{Y}_m)}{\sigma} + \frac{\sqrt{n}(F^{-1}(AQL) - \mu)}{\sigma}\frac{S_m}{\sigma} \right].$$

Clearly, $\sigma/S_m = 1 + (\sigma - S_m)/S_m \to 1$, as $m \to \infty$, a.s. Thus, it suffices to show that $W_{n,m}$ is $o_P(1)$. Using $S_m/\sigma = 1 + (S_m - \sigma)/\sigma$, we see that the second summand of $W_{n,m}$ can be written as

$$\frac{\sqrt{n}(F^{-1}(AQL) - \mu)}{\sigma} + \frac{F^{-1}(AQL) - \mu}{\sigma}\sqrt{n}\frac{S_m - \sigma}{\sigma}.$$

Rearranging terms, we see that $W_{n,m}$ can be written as

$$\sqrt{\frac{n}{m}}\left[-\sqrt{m}\frac{F_m^{-1}(AQL) - F^{-1}(AQL)}{\sigma} + \sqrt{m}\frac{\overline{Y}_m - \mu}{\sigma}\right.$$
$$\left.+\frac{F^{-1}(AQL) - \mu}{\sigma}\sqrt{m}\frac{S_m - \sigma}{\sigma}.\right]$$

By Theorem 6 (ii),

$$\sqrt{m}(F_m^{-1}(AQL) - F^{-1}(AQL)) = O_P(1).$$

Further, since $EY_1^4 < \infty$ by assumption, we also have

$$\sqrt{m}(\overline{Y}_m - \mu) = O_P(1) \quad \text{and} \quad \sqrt{m}(S_m - \sigma) = O_P(1),$$

as $m \to \infty$; for the second statistic also confer the proof of Theorem 5. Since by Assumption (A) $n/m = o(1)$, as $n, m \to \infty$, we obtain $W_{n,m} = o_P(1)$ follows, which completes the proof. □

Proof. (*Theorem 5*)
For brevity of notation, we put $p_\alpha = AQL$ and $p_\beta = RQL$. The Bahadur representation yields for $p \in \{p_\alpha, p_\beta\}$

$$\sqrt{m}[F_m^{-1}(p) - F^{-1}(p)] = \frac{1}{\sqrt{m}}\sum_{i=1}^{m}\frac{p - 1(Y_i \le F^{-1}(p))}{F'(F^{-1}(p))} + o_P(1).$$

Further, with $\widetilde{Y}_i = Y_i - \mu$ and $\overline{\widetilde{Y}} = m^{-1}\sum_{i=1}^{m}\widetilde{Y}_i$ we obtain the asymptotic linearization

$$\frac{1}{\sqrt{m}}\sum_{i=1}^{m}[(Y_i - \overline{Y})^2 - \sigma^2] = \frac{1}{\sqrt{m}}\sum_{i=1}^{m}(\widetilde{Y}_i^2 - \sigma^2) - 2\overline{\widetilde{Y}}\sqrt{m}\overline{\widetilde{Y}} + \sqrt{m}(\overline{\widetilde{Y}})^2$$
$$= \frac{1}{\sqrt{m}}\sum_{i=1}^{m}(\widetilde{Y}_i^2 - \sigma^2) + o_P(1),$$

since $\sqrt{m}\overline{\widetilde{Y}} = O_P(1)$ and $\overline{\widetilde{Y}} = o_P(1)$, which, of course, carries over to $\sqrt{m}(S_m^2 - \sigma^2)$. This gives

$$U_m = \sqrt{m}\begin{pmatrix} F_m^{-1}(p_\alpha) - F^{-1}(p_\alpha) \\ F_m^{-1}(p_\beta) - F^{-1}(p_\beta) \\ S_m^2 - \sigma^2 \end{pmatrix} = \frac{1}{\sqrt{m}}\sum_{i=1}^{m} Z_i + o_P(1),$$

where for $i = 1, \dots, m$

$$Z_i = \begin{pmatrix} (p_\alpha - 1(Y_i \le F^{-1}(p_\alpha)))/F'(F^{-1}(p_\alpha)) \\ (p_\beta - 1(Y_i \le F^{-1}(p_\beta)))/F'(F^{-1}(p_\beta)) \\ (Y_i - \mu)^2 - \sigma^2 \end{pmatrix}.$$

Notice that $Z_1, \dots, Z_m$ are i.i.d. with a finite second moment. Thus, an application of the multivariate central limit theorem yields

$$U_m \xrightarrow{d} U \sim N(\mathbf{0}, \Sigma), \text{ with } \Sigma = E(Z_1^2) = \begin{pmatrix} \Sigma_{\alpha\alpha} & \Sigma_{\alpha\beta} & \zeta_{Y\alpha} \\ \Sigma_{\alpha\beta} & \Sigma_{\beta\beta} & \zeta_{Y\beta} \\ \zeta_{Y\alpha} & \zeta_{Y\beta} & \zeta_{YY} \end{pmatrix},$$

as $m \to \infty$, where the entries are as given in the theorem. Now consider

$$\sqrt{m}(n_m - n_\infty) = \sqrt{m}\left(\frac{S_m^2 \cdot (\Phi^{-1}(\alpha) - \Phi^{-1}(1-\beta))^2}{(F_m^{-1}(p_\alpha) - F_m^{-1}(p_\beta))^2} - \frac{\sigma^2 \cdot (\Phi^{-1}(\alpha) - \Phi^{-1}(1-\beta))^2}{(F^{-1}(p_\alpha) - F^{-1}(p_\beta))^2}\right). \tag{14}$$

Observe that we may write

$$\sqrt{m}(n_m - n_\infty) = \sqrt{m}[g(F_m^{-1}(p_\alpha), F_m^{-1}(p_\beta), S_m^2) - g(F^{-1}(p_\alpha), F^{-1}(p_\beta), \sigma^2)]$$

where the function $g : D \to \mathbb{R}$, $D = \{(x, y, z) \in \mathbb{R}^3 : x \neq y\}$, is given by

$$g(x, y, z) = C(\alpha, \beta) \cdot \frac{z}{(x-y)^2}, \quad C(\alpha, \beta) = (\Phi^{-1}(\alpha) - \Phi^{-1}(1-\beta))^2$$

for $(x, y, z) \in D$. The function g is differentiable with

$$\dot{g}(x, y, z) = C(\alpha, \beta)(-2z(x-y)^{-3}, 2z(x-y)^{-3}, (x-y)^{-2}),$$

such that

$$\begin{aligned} \mathbf{g} &= \dot{g}(F^{-1}(p_\alpha), F^{-1}(p_\beta), \sigma^2) \\ &= \frac{C(\alpha, \beta)}{(F^{-1}(p_\alpha) - F^{-1}(p_\beta))^3} \cdot (-2\sigma^2, 2\sigma^2, (F^{-1}(p_\alpha) - F^{-1}(p_\beta)). \end{aligned}$$

The delta method now implies that

$$\sqrt{m}(n_m - n_\infty) \xrightarrow{d} \mathbf{g}U \sim N(\mathbf{0}, \eta^2),$$

as $m \to \infty$, where

$$\eta^2 = \mathbf{g}\Sigma\mathbf{g}'.$$

This completes the proof. □

References

Bahadur, R. R. (1966) A note on quantiles in large samples. *Annals of Mathematical Statistics, 37*, 577–580.

Herrmann, W., & Steland, A. (2010) Evaluation of photovoltaic modules based on sampling inspection using smoothed empirical quantiles. *Progress in Photovoltaics, 18*(1), 1–9.

Herrmann, W., Althaus, J., Steland, A., & Zaehle, H. (2006) Statistical and experimental methods for assessing the power output specification of PV modules. In *Proceedings of the 21st European photovoltaic solar energy conference*, Dresden (pp. 2416–2420).

Herrmann, W., Steland, A., & Herff, W. (2010) Sampling procedures for the validation of PV module output specification. In *Proceedings of the 24th European photovoltaic solar energy conference*, Hamburg (pp. 3540–3547). ISBN: 3-936338-25-6, doi: 10.4229/24thEUPVSEC2009-4AV.3.70.

Hyndman R. J., & Fan Y. (1996) Sample quantiles in statistical packages. *The American Statistician, 50*(4), 361–364.

Kiefer, J. (1967) On Bahadurs representation of sample quantiles. *Annals of Mathematical Statistics, 38*, 1323–1342.

Kuurne, J., Tolvanen, A., Hyvärinen, J., & Oy, E. (2008) Sweep time, spectral mismatch and ligh soaking in thin film module measurements. In *PVSC '08. 33rd IEEE*, San Diego (pp. 1–3). ISSN 0160-8371.

Meisen, S. (2010) Ein nichtparametrisches Verfahren zur Annahmeprüfung. Diploma thesis, Institute of Statistics, RWTH Aachen University.

Rau, U., Schmitt, M., & Parisi, J. (1989) Persistent photoconductivity in Cu(In,Ga)Se_2 heterojunctions and thin films prepared by sequential deposition. *Journal of Applied Physics, 73*(2), 223–225.

Roy, J.N., Gariki, G.R., & Nagalakhsmi, V. (2010) Reference module selection criteria for accurate testing of photovoltaic (PV) panels. *Solar Energy, 84*, 32–36.

Ruberto, M.N., & Rothwarf, J. (1987) Time-dependent open-circuit voltage in $CuInSe_2$/CdS solar cells: Theory and experiment. *Journal of Applied Physics, 61*(9), 4662–4669.

Schilling, D.G., & Neubauer, D.V. (2009) *Acceptance sampling in quality control*. Boca Raton: Chapman & Hall/CRC.

Steland, A., & Zaehle, H. (2009) Sampling inspection by variables: Nonparametric setting. *Statistica Neerlandica, 63*(1), 101–123.

Steland, A., Padmanabhan, A.R., & Akram, M. (2011) Resampling methods for the nonparametric and generalized Behrens-Fisher problems. *Sankhya A, 73*, 267–302.

Virtuani, A., Muellejans, H., Ponti, F., & Dunlop, E. (2010) Comparison of indoor and outdoor performance measurements of recent commercially available solar modules. In *Proceedings of the 23th European photovoltaic solar energy conference*, Hamburg (pp. 2379–2385). ISBN: 3-936338-25-6, doi: 10.4229/24thEUPVSEC2009-3CO.12.1.

On Practical Uses of ISO Standards: Two Case Studies

Jørgen Iwersen

Abstract Statistical ISO-standards such as ISO 2859-1 and ISO 3951-1 are still used in many organizations for many different purposes, from incoming control to release of final product. In the current era of optimization everywhere there is, however, a new focus on sampling and use of resources. In many organizations new technologies like Lean, Six Sigma, etc. are widely used, and hence the sample sizes proposed by the ISO standards are often challenged with the purpose of reducing sampling to reduce cost both directly in terms of fewer samples but also indirectly in terms of saved costs in the labs.

In this paper I will present two very different examples from the pharmaceutical world. The first example is an algorithm to lower the sample size in In-Process Control (IPC). In this case, we wanted to do some testing in real time as IPC controls, to be able to react in time if these parameters show a trend or go out of control. These plans are based on ISO standards in the sense that we retain the Consumer Risk Quality (CRQ) at the levels of the standard, but we then reduce the sample size to save costs and reduce the workload of sampling and analyzing samples in production.

The second case study is from the other end of the spectrum. In this case we look at a vision system that enables us to do 100% control for many parameters. We also retain the CRQ from ISO, but we rewrite the acceptance criterion to allow for 100% control of outgoing batches.

Keywords ISO standards • In-process control • Vision systems • AQL • LQ • CRQ

J. Iwersen (✉)
Novo Nordisk A/S, Novo Alle, DK-2880 Bagsvaerd, Denmark
e-mail: jiw@novonordisk.com

H.-J. Lenz et al. (eds.), *Frontiers in Statistical Quality Control 10*,
Frontiers in Statistical Quality Control 10, DOI 10.1007/978-3-7908-2846-7_18,

1 Introduction

Statistical ISO-standards such as ISO 2859–1 (1999), ISO 2859–2 (1985) and ISO 3951–1 (2005) are still used in many organizations for many different purposes, from incoming control to release of final product. In the current era of optimization in every step of production processes there is, however, an increased focus on sampling sizes and use of resources. In many organizations new technologies like Lean, Six Sigma, etc. are widely used, and hence the sample sizes proposed by the ISO standards are often challenged with the purpose of reducing sampling to reduce cost – both directly in terms of smaller samples but also indirectly in terms of saved costs in the laboratories.

This paper presents two very different examples from the pharmaceutical world. The first example is an algorithm to use ISO 2859-1 for In-Process Control (IPC). In this case, we wanted to do some testing in real time as IPC controls, to be able to react in time if important quality characteristics show a trend or go out of control. These plans are based on ISO standards in the sense that we retain the Consumer Risk Quality (CRQ) at the levels of the standard, but we then change the sample size to fit the application. It should be noted that the sample size may be lower than suggested by ISO 2859-1 if a procedure with a low acceptance number is chosen, whereas the sample size will increase if a more elaborate procedure with a larger acceptance number is chosen.

The second case study is from the other end of the spectrum. In this case we look at a vision system that enables us to do 100% control for many quality characteristics simultaneously. In this case we also retain the CRQ from ISO, but we rewrite the acceptance criterion to allow for 100% control of outgoing batches.

2 Sampling Plans from ISO 2859-1

ISO 2859-1 is a foundation for sampling from a series of batches operating on roughly the same quality level, i.e. the proportion of defects is fairly constant. The sampling plans are indexed according to AQL, i.e. Acceptance Quality Limit. Note that processes are expected to have averages better than AQL to avoid rejections under the sampling systems provided in the standard.

When sampling is performed according to ISO 2859-1, the sampling procedure includes at least sampling at the normal inspection level and the tightened inspection level. It is possible to include the reduced inspection level, if the process has proven to be stable and operates at a low proportion of defects compared to AQL.

When two of five consecutive batches are rejected on normal inspection, inspections must continue at the tightened inspection level. Inspection may resume on normal inspection level, if five consecutive batches are accepted at the tightened level. In most cases, the impact of tightened inspection compared to normal inspection is that the sample size is the same, but the acceptance number is reduced by one, which is a more demanding acceptance criterion. In most cases this corresponds to inspection at an AQL level one step lower. Hence, the process needs to operate at defect levels lower than the one step lower AQL for five consecutive batches to be accepted at tightened inspection in order for inspection to resume at the normal inspection level.

Finally, if five batches are rejected at the tightened inspection level, production must be terminated, and the process improved. Then one may start all over again, but inspection resumes at the tightened inspection level.

In the two cases in this paper, it would be rather confusing to operate with different sampling plans. To avoid this, without being out of compliance with the standard, we only employ plans on the tightened inspection level. This choice allows us to use one sampling plan for each AQL-value (corresponding to tightened inspection), but the termination rule still applies. For this concept to work, all parameters have to operate at a quality level better than their respective AQL-values.

3 IPC-Sampling on a Medical Filling Process Based on ISO 2859-1

In this case, we consider filling of liquid medical product. Such products are usually going through preparation, where raw materials and active pharmaceutical ingredients are mixed. Then product is filled into containers on a filling line, and afterwards the containers are inspected, labeled and packaged.

Until recently, sampling was performed after the product had been filled. This is efficient from a sampling point of view, since operators may just collect random samples during filling, and send them to a laboratory afterwards.

In reality, a number of potential defects may be or are entirely caused by the filling equipment. In such cases, it is not efficient to do the testing afterwards, since it is not possible to adjust the filling process if this is the case. Hence, it was decided to move the control for these types of defects from QC-labs into running production, and introduce In Process Control (IPC).

When the inspection is moved from sampling after filling to sampling during filling, it is important that the sample is still applicable for batch release. It would be a waste of time and money to do IPC sampling and then draw a sample for batch release afterwards. Hence, the sample size needs to be sufficient for batch release.

The parameters (quality characteristics) included will mostly be physical attributes of the product and primary packaging materials. Potential defects in the product itself are mostly the presence of various particles in the filled product. Particles may be either foreign material from e.g. the filling line or ingredients crystallizing on the container walls. From primary packaging a number of important parameters include missing glass beads in suspension products, chips and cracks in the container and "black" spots on the container or piston.

Usually, the potential defects moved to IPC do not include the chemical parameters such as potency, degradation products, conservatives, etc. These parameters are usually determined by batch preparation and raw materials and are not expected to change significantly during filling, unless filling tubes are open, product is subjected to excess heat or the process is stopped for a prolonged period of time.

The chemical parameters are tested during preparation, and it is not permissable to make changes during filling, if e.g. potency is too high, it would be a major GMP violation to try to add water to dilute it during filling. GMP – Good Manufacturing Practice – is a regulatory requirement for production, to ensure product safety and quality. Hence, the chemical parameters will usually remain fairly constant during filling, and are not considered in IPC controls at the moment. In the future when fast accurate measurement methods are available for potency, it will probably be included in IPC too.

3.1 Determination of Sample Size

When confronted with the prescribed sample sizes from ISO 2859-1, the production department initially decided that these sample sizes are too high for IPC. Hence, a method to determine smaller sample sizes without violating ISO 2859-1 requirements was needed. Furthermore, the nature of the sampling, as well as handling and analysis of samples, makes it highly problematic to change sample sizes and acceptance criteria between batches.

To ensure compliance with ISO 2859-1 without alternate sampling plans, we only have one solution, i.e. only to operate on the tightened inspection level. Hence, the parameters considered have to be operating at defect levels that are much better than the AQL, in most cases they need to be well below the next lower preferred AQL in the ISO tables, e.g. if AQL is 1.0%, the parameter has to operate at a defect level under 0.65%. If the proportion of defects does not remain under 0.65% in this case, there is a high probability of rejecting five batches over time, and hence the termination rule will apply. Note that this principle only eliminates the need for using the shifting rules; the termination rule still applies.

To assist the production department in lowering the sample sizes, we use the following algorithm.

1. Determine the relevant sampling plan from ISO 2859-1, based on AQL, batch size and tightened inspection.
2. Calculate the consumer risk quality, i.e. the proportion of defects with an OC-value of 10%.
3. For acceptance numbers 0–3, determine the sample size that satisfies the CRQ-value, i.e. the sample size has to be sufficiently large to ensure that the actual CRQ-value does not exceed the CRQ-value proposed by the ISO-plan under 1.

This procedure is efficient, if the processes operate well below the AQL-values of the parameters. Otherwise it will terminate rapidly. In the following example, an

Table 1 AQL-values for IPC

Parameter	AQL(%)
Pressure test	0.1
Defect capsule	0.4
Height of container including capsule	1.0
Cosmetics	2.5

Table 2 Sampling plans from ISO 2859-1 and corresponding CRQ-values

Parameter	AQL(%)	Sampling plan, (n,Ac)	CRQ (%)
Pressure test	0.1	(800,1)	0.485
Defect capsule	0.4	(800,5)	1.16
Height of container including capsule	1.0	(800,12)	2.21
Cosmetics	2.5	(500,18)	4.92

n is the sample size and Ac is the acceptance number

table is shown. The last column is the probability of accepting a batch, if the defect rate is at AQL. It is seen from the tables that accept zero plans are only efficient, if the defect rate is fairly close to 0.

3.2 Example

In this case the company needs to inspect and release batches of a medical product filled in glass containers. It is decided to do the inspection as IPC sampling and use the IPC-data for batch release as well.

The parameters and AQL-levels are shown in Table 1.

The shown AQL-values are for illustration only, not necessarily the values in practical applications.

In this case LQ/CRQ-values have not been determined. Hence, to assure the same risk to the customer as would be the case using ISO 2859-1, we use the 10%-point on the OC-curve, i.e. the proportion of defects corresponding to OC = 10%, to determine sampling plans with lower acceptance numbers.

Suppose we have batch sizes of 300,000 units and use inspection level II. In this case, we get the following sampling plans at tightened inspection having the same CRQ as ISO 2859-1 (Table 2).

The procedure is now controlled by the lowest AQL, in the sense that the sampling plan for the lowest AQL-value will determine the sample size. The other sampling plans for the higher AQL-values would allow for lower samples sizes, but since we are performing IPC controls, it will often be easier to run the IPC if we use the same sample size for all parameters. Hence, we will just adjust the acceptance numbers for the other AQL-values accordingly.

Table 3 Sampling plans for IPC

	AQL = 0.1%		AQL = 0.4%		AQL = 1.0%		AQL = 2.5%	
Sample size	Ac	PR(%)	Ac	PR(%)	Ac	PR(%)	Ac	PR(%)
475	0	37.8	2	29.6	6	20.1	16	9.2
805	1	19.3	5	10.7	12	6.5	31	0.8
1,100	2	9.9	7	7.8	17	3.1	44	0.1
1,380	3	5.1	10	2.6	23	0.8	57	0.0

To make a decision on IPC-samples, we usually determine the plans for the lowest AQL value with Ac $= 0, 1, 2, 3$. First we determine the sample size for each acceptance number, $n^\star$:

$$n^* = \min[P\{Bin(n, CRQ = 0.485\%) \leq Ac\} \leq 0.10] \tag{1}$$

Then we determine the acceptance numbers for all AQL-values above 0.1%, Ac*, as

$$Ac^* = \max[Ac|P\{Bin(n, CRQ_{AQL}) \leq Ac\} \leq 0.10], \tag{2}$$

where CRQ$_{AQL}$ is the CRQ-value corresponding to the respective AQL-values.

To enable reasonable decisions on sample sizes, we compute the OC-value at p = AQL, for all the suggested sampling plans. In the above case, the decision table is given as Table 3.

In Table 3, PR is the producers risk, i.e. the probability of rejecting the batch, given that the proportion of defective units equals the AQL.

The above results may not be a surprise to statisticians, but they do surprise production people. It is obvious from the table, that unless the parameter with an AQL of 0.1% operates at a defect level much lower than AQL, the accept-zero plan is useless, since the probability of rejecting a process with a defect rate at AQL is 37.8%. To a lesser degree, this conclusion applies to the other three AQL-values.

Depending on historical data, the plans for acceptance numbers 1, 2 and 3 may be considered. But if production periodically come close to the AQL in terms of defective units, we have to use the plan $(n, \mathrm{Ac}) = (1,380, 3)$ for AQL $= 0.1\%$. This is a fairly dramatic conclusion, since it is a 72.5% increase in sampling size compared to the standard plan ($n = 800$), but it is necessary to reduce the producer's risk if the process quality is not better than specified by the AQL.

From an applied view, however, the above table is very useful. It makes it much easier for engineers and chemists to understand the ramifications of their decisions, and tends to speed up decision-making.

This example suggests that we might often increase sample sizes instead of reducing them when basing IPC sampling on ISO standards. However, this is not a general property of the method. The reason for the larger sample sizes is that we operate on low AQL-values, which in turn lead to low acceptance numbers at the lowest AQL values. Hence, there is no room to in which to "optimize". If the smallest AQL had been higher, reductions might have been possible. Hence, the example illustrates a very important property of the sampling plans in ISO 2859-1. When the chosen sampling plan has low acceptance numbers, 0 or 1, it is critical to be operating at a better level than AQL to pass inspection, whereas processes may operate closer to AQL with higher acceptance numbers. On the rare occasion where the acceptance number is high, we may operate closer to the AQL. Using the above procedure, where all plans are on the tightened inspection level, the probability of rejecting a batch at the AQL will always be 5.1% or higher, hence we will never be able to operate at AQL with more than one parameter, without causing an unacceptable high overall probability of rejecting a batch that is basically satisfactory.

4 Determination of Release Criterion for 100% Inspection of Batches

During the inspection phase, many physical parameters are inspected using a visual inspection machine. In practice, all units in a batch are going through the inspection machine on a conveyer belt, a digital camera takes photos of the units and, based on the pictures, units are then classified as "go" or "no go" in real time using advanced software.

Due to the nature of this process, all units go through the vision system, and hence we do 100% control of submitted batches, contrary to standard sampling inspection. Since many sampling and other quality procedures are based on ISO standards, we would prefer this inspection to be similarly based.

Physical parameters inspected in this manner, include particles in product, glass beads (in suspension products), cosmetics, etc.

4.1 Determination of Acceptance Criterion

For this inspection, the sample size is the batch size. The unknown is the proportion of defects we are allowed to accept.

When using vision systems, the acceptance criterion is often related to detection rates obtained from human inspection also known as manual inspection. Clearly, the detection rates from a vision system should at least equal the detection rate from manual inspection.

To assure the consumer that quality is satisfactory, the proportion of undetected defects must not exceed the AQL value of the corresponding sampling plan from ISO 2859-1. Since we inspect all units, we may use the AQL as the limit in this application.

Now, let

DRV = Detection Rate for Vision system
DRM = Detection Rate for Manual inspection
FAR = False Alarm Rate
p = proportion of defects in a batch

Hence for the vision system, we obtain,

$$\text{Proportion of Detected Defects} = \text{p} \times \text{DRV} + (1\text{-p}) \times \text{FAR} \tag{3}$$

$$\text{Proportion of Undetected Defects} = \text{p} \times (1\text{-DRV}) \leq \text{AQL} \tag{4}$$

To assure the customer, we insert the maximum Eq. 4 into Eq. 3:
Proportion of Detected Defects = p × DRV

$$\leq \text{AQL} \times \text{DRV} \,/\, (1\text{-DRV}) \tag{5}$$

Hence, we denote the maximum value MPDD, i.e. Maximum Proportion of Detected Defects,

$$\text{MPDD} = \text{AQL} \times \text{DRV} \,/\, (1\text{-DRV}) \tag{6}$$

For practical purposes, the real issue in Eq. 6 is the False Alarm Rate. Since a vision system is based on digital images from a camera, it will depend on calibration and the environment in the production facility, and hence it will not be constant over time. It is time consuming (and sometimes costly) to evaluate all rejects from all batches and it would be contrary to the purpose of the procedure. Hence, a simple approach is to assume that FAR is 0. If a given batch yields too many rejects, we may determine FAR afterwards by inspection of all rejected items from that particular batch.

In the real world we do not observe the proportion of detected defect, but instead the proportion of perceived defects:

$$\text{Proportion of Perceived Defects} = \text{Proportion of Detected Defects} + (1\text{-p}) * \text{FAR} \tag{7}$$

If possible, the production will estimate the proportion of detected defects by the proportion of perceived defects. The problem is when a batch get rejected the reason might be the false alarm rate. Hence, if for whatever reason the batch is valuable to the company, we may go back and evaluate the false alarm rate by manual (or other) inspection of the rejected units, and reevaluate the proportion of defected defects. This way, good batches may be saved from scrap, which would be caused by a posi-tive false alarm rate.

This criterion should not stand alone – it would be advisable to run a control chart to monitor batch data. Otherwise we have a risk of the process drifting or shifting towards AQL, and ultimately exceeding it without prior warning. In contrast to the previous example, this monitoring scheme allows the process to operate close to the AQL without rejection.

Table 4 AQL-values for control by vision system

Parameter	AQL(%)
Defect capsule	0.1
Plunger orientation	0.4
Height of container including capsule	1.0
Cosmetics	2.5

Table 5 Acceptance criterion for control by vision system

Parameter	AQL (%)	Manual detection rate(%)	MPDD (%)
Defect capsule	0.1	50	0.1
Plunger orientation	0.4	90	3.6
Height of container including capsule	1.0	60	1.5
Cosmetics	2.5	90	22.5

In practice, this criterion is often used in a more realistic setting, using the detection rate for manual inspection – which is often much smaller – since the detection rate for the vision system is not known very well, and may fluctuate over time. Furthermore, when comparing data from different production sites it is beneficial to operate on a common threshold, i.e. manual detection rate, than having different detection rates from various production sites go into the systems.

It always has to be validated that the detection rate for the vision system is better than manual inspection, but since vision systems are complex devices that are calibrated between batches and so on, the detection rate is not necessarily constant for all parameters, and it is not necessarily well established.

4.2 Example

In this case the company needs to inspect and release batches of a medical product filled in glass containers. It is decided to do the inspection as IPC sampling and use the IPC-data for batch release as well. The parameters and AQL-levels are as follows (Table 4).

The shown AQL-values are for illustration only, not necessarily the values in practical applications.

As above the batch size is 300,000 units, and the AQL-values are the same as in Table 2. Then we apply Eq. 6 with FAR $= 0$ to obtain the acceptance criterion for each parameter (Table 5).

This method is still in development, so there are no real batch data available at this time. The procedure is conservative, in the sense that a huge problem in visual inspection is the presence of false rejects. Hence the proportion of defects from an inspection will be inflated by false rejects, and depending on the setting of the vision system this component may dominate in some scenarios.

We may, however, determine the actual FAR for a given batch afterwards by manual inspection of all rejected items. Then we may apply Eq. 7 with the actual FAR value and perform an improved release test of the given batch.

Batch results should be plotted in control charts and monitored for trends. Furthermore it is easy to ensure that the detection rate for the vision system is always at least as good as the manual reference or better.

References

ISO 2859–1 (1999) *Sampling procedures for inspection by attributes – Part 1: Sampling schemes indexed by acceptance quality limit (AQL) for lot-by-lot inspection*. Geneva: International Standardisation Organisation.

ISO 2859–2 (1985) *Sampling procedures for inspection by attributes – Part 2: Sampling plans indexed by limiting quality (LQ) for isolated lot inspection*. Geneva: International Standardisation Organisation.

ISO 3951–1 (2005) *Sampling procedures for inspection by variables – Part 1: Specification for single sampling plans indexed by acceptance quality limit (AQL) for lot-by-lot inspection – Single quality characteristic and single AQL*. Geneva: International Standardisation Organisation.

Part II
Off-line Control

Hybrid Space-Filling Designs for Computer Experiments

Rachel T. Johnson, Douglas C. Montgomery, and Kathryn S. Kennedy

Abstract Computer models play an increasingly important role in engineering design and in the study of complex systems, where physical experiments on the real system or even a prototype are prohibitively expensive. Both deterministic and stochastic computer models are used in these situations. A deterministic computer model is a set of complex equations whose solution depends on the input conditions and the levels of design factors or parameters but not on random elements. Examples include finite element models and computational fluid dynamics models. Space-filling designs are usually employed to study these deterministic computer models and often the modeling strategy involves fitting a spatial correlation or Kriging model (the Gaussian stochastic process model) to the data, because this model interpolates the experimental data exactly. We provide a survey of these designs and the modeling strategy, and propose a new type of hybrid space-filling design. The new design is a hybrid consisting of design points from a traditional space-filling design augmented by runs from a near saturated I-optimal design for a polynomial. We illustrate the construction of these designs with examples, and demonstrate their performance in response prediction for several situations. A comparison with standard space-filling designs is provided.

Keywords Gaussian process model • Optimal design • Linear regression • Response surface

---

R.T. Johnson
Operations Research Department, Naval Postgraduate School, Monterey, CA 93943, USA

D.C. Montgomery (✉) · K.S. Kennedy
School of Computing, Informatics and Decision Systems Engineering, Arizona State University, Tempe, AZ 85282, USA
e-mail: doug.montgomery@asu.edu; kathryn.kennedy@gmail.com

H.-J. Lenz et al. (eds.), *Frontiers in Statistical Quality Control 10*,
Frontiers in Statistical Quality Control 10, DOI 10.1007/978-3-7908-2846-7_19,

1 Introduction

Computer simulation models are often used in place of or in conjunction with physical experiments. Computer experiments can be computationally expensive in terms of time required to run an experiment on the simulation model. Therefore, surrogate models or metamodels are often used to mimic the input – output relationship in the form of a simpler mathematical expression that can be quickly computed. Surrogate models encompass a broad range of techniques ranging from parametric to nonparametric analysis.

The response surface in simulation model output can be very complex and determining which model fitting technique to use is dependent upon several factors such as the problem and the goal of the model. Two popular choices for surrogate models are linear regression models (polynomials) and Gaussian Process (GASP) models. See Santner et al. (2003) and Feng et al. (2006) for a review of using both models for computer simulation output. Ankenman et al. (2008) discusses the application of the GASP model to stochastic computer output.

Determining which experimental design technique should be used based on the choice of surrogate model is an important decision. There are several publications that evaluate experimental designs for computer simulations. Hussain et al. (2002) present seven two-dimensional functions that were used to test two metamodels. The metamodels tested were a radial basis function, which was originally developed to fit irregular topographic contours of geographical data, and quadratic polynomial models. Allen et al. (2003) compares combinations of experimental design classes with respect to second-order response surfaces and Kriging modeling methods. Bursztyn and Steinberg (2004) develop a new method of design comparison based on a Bayesian interpretation of an alias matrix. Chen et al. (2008) discuss various designs used for computer simulation models and various surrogate model choices.

Johnson et al. (2010) evaluate space-filling designs and optimal designs with respect to their performance when used to fit linear regression models. They compare designs based on their prediction variance. Their conclusions indicate that: (1) space-filling designs do not perform as well as optimal designs with respect to a linear regression model, (2) of the space-filling designs sphere packing designs generally have the lowest prediction variance followed closely by the Latin Hypercube designs, and (3) augmentation of space-filling designs with I-optimal points is suggested whenever initial modeling indicates that the computer simulation model can be adequately approximated by a polynomial. Their last point suggests that hybrid designs, which combine both optimal points and space-filling points, have the potential to be powerful designs. We compare hybrid space-filling experimental designs based on their prediction variance with respect to the Gaussian process model and linear regression models both theoretically and empirically.

2 Hybrid Space-Filling Designs

Building up information sequentially through design augmentation is efficient and economical. Montgomery (2009) points out that it is almost always preferable to run a fractional design, analyze the results, and then decide on the best set of runs to perform next. We believe that design augmentation can also be used in computer simulation modeling. Johnson et al. (2010) demonstrates that augmenting a space-filling design with optimal points can be effective in improving the prediction variance across the design region. This is important since the goal of a surrogate model is to closely approximate the computer model. In this paper we introduce augmentation of space-filling designs and referred to them as hybrid space-filling designs. The next subsections describe space-filling designs and optimal designs and then introduce the hybrid designs.

2.1 Space-Filling and Optimal Designs

Space-filling designs attempt to fill the interior portion of the design space. We chose the Latin hypercube to evaluate, which is one of the most popular designs for deterministic models. The Latin hypercube design was developed by McKay et al. (1979). The Latin Hypercube design is defined in Fang et al. (2006) as, "A Latin Hypercube design (LHD) with n runs and s input variables, denoted by LHD(n, s), is an $n \times s$ matrix, in which each column is a random permutation of $\{1, 2, \ldots, n\}$." Examples of applications of LHDs can be found in Welch et al. (1992) and Storlie and Helton (2008).

Special versions of the Latin Hypercube allow for the specification of additional criterion. One such specification is maximin distance. The maximin distance criterion maximizes the minimum inter-site distance and is specified by

$$\max_{D} \min_{u,v \in D} d(u,v) = \min_{u,v \in D_*} d(u,v),$$

where d(u,v) is a positive distance between two points within the design space D, and D_* represents the final design matrix. The maximin criterion is used as a secondary criterion for creating the LHDs. All of the LHD generated in this paper are maximin Latin Hypercube designs.

In this paper, we augment the LHDs with I-optimal points. Strictly I-optimal designs, or integrated variance designs, minimize the average scaled prediction variance over the design region with respect to a pre-specified model. Myers et al. (2009) discuss the I-optimal design with respect to the linear regression model in detail.

Table 1 Minimum number of design points needed ($n = p$)

	Order of polynomial			
Factors	2	3	4	5
2	6	10	15	21
3	10	20	35	56
4	15	35	70	126
5	21	56	126	252

2.2 Hybrid Space-Filling Designs

Hybrid space-filling designs are created by generating a space-filling LHD, in n points and then augmenting that design with m I-optimal points. The I-optimal points are constructed assuming that the model the experimenter plans to fit is a polynomial of a specified order. For this paper, designs were created to be saturated designs when fully completed, ranging from a full LHD to a full I-optimal design.

In order to test the predictive capabilities of space-filling designs and optimal designs when fitting a linear regression model, we generated designs ranging from two to five factors and used second order to fifth order polynomials to generate the $\mathbf{X}$ model matrix. To generate a space-filling design, no model specification is necessary, only the required number of points (sample size) is needed. To generate an optimal design, a model must be specified as well as the number of design points required. Table 1 illustrates the minimum number of design points (n) needed to fit a given polynomial with two to five factors (the number of parameters in the intended model is represented by p).

Four to seven designs were generated for each combination of number of factors and polynomial order, and sample size was held constant at the minimum number of design points needed so that designs within each category could be compared directly. Designs ranged from a full Latin hypercube to a full I-optimal design. For each combination, two to five smaller Latin hypercubes were augmented with I-optimal points. The notation used to identify designs throughout this work is LwIz_xF_yO, where w is the number of initial Latin hypercube points, z is the number of I-optimal points augmented to the Latin hypercube, x represents the number of factors, and y represents the order of the intended polynomial specified for the augmentation.

3 Results

3.1 Theoretical Prediction Variance

Here we compare our hybrid space-filling designs with strictly space-filling designs and strictly optimal designs using prediction variance over the design region.

The prediction variance is a standard criterion for comparing designs when modeling physical systems. The scaled prediction variance (SPV) can be evaluated at any point, x_0, within the design space as:

$$\frac{NV[\hat{y}(x_0)]}{\sigma^2} = Nx_0'(X'X)^{-1}x_0$$

where **X** represents the design matrix expanded to model form. For the SPV, the variance is multiplied by the number of runs, N, in order to penalize designs with larger sample size. However, no penalty is necessary when comparing designs of the same sample size. Since deterministic computer experiments have no stochastic component it is necessary to justify the use of scaled prediction variance as a performance criterion. Suppose that a given computer experiment is adequately modeled using a polynomial fit. The difference between the observed and fitted values in a deterministic computer model, however, is not stochastic error, but rather is model bias. If the polynomial model adequately describes the response surface of the true underlying function, the model bias of the fitted coefficients is negligible. The model bias of an individual prediction is also fairly small because the fit is adequate. Assume that the source of this bias is due to multiple high order terms. Thus deviations between the observed and predicted values will behave like the sum of a number of independent small quantities. Appealing to the central limit theorem, as the number of these bias quantities gets large, these deviations will converge to the normal distribution. We then justify the prediction variance criterion as a measure of the sum of a large number of small biases.

In order to compare the various designs, test spaces with 10,000 uniformly distributed points were generated. The prediction variance was then calculated over the entire design space for each design, assuming the full form of the polynomial model. Comparing the designs based on summary statistics can be problematic, since designs with the same mean prediction variance could have very different profiles. Hence, rather than trying to balance comparisons of the mean or maximum prediction variances, the prediction variances were sorted from smallest to largest and plots similar to Fraction of Design Space (FDS) plots (Zaharan et al. 2003) were generated. FDS plots graph the empirical distribution function of the prediction variance over the design space. They efficiently present a large amount of information, allowing for comparisons of design performance over the whole design space rather than simply comparing designs based on summary statistics such as the mean or maximum. Because the plot does not address location within the design space, it is only appropriate in cases where the entire design space is of equal importance. If some regions are more interesting than others, a weighting scheme or partitioning should be applied. The FDS plot for the four factor, second order polynomial case is presented in Fig. 1 as an example, assuming the full form of the model.

As expected, the full I-optimal designs performed best in terms of prediction variance, since the I-optimal criteria minimizes the average variance of prediction over the design region (with respect to the hypothesized model form). The full Latin

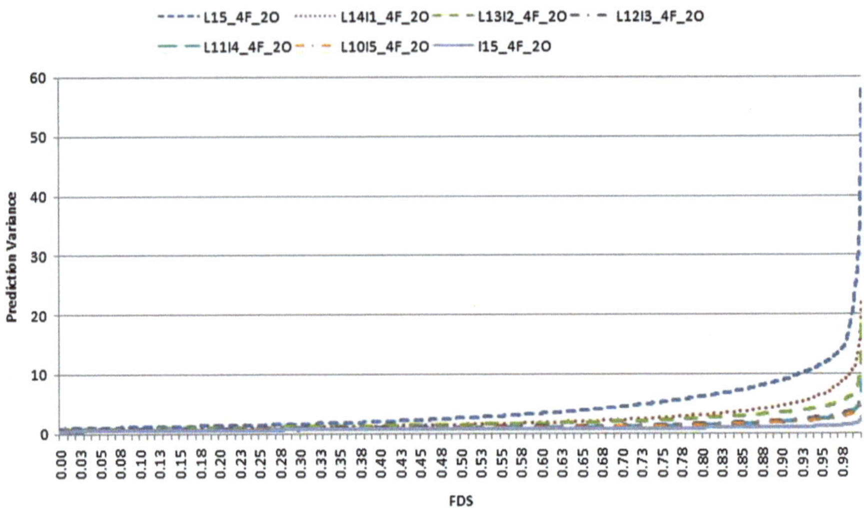

Fig. 1 FDS plot for four factor, second order designs

hypercube has the highest prediction variance across the whole design space, and the prediction variance reduces as more I-optimal points are added to the designs. The hybrid designs in the other design categories perform similarly.

3.2 *Empirical Mean Squared Error*

To evaluate the prediction properties of the GASP model and polynomials for the hybrid designs, a response variable was created for each of the designs using a test function. The designs were then analyzed using both a GASP model and a polynomial. There is likely strong multicollinearity present amongst the factors in the fitting of the polynomial models, particularly for the higher order models. While this means that the individual model parameters are likely to be poorly estimated, it does not necessarily imply that the full model will have poor predictive capability if used within the same design space. For the purpose of this work, the emphasis is not on the model fitting itself but rather to compare the predictive ability of the full models to that of the GASP models. The generated models were then used to predict the response values for 10,000 randomly generated test points, and the error calculated as the difference from the values predicted by the test function. For each of the test functions used, the function and its source is described, a response surface varying two of the input factors is shown, and results pertaining to root mean squared error (RMSE) for the linear regression models (polynomials) and GASP models is provided. Descriptions of the results are also included.

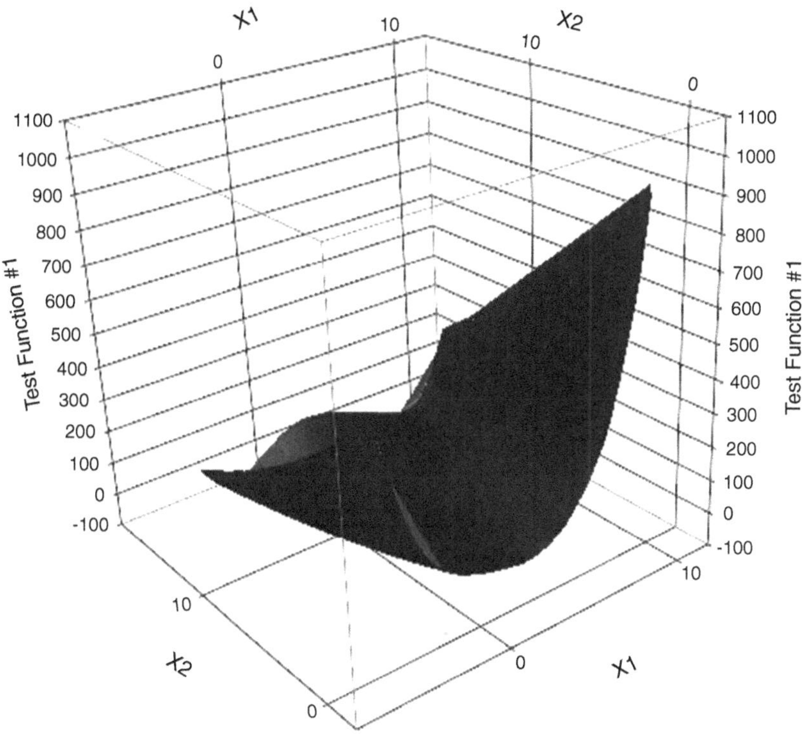

Fig. 2 Surface plot of test function 1

Test Function 1: The first test function was used by Santner et al. (2003), and first appeared in Brainin (1972). The function is

$$y = \left(x_2 - \frac{5.1}{4\pi^2}x_1^2 - \frac{5}{\pi}x_1 - 6\right)^2 + 10\left(1 - \frac{1}{8\pi}\right)\cos(x_1) + 10$$

$$x_1 \in [-5, 10], x_2 \in [0, 15]$$

A surface plot of test function 1 is presented in Fig. 2. The sample size (n) for the designs increases as the intended polynomial order increases, since the number of design points was held equal to the number of terms in the linear regression model. The fitting of the GASP model is also dependent on the sample size, since the fitted model will interpolate the design points. In Fig. 3, it can be seen that the RMSE for both models is reduced as the number of design points increases. There does not seem to be a tractable pattern of how the RMSE varies depending on the design composition (ratio of space-filling to I-optimal points). Because the location of the design points is a factor in both models, the lack of a defined pattern may be related to the fact that only one design was generated for each composition.

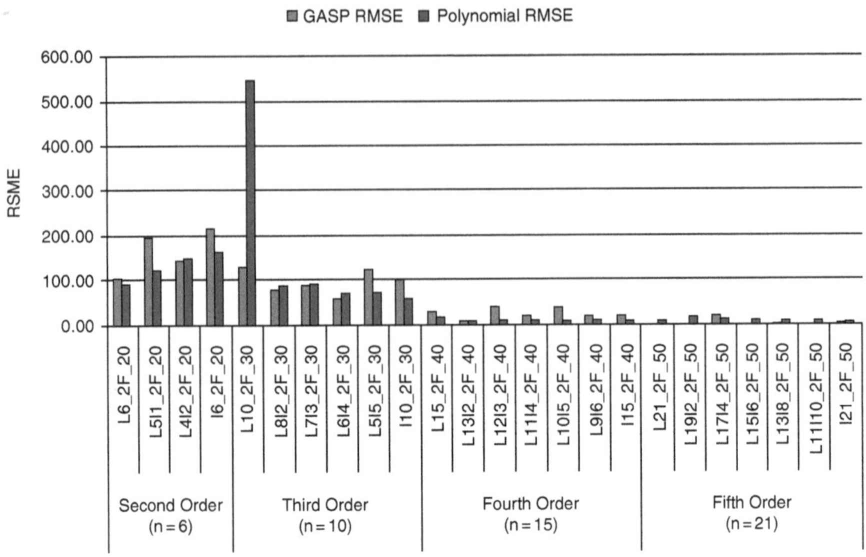

Fig. 3 RMSE for two-factor designs

Test Function 2: The second test function is found in Allen et al. (2003) and is designed to act as a surrogate model for a plastic seal design. The approximate analytical function is given as

$$y(x_1, x_2, x_3) = (105[0.58(x_2 + x_3 - 0.85) + 3.0]^3 x_2) \times \left(\frac{\sin\left[\dfrac{1.5x_3}{x_1 - 2.0}\right]}{(x_1 - 2.0)^2} \right)$$

where x_1, x_2 and x_3 represent input parameter dimensions on the plastic seal. The bounds for the parameters are (in millimeters): $4 \leq x_1 \leq 7$, $0.7 \leq x_2 \leq 1.7$, and $0.055 \leq x_3 \leq 0.500$. A surface plot of test function 2 is shown in Fig. 4, holding $x_3 = 0.2225$.

As with the two factor designs, in the three factor case both model types' RMSE improves as the sample size increases (Fig. 5). The models perform comparably, and there does not seem to be a tractable pattern of how the RMSE varies depending on the design composition (ratio of space-filling to I-optimal points).

Test Function 3: Our final test function was first published in Morris et al. (1993) and subsequently used for comparing metamodels in Allen et al. (2003). The function is

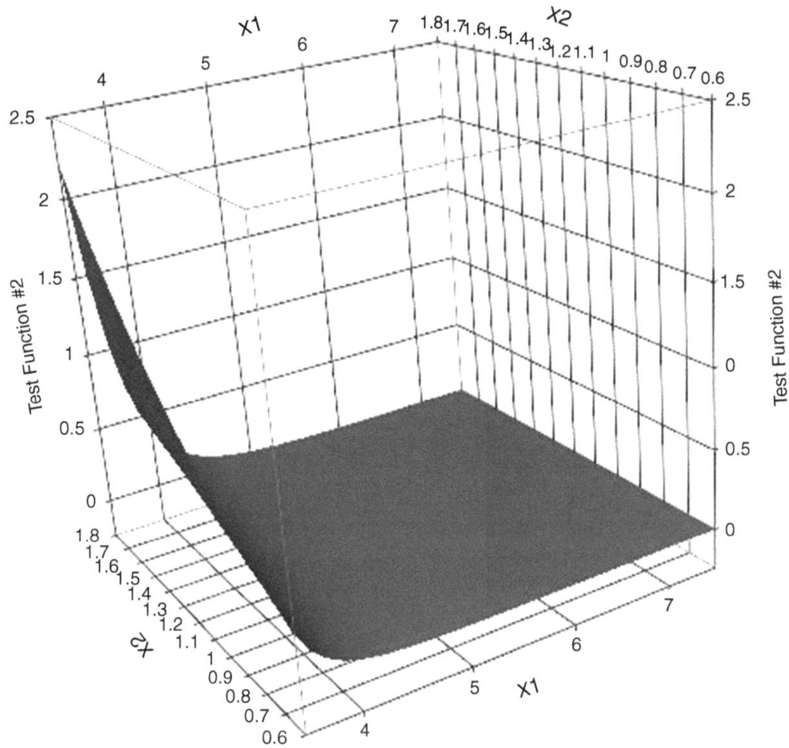

Fig. 4 Surface plot of test function 2 with $x_3 = 0.2225$

$$y(x_1, x_2, x_3, x_4, x_5, x_6, x_7, x_8) = \frac{2\pi x_3(x_4 - x_6)}{\ln\left(\dfrac{x_2}{x_1}\right)\left[1 + \dfrac{2x_7x_3}{\ln\left(\dfrac{x_2}{x_1}\right)x_1^2x_8} + \dfrac{x_3}{x_5}\right]}$$

where y predicts water flow – in cubic meters per year – as a function of eight design dimensions. As in Allen et al. (2003), we only vary x_1, x_4, x_6, and x_7 and set the other four variables at their midpoint of the specified ranges from the experiment demonstrated in Morris et al. (1993). The ranges and fixed values for each of the variables are presented in Table 2.

For the four factor designs, the polynomials' performance improves demonstrably as the number of design points and correspondingly the number of terms in the model increase. The GASP models exhibit the lowest RMSE values when n = 35, which corresponds well to work by Loeppky et al. (2008) that indicates that the GASP model works well given ten times the number of factors' worth of runs (Fig. 6).

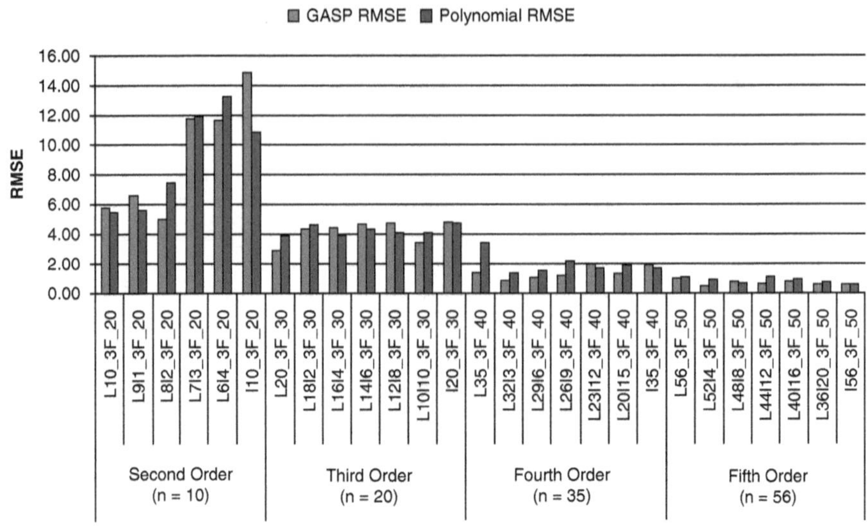

Fig. 5 RMSE for three factor designs

Table 2 The ranges and fixed values for the experimental and fixed variables in test function 3

Experimental variables			Fixed variables	
Variable	Low	High	Variable	Fixed value
*x*1	0.05	0.15	*x*2	25,050
*x*4	990	1,110	*x*3	89,335
*x*6	700	820	*x*5	89.6
*x*7	1,120	1,680	*x*8	9,855

Test function 3 was also used to evaluate the five factor designs. Factor x_2 was added to the factors that were varied, and x_3, x_5, and x_8 were all held constant at the same levels (Fig. 7).

Similar results were seen in the five factor designs and models as were evidenced in four factors. Polynomial model performance improves with the addition of more design points and model terms, while the GASP models exhibit the lowest RMSE values when n is approximately ten times the number of factors ($n = 56$).

3.3 *Design Variability*

As noted earlier, there is also variability imparted on these summary statistics based on the exact design employed. To assess how that variability affected the comparison of the designs, four additional designs were created for each of the design combinations for two factors.

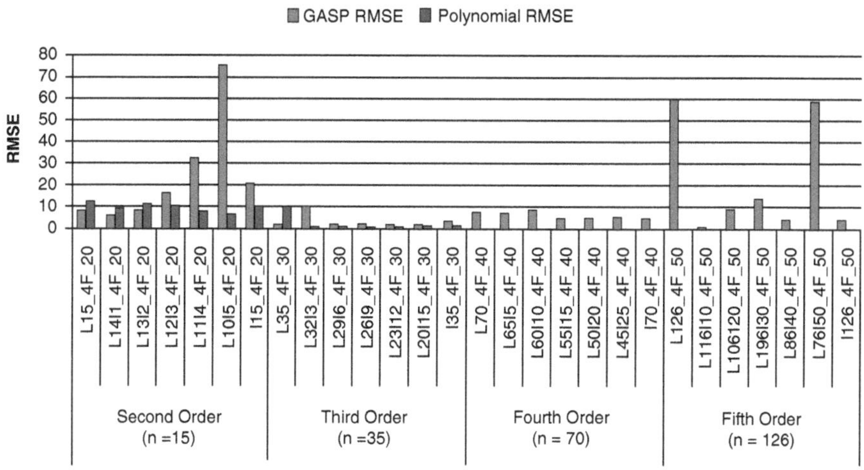

Fig. 6 RMSE for four factor designs

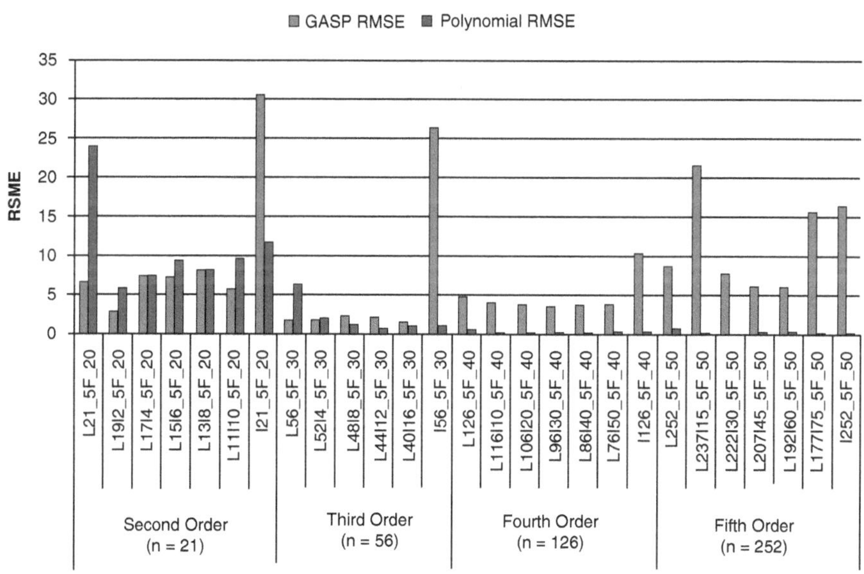

Fig. 7 RMSE for five factor designs

Figure 8 illustrates design variability. Each design was created by generating a Latin hypercube design in two factors with nine runs. Then each design was augmented with six I-optimal points, specifying a fourth order polynomial as the intended response function. The design space and sample size are the same, yet the points are placed in different locations.

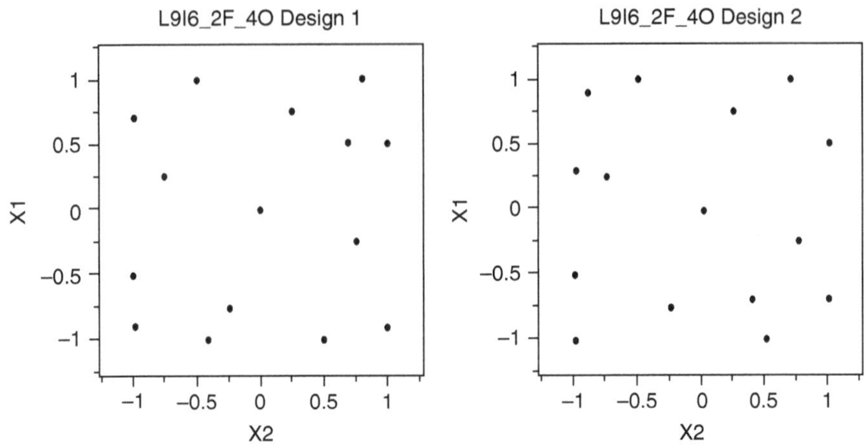

Fig. 8 Two factor, fourth order designs

The designs were then analyzed under two scenarios – one in which test function 1 was used in a deterministic fashion, and another in which normally distributed random error was added to make the process stochastic. In both cases, deterministic and stochastic, the designs perform as expected for the polynomials. The full I-optimal designs consistently have the lowest RMSE, while the full Latin hypercube designs consistently have the highest RMSE. As the number of I-optimal points in the design increases, the RMSE decreases. There was no apparent relationship between the mixture of design points and the RMSE for the GASP model, differences seemed to be solely related to sample size (as sample size increased, RMSE decreased) and whether the response was deterministic or stochastic (higher RMSE were evidenced in the stochastic case).

In general, the fitted error for the GASP models was higher than that of the polynomials for small designs ($n = 6$ and 10). As the number of design points increases, the GASP models begin to perform comparably to the polynomials in terms of fitted error, which corresponds to work by Loeppky et al. (2008) that indicates that the GASP model works well given ten times the number of factors' worth of runs. The GASP models also begin to perform comparably or better than the polynomials as error is introduced into the system.

As an example, the results for the two factor, fourth order designs are presented in the form of box plots. Figure 9 displays the RMSE values for each of the augmented space-filling designs fit to the responses with no random error, while Fig. 10 includes random error in the test function. Results from the GASP models and polynomials are presented side by side for comparison.

As can be seen by comparing Figs. 9 and 10, although the error in the stochastic case is higher for both GASP models and the polynomials, the GASP models are performing as well or better than the polynomials in the stochastic case. In the deterministic case, the RMSE for the polynomial models was much smaller than that of the GASP models.

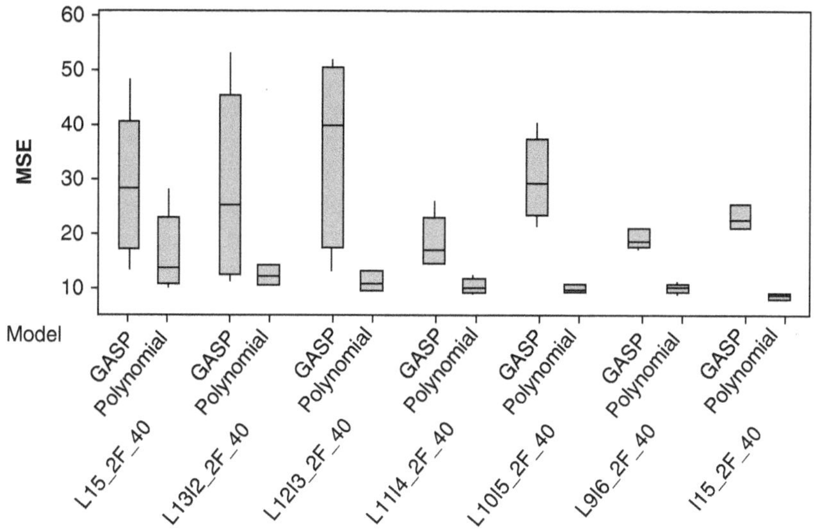

Fig. 9 Deterministic RMSE for two factor, fourth order designs

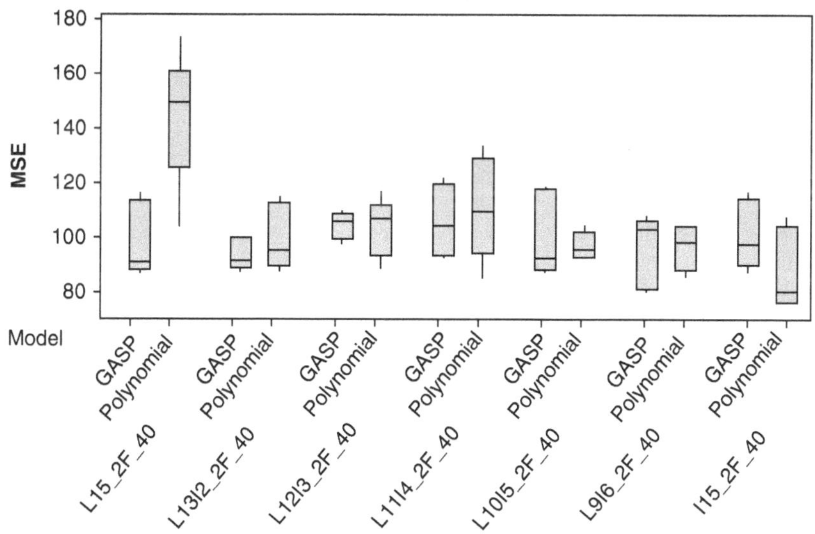

Fig. 10 Stochastic RMSE for two factor, fourth order designs

4 Conclusions

The results presented in this paper give insight into how hybrid space-filling designs perform with respect to prediction variance properties for the linear regression model and the GASP model. The designs are compared to both solely space-filling and solely optimal designs. Follow up work will be done to evaluate how the hybrid designs perform at larger sample sizes, and investigate further how design variability affects choice of model.

One of the benefits of a computer simulation models is the ability to build up a design sequentially, without concern for blocking or randomization. Note that in deterministic models replication and randomization are not needed and in stochastic models randomization can be controlled through the random number generator. Either way, in computer simulation experiments the space-filling-hybrid design is an excellent choice. After running a preliminary set of runs, the experimenter has a better idea of what modeling strategy to use. At this point the design can be augmented with a criterion that is optimal for that strategy.

While some might question the use of the space-filling design for polynomials at all, it is important to remember that in advance of any experimentation it is impossible to know whether a polynomial model of any order will prove to be adequate. Using a space-filling design for initial exploration makes considerable practical sense.

References

Allen, T. T., Bernshteyn, M. A., & Kabiri-Bamoradian, K. (2003). Constructing meta-models for computer experiments. *Journal of Quality Technology, 35*(3), 264–274.

Ankenman, B., Nelson, B. L., & Staum, J. (2008). Stochastic kriging for simulation metamodeling. In *Proceedings of the 2008 winter simulation conference*, Miami (pp. 362–370).

Bursztyn, D., & Steinberg, D. M. (2006). Comparison of designs for computer experiments. *Journal of Statistical Planning and Inference, 136*, 1103–1119.

Chen V., Tsui, K-L., Barton, R., & Meckensheime, M. (2006). A review on design, modeling and applications of computer experiments. *IEE Transactions, 38*, 273–291.

Fang, K. T., Li, R., & Sudjianto, A. (2006). *Design and modeling for computer experiments*. Boca Raton: Taylor & Francis.

Hussain, M. F., Barton, R. R., & Joshi, S. B. (2002). Metamodeling: Radial basis functions, versus polynomials. *European Journal of Operational Research, 138*, 142–154.

Johnson, R. T., Montgomery, D. C., Jones, B., & Parker, P. A. (2010). Comparing computer experiments using high order polynomial metamodels. *Journal of Quality Technology, 42*(1), 86–102.

Johnson, R. T., Montgomery, D. C., & Jones, B. (2010). An empirical study of the prediction performance of space-filling designs. *International Journal of Experimental Design and Process Optimisation* (to appear).

Jones, B., & Johnson, R. T. (2009). The design and analysis of the gaussian process model. *Quality and Reliability Engineering International, 25*, 515–524.

Loeppky, J.L., Sacks, J., & Welch, W. (2008). Choosing the sample size of a computer experiment: A practical guide. Technical Report Number 170, National Institute of Statistical Sciences.

McKay, N. D., Conover, W. J., & Beckman, R. J. (1979). A comparison of three methods for selecting values of input variables in the analysis of output from a computer code. *Technometrics, 21*, 239–245.

Montgomery, D. C. (2009). *The design and analysis of experiments*, 7th ed. New York: Wiley.

Morris, M. D., Mitchell, T. J., & Ylvisaker, D. (1993). Bayesian design and analysis of computer experiments: Use of derivatives in surface prediction. *Technometrics, 35*, 243–255.

Myers, R. H., Montgomery, D. C., & Anderson-Cook, C. M. (2002). *Response surface methodology: process and product optimization using designed experiments*, 3rd ed. New York: John.

Santner, T. J., Williams, B. J., & Notz, W. I. (2003). *The design and analysis of computer experiments*. Springer series in statistics. New York: Springer.

Storlie, C. B., & Helton, J. C. (2008). Multiple predictor smoothing methods for sensitivity analysis: Example results. *Reliability Engineering & System Safety, 93*, 55–77.

Welch, W. J., Buck, R. J., Sacks, J., Wynn, H. P., Mitchell, T. J., & Morris, M. D. (1992). Screening, predicting, and computer experiments. *Technometrics, 34*(1), 15–25.

Zahran, A., Anderson-Cook, C. M., & Myers, R. H. (2003). Fraction of design space to assess the prediction capability of response surface designs. *Journal of Quality Technology, 35*, 377–386.

Optimal Design for Multifactor Life Testing Experiments for Exponentially Distributed Lifetimes

Brandon R. Englert, Steven E. Rigdon, Connie M. Borror, Douglas C. Montgomery, and Rong Pan

Abstract Life testing experiments differ from most experiments in a number of ways. Instead of assuming a normal distribution for the response, we often assume a distribution such as the exponential or Weibull. Also, censoring, the termination of a life test before all units have failed, is common in life testing experiments. We investigate algorithms to obtain optimal, or near-optimal designs for multifactor experiments assuming an exponential distribution for the lifetimes.

Keywords A-optimality • D-optimality • Bayesian optimal design • Genetic algorithm

1 Introduction

There are a number of aspects that make design of experiments for life testing different from most traditional designs.

First, while most work on optimal design assumes a normally distributed response, the distribution of lifetimes is usually not normal. Rather, the distribution of life times is usually skewed to the right, so distributions such as the exponential, gamma and Weibull are often used.

Second, it is likely that several of the items placed on test have very long lives that last well beyond the planned time of the experiment. Thus, in order to terminate

---

B.R. Englert · S.E. Rigdon (✉)
Department of Mathematics and Statistics, Southern Illinois University Edwardsville, 1333 Science Building, Edwardsville, IL, 62026-1653, USA
e-mail: bfish@siue.edu; srigdon@siue.edu

C.M. Borror · D.C. Montgomery · R. Pan
School of Computing, Informatics and Decision Systems Engineering, Arizona State University, Tempe, AZ 85281, USA
e-mail: conni@asu.edu; doug.montgomery@asu.edu; Rong.Pan@asu.edu

H.-J. Lenz et al. (eds.), *Frontiers in Statistical Quality Control 10*,
Frontiers in Statistical Quality Control 10, DOI 10.1007/978-3-7908-2846-7_20,

the life test in a reasonable amount of time, life tests are often *censored*. The most common censoring plans are *time censoring* (also called Type I censoring), where the test is terminated at a predetermined time, and *failure censoring* (also called Type II censoring), where the test is terminated after a predetermined number of failures. There are other types of censoring, including hybrids of these. A good design for life testing must account for the possibility that some items will be censored. In fact, it is possible to include the censoring time as an input variable, and therefore choose an optimal censoring time.

Third, life tests can have an accelerating factor, that is, a variable designed to make the items fail sooner. Life tests, by their very nature, take time, often a long time. In order to complete the experiment in a reasonable amount of time a variable such as temperature (or voltage, vibration, etc.) is used to induce early failures. By using varying levels of this accelerating factor, it is possible to extrapolate back to the unaccelerated use condition. For example, by running a test on the lifetime of tires at 200°C, 150°C, and 100°C, users could extrapolate to the actual use condition of say 50°C. Running the experiment at 50°C would likely lead to a very long test, much longer that the test at these more extreme values. Thus, the design of the acceleration plan can be part of the overall design for the life test.

For a normally distributed response, most measures of design optimality, e.g., D-, A-, and I-optimality, do not depend on the unknown parameters. For life testing distributions like the Weibull or gamma, these criteria do depend on unknown parameters. This creates the dilemma that we must know the unknown parameters in order to construct an optimal design, but if we knew the parameters, we wouldn't have to bother with designing an experiment so we could learn about the parameters.

One approach to this dilemma is to guess the value of the parameter, and then construct a design to be optimal for this choice. But if our guess for the parameter is far from its true value, then the design may be far from optimal. The approach taken by Gotwalt et al. (2009) was to propose a prior for the parameter θ and then average over the prior. In this case the objective is to choose the design to maximize or minimize

$$v(\mathbf{d}) = \int_{\Theta} v(\boldsymbol{\theta}, \mathbf{d})\, p(\boldsymbol{\theta})\, d\boldsymbol{\theta}.$$

This method involves a mix of classical (the design criterion $v(\boldsymbol{\theta}, \mathbf{d})$) and Bayesian (the prior $p(\boldsymbol{\theta})$) approaches, so we call this "pseudo-Bayes" design optimality.

In this paper, we take a purely Bayesian approach for constructing optimal designs for life testing experiments.

2 Bayesian Design

The Bayesian approach to design involves the following aspects of the design and the data.

- **d** is the design. This will include the values for the predictor variables that will be used in the experiment. It could also include other aspects of the experiment, such as the censoring scheme. Each run in the experiment consists of a set of values of the predictor variables selected from the design space **D**. What is often called the design matrix **X** is obtained from **d** by adding a column of ones (if there is a constant term in the model) and additional terms corresponding to higher-order terms in the model.
- $\boldsymbol{\theta}$ is the parameter, which is assumed to come from the parameter space Θ.
- **y** is the observed data, whose distribution depends on the parameter $\boldsymbol{\theta}$ and the design **d**. The sample space $\mathcal{Y}$ consists of all possible outcomes for **y**.
- The prior distribution $p(\boldsymbol{\theta})$ is independent of the design **d**. The distribution for **y**, $p(\mathbf{y}|\boldsymbol{\theta},\mathbf{d}) = L(\boldsymbol{\theta}|\mathbf{y},\mathbf{d})$, depends on both the parameter $\boldsymbol{\theta}$ and the design **d**. The posterior of $\boldsymbol{\theta}$ also depends on both **y** and **d**. The posterior of $\boldsymbol{\theta}$ is obtained from Bayes theorem

$$p(\boldsymbol{\theta}|\mathbf{y},\mathbf{d}) = \frac{p(\boldsymbol{\theta})\, p(\mathbf{y}|\boldsymbol{\theta},\mathbf{d})}{\int_{\Theta} p(\boldsymbol{\theta})\, p(\mathbf{y}|\boldsymbol{\theta},\mathbf{d})\, d\boldsymbol{\theta}}.$$

- $u(\mathbf{d},\boldsymbol{\theta},\mathbf{y})$ is a utility function that incorporates the utility of using design **d** when the true parameter is $\boldsymbol{\theta}$ and the data are given by the vector **y**.

The optimal design is then the one that maximizes expected utility, that is, the utility integrated over both the parameter space Θ and the sample space $\mathcal{Y}$:

$$\begin{aligned} u(\mathbf{d}) &= \int_{\mathcal{Y}}\int_{\Theta} u(\mathbf{d},\boldsymbol{\theta},\mathbf{y})\ p(\boldsymbol{\theta},y|\mathbf{d})\ d\boldsymbol{\theta}\ d\mathbf{y} \\ &= \int_{\mathcal{Y}}\int_{\Theta} u(\mathbf{d},\boldsymbol{\theta},\mathbf{y})\ p(\boldsymbol{\theta})\ p(\mathbf{y}|\mathbf{d},\boldsymbol{\theta})\ d\boldsymbol{\theta}\ d\mathbf{y} \end{aligned}$$

There are a number of choices for the utility function $u(\mathbf{d},\boldsymbol{\theta},\mathbf{y})$. Two of the most common choices are based on the Shannon information or on quadratic loss.

The expected gain in information from an experiment is the expected logarithm of the ratio of the posterior and the prior; that is

$$\begin{aligned} U(\mathbf{d}) &= \int_{\mathcal{Y}}\int_{\Theta} \log\left(\frac{p(\boldsymbol{\theta}|\mathbf{y},\mathbf{d})}{p(\boldsymbol{\theta})}\right)\ p(\boldsymbol{\theta})\ p(\mathbf{y}|\mathbf{d},\boldsymbol{\theta})\ d\boldsymbol{\theta}\ d\mathbf{y} \\ &= \int_{\mathcal{Y}}\int_{\Theta} \log\left[p(\boldsymbol{\theta}|\mathbf{y},\mathbf{d})\right]\ p(\boldsymbol{\theta})\ p(\mathbf{y}|\mathbf{d},\boldsymbol{\theta})\ d\boldsymbol{\theta}\ d\mathbf{y} \\ &\quad - \int_{\Theta} \log\left[p(\boldsymbol{\theta})\right]\ p(\boldsymbol{\theta}) \left(\int_{\mathcal{Y}} p(\mathbf{y}|\mathbf{d},\boldsymbol{\theta})\ d\mathbf{y}\right) d\boldsymbol{\theta} \\ &= \int_{\mathcal{Y}}\int_{\Theta} \log\left[p(\boldsymbol{\theta}|\mathbf{y},\mathbf{d})\right] p(\boldsymbol{\theta})\, p(\mathbf{y}|\mathbf{d},\boldsymbol{\theta})\, d\boldsymbol{\theta}\, d\mathbf{y} - \int_{\Theta} \log\left[p(\boldsymbol{\theta})\right] p(\boldsymbol{\theta})\,(1)\, d\boldsymbol{\theta} \end{aligned}$$

Since the second term in this last expression is independent of **d**, maximizing the expected gain in information is equivalent to maximizing the Shannon information in the posterior (SIP):

$$U_{\mathrm{SIP}}(\mathbf{d}) = \int_{\mathcal{Y}} \int_{\Theta} \log\left[p\left(\boldsymbol{\theta} \mid \mathbf{y}, \mathbf{d}\right)\right] \; p\left(\boldsymbol{\theta}\right) \; p\left(\mathbf{y} \mid \mathbf{d}, \boldsymbol{\theta}\right) \; d\boldsymbol{\theta} \; d\mathbf{y}.$$

Another approach is to maximize the negative or the reciprocal of the squared error loss (NSEL and RSEL, respectively); for example,

$$\begin{aligned} U_{\mathrm{NSEL}}(\mathbf{d}) &= -E\left[\left(\boldsymbol{\theta} - \hat{\boldsymbol{\theta}}\right)' A \left(\boldsymbol{\theta} - \hat{\boldsymbol{\theta}}\right)\right] \\ &= -\int_{\mathcal{Y}} \int_{\Theta} \left(\boldsymbol{\theta} - \hat{\boldsymbol{\theta}}\right)' A \left(\boldsymbol{\theta} - \hat{\boldsymbol{\theta}}\right) \; p\left(\boldsymbol{\theta}\right) \; p\left(\mathbf{y} \mid \mathbf{d}, \boldsymbol{\theta}\right) \; d\boldsymbol{\theta} \; d\mathbf{y} \end{aligned}$$

for some matrix A. Here, $\hat{\boldsymbol{\theta}} = E\left[\boldsymbol{\theta} \mid \mathbf{y}, \mathbf{d}\right]$ is the posterior mean of $\boldsymbol{\theta}$. If $A = I$, then this approach corresponds to minimizing the sum of the expected posterior variances for the parameters in the vector $\boldsymbol{\theta}$. See Chaloner and Verdinelli (1995) for a thorough discussion of the choice of a utility function.

Unfortunately, some of the simulation based optimal design approaches described in the next section require a nonnegative utility function $u\left(\mathbf{d}, \boldsymbol{\theta}, \mathbf{y}\right)$, and both of the utility functions u_{SIP} and u_{NSEL} are sometimes, or always, nonpositive. We could add a constant to the values of u_{NSEL} or u_{SIP} to ensure that all values are positive. As alternatives to u_{NSEL} and u_{RSEL}, we could choose as the utility function

$$U_{\mathrm{N}}(\mathbf{d}) = \exp\left[-E\left(\left(\boldsymbol{\theta} - \hat{\boldsymbol{\theta}}\right)' A \left(\boldsymbol{\theta} - \hat{\boldsymbol{\theta}}\right)\right)\right], \tag{1}$$

or

$$U_{\mathrm{C}}(\mathbf{d}) = \frac{1}{1 + E\left(\left(\boldsymbol{\theta} - \hat{\boldsymbol{\theta}}\right)' A \left(\boldsymbol{\theta} - \hat{\boldsymbol{\theta}}\right)\right)}. \tag{2}$$

Because the functional forms in Eqs. 1 and 2 are similar to the normal and Cauchy distributions, we refer to these as the *normal* loss function and the *Cauchy* loss function, respectively. As an alternative to $U_{\mathrm{SIP}}(\mathbf{d})$, we could use

$$U_{\mathrm{ESI}}(\mathbf{d}) = \exp\left[-\int_{\mathcal{Y}} \int_{\Theta} \log\left[p\left(\boldsymbol{\theta} \mid \mathbf{y}, \mathbf{d}\right)\right] \; p\left(\boldsymbol{\theta}\right) \; p\left(\mathbf{y} \mid \mathbf{d}, \boldsymbol{\theta}\right) \; d\boldsymbol{\theta} \; d\mathbf{y}\right]. \tag{3}$$

We call this the exponential Shannon information utility function.

Many other utility functions are possible. In the reliability context, we could base the utility function on estimation of a fixed percentile of the lifetime distribution or the reliability at some fixed time for some level of the input factors.

3 Optimization Algorithms

Once a utility function is selected, we must choose an algorithm to perform the maximization of $u(\mathbf{d})$. We consider three algorithms: a genetic algorithm and two simulation based optimization algorithms.

3.1 Genetic Algorithms

The concept of using a heuristic algorithm to mimic biological evolution is usually attributed to Holland (1975). The ideas behind genetic algorithms have been broadened, and many similar algorithms are now called evolutionary algorithms (Eiben and Smith 2003). Heredia-Langner et al. (2003) used a genetic algorithm to find near D-optimal designs assuming a normally distributed response.

Genetic algorithms begin with a collection of candidate designs and then proceed iteratively through three steps: recombination, mutation, and selection. A genetic algorithm for selecting an optimal design, given a utility function $u(\mathbf{d}, \boldsymbol{\theta}, \mathbf{y})$, would follow these steps. Begin with a collection $\mathcal{D}$ of possible designs; usually the number N of designs in $\mathcal{D}$ is rather large, often in the hundreds. Then, select M pairs of designs from $\mathcal{D}$ and have each pair exchange some common features. This could mean creating new designs by selecting some components at random from each of the two designs. This will thus create M new designs that are added to the collection $\mathcal{D}$. This step is called recombination, or crossover. Next, select K designs from $\mathcal{D}$ and impose slight random changes to some of the elements in the selected designs. This is the mutation step and it adds an additional K designs to the collection $\mathcal{D}$. Finally, the utility of each of the $N+M+K$ designs that now comprise $\mathcal{D}$ are computed. The designs with the highest utility have the greatest chance of being selected to survive to the next generation. Various strategies can be applied here, including selecting the N designs with the highest utility, or making the probability of survival proportional to the utility function. Hybrids of these strategies can also be used. These steps are repeated through a number of generations. We always keep track of the best design in each generation.

3.2 Müller's Markov Chain Simulation Algorithm

Müller (1999) and Müller et al. (2004) proposed the use of an augmented probability model together with Markov Chain Monte Carlo (MCMC) and simulated annealing. The basic idea behind their algorithm is to place an auxiliary probability distribution on the design $\mathbf{d}$ together with the parameter $\boldsymbol{\theta}$ and the data $\mathbf{y}$. If we choose

$$h(\mathbf{d}, \boldsymbol{\theta}, \mathbf{y}) \propto u(\mathbf{d}, \boldsymbol{\theta}, \mathbf{y})\ p(\boldsymbol{\theta}, \mathbf{y}|\mathbf{d}), \tag{4}$$

that is, the PDF is proportional to the utility function, then the marginal of **d** is then

$$h(\mathbf{d}) \propto \int_{\mathcal{Y}} \int_{\Theta} u(\mathbf{d}, \boldsymbol{\theta}, \mathbf{y})\, p(\boldsymbol{\theta}, \mathbf{y}|\mathbf{d})\; d\boldsymbol{\theta}\; d\mathbf{y} = U(\mathbf{d}).$$

Thus, the optimal design problem reduces to the problem of finding the mode for the distribution of the marginal of **d**.

The marginal of **d** is, however, often very flat, making it difficult to find the mode. Müller (1999) and Müller et al. (2004) suggest using J copies of $(\boldsymbol{\theta}, \mathbf{y})$ instead of just one, as in Eq. 4. Thus,

$$h(\mathbf{d}, (\boldsymbol{\theta}_1, \mathbf{y}_1), (\boldsymbol{\theta}_2, \mathbf{y}_2), \ldots, (\boldsymbol{\theta}_J, \mathbf{y}_J)) \propto \prod_{j=1}^{J} u\left(\mathbf{d}, \boldsymbol{\theta}_j, \mathbf{y}_j\right)\; p\left(\boldsymbol{\theta}_j, \mathbf{y}_j|\mathbf{d}\right)$$

which leads to

$$h_J(\mathbf{d}) \propto \left[\int_{\mathcal{Y}} \int_{\Theta} u(\mathbf{d}, \boldsymbol{\theta}, \mathbf{y})\, p(\boldsymbol{\theta}, \mathbf{y}|\mathbf{d})\; d\boldsymbol{\theta}\; d\mathbf{y}\right]^J = [U(\mathbf{d})]^J.$$

Raising $U(\mathbf{d})$ to the power J serves to "sharpen" the PDF near the mode, making it easier for the Markov chain simulation to find it. Note that all along we have been writing $p(\boldsymbol{\theta}, \mathbf{y}|\mathbf{d})$ as if we were conditioning on **d**; it would have been more accurate to write $p_{\mathbf{d}}(\boldsymbol{\theta}, \mathbf{y})$ to indicate the dependence on **d**. Now, however, we are conditioning on **d** in the probabilistic sense, so we write $p(\boldsymbol{\theta}, \mathbf{y}|\mathbf{d})$. The value of J is chosen in such a way that $J \to \infty$ in the Markov chain simulation, although in practice, J usually takes on moderately small values and grows linearly or logarithmically. The reciprocal of J can be thought of as "temperature" in a simulated annealing algorithm. As $J \to \infty$, $T = 1/J \to 0$, a requirement of simulated annealing algorithms.

The algorithm of Müller et al. (2004) uses a Metropolis-Hastings algorithm to move through the design space in search of the mode for $h_J(\mathbf{d})$. Their algorithm can be described as follows.

3.2.1 Markov Chain Simulation Algorithm of Müller

Step 0: Set $t = 0$ and begin with an initial design $\mathbf{d}^{(0)}$. Set $J = J(t)$. Then for $j = 1, 2, \ldots, J$, simulate $\boldsymbol{\theta}_j \sim p(\boldsymbol{\theta})$ and $\mathbf{y}_j \sim p\left(\mathbf{y}|\boldsymbol{\theta}, \mathbf{d}^{(0)}\right)$, and compute

$$u^{(0)} = \prod_{j=1}^{J} u\left(\mathbf{d}, \boldsymbol{\theta}_j, \mathbf{y}_j\right).$$

Step 1: Generate a candidate design $\tilde{\mathbf{d}}$ by some mutation of $\mathbf{d}$. Specifically, simulate

$$\begin{aligned} \tilde{\mathbf{d}} &\sim q\left(\tilde{\mathbf{d}}|\mathbf{d}\right) && \text{[the proposal distribution]} \\ \boldsymbol{\theta}_j &\sim p\left(\boldsymbol{\theta}\right) && \text{[the prior for } \boldsymbol{\theta}\text{]} \\ \mathbf{y}_j &\sim p\left(\mathbf{y}|\boldsymbol{\theta},\tilde{\mathbf{d}}\right) && \text{[the likelihood]} \end{aligned}$$

and compute

$$\tilde{u} = \prod_{j=1}^{J} u\left(\tilde{\mathbf{d}},\boldsymbol{\theta}_j,\mathbf{y}_j\right).$$

Step 2: We must choose whether to move from $\mathbf{d}^{(0)}$ to the proposed design $\tilde{\mathbf{d}}$, so we compute the acceptance probability

$$\alpha = \min\left(1, \frac{\tilde{u}\, q\left(\mathbf{d}^{(0)}|\tilde{\mathbf{d}}\right)}{u^{(0)}\, q\left(\tilde{\mathbf{d}}|\mathbf{d}^{(0)}\right)}\right).$$

Note that if q is a symmetric function, then α reduces to $\alpha = \min\left(1, \tilde{u}/u^{(0)}\right)$.

Step 3: Assign $\left(\mathbf{d}^{(1)}, u^{(1)}\right) = \left(\tilde{\mathbf{d}}, \tilde{u}\right)$ with probability α, and $\left(\mathbf{d}^{(1)}, u^{(1)}\right) = \left(\mathbf{d}^{(0)}, u^{(1)}\right)$ with probability $1-\alpha$. In other words, accept the move to $\left(\mathbf{d}^{(1)}, u^{(1)}\right)$ with probability α, and remain in state $\left(\mathbf{d}^{(0)}, u^{(0)}\right)$ with probability $1-\alpha$.

Step 4: Set $t = t+1$, and $J = J(t)$. Next, set $\left(\mathbf{d}^{(0)}, u^{(0)}\right) = \left(\mathbf{d}^{(1)}, u^{(1)}\right)$.

Step 5: Repeat Steps 1–4 until the Markov chain has practically converged.

When the Markov chain $\left(\mathbf{d}^{(t)}, u^{(t)}\right)$ has settled down to its steady state (for all practical purposes), the values of $\mathbf{d}^{(t)}$ should be near the mode of h. We could then take as our near-optimal design the value of $\mathbf{d}$ that produced the largest $u^{(t)}$ after allowing for the Markov chain to burn in.

3.3 Interacting Particle Algorithm

Amzal et al. (2006) proposed a generalization of the Markov chain simulation algorithm described in the previous subsection. They suggest simulating M parallel Markov chains, so that at each time step t, the Markov chain state is $\left(\left(\mathbf{d}_1^{(t)}, u_1^{(t)}\right), \left(\mathbf{d}_2^{(t)}, u_2^{(t)}\right), \ldots, \left(\mathbf{d}_M^{(t)}, u_M^{(t)}\right)\right)$. Each of the pairs $\left(\mathbf{d}_i^{(t)}, u_i^{(t)}\right)$ is called a particle. The idea of carrying many designs at a time is reminiscent of genetic algorithms. The algorithm then proceeds as follows.

3.3.1 Algorithm of Amzal et al. (2006)

Step 0: Let $\left(\mathbf{d}_1^{(0)}, \mathbf{d}_2^{(0)}, \ldots, \mathbf{d}_M^{(0)}\right)$ be a collection of M designs. Set $J = J(0)$.
Step 1: For $i = 1, 2, \ldots, M$:

1a. Use importance sampling (IS) to simulate the proposal design $\tilde{\mathbf{d}}_i^{(1)}$ given $\tilde{\mathbf{d}}_i^{(0)}$. This could be done using a random walk, or $\mathbf{d}_i^{(1)}$ could be chosen to be independent of $\tilde{\mathbf{d}}_i^{(0)}$. Call this proposal density $q_{\text{IS}}\left(\tilde{\mathbf{d}}|\mathbf{d}_i^{(0)}\right)$.

1b. For $j = 1, 2, \ldots, J$, simulate $\tilde{\boldsymbol{\theta}}_{ij}^{(1)}$ from $p(\boldsymbol{\theta})$ and $\tilde{\mathbf{y}}_{ij}$ from $p\left(\mathbf{y}|\tilde{\boldsymbol{\theta}}_{ij}^{(1)}, \mathbf{d}_i^{(1)}\right)$.

1c. Compute

$$\tilde{u}_i^{(1)} = \prod_{j=1}^{J} u\left(\mathbf{d}_i^{(1)}, \tilde{\boldsymbol{\theta}}_{ij}^{(1)}, \tilde{\mathbf{y}}_{ij}\right) \quad \text{and} \quad w_i^{(1)} = \frac{\tilde{u}_i^{(1)}}{q_{\text{IS}}\left(\tilde{\mathbf{d}}_i^{(1)}|\tilde{\mathbf{d}}_i^{(0)}\right)}.$$

1d. Normalize the weights $w_i^{(1)}$ so that they sum to 1.

Step 2: Use multinomial resampling to select $\left(\hat{\mathbf{d}}_1^{(1)}, \hat{\mathbf{d}}_2^{(1)}, \ldots, \hat{\mathbf{d}}_M^{(1)}\right)$ from $\left(\tilde{\mathbf{d}}_1^{(1)}, \tilde{\mathbf{d}}_2^{(1)}, \ldots, \tilde{\mathbf{d}}_M^{(1)}\right)$ using weights $w_i^{(1)}$ and utilities $\tilde{u}_i^{(1)}$.
Step 3: For $i = 1, 2, \ldots, M$

3a. Use the Metropolis-Hastings (MH) algorithm to simulate $\bar{\mathbf{d}}_i^{(1)}$ from some proposal distribution $q_{\text{MH}}\left(\mathbf{d}|\hat{\mathbf{d}}_i^{(1)}\right)$.

3b. For $j = 1, 2, \ldots, J$, simulate $\bar{\boldsymbol{\theta}}_{ij}^{(1)}$ from $p(\boldsymbol{\theta})$ and $\bar{\mathbf{y}}_{ij}$ from $p\left(\mathbf{y}|\bar{\boldsymbol{\theta}}_{ij}^{(1)}, \bar{\mathbf{d}}_i^{(1)}\right)$.

3c. Compute

$$\bar{u}_i = \prod_{j=1}^{J} u\left(\bar{\mathbf{d}}_i^{(1)}, \bar{\boldsymbol{\theta}}_{ij}^{(1)}, \bar{\mathbf{y}}_{ij}^{(1)}\right) \quad \text{and} \quad \alpha_i = \min\left(1, \frac{\bar{u}_i^{(1)} q_{\text{MH}}\left(\hat{\mathbf{d}}_i^{(1)}|\bar{\mathbf{d}}_i^{(1)}\right)}{\hat{u}_1 q_{\text{MH}}\left(\bar{\mathbf{d}}_i^{(1)}|\hat{\mathbf{d}}_i^{(1)}\right)}\right).$$

Step 4: Set $\mathbf{d}_i^{(1)} = \bar{\mathbf{d}}_i^{(1)}$ with probability α_i, and $\mathbf{d}_i^{(1)} = \hat{\mathbf{d}}_i^{(1)}$ with probability $1-\alpha_i$.
Step 5: Set $t = t + 1$, and $J = J(t)$. Next, set $\mathbf{d}^{(0)} = \mathbf{d}^{(1)}$.
Step 6: Repeat Steps 1–5 until the Markov chain has practically converged.

4 Optimal Design for the Exponential Distribution

4.1 First-Order Model in $p = 2$ Variables

If we assume that the distribution of the lifetime Y_i is exponential with mean $\exp\left[\boldsymbol{\beta}'\mathbf{x}_i\right]$, where $\boldsymbol{\beta} = [\beta_0, \beta_1, \ldots, \beta_p]'$ and $\mathbf{x} = [1, x_{i1}, \ldots, x_{ip}]'$, then the likelihood function is

$$L(\boldsymbol{\beta}|\mathbf{y}) = \prod_{i=1}^{n} \frac{1}{\exp[\boldsymbol{\beta}'\mathbf{x}_i]} \exp\left\{\exp\left[\boldsymbol{\beta}'\mathbf{x}_i\right]\right\}.$$

We take as prior the normal distribution

$$\boldsymbol{\beta} \sim N\left([\mu_0, \mu_1, \ldots, \mu_p]', \mathrm{diag}\left(\sigma_0^2, \sigma_1^2, \ldots, \sigma_p^2\right)\right).$$

For a relatively noninformative prior distribution the values of σ_j^2 can be taken to be large. The log posterior for $p = 2$ predictor variables, is then

$$\ell(\beta) = \log p(\boldsymbol{\beta}|\mathbf{y}) = \log c - \frac{1}{2\sigma_0^2}(\beta_0 - \mu_0)^2 - \frac{1}{2\sigma_1^2}(\beta_1 - \mu_1)^2 - \frac{1}{2\sigma_2^2}(\beta_2 - \mu_2)^2$$
$$- \beta_0 x_{\cdot 0} - \beta_1 x_{\cdot 1} - \beta_2 x_{\cdot 2} - \sum_{i=1}^{n} y_i \exp\left[-\beta_0 - \beta_1 x_{i1} - \beta_2 x_{i2}\right]$$

where

$$x_{\cdot 0} = \sum_{i=1}^{n} x_{i0} = n, \quad \text{and} \quad x_{\cdot j} = \sum_{i=1}^{n} x_{ij}, \quad j = 1, 2.$$

The extension to p variables should be clear. The first partial derivatives are

$$\frac{\partial \ell}{\partial \beta_j} = -\frac{\beta_j - \mu_j}{\sigma_j^2} - x_{\cdot 0} + \sum_{i=1}^{n} x_{ij} y_i \exp\left[-\beta_0 - \beta_1 x_{i1} - \beta_2 x_{i2}\right], \quad j = 0, 1, 2 \quad (5)$$

and the second partials are

$$\frac{\partial^2 \ell}{\partial \beta_j^2} = -\frac{1}{\sigma_j^2} - \sum_{i=1}^{n} x_{ij}^2 y_i \exp\left[-\beta_0 - \beta_1 x_{i1} - \beta_2 x_{i2}\right]$$
$$\frac{\partial^2 \ell}{\partial \beta_j \partial \beta_k} = -\sum_{i=1}^{n} x_{ij} x_{ik} y_i \exp\left[-\beta_0 - \beta_1 x_{i1} - \beta_2 x_{i2}\right].$$

With these partial derivatives, we can use the Newton-Raphson method to find the posterior mode. At the posterior mode $\hat{\boldsymbol{\beta}}$, we can estimate the covariance matrix of the posterior distribution for $\boldsymbol{\beta}$ (Gelman et al. 2004, Sect. 4.1):

$$V \approx \left[-\ell''\left(\hat{\boldsymbol{\beta}}\right)\right]^{-1}. \tag{6}$$

In order to apply a genetic algorithm, we must specify a utility function $U(\mathbf{d})$ to be maximized. We have considered the following:

$$U_1\left(\mathbf{d}\right) = \left(\sum_{i=0}^{2} V_{ii}\right)^{-1} \quad \text{[sum of posterior variance of all parameters]}$$

$$U_2\left(\mathbf{d}\right) = \left(\sum_{i=1}^{2} V_{ii}\right)^{-1} \quad \text{[sum of posterior variance of slope parameters]}$$

$$U_3\left(\mathbf{d}\right) = \frac{1}{\det\left(V\right)} \quad \text{[proportional to volume of posterior probability region]}$$

We applied a genetic algorithm that proceeds as follows:

Step 0: Begin with a collection $\mathcal{D}$ of N designs selected at random (uniformly across the design space). Set $k = 1$.

Step 1: (Recombination or cross-over) Select M pairs of designs at random. Each pair of designs exchanges some information to create a new design. Add these M new designs to the collection $\mathcal{D}$.

Step 2: (Mutation) Select K designs from $\mathcal{D}$ and impose random variation in some of the components. Add these new designs to the collection $\mathcal{D}$.

Step 3: (Computing the fitness) For each of the $N + M + K$ designs that comprise $\mathcal{D}$, compute the utility by repeating Steps 3a–e for $j = 1, 2, \ldots, J$.

Step 3a: Simulate $\boldsymbol{\beta}^{(j)} = (\beta_0, \beta_1, \beta_2)'$ from the prior distribution $p\left(\beta_0, \beta_1, \beta_2\right)$.

Step 3b: Simulate observations $\mathbf{y}^{(j)} = (y_1, y_2, \ldots, y_n)$ from $p\left(\mathbf{y}|\boldsymbol{\beta}^{(j)}, \mathbf{d}\right)$.

Step 3c: Find the posterior mode $\hat{\boldsymbol{\beta}}^{(j)}$.

Step 3d: Approximate the posterior covariance matrix $V^{(j)}$ using Eq. 6.

Step 3e: Using the above information, compute $U\left(\mathbf{d}, \boldsymbol{\beta}^{(j)}, \mathbf{y}^{(j)}\right)$.

Step 3f: To estimate the average fitness across these simulations, take

$$U(\mathbf{d}) = \frac{1}{J}\sum_{j=1}^{J} U\left(\mathbf{d}, \boldsymbol{\beta}^{(j)}, \mathbf{y}^{(j)}\right).$$

Step 4: (Selection) Rank all the designs in $\mathcal{D}$ according to fitness. Select the very best designs in $\mathcal{D}$ for sure, and randomly select other designs based on proportional selection from the remaining designs, to determine which designs survive to the next generation. The probability of selection is greater for larger values of $U_m(\mathbf{d})$. At this stage, N designs have been selected to survive to the next generation.

Step 5: Set $k = k + 1$ and repeat Steps 1–4 until the desired number of generations has been computed.

4.2 Second-Order Model in $p = 2$ Variables

Suppose now that we have a second-order model in $p = 2$ variables; that is,

$$Y_i \sim \text{EXP}\left(\beta_0 + \beta_1 x_{i1} + \beta_2 x_{i2} + \beta_{11} x_{i1}^2 + \beta_{22} x_{i2}^2 + \beta_{12} x_{i1} x_{i2}\right)$$

The logarithm of the posterior distribution is then

$$\begin{aligned}\ell(\beta) = \log c &- \frac{1}{2\sigma_0^2}(\beta_0 - \mu_0)^2 - \frac{1}{2\sigma_1^2}(\beta_1 - \mu_1)^2 - \frac{1}{2\sigma_2^2}(\beta_2 - \mu_2)^2 \\ &- \frac{1}{2\sigma_{11}^2}(\beta_{11} - \mu_{11})^2 - \frac{1}{2\sigma_{22}^2}(\beta_{22} - \mu_{22})^2 - \frac{1}{2\sigma_{12}^2}(\beta_{12} - \mu_{12})^2 \\ &- \beta_0 x_{\cdot 0} - \beta_1 x_{\cdot 1} - \beta_2 x_{\cdot 2} - \beta_{11} x_{\cdot ,11} - \beta_{22} x_{\cdot ,22} - \beta_{12} x_{\cdot ,12} \\ &- \sum_{i=1}^{n} y_i \exp\left[-\beta_0 - \beta_1 x_{i1} - \beta_2 x_{i2} - \beta_{11} x_{i1}^2 - \beta_{22} x_{i2}^2 - \beta_{12} x_{i1} x_{i2}\right].\end{aligned}$$

The first and second partial derivatives can be computed as in Eq. 5. The posterior mode $\hat{\beta}$ can then be quickly determined by Newton-Raphson, and the posterior covariance matrix can be estimated by Eq. 6. The GA for the second-order model proceeds through essentially the same steps described in the previous subsection.

5 Results

To illustrate the genetic algorithm described in the previous section, we have applied it to the problem of finding a 16-run design that minimizes the sum of the posterior variances of the parameters for first-order and second-order models in $p = 2$ variables. This is similar to A-optimality in classical design of experiments.

Suppose first that the design region is $[-1, 1] \times [-1, 1]$. Based on experience with optimal designs for normally distributed responses, we would expect that the optimal design would place points near the corners of the design region for a first-order model. After 1,000 generations, the best design among all of the designs in $\mathcal{D}$ is shown in Fig. 1. However, this is not necessarily the best design that has occurred

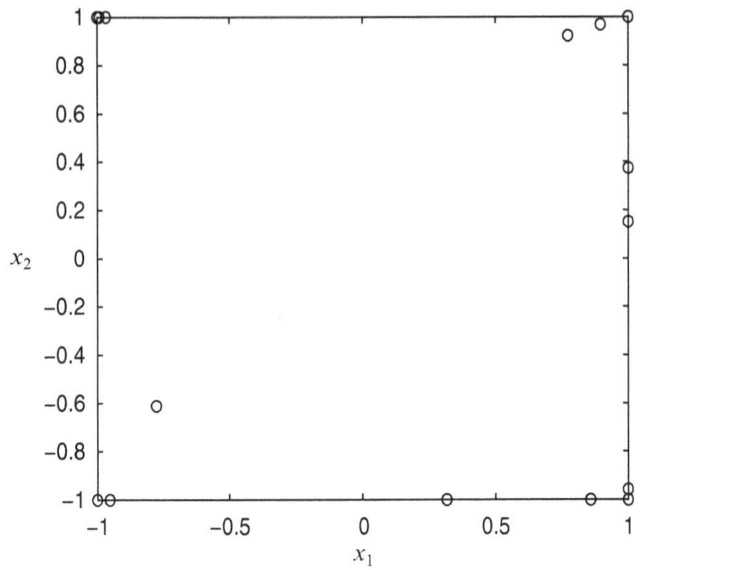

Fig. 1 Best design after 1,000 generations of GA for first-order exponential regression model on square design region

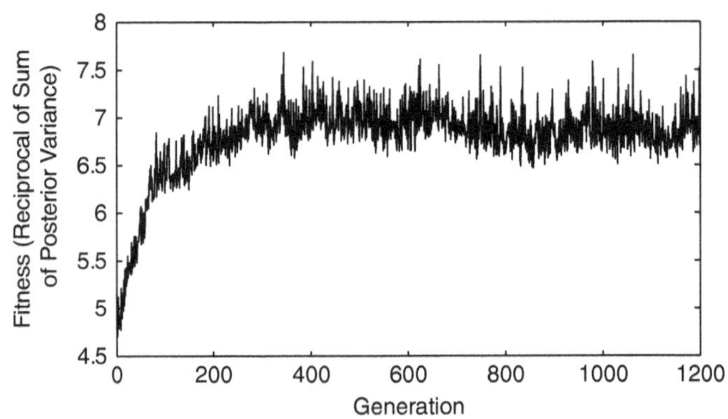

Fig. 2 Fitness across many generations of GA for first-order exponential regression model on square design region

in the history of the genetic algorithm. In the context of optimal design, we define the "fitness" of a design to be equal to the Bayesian A-optimality criterion (the sum of the posterior variances), or whatever criterion is used for choosing an optimal design. The plot of the fitness of the best design in each generation is shown in Fig. 2. As expected, most of the design points cluster around the corners of the design region. The fitness seems to increase, with some randomness, due to the estimation (not evaluation) of the fitness for each design in Step 3f above.

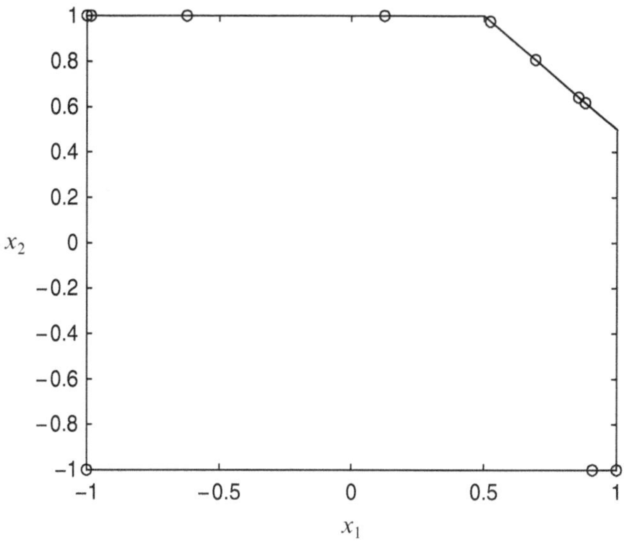

Fig. 3 Best design after 1,000 generations of GA for first-order exponential regression model on irregular design region

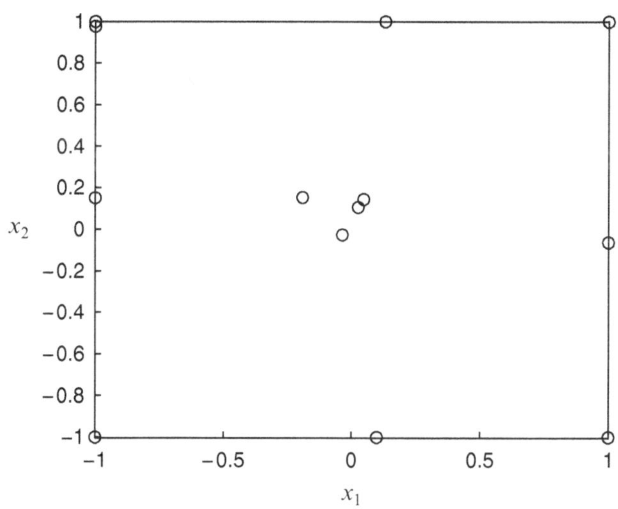

Fig. 4 Best design after 1,000 generations of GA for second-order exponential regression model on square design region

To see what happens to an irregular design region, we considered the design region $[-1, 1] \times [-1, 1]$ with the additional constraint that $x_1 + x_2 \leq 1.5$; this effectively cuts off a triangular region in the upper right corner of the design region. Figure 3 shows the best design after 1,000 generations.

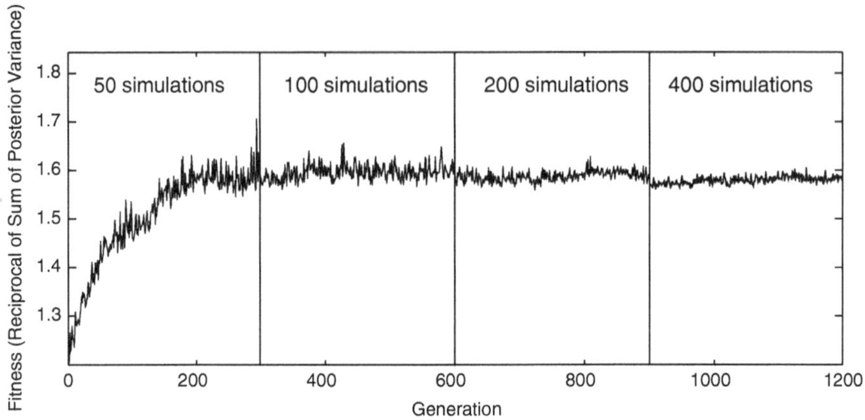

Fig. 5 Fitness across generations of GA for second-order exponential regression model. The numbers above the curves indicate the number of simulations used to estimate the posterior variance

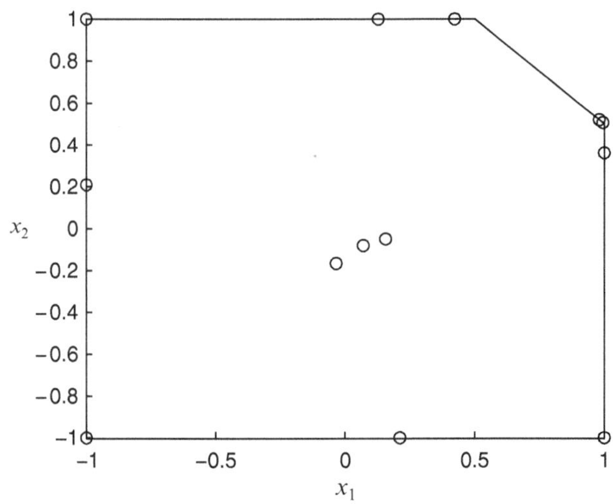

Fig. 6 Best design after 1,000 generations of GA for second-order exponential regression model on irregular design region

We also ran the GA to find a near optimal design for a second-order model with $p = 2$ variables. The best design after 1,200 generations is shown in Fig. 4. This design has a structure that is similar to a central composite design, with two runs at each corner, one axial point along each axis (both positive and negative) and several points near the center. Figure 5 shows a trace plot of the best design's fitness across all 1,200 generations. There is a lot of noise in the estimation of the fitness, due to the simulation that is required. We applied an increasing number of simulations as the GA progressed, much like "temperature" increases in a simulated annealing algorithm. Figure 6 shows the best design for a second-order model in two variables with the same irregular region described above.

6 Conclusions

We have shown how to apply genetic algorithms to find a near optimal design for life testing experiments when there are two predictor variables and the response is exponential. The genetic algorithms move very quickly away from a set of designs chosen at random toward designs that are near optimal. Because we are using simulation to estimate the value of the fitness (the sum of the posterior variances), there is considerable noise in the successive values of the fitness. The near-optimal Bayesian designs obtained by the genetic algorithm are similar to classical designs such as the 2-level factorial design for a first-order design and a central composite design for a second-order model. While the exponential distribution is somewhat unrealistic in practice, we expect that the basic idea of the genetic algorithm for optimal design will carry over to more realistic distributions, such as the Weibull, which we will address in future studies. The issue of censoring, which we have ignored here, will also be addressed in future work.

Acknowledgements This research was supported in part by the National Science Foundation under contract CMMI-0928046.

References

Amzal, B., Bois, F. Y., Parent, E., & Robert, C. P. (2006). Bayesian-optimal design via interacting particle systems. *Journal of the American Statistical Association, 101*, 773–785.

Chaloner, K., & Verdinelli, I. (1995). Bayesian experimental design: A review. *Statistical Science, 10*, 273–304.

Eiben, A. E., & Smith, J. E. (2003). *Introduction to evolutionary computing*, Berlin: Springer.

Gelman, A., Carlin, J. B., Stern, H. S., & Rubin, D. B. (2004). *Bayesian data analysis* (2nd ed.). Boca Raton: Chapman & Hall/CRC.

Gotwalt, C. M., Jones, B. A., & Steinberg, D. M. (2009). Fast computation of designs robust to parameter uncertainty for nonlinear settings. *Technometrics, 51*, 88–95.

Heredia-Langner, A., Carlyle, W. M., Montgomery, D. C., Borror, C. M., & Runger, G. C. (2003). Genetic algorithms for the construction of D-optimal designs. *Journal of Quality Technology, 35*, 28–46.

Holland J. H (1975). *Adaptation in natural and artificial systems*. Ann Arbor: University of Michigan Press.

Müller, P. (1999). Simulation-based optimal design, with discussion. In J. M. Bernardo, J. O. Berger, A. P. Dawid, & A. F. M. Smith (Eds.), *Bayesian statistics 6* (pp. 459–474). Oxford: Clarendon Press.

Müller, P., Sansó, B., & De Iorio, M. (2004). Optimal Bayesian design by inhomogeneous Markov chain simulation. *Journal of the American Statistical Association, 99*, 788–798.

Accelerated Lifetime Testing of Thermal Insulation Elements

Rainer Göb, Kristina Lurz, and Ulrich Heinemann

Abstract Thermal insulation materials play an important role in the area of energy technology. Thermal insulation elements (TIEs) are used in fields where high quality and high convenience insulation is required as e.g. in buildings. The TIE manufacturing sector is evolving, but not completely mature. Producers and users have an urgent demand for quality control techniques. Hitherto, quality control and service lifetime prediction for TIEs have mainly been considered from a physical viewpoint with strong emphasis on measurement issues. Rigorous statistical approaches are still missing. From a review of the physical models for TIE degradation over time we build a mixed nonlinear regression model of degradation as a function of time and ambient temperature. The model accounts for measurement-to-measurement and for unit-to-unit variation. We investigate inferential techniques for model parameter estimation and lifetime prediction, and we study the design of accelerated experiments on TIEs.

Keywords Accelerated lifetime testing • Nonlinear regression • Linear mixed regression model • Lifetime prediction • Prediction interval • Design of experiments

R. Göb (✉) · K. Lurz
Institute for Applied Mathematics and Statistics, University of Würzburg, Sanderring 2, D-97070 Würzburg, Germany
e-mail: goeb@mathematik.uni-wuerzburg.de; kristina.lurz@uni-wuerzburg.de

U. Heinemann
Bavarian Center for Applied Energy Research, Am Hubland, D-97074 Würzburg, Germany
e-mail: Ulrich.Heinemann@zae.uni-wuerzburg.de

H.-J. Lenz et al. (eds.), *Frontiers in Statistical Quality Control 10*,
Frontiers in Statistical Quality Control 10, DOI 10.1007/978-3-7908-2846-7_21,

1 Physical Background

Thermal insulation elements (TIEs) are used for high quality and high convenience insulation purposes. The essential quality characteristic of a thermal insulation element is *thermal conductivity*. Thermal conductivity at a fixed temperature, measured in units of Watt per meter times Kelvin (W/ (m K)) is a function of two main components: (1) The gaseous thermal conductivity λ_{gas}, and (2) the thermal conductivity λ_w due to water content. The effect of these components on total thermal conductivity $\lambda = \lambda_{total}$ is analysed by the additive model

$$\lambda_{total} = \lambda_{ev} + \lambda_{gas} + \lambda_w + \lambda_c, \tag{1}$$

see Caps et al. (1997), Heinemann (2008). In the latter equation, λ_{ev} is the thermal conductivity in the evacuated and dry state which depends on the solid conduction and radiative transfer in the core material, and λ_c is a coupling term which becomes noticeable at high gas pressure when the contact resistances between the grains are thermally shorted by the gas molecules. The thermal conductivity of a TIE is dependent on the thickness of the element. The thermal conductivity divided by the thickness of an insulation element in units of meter leads to the *heat transfer coefficient*, measured in units of $W/\left(m^2\,K\right)$, see e.g. Kreith and Bohn (2000).

The thermal performance of the insulation elements is very slowly degrading. Gas and water vapour are gradually penetrating over time, and the thermal conductivity is slowly, but steadily increasing. Regulations as the Energy Saving Regulation (EnEV) given by the German Federal Government in 2009, see Bundesregierung der Bundesrepublik Deutschland (2007, 2009), impose restrictions on the thermal performance of insulation material, which are often given in terms of upper specification limits for the heat transfer coefficient. For instance, the heat transfer coefficient of outer walls of newly installed residential buildings is not to exceed $0.28\ W/\left(m^2\,K\right)$. On the basis of such regulations, the TIE industry derives upper specification limits of the thermal conductivity for the long-term performance. To verify compliance to such specifications, producers and users need long-term forecasts on the thermal conductivity.

The evolution of an elements' thermal conductivity depends on endogenous characteristics of the element and exogenous factors. The endogenous factors are related to the elements' technical design and to manufacturing factors. The three main exogenous factors are environmental temperature, environmental air humidity, usage time. The physical laws which express the dependence of the additive components in Eq. 1 on these factors are nonlinear in nature. Hence the corresponding statistical analysis is based on nonlinear regression models.

In experimental practice on TIEs, the readily accessible approach is to measure the internal gas pressure and the water content, and to infer on the thermal conductivity via the additive Eq. 1. An increase in the water content can be measured by simple weighing.

2 Statistical Issues

Scientific research on TIEs and the development of TIE industrial manufacturing technology are in a phase of rapid evolution. The TIE manufacturing sector is growing at remarkable annual rates, in particular in emerging industrial countries like China. TIE manufacturing technology is not completely mature. Producers and users face considerable quality problems. Since TIEs are primarily used for durable installations, the prediction of service lifetime, i.e., the service time until violation of specifications on thermal conductivity, is a paramount issue in the TIE industry. Predictions for the service lifetime of the TIEs follow typical guarantee periods in the German building industry, which are 2, 4 or 5 years or, if a long-term use is of interest, even 10, 25 or 50 years.

Quality control and service lifetime prediction for TIEs have mainly been considered from a physical viewpoint with strong emphasis on measurement issues. A thorough statistical analysis is still missing.

In industrial contexts, service lifetime prediction has to be rapid. However, even under accelerated levels of the influential factors temperature and humidity, the increase of thermal conductivity is slow. Critical upper limits for the thermal conductivity in TIEs are never attained over reasonably short experimental periods. Lifetime distribution analysis, albeit based on accelerated experiments, cannot be used. Service lifetime prediction has to be based on regression models. The statistical analysis of experiments on TIEs has to account for three levels of variation: (1) Measurement-to-measurement variation in runs of repeated measurements, (2) run-to-run variation due to set-ups in measurement instruments and laboratory environment, (3) unit-to-unit variation resulting from manufacturing instability. Because of the immaturity of the TIE manufacturing sector, unit-to-unit variation is particularly strong and serious.

The subsequent study accounts for the above requirements by analysing experiments on TIEs by a *nonlinear mixed regression model*. Such models, also referred to as *hierarchical nonlinear regression models*, have recently received considerable interest in the literature, see the survey Davidian and Giltinian (2003). The interest in design requires explicit expressions for the variance components associated to the different levels of variation. This is best achieved by partially linearising the model as suggested in Göb and Lurz (2010). Section 3 describes the structure of TIE testing experiments. Section 4 transfers the existing physical modeling approaches into a statistical model which is identified as special case of a general mixed model scheme in Sect. 5. The empirical analysis of the general mixed model scheme is developed by Sect. 6. Section 7 specialises to the general results to the specific TIE model, and Sect. 8 exemplifies by applying to data from a laboratory experiment on TIEs. Issues of optimal design are considered in Sect. 9.

test unit	exposure	measurements of thermal conductivity
1	non-accelerated	$Y_{11}, Y_{12}, \ldots, Y_{1,n_1}$
2	non-accelerated	$Y_{21}, Y_{22}, \ldots, Y_{2,n_2}$
$\vdots$	$\vdots$	$\vdots$
m_1	non-accelerated	$Y_{m_1 1}, Y_{m_1 2}, \ldots, Y_{m_1 n_{m_1}}$
m_1+1	accelerated	$Y_{m_1+1\ 1}, Y_{m_1+1\ 2}, \ldots, Y_{m_1+1\ n_{m_1+1}}$
m_1+2	accelerated	$Y_{m_1+2\ 1}, Y_{m_1+2\ 2}, \ldots, Y_{m_1+2\ n_{m_1+2}}$
$\vdots$	$\vdots$	$\vdots$
$m = m_1 + m_2$	accelerated	$Y_{m_1+m_2\ 1}, Y_{m_1+m_2\ 2}, \ldots, Y_{m_1+m_2\ n_{m_1+m_2}}$

arrangement of measurement time points

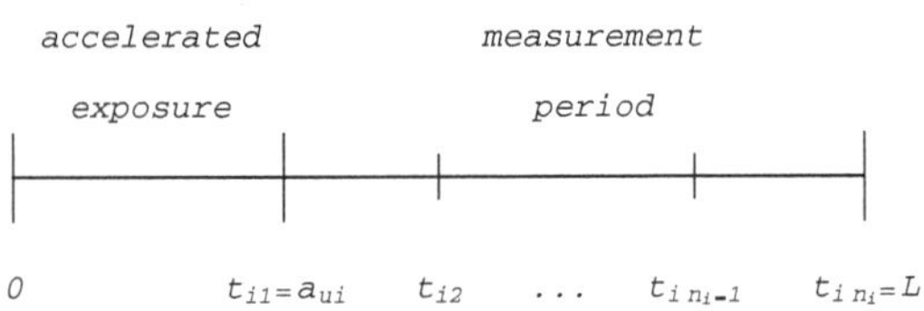

Fig. 1 Testing experiment

3 TIE Testing Experiments

Figure 1 outlines a common scheme for accelerated testing experiments for TIEs. Consider a set of m test units. The units $1, \ldots, m_1$ are held at non-accelerated normal usage conditions over the total experimental period of L days. Each unit $i = m_1 + 1, \ldots, m_1 + m_2 = m$ is exposed to accelerated conditions with respect to temperature and/or humidity in climatic chambers over $a_{ui} < L$ days. On each test unit $i = 1, \ldots, m$, n_i measurements are taken at times $a_{ui} = t_{i1} < t_{i2} < \ldots < t_{in_i} = L$ where $a_{ui} = 0$ for units $i = 1, \ldots, m_1$.

We consider an instance of the above general experimental scheme, specialised in several respects. The thermal conductivity is measured directly. Precise measurements, so-called *absolute value measurements*, are expensive in time and equipment. A large number of repetitions is unfeasible for an industrial company, the sample sizes n_i on the test units usually remain at the lowest supportable level. The design developed in Sect. 9, below, assumes measurement samples of size $n_i = 4$ on each test unit.

It is assumed that the elements under investigation contain desiccants and getter materials to absorb penetrating water vapour, so that the influence of water vapour on the element's degradation is negligible. This has consequences for the experimental design and the experimental analysis. The experimental design controls the factor *temperature* only, whereas humidity remains uncontrolled at ambient levels. Non-accelerated exposure is equivalent to storage at a lower temperature T_1, accelerated exposure is equivalent to storage at a higher temperature T_u for all units $i = 1, \ldots, m$. In the model Eq. 1, the water vapour component λ_w can be neglected.

4 The Model

The statistical model is based on a special and simplified case of Eq. 1.

The gaseous thermal conductivity as a function $\lambda_{gas} = \lambda_{gas}(p_{gas})$ of the gas pressure p_{gas}, measured in bar, in Eq. 1 is modeled by the Knudsen equation

$$\lambda_{gas}(p_{gas}) = \frac{\lambda_{fg}}{1 + \frac{p_{1/2}}{p_{gas}}}, \tag{2}$$

see Caps et al. (1997). Here λ_{fg} is the thermal conductivity of the free and still gas, and $p_{1/2}$ the pressure at which the thermal conductivity of the gas equals one half of λ_{fg}. The application of the Knudsen equation in describing the gaseous thermal conductivity of TIEs traces back to Kaganer (1969). Recent research has extensively investigated and validated the use of the equation for modeling the gaseous thermal conductivity of various core materials, see Caps et al. (1996, 1997) and Heinemann et al. (1999).

The dependence of gas pressure p_{gas} on time and temperature is expressed by an Arrhenius-type function. Let T_0 be a reference temperature, and let a test unit be stored x time units at a temperature level T. Then the increase Δp_{gas} in pressure over the x time units is

$$\Delta p_{gas} = c \cdot \exp\left(\frac{\delta}{R}\left[\frac{1}{T_0} - \frac{1}{T}\right]\right) x. \tag{3}$$

In Eq. 3, the *acceleration factor* c depends on general material properties and on characteristics of an individual test unit, the *activation energy* δ, measured in J/mol, depends on general material properties, and $R = 8.314472\,\text{J}/\,(\text{mol K})$ is the universal gas constant.

For the purposes of experimental analysis and forecasting, the coupling term λ_c in Eq. 1 can be ignored. The coupling effect becomes noticeable for high gas pressures p_{gas} only. At $p_{gas} = 10$ mbar, the contribution of λ_c to total thermal conductivity $\lambda = \lambda_{total}$ is approximately 4% only. Ten mbar already lead to a thermal conductivity far beyond criticial specification limits which are not attained in relatively short-lasting experiments.

The thermal conductivity λ_w due to water content is ignored by assuming the presence of efficient getter and absorption materials in the TIE, see Sect. 3, above.

Summarising all assumptions and Eqs. 1, 2, and 3, we obtain the following equation for a measurement of total thermal conductivity λ_{total} after x time units exposure at non-accelerated temperature T_l and a time units exposure at accelerated temperature T_u:

$$\begin{aligned} \lambda_{\text{total}} &= \lambda_{\text{ev}} + \lambda_{\text{gas}}(p_{\text{gas}}) = \lambda_{\text{ev}} + \frac{\lambda_{\text{fg}}}{1 + \frac{p_{1/2}}{p_{\text{gas}}}} = \\ &\lambda_{\text{ev}} + \frac{\lambda_{\text{fg}}}{1 + \frac{1}{\frac{p_0}{p_{1/2}} + \frac{c}{p_{1/2}}\left\{\exp\left(\frac{\delta}{R}\left[\frac{1}{T_0} - \frac{1}{T_1}\right]\right)x + \exp\left(\frac{\delta}{R}\left[\frac{1}{T_0} - \frac{1}{T_{\text{u}}}\right]\right)a\right\}}}, \end{aligned} \tag{4}$$

where p_0 is the initial gas pressure in the TIE.

Among the three levels of variation discussed in Sect. 2, the model will not account for run-to-run variation due to set-ups in measurement instruments and laboratory environment. For the considered absolute value measurements of thermal conductivity such effects are basically negligible. Nevertheless, the model has to account for measurement-to-measurement variation and for unit-to-unit variation.

Unit-to-unit differences due to inhomogeneous manufacturing become noticeable in gas pressure, in particular, in the parameters p_0 (initial gas pressure) and c (acceleration factor). Accordingly, for unit i, we introduce a random deviation b_{i1} of the relative initial gas pressure $p_0/p_{1/2}$ from the population average, and a random deviation b_{i2} of the relative acceleration factor $c/p_{1/2}$ from the population average. Consider the thermal conductivity measurement Y_{ij} obtained on unit i at the jth measurement time in a testing experiment as described by Fig. 1. In Eq. 4, we introduce random deviations b_{i1}, b_{i2} of unit i from the population average, and rename the parameters in a manner more familiar in regression analysis. Then we can model the conditional expectation $E[Y_{ij}|b_{i1}, b_{i2}]$ by the nonlinear mixed model regression function $E[Y_{ij}|b_{i1}, b_{i2}] = f(x_{ij}, \boldsymbol{a}_i, \boldsymbol{\beta}, \delta, \boldsymbol{b}_i)$, where

$$\begin{aligned} E[Y_{ij}|b_{i1}, b_{i2}] &= f(x_{ij}, \boldsymbol{a}_i, \boldsymbol{\beta}, \delta, \boldsymbol{b}_i) = \\ \lambda_{\text{ev}} &+ \lambda_{\text{fg}} \times \left\{ 1 - \left[1 + \beta_1 + b_{i1} + (\beta_2 + b_{i2}) \cdot \right.\right. \\ &\left.\left. \left(\exp\left(\frac{\delta}{R} a_{T_1}\right) x_{ij} + \exp\left(\frac{\delta}{R} a_{T_{\text{u}}}\right) a_{\text{u}i} \right) \right]^{-1} \right\}. \end{aligned} \tag{5}$$

Table 1 explains and relates the parameters of the regression function Eq. 5 to the parameters in Eq. 4.

In Göb and Lurz (2010) it is shown that the direct analysis of the nonlinear mixed regression function Eq. 5 runs into numerical and conceptual problems. Instead, it is recommended to linearise Eq. 5 by the transformation

$$\mathcal{T}(y) = \frac{\lambda_{\text{fg}}}{\lambda_{\text{fg}} - y + \lambda_{\text{ev}}}. \tag{6}$$

Applying $\mathcal{T}$ both to the measurements Y_{ij} and to the conditional mean $E[Y_{ij}|b_{i1}, b_{i2}] = f(\boldsymbol{x}_{ij}, \boldsymbol{a}_i, \boldsymbol{\beta}, \delta, \boldsymbol{b}_i)$ leads to considering the mixed model

Table 1 Parameters of model Eq. 5

λ_{ev}	Thermal conductivity in the evacuated and dry state of the TIE
β_1	$\beta_1 = p_0/p_{1/2}$ average initial gas pressure in a TIE at time 0
β_2	$\beta_2 = c/p_{1/2}$ average acceleration factor of a TIE
δ	Activation energy
$\boldsymbol{\beta}$	Vector $\boldsymbol{\beta} = (\beta_1, \beta_2)^\top$ of β-regression coefficients
b_{i1}	Random deviation of the relative initial gas pressure of unit i from the population average
b_{i2}	Random deviation of the relative acceleration factor of unit i from the population average
$\boldsymbol{b}_i$	Vector $\boldsymbol{b}_i = (b_{i1}, b_{i2})^\top$ of random deviations of unit i from the population average
x_{ij}	Total non-accelerated exposure time of unit i until measurement j
T_{l}	Non-accelerated temperature in K
T_{u}	Accelerated temperature in K
ϑ_{l}	$\vartheta_{\text{l}} = T_{\text{l}} - 273.15$ non-accelerated temperature in °C
ϑ_{u}	$\vartheta_{\text{u}} = T_{\text{u}} - 273.15$ accelerated temperature in °C
$a_{T_{\text{l}}}$	Transformed non-accelerated temperature $a_{T_{\text{l}}} = \frac{1}{T_0} - \frac{1}{T_{\text{l}}}$
$a_{T_{\text{u}}}$	Transformed accelerated temperature $a_{T_{\text{u}}} = \frac{1}{T_0} - \frac{1}{T_{\text{u}}}$
$a_{\text{u}i}$	Total accelerated exposure time of unit i
R	Universal gas constant $R = 8.314472$ J/ (mol K)
λ_{fg}	Thermal conductivity $\lambda_{\text{fg}} = 26 \cdot 10^{-3}$ W/ (m K) of the free and still gas

$$\mathcal{T}(Y_{ij}) \;=\; Z_{ij} \;=\; g(x_{ij}, \boldsymbol{a}_i, \boldsymbol{\beta}, \delta, \boldsymbol{b}_i) + \varepsilon_{ij}, \tag{7}$$

$$\begin{aligned} g(x_{ij}, \boldsymbol{a}_i, \boldsymbol{\beta}, \delta, \boldsymbol{b}_i) \;&=\; \mathcal{T}\Big(f(x_{ij}, \boldsymbol{a}_i, \boldsymbol{\beta}, \delta, \boldsymbol{b}_i)\Big) \;= \\ 1 + \beta_1 + b_{i1} + (\beta_2 + b_{i2}) &\left\{\exp\left(\frac{\delta}{R} a_{T_{\text{l}}}\right) x_{ij} \;+\; \exp\left(\frac{\delta}{R} a_{T_{\text{u}}}\right) a_{\text{u}i}\right\}. \end{aligned} \tag{8}$$

The parameter λ_{ev} is integrated into the transformed observations and cannot be analysed empirically on grounds of Eqs. 7 and 8. Section 7 explains how to handle the parameter λ_{ev}. The assumptions on the stochastic parameters $\boldsymbol{b}_i$ and ε_{ij} are discussed in the more general framework established by the subsequent Sect. 5.

5 A General Mixed Model Scheme

The model defined by Eqs. 7 and 8 is a special case of a more general mixed model scheme.

We consider a group structured sampling model. Let $1, \ldots, m$ be independent groups with n_i observations $Z_{i1}, \ldots, Z_{in_i}$ in group i. The sampling vector $\boldsymbol{Z}_i = (Z_{i1}, \ldots, Z_{in_i})^\top$ in group i follows the model

$$\boldsymbol{Z}_i = \mathbf{X}_i \boldsymbol{\alpha}_i + \boldsymbol{\varepsilon}_i . \tag{9}$$

$\boldsymbol{\varepsilon}_i$ is a vector of residuals with $E[\boldsymbol{\varepsilon}_i] = \mathbf{0}$, $\mathrm{Cov}[\boldsymbol{\varepsilon}_i] = \sigma^2 \mathbf{I}_{n_i}$. $\mathbf{X}_i$ is an $n_i \times q$ design matrix of known regressor values. The q-dimensional random parameter $\boldsymbol{\alpha}_i$ follows the equation

$$\boldsymbol{\alpha}_i = \boldsymbol{\Lambda}_i \boldsymbol{\beta}_i + \boldsymbol{d}_i , \tag{10}$$

where $\boldsymbol{\Lambda}_i = \boldsymbol{\Lambda}_i(\boldsymbol{\delta})$ is a nonsingular $q \times q$ matrix depending on further parameters $\boldsymbol{\delta}$, and $\boldsymbol{d}_i$ is a fixed and known vector in $\mathbb{R}^q$. The q-dimensional random parameter $\boldsymbol{\beta}_i$ follows the equation

$$\boldsymbol{\beta}_i = \boldsymbol{\beta} + \boldsymbol{b}_i , \tag{11}$$

where $\boldsymbol{\beta}$ is an unknown vector in $\mathbb{R}^q$, and $\boldsymbol{b}_i$ is a q-dimensional random vector with the properties

$$E[\boldsymbol{b}_i] = \mathbf{0}, \quad \mathrm{Cov}[\boldsymbol{b}_i] = E[\boldsymbol{b}_i \boldsymbol{b}_i^\top] = \mathbf{B}, \quad \mathrm{Cov}[\boldsymbol{b}_i, \boldsymbol{\varepsilon}_i] = E[\boldsymbol{b}_i \boldsymbol{\varepsilon}_i^\top] = \mathbf{O}. \tag{12}$$

The variance-covariance matrix of the sampling vector is

$$\mathrm{Cov}[\boldsymbol{Z}_i] = \boldsymbol{\Lambda}_i \mathbf{X}_i \mathbf{B} (\boldsymbol{\Lambda}_i \mathbf{X}_i)^\top + \sigma^2 \mathbf{I}_{n_i} . \tag{13}$$

The model defined by Eqs. 7 and 8 is obtained with the following definitions:

$$\boldsymbol{\beta} = \begin{pmatrix} \beta_1 \\ \beta_2 \end{pmatrix}, \quad \boldsymbol{b}_i = \begin{pmatrix} b_{i1} \\ b_{i2} \end{pmatrix}, \quad \boldsymbol{d}_i = \begin{pmatrix} 1 \\ 0 \end{pmatrix}, \quad \boldsymbol{\delta} = \delta, \tag{14}$$

$$\boldsymbol{\Lambda}_i = \boldsymbol{\Lambda}_i(\delta) = \begin{pmatrix} 1 & \exp\left(\frac{\delta}{R} a_{T_\mathrm{u}}\right) a_{\mathrm{u}i} \\ 0 & \exp\left(\frac{\delta}{R} a_{T_\mathrm{l}}\right) \end{pmatrix}, \quad \boldsymbol{X}_i = \begin{pmatrix} 1 & x_{i1} \\ \vdots & \vdots \\ 1 & x_{in_i} \end{pmatrix}. \tag{15}$$

In the variance-covariance matrix Eq. 13, the component $\boldsymbol{\Lambda}_i \mathbf{X}_i \mathbf{B} (\boldsymbol{\Lambda}_i \mathbf{X}_i)^\top$ corresponds to unit-to-unit variation, and $\sigma^2 \mathbf{I}_{n_i}$ corresponds to measurement-to-measurement variation.

6 Empirical Analysis of the General Mixed Model Scheme

Consider the general mixed model scheme introduced in Sect. 5 in the following knowledge context. (1) Known: design matrices $\mathbf{X}_i$, $\boldsymbol{d}_i$, matrices $\boldsymbol{\Lambda}_i(\cdot)$. (2) Unknown: regression parameter vectors $\boldsymbol{\beta}$ and $\boldsymbol{\delta}$, random effect vectors $\boldsymbol{b}_i$, residual variance σ^2, variance-covariance matrix $\text{Cov}[\boldsymbol{b}_i] = \mathbf{B}$.

Many approaches to the analysis of linear mixed models have been discussed in literature, see Christensen (1996) for a survey. Throughout, these approaches need additional assumptions, generally stipulating a diagonal random effects variance-covariance matrix $\mathbf{B}$. This assumption is not necessarily valid in the application to TIEs. The situation is further complicated by the nonlinear component $\boldsymbol{\Lambda}_i(\delta)$. Methods for the estimation of δ are considerably simplified if explicit expressions for the estimates of the remaining unknown parameters are available.

The theory of the method suggested by Göb and Lurz (2010) proceeds in two steps. Step I: Assume known $\boldsymbol{\delta}$, and reduce the analysis of all further parameters to standard OLS methods based on minimising the estimated residual variance $\widehat{\sigma}^2$. $\mathbf{B}$ is made estimable by imposing the intuitively plausible ANOVA-like restriction $\sum_i \widehat{\boldsymbol{b}}_i = \mathbf{0}$ on the estimates of the random deviations. Step II: $\boldsymbol{\delta}$ is estimated by minimising $\sum_i \|\widehat{\boldsymbol{b}}_i(\boldsymbol{\delta})\|^2$ in $\boldsymbol{\delta}$.

6.1 Step I Estimators and Predictors Under $\boldsymbol{\delta}$ Known

Step I provides the following results, depending on the argument $\boldsymbol{\delta}$. The estimator of $\boldsymbol{\alpha}_i$ and the unbiased estimator of $\boldsymbol{\beta}$ are

$$\widehat{\boldsymbol{\alpha}}_i = (\mathbf{X}_i^\top \mathbf{X}_i)^{-1}\mathbf{X}_i^\top \boldsymbol{Z}_i, \quad \widehat{\boldsymbol{\beta}} = \frac{1}{m}\sum_{i=1}^{m} \boldsymbol{\Lambda}_i^{-1}(\widehat{\boldsymbol{\alpha}}_i - \boldsymbol{d}_i), \tag{16}$$

where

$$E[\widehat{\boldsymbol{\alpha}}_i] = \boldsymbol{\Lambda}_i\boldsymbol{\beta} + \boldsymbol{d}_i, \quad \text{Cov}[\widehat{\boldsymbol{\alpha}}_i] = \sigma^2(\mathbf{X}_i^\top \mathbf{X}_i)^{-1} + \boldsymbol{\Lambda}_i \mathbf{B} \boldsymbol{\Lambda}_i^\top. \tag{17}$$

The estimator of $\boldsymbol{b}_i$ is

$$\widehat{\boldsymbol{b}}_i = \boldsymbol{\Lambda}_i^{-1}(\widehat{\boldsymbol{\alpha}}_i - \boldsymbol{d}_i) - \widehat{\boldsymbol{\beta}} = \boldsymbol{\Lambda}_i^{-1}(\widehat{\boldsymbol{\alpha}}_i - \boldsymbol{d}_i) - \frac{1}{m}\sum_{l=1}^{m} \boldsymbol{\Lambda}_l^{-1}(\widehat{\boldsymbol{\alpha}}_l - \boldsymbol{d}_l). \tag{18}$$

In each group i, the residual vector $\boldsymbol{\varepsilon}_i$ is estimated by

$$\widehat{\boldsymbol{\varepsilon}}_i = \boldsymbol{Z}_i - \mathbf{X}_i\widehat{\boldsymbol{\alpha}}_i. \tag{19}$$

Let $N = n_1 + \ldots + n_m$ be the total sample size. An unbiased pooled variance estimator for the residual variance σ^2 is

$$S_Z^2 = \frac{1}{N - mq} \sum_{i=1}^{m} \widehat{\boldsymbol{\varepsilon}}_i^\top \widehat{\boldsymbol{\varepsilon}}_i. \tag{20}$$

Unbiased estimators for $\mathbf{B}$ and for $\mathrm{Cov}[\widehat{\boldsymbol{\beta}}]$ are

$$\widehat{\mathbf{B}} = \frac{1}{m-1} \sum_{i=1}^{m} \widehat{\boldsymbol{b}}_i \widehat{\boldsymbol{b}}_i^\top - \frac{S_Z^2}{m} \sum_{l=1}^{m} \boldsymbol{\Lambda}_l^{-1} (\mathbf{X}_l^\top \mathbf{X}_l)^{-1} (\boldsymbol{\Lambda}_l^{-1})^\top, \tag{21}$$

$$\widehat{\mathrm{Cov}[\widehat{\boldsymbol{\beta}}]} = \frac{1}{m} \left\{ \widehat{\mathbf{B}} + \frac{S_Z^2}{m} \sum_{l=1}^{m} \boldsymbol{\Lambda}_l^{-1} (\mathbf{X}_l^\top \mathbf{X}_l)^{-1} (\boldsymbol{\Lambda}_l^{-1})^\top \right\}. \tag{22}$$

An unbiased estimator for $\mathrm{Cov}[\widehat{\boldsymbol{b}}_i] = \mathrm{Cov}[\widehat{\boldsymbol{b}}_i, \widehat{\boldsymbol{b}}_i]$ is

$$\widehat{\mathrm{Cov}[\widehat{\boldsymbol{b}}_i]} = \widehat{\mathrm{Cov}[\widehat{\boldsymbol{\beta}}]} + \left(1 - \frac{2}{m}\right) \left\{ S_Z^2 \boldsymbol{\Lambda}_i^{-1} (\mathbf{X}_i^\top \mathbf{X}_i)^{-1} (\boldsymbol{\Lambda}_i^{-1})^\top + \widehat{\mathbf{B}} \right\}. \tag{23}$$

An unbiased estimator for $\mathrm{Cov}[\widehat{\boldsymbol{b}}_i, \widehat{\boldsymbol{b}}_j]$, $i \neq j$, is

$$\begin{aligned} \widehat{\mathrm{Cov}[\widehat{\boldsymbol{b}}_i, \widehat{\boldsymbol{b}}_j]} &= \frac{S_Z^2}{m^2} \sum_{l=1}^{m} \boldsymbol{\Lambda}_l^{-1} (\mathbf{X}_l^\top \mathbf{X}_l)^{-1} (\boldsymbol{\Lambda}_l^{-1})^\top - \frac{1}{m} \widehat{\mathbf{B}} \\ &\quad - \frac{S_Z^2}{m} \left(\boldsymbol{\Lambda}_i^{-1} (\mathbf{X}_i^\top \mathbf{X}_i)^{-1} (\boldsymbol{\Lambda}_i^{-1})^\top + \boldsymbol{\Lambda}_j^{-1} (\mathbf{X}_j^\top \mathbf{X}_j)^{-1} (\boldsymbol{\Lambda}_j^{-1})^\top \right). \end{aligned} \tag{24}$$

Let $\boldsymbol{\Lambda}$ be a $q \times q$ matrix, and let $\boldsymbol{x}$ be a vector in $\mathbb{R}^q$. Consider the response $Z(\boldsymbol{x}) = (\boldsymbol{\Lambda}(\boldsymbol{\beta} + \boldsymbol{b}) + \boldsymbol{d})^\top \boldsymbol{x} + \varepsilon$ under the regressor $\boldsymbol{x}$ where the random effect $\boldsymbol{b}$ and the residual ε are independent of all variables Z_{ij} used for parameter estimation. A predictor is given by

$$\widehat{Z}(\boldsymbol{x}) = (\boldsymbol{\Lambda} \widehat{\boldsymbol{\beta}} + \boldsymbol{d})^\top \boldsymbol{x} \tag{25}$$

with $E[\widehat{Z}(\boldsymbol{x})] = E[Z(\boldsymbol{x})]$. An unbiased estimator of the prediction error variance is

$$\begin{aligned} &\widehat{\sigma}^2_{\widehat{Z}(x) - Z(x)} = \\ &S_Z^2 + \boldsymbol{x}^\top \boldsymbol{\Lambda} \left(\frac{m+1}{m} \widehat{\mathbf{B}} + \frac{S_Z^2}{m^2} \sum_{l=1}^{m} \boldsymbol{\Lambda}_l^{-1} (\mathbf{X}_l^\top \mathbf{X}_l)^{-1} (\boldsymbol{\Lambda}_l^{-1})^\top \right) \boldsymbol{\Lambda}^\top \boldsymbol{x}. \end{aligned} \tag{26}$$

6.2 Distribution Analysis for Step I Estimators and Predictors

We consider two additional distribution assumptions: (1) $\boldsymbol{b}_i$ has a q-dimensional normal distribution $N(\mathbf{0}, \mathbf{B})$. (2) The residual vector $\boldsymbol{\varepsilon}_i$ has a normal distribution $N(\mathbf{0}, \sigma^2\mathbf{I})$. In Göb and Lurz (2010), the subsequent results on the distribution of estimators and predictors are established.

The random variables $\sum_{i=1}^{m} \widehat{\boldsymbol{b}}_i \widehat{\boldsymbol{b}}_i^{\top}$ and S_Z^2 are independent. $\frac{N-mq}{\sigma^2} S_Z^2$ has the χ^2-distribution $\chi^2(N-mq)$.

The prediction error has a univariate normal distribution with mean 0 and variance $\sigma^2_{\widehat{Z}(x)-Z(x)}$.

Under small residual variance σ^2, such that $\sigma^2 \sum_{i=1}^{m} \boldsymbol{\Lambda}_i^{-1}(\mathbf{X}_i^{\top}\mathbf{X}_i)^{-1}(\boldsymbol{\Lambda}_i^{-1})^{\top}$ is negligible in comparison with $\mathbf{B}$, $\widehat{\boldsymbol{\beta}}$ and $\sum_{i=1}^{m} \widehat{\boldsymbol{b}}_i \widehat{\boldsymbol{b}}_i^{\top}$ are approximately independent. Hence the prediction error $\widehat{Z}(\boldsymbol{x}) - Z(\boldsymbol{x}) = (\boldsymbol{\Lambda}(\widehat{\boldsymbol{\beta}} - \boldsymbol{\beta} - \boldsymbol{b}))^{\top}\boldsymbol{x}$ and the estimator $\widehat{\sigma}^2_{\widehat{Z}(x)-Z(x)}$ of the prediction error variance are approximately independent. In Göb and Lurz (2010), the method of Satterthwaite (1941, 1946) and Welch (1947), is applied to obtain for the ratio

$$\frac{\widehat{Z}(\boldsymbol{x}) - Z(\boldsymbol{x})}{\widehat{\sigma}_{\widehat{Z}(x)-Z(x)}} \tag{27}$$

an approximate central t-distribution $t(\nu)$ with estimated degree of freedom

$$\nu = \frac{\widehat{\sigma}^4_{\widehat{Z}(x)-Z(x)}}{\frac{S_Z^4 \Delta^2}{N-mq} + \left(\frac{m+1}{m(m-1)}\right)^2 \sum_{1 \le i,j \le m} \left(\boldsymbol{x}^{\top}\boldsymbol{\Lambda}\widehat{\mathrm{Cov}[\widehat{\boldsymbol{b}}_i, \widehat{\boldsymbol{b}}_j]}\boldsymbol{\Lambda}^{\top}\boldsymbol{x}\right)^2}, \tag{28}$$

where

$$\Delta = 1 - \frac{1}{m}\boldsymbol{x}^{\top}\boldsymbol{\Lambda}\sum_{l=1}^{m} \boldsymbol{\Lambda}_l^{-1}(\mathbf{X}_l^{\top}\mathbf{X}_l)^{-1}(\boldsymbol{\Lambda}_l^{-1})^{\top}\boldsymbol{\Lambda}^{\top}\boldsymbol{x}. \tag{29}$$

Hence approximate bounds for prediction intervals for $Z(\boldsymbol{x})$ are of the form

$$\widehat{Z}(\boldsymbol{x}) -/+ z\widehat{\sigma}_{\widehat{Z}(x)-Z(x)}, \tag{30}$$

where z is an appropriate quantile of the central t-distribution $t(\nu)$.

The results obtained are valid in the case that the assumption of the normality of the q-dimensional normal distribution of $\boldsymbol{b}_i$ and the residual vector $\boldsymbol{\varepsilon}_i$ holds. The assumption of normally distributed residuals $\boldsymbol{\varepsilon}_i$ is well-supported. The residuals represent measurement-to-measurement error. Following Gauss' derivation of the normal distribution, there are good reasons to assume measurement errors to be normally distributed. The normality assumption for the random regression parameters $\boldsymbol{b}_i$ is more questionable. These parameters represent item-to-item variation due to manufacturing or handling. Though many manufacturing processes exhibit normally distributed deviations from target, there is clearly no general law to support this assumption in any particular case. In the presented context of the analysis of TIEs usually small sample sizes occur, so that the assumptions cannot be supported by asymptotic results. Results obtained in course of the empirical study presented in Sect. 8, like residual plots and quantile plots, do not contradict the normality assumption. However, the validity of the normality assumption cannot thoroughly be verified at the moment and has to be postponed to future investigations.

6.3 Step II Estimation of $\boldsymbol{\delta}$

$\boldsymbol{\delta}$ is estimated by minimising $O(\boldsymbol{\delta}) = \sum_i \|\widehat{\boldsymbol{b}}_i(\boldsymbol{\delta})\|^2$ in $\boldsymbol{\delta}$, where $\widehat{\boldsymbol{b}}_i(\boldsymbol{\delta})$ is taken from Eq. 18. In Göb and Lurz (2010), the problem is reduced to a multiresponse regression problem. A simple solution is obtained if one component $\sum_i \widehat{b}_{ij}(\boldsymbol{\delta})^2$ is dominating in $O(\boldsymbol{\delta})$. This is the case for the application of the general scheme, Eqs. 7 and 8, to TIE analysis, where $O(\delta) \approx \sum_i \widehat{b}_{i1}(\delta)^2$. In this case, point estimates and confidence regions are easily obtained from classical linear or nonlinear regression analysis.

For prediction purposes, it is suggested in Göb and Lurz (2010) to calculate a confidence region C for $\boldsymbol{\delta}$ and to use the value $\boldsymbol{\delta} \in C$ which maximises the length of a two-sided prediction interval of the type Eq. 30. A prescribed level γ in the final prediction interval is obtained via the Bonferroni inequality by choosing partial levels $\gamma_i = 0.5(1+\gamma)$ for C and for the prediction interval depending on δ.

7 Estimation of Mixed Model Parameters from TIE Experiments

The linearised TIE model defined by Eqs. 7 and 8 can be analysed empirically by applying the results of Sect. 6, inserting the model quantities according to the definitions in Eqs. 14 and 15.

Being a parameter of the linearising transformation Eq. 6, the parameter λ_{ev} (thermal conductivity in the evacuated and dry state of the TIE) cannot be estimated on grounds of the linearised model. For some types of TIEs, there is reliable expert information on λ_{ev}, e.g. $1\,\mathrm{W}/(\mathrm{m\,K}) \leq \lambda_{ev} \leq 4\,\mathrm{W}/(\mathrm{m\,K})$. Furthermore a negative relation between λ_{ev} and the parameter β_1 exists due to the nature of the two parameters. Keeping this relation in mind, it has to be ensured that β_1 is a positive real number. The subsequent empirical study uses the *worst forecast estimator* for λ_{ev} suggested in Göb and Lurz (2010): For each value λ_{ev} from a prior information set D, all other parameters are estimated from the linearised model. The estimator $\widehat{\lambda}_{ev}$ is the value from D which leads to the less favourable, i.e., highest, forecast of total thermal conductivity for a specified forecasting time, under the restriction $\beta_1 > 0$.

8 Empirical Analysis of a TIE Experiment

We apply the method described in Sect. 7 to experimental data obtained at the Bavarian Center for Applied Energy Research, Würzburg, Germany. $m = 9$ test units were subject to an experiment as described in Sect. 3. $m_1 = 4$ test units were stored at $\vartheta_l = (T_l - 273.15)°C = 20°C$ (non-accelerated exposure) permanently, $m_2 = 5$ test units were initially exposed to $\vartheta_u = (T_u - 273.15)°C = 70°C$ (accelerated exposure) temporarily, and then measured at the non-accelerated temperature of $\vartheta_l = 20°C$. The sample sizes were very small: Three absolute value measurements were taken on each of the non-accelerated units, two measurements on each of the accelerated units. The following estimates were obtained:

$$\widehat{\lambda}_{ev} = 3.45, \quad \widehat{\beta}_1 = 0.000403441, \quad \widehat{\beta}_2 = 0.000604925, \quad \widehat{\delta} = 42\,879.7,$$

$$\widehat{\sigma}^2 = 0.0000143776, \quad \widehat{\mathbf{B}} = \begin{pmatrix} 0.0002444814 & 1.20215 \times 10^{-6} \\ 1.20215 \times 10^{-6} & 3.61278 \times 10^{-8} \end{pmatrix}.$$

The interest from the TIE producer's perspective is to predict the thermal conductivity $\lambda(x)$ at the end of the TIE's guarantee period, which ranges between 2, 4 or 5 years and 10, 25 or 50 years for long-time installations in buildings. We exemplarily demonstrate the prediction of the service lifetime for a guarantee period of 10 years ($x = 3{,}650$ days) under 20°C ambient temperature, which is considered a conservative upper bound for the average daily temperature in Germany over the course of a year. On grounds of the linearised model, the prediction $\widehat{Z}(x)$ by Eq. 25 and the two-sided prediction limits $Z_L(x) < Z_U(x)$ by Eq. 30 at a confidence level of 0.95 are

$$\widehat{Z}(x) = 1.36992, \quad Z_L(x) = 1.16049, \quad Z_U(x) = 1.57934.$$

The estimated prediction error variance is $\widehat{\sigma}_{\widehat{Z}(x)-Z(x)} = 0.00440372$ where the three components in Eq. 26 are

$$S_Z^2 = 0.0000143776, \quad \boldsymbol{x}^\top \boldsymbol{\Lambda}\Big(\frac{m+1}{m}\widehat{\mathbf{B}}\Big)\boldsymbol{\Lambda}^\top \boldsymbol{x} = 0.00434685,$$

$$\boldsymbol{x}^\top \boldsymbol{\Lambda}\Big(\frac{S_Z^2}{m^2}\sum_{l-1}^{m} \boldsymbol{\Lambda}_l^{-1}(\mathbf{X}_l^\top \mathbf{X}_l)^{-1}(\boldsymbol{\Lambda}_l^{-1})^\top\Big)\boldsymbol{\Lambda}^\top \boldsymbol{x} = 0.0000424975.$$

The component $\boldsymbol{x}^\top \boldsymbol{\Lambda}(\frac{m+1}{m}\widehat{\mathbf{B}})\boldsymbol{\Lambda}^\top \boldsymbol{x}$ corresponding to unit-to-unit variation strongly dominates the total prediction variance, the components in S_Z^2 corresponding to measurement-to-measurement variation are very small.

Both the point prediction $\widehat{Z}(x)$ and the limits of the prediction interval have to be retransformed by applying the inverse $\mathcal{T}^{-1}$ of the linearising transformation Eq. 6. The predicted thermal conductivity and the corresponding prediction limits at a confidence level of $\gamma = 0.95$ are

$$\widehat{\lambda}(x) = 0.0104707, \quad \lambda_L(x) = 0.0070458, \quad \lambda_U(x) = 0.0129874.$$

9 Experimental Design

Long-term predictions on thermal conductivity are the primary interest of the statistical inference from experiments on TIEs. Hence the quality of experimental design has to be evaluated with respect to forecasting accuracy, i.e., the prediction error variance. Designs minimising the prediction error variance for a prescribed forecasting horizon are a special case of *C-optimal* designs, see Atkinson et al. (2007). Subsequently, predictions are considered for a guarantee period of 10 years, i.e., $x = 3{,}650$ days, under the non-accelerated temperature 20°C.

The experimental design has to specify the following issues: (1) Levels of temperature T_l, T_u at normal usage conditions and at accelerated conditions. (2) Total number m of testing units, allocation numbers m_1, m_2 of units over non-accelerated and accelerated conditions. (3) Total length L of experimental period. (4) For each testing unit i, the length a_{ui} of exposure to accelerated conditions. (5) For each testing unit i, total number n_i of measurements and times $t_{i1}, \ldots, t_{in_i}$ of measurements. Restrictions of industrial practice lead to the following simplifications: (1) The accelerated exposure length is either $a_{ui} = 0$ or $a_{ui} = a_u$ identical for all units in accelerated exposure. (2) Equally distanced measurement times $t_{i1}, \ldots, t_{in_i}$ over the measurement period $L - a_{ui}$. (3) Cost restrictions require a small number of measurements $n_i = 4$ for each testing unit i. The result of the experiment in Sect. 8 shows that measurement-to-measurement variation is very small so that a small number of measurements is tolerable.

The empirical basis for constructing an optimum design are the results of the experiment considered in Sect. 8, in particular the estimates $\widehat{\boldsymbol{\beta}}$, $\widehat{\sigma}^2 = S_Z^2$, $\widehat{\mathbf{B}}$. The interdependence of the parameter estimates is too involved for determining an optimum design analytically. Instead, we use a simulation approach, based on the normality assumptions of Sect. 6.2. All parameter estimators depend on the estimates $\widehat{\boldsymbol{\alpha}}_i$. Because of the independence of the groups $\widehat{\boldsymbol{\alpha}}_1, \ldots, \widehat{\boldsymbol{\alpha}}_m$ are independent. Each $\widehat{\boldsymbol{\alpha}}_i$ has a bivariate normal distribution with parameters from Eq. 17. Inserting the estimators obtained in Sect. 8 into Eq. 17, vectors $(\widehat{\boldsymbol{\alpha}}_1, \ldots, \widehat{\boldsymbol{\alpha}}_m)$ can be simulated for each interesting design constellation. For each simulated vector $(\widehat{\boldsymbol{\alpha}}_1, \ldots, \widehat{\boldsymbol{\alpha}}_m)$, all parameters are re-estimated and the two-sided level 0.95 prediction interval for a 10-year horizon under the non-accelerated temperature $\vartheta_l = 20$°C is calculated. Each particular design constellation is evaluated by the average length of intervals from 10,000 simulation runs.

Due to the presence of a mixed model, the construction of an experimental design minimising the prediction error variance requires to distinguish between the uncertainty in the prediction caused by unit-to-unit variation and the part of the prediction error variance associated to measurement error. In the present industrial environment, unit-to-unit variation will always be strongly dominating, see the experimental results in Sect. 8, and the accuracy of the prediction is primarily influenced by the number of testing units. The temperature levels, the length of the accelerated exposure period, and the length of the total experimental period have only small impact on the total length of the prediction interval, but rather influence on coefficients of the estimated measurement variance. Therefore, the design analysis proceeds as follows:

- The total length of the prediction interval

$$2 \cdot z \cdot \widehat{\sigma}_{\widehat{Z}(x)-Z(x)}, \tag{31}$$

 is used to assess the influence of the number m of experimental units.
- The part

$$\Phi \quad := \quad x^\top \Lambda \left(\frac{S_Z^2}{m^2} \sum_{l=1}^{m} \Lambda_l^{-1} (\mathbf{X}_l^\top \mathbf{X}_l)^{-1} (\Lambda_l^{-1})^\top \right) \Lambda^\top x \tag{32}$$

 of the prediction error variance is considered to assess the influence of the following design components: temperature levels T_l, T_u, length a_{ui} of the accelerated exposure period, length of the total experimental period, allocation numbers m_1 (number of units without accelerated exposure) and m_2 (number of units with accelerated exposure).

9.1 Choice of the Numbers m, m_1, m_2 of Testing Units

To demonstrate the influence of the number of testing units m on the length of the prediction interval Eq. 31, m is varied while all other experimental factors are kept constant.

Figure 2 shows that a huge improvement in the length of the prediction interval can be achieved by increasing the number of testing units involved in the experiment up to a number of about 24. Only minor improvement can be observed when including more than about 24 testing units.

To receive information about the influence of the allocation numbers m_1 and m_2, we compare the term Φ, see Eq. 32, for the two cases $m_1 = m_2$ and $m_1 = 2m_2$. As Fig. 2 reveals, the case $m_1 = m_2$, i.e., storing the same number of units under accelerated and under non-accelerated temperature, shows better results than the set-up $m_1 = 2m_2$.

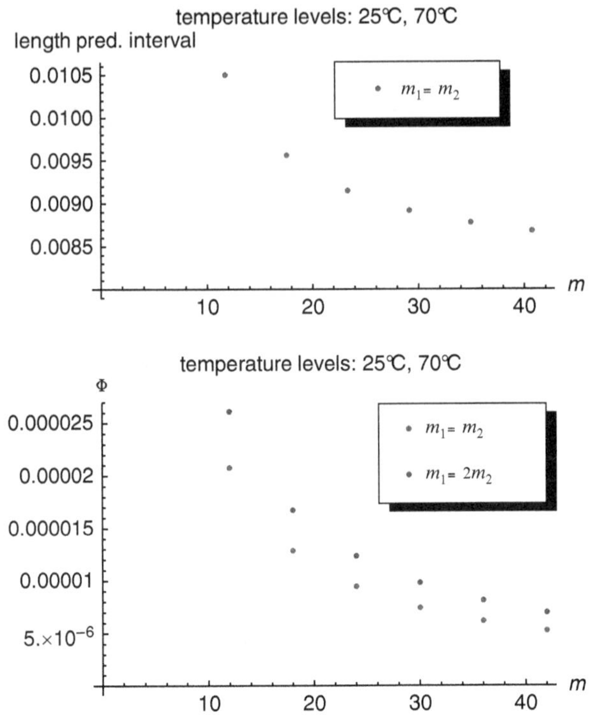

Fig. 2 Length of the prediction interval (unit: W/ (m K)) and term Φ as a function of the number $m = m_1 + m_2$ of testing units. m_1, m_2: allocation numbers of units over non-accelerated and accelerated conditions. Levels of the experimental factors: Temperatures: $\vartheta_l = 25°C$, $\vartheta_u = 70°C$, $L = 120$ days, $a_u = 60$ days, $n_i = 4$, equally distanced $t_{i1}, \ldots, t_{in_4}$ over the measurement period $L - a_{ui}$

9.2 Choice of the Length L of the Experimental Period

To analyse the influence of extending the length L of the experimental period, we vary L while keeping all other experimental factors constant and observe the term Φ, see Eq. 32.

Figure 3 shows the term Φ as a function of the length L of the experimental period in days. Strong improvement in Φ can be achieved when choosing $L \geq 150$ days.

9.3 Choice of Temperature Levels T_l, T_u and of the Length a_u of the Accelerated Exposure Period

The non-accelerated temperature T_l is prescribed at (273.15 + 25) K, which is considered to be an upper bound for the average temperature stress in applications.

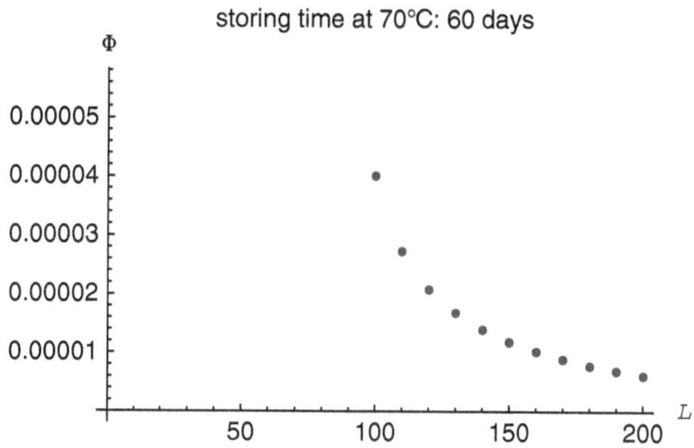

Fig. 3 Φ as a function of the length L of the experimental period in days. Levels of the experimental factors: Temperatures: $\vartheta_l = 25°C$, $\vartheta_u = 70°C$, $a_u = 60$ days, $n_i = 4$, equally distanced $t_{i1}, \ldots, t_{in_4}$ over the measurement period $L - a_{ui}$, $m_1 = 6 = m_2 \Rightarrow m = 12$.

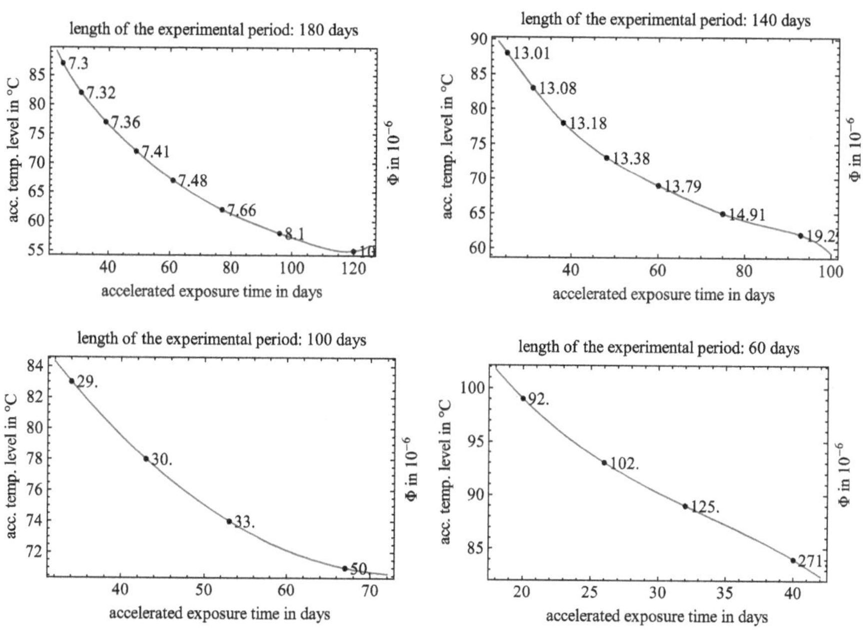

Fig. 4 Optimum combinations of accelerated exposure time a_u and accelerated temperature ϑ_u under four levels of the total length L of the experimental period. $n_i = 4$, measurement times equally distanced $t_{i1}, \ldots, t_{in_4}$ over the measurement period $L - a_{ui}$, $m_1 = 6 = m_2 \Rightarrow m = 12$

In industrial applications, the interest is to reduce the length of the experimental period L, in particular the accelerated exposure time a_u, so as to reduce costs and achieve rapid decisions, without a significant loss in prediction precision. Under a prescribed level of forecasting reliability γ, a reduced experimental period a_u can

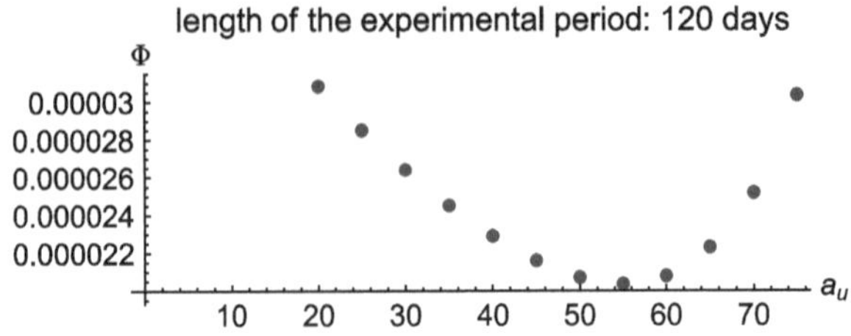

Fig. 5 Φ as a function of the accelerated exposure period a_{u} in days. Levels of the experimental factors: Temperatures: $\vartheta_{\mathrm{l}} = 25°\mathrm{C}$, $\vartheta_{\mathrm{u}} = 70°\mathrm{C}$, $L = 120\,\mathrm{days}$, $n_i = 4$, equally distanced $t_{i1}, \ldots, t_{in_4}$ over the measurement period $L - a_{\mathrm{u}i}$, $m = 12$, $m_1 = 6 = m_2$

be compensated by an appropriately increased accelerated experimental temperature T_{u}. Figure 4 considers four values of the total length L, in days, of the experimental period: $L = 180$, $L = 140$, $L = 100$, $L = 60$. The horizontal axis is indexed by the accelerated exposure time a_{u}, in days. In each block, the left-hand vertical axis is indexed by the accelerated temperature level $\vartheta_{\mathrm{u}} = T_{\mathrm{u}} - 273.15$, the right-hand vertical axis by the variance component Φ, which expresses the design effect. The curve associates each accelerated exposure time a_{u} with the corresponding accelerated temperature level $\vartheta_{\mathrm{u}} = T_{\mathrm{u}} - 273.15$ which provides the minimum value of Φ. Increased accelerated temperature goes along with decreased exposure time, and also with a decrease in Φ. However, this effect cannot be exploited ad libitum. The practically maximum accelerated temperature ϑ_{u} is around 80°C. Exposure at higher temperatures causes immediate damages in the TIEs. Decreasing the total length L inflates the variance component Φ. Small L below 100 days implies very high Φ, and leads to practically unfeasible accelerated temperatures. The experimenter can use Fig. 4 to choose the appropriate design under given restrictions on time, temperature, and desired variance component Φ.

Figure 5 considers the choice of the optimal length a_{u} of the accelerated exposure time under fixed total length L and a prescribed accelerated temperature level ϑ_{u}. We consider $\vartheta_{\mathrm{l}} = 25°\mathrm{C}$, $\vartheta_{\mathrm{u}} = 70°\mathrm{C}$, and, in view of the preceding results, we consider $L = 120$ days. Figure 5 shows that Φ as a function of a_{u} adopts a minimum at approximately $a_{\mathrm{u}} = 55$ days.

10 Conclusion and Outlook

Starting from physical models, we have established a statistical nonlinear mixed regression model for the thermal conductivity of TIEs. The model accounts for two levels of variation: measurement-to-measurement variation, and, important in view of the present immaturity of the TIE manufacturing sector, unit-to-unit variation. Techniques for model estimation, prediction, and experimental design have been developed. However, there are issues remaining for future research. (1) The present model covers the factor time and temperature, but excludes the ambient factor humidity. This is appropriate only for TIEs equipped with appropriate humidity getters. To account for a larger class of TIEs, humidity should be integrated in a future model. (2) The validity of the method for constructing approximate prediction intervals strongly depends on the assumption that the measurement variance component is small in comparison with the variance component due to the TIEs. This assumption widely holds in the present state of not completely mature TIE manufacturing. However, a valid technique should also be developed for applications to TIEs from more homogeneous manufacturing likely to evolve over the next years.

References

Atkinson, A. C., Donev, A. N., & Tobias, R. D. (2007). *Optimum experimental designs, with SAS*. New York: Oxford University Press.

Bundesregierung der Bundesrepublik Deutschland (2007). Verordnung über energiesparenden Wärmeschutz und energiesparende Anlagentechnik bei Gebäuden (Energieeinsparverordnung - EnEV) vom 24. Juli 2007. Bundesgesetzblatt Jahrgang 2007, Teil I, Nr. 34, pp. 1519–1563.

Bundesregierung der Bundesrepublik Deutschland (2009). Verordnung zur Änderung der Energieeinsparverordnung vom 29. April 2009. Bundesgesetzblatt Jahrgang 2009, Teil I, Nr. 23, pp. 954–989.

Caps, R., Hetfleisch, J., Rettelbach, T., & Fricke, J. (1996). Thermal conductivity of spun glass fibers as filler material for vacuum insulations. *Thermal Conductivity, 23*, 373–382.

Caps, R., Rettelbach, T., Ehrmanntraut, M., Korder, S., & Fricke, J. (1997). Development of vacuum super insulations with glass cover and powder filling. In R. S. Graves & R. R. Zarr (Eds.), *Insulation materials: Testing and applications* (Vol. 3. ASTM STP 1320) (pp. 270–282). West Conshohocken: American Society for Testing and Materials.

Christensen, R. (1996). *Plane answers to complex questions: The theory of linear models* (2nd ed.). New York/Berlin/Heidelberg: Springer.

Davidian, M., & Giltinian, D. M. (2003). Nonlinear models for repeated measurement data: An overview and update. *Journal of Agricultural, Biological, and Environmental Statistics, 8*(4), 387–419.

"Verordnung über energiesparenden Wärmeschutz und energiesparende Anlagentechnik bei Gebäuden (Energieeinsparverordnung – EnEV)" (2007, 2009). http://www.gesetze-im-internet.de/bundesrecht/enev_2007/gesamt.pdf.

Göb, R., & Lurz, K. (2010). A mixed effects model for experiments on thermal insulation elements. Technical Reports of the Institute of Mathematics of the University of Würzburg, 2010.

Heinemann, U. (2008). Influence of water on the total heat transfer in evacuated insulations. *International Journal of Thermophysics, 29*(2), 735–749.

Heinemann, U., Caps, R., & Fricke, J. (1999). Characterization and optimization of filler materials for vacuum super insulations. *Vuoto scienza e tecnologia, XXVIII*(1–2), 43–46.

Kaganer, M.G. (1969). *Thermal insulation in cryogenic engineering*. Jerusalem: Israel program for scientific translations.

Kreith, F., & Bohn M. S. (2000). *Principles of heat transfer*. USA: Thomson Press.

Satterthwaite, F. E. (1941). Synthesis of variance. *Psychometrika, 6*, 309–316.

Satterthwaite, F. E. (1946). An approximate distribution of estimates of variance components. *Biometrics Bulletin, 2*(6), 110–114.

Welch, B. L. (1947). The generalization of "student's" problem when several different population variances are involved. *Biometrika, 34*, 28–35.

Proposal of Advanced Taguchi's Linear Graphs for Split-Plot Experiments

Tomomichi Suzuki, Hironobu Kawamura, Seiichi Yasui, and Yoshikazu Ojima

Abstract Taguchi's orthogonal arrays and linear graphs are convenient tools for the design of fractional factorial experiments, especially for practitioners. Taguchi also proposed how to use them in split-plot designs and prepared linear graphs for split-plot designs. For the orthogonal array of order 16, Taguchi proposed one which is called L_{16} orthogonal array. Taguchi presented 18 linear graphs when a L_{16} orthogonal array is used in split-plot designs. Those linear graphs are capable of showing main effects of whole plots, subplots, sub-subplots, and so on, but they are not capable of showing interaction effects of plots of different levels. Also, those linear graphs do not cover all the possible designs, and there exist a lot of other linear graphs that can be applied when using L_{16} orthogonal arrays. The primary objective of this paper is to propose an improved version of linear graphs. Another purpose of this paper is to investigate how to list all the possible linear graphs that can be applied when using L_{16} orthogonal arrays. A proposal is made and many new linear graphs are presented.

Keywords Fractional factorial design • Two-factor interaction • Orthogonal array

1 Introduction

Fractional factorial designs are popularly used in screening stages. Orthogonal arrays (Hedayat et al. 1999) are effectively applied in those designs. The widely used approach for designing those designs is described in many text books such as Box et al. (2005) and Wu and Hamada (2000). There is another approach

T. Suzuki (✉) · H. Kawamura · S. Yasui · Y. Ojima
Department of Industrial Administration, Tokyo University of Science, 2641 Yamazaki, Noda, Chiba, 278-8510 Japan
e-mail: suzuki@ia.noda.tus.ac.jp; kawamura@ia.noda.tus.ac.jp; yasui@ia.noda.tus.ac.jp; ojima@ia.noda.tus.ac.jp

H.-J. Lenz et al. (eds.), *Frontiers in Statistical Quality Control 10*,
Frontiers in Statistical Quality Control 10, DOI 10.1007/978-3-7908-2846-7_22,

proposed by Taguchi (1976). Taguchi's orthogonal arrays and linear graphs are convenient tools for the design of fractional factorial experiments, especially for practitioners. Taguchi also proposed how to use them in split-plot designs and prepared linear graphs for split-plot designs. For the orthogonal arrays of order 16, Taguchi proposed one which is called L_{16} orthogonal array. Taguchi presented 18 linear graphs when a L_{16} orthogonal array is used in split-plot designs. Those linear graphs are capable of showing main effects of whole plots, subplots, sub-subplots, and so on, but they are not capable of showing interaction effects of plots of different levels. Also, those linear graphs do not cover all the possible designs, and there exist a lot of other linear graphs that can be applied when using L_{16} orthogonal arrays.

The primary objective of this paper is to propose an improved version of linear graphs. The secondary purpose of this paper is to investigate how to list all the possible linear graphs that can be applied when using L_{16} orthogonal arrays. A proposal is made and many new linear graphs are presented.

2 Designing Fractional Factorial Experiments

In this chapter, two approaches of designing fractional factorial designs using orthogonal arrays are explained. One is the conventional approach and the other one is the Taguchi approach.

2.1 Conventional Approach to Fractional Factorial Experiments

The conventional approach to fractional factorial experiments is to use defining relations. Generating relations (generators) are chosen to run an experiment. All the alias relationships are derived using the generators. For example, consider the example of Table 1 presented in Box et al. (2005).

In this case, the generating relation of the design is

$$\boxed{\mathbf{D} = \mathbf{ABC}}$$

and all the alias relations are

$$\mathbf{I} = \mathbf{ABCD}, \mathbf{A} = \mathbf{BCD}, \mathbf{B} = \mathbf{ACD}, \mathbf{C} = \mathbf{ABD}, \mathbf{D} = \mathbf{ABC}$$
$$\mathbf{AB} = \mathbf{CD}, \mathbf{AC} = \mathbf{BD}, \mathbf{AD} = \mathbf{BC}$$

Table 1 An eight-run experiment

Factor	a	b	c	ab	ac	bc	abc
Run#	A	B	C				D
1	−	−	−	+	+	+	−
2	+	−	−	−	−	+	+
3	−	+	−	−	+	−	+
4	+	+	−	+	−	−	−
5	−	−	+	+	−	−	+
6	+	−	+	−	+	−	−
7	−	+	+	−	−	+	−
8	+	+	+	+	+	+	+

This is a 2^{4-1}_{IV} design. The effects, for example the main effect of **A** and the three factor interaction effect **BCD**, are confounded. So, we cannot estimate all the effects independently. Even if we can assume that the three or higher factor interactions are negligible, two factor interaction effects are still confounded. If we want to estimate a particular two factor interaction effects, we have to ignore the other, or to choose other designs, namely a larger experiment.

The merit of this conventional approach is that once one masters it, various experimental designs can be designed. The drawback, on the other hand, is that this approach requires the knowledge about related DOE topics. Also, we have to consider that there will be many interaction effects that are confounded.

2.2 *Taguchi's Approach to Fractional Factorial Experiments*

Taguchi proposed to use linear graphs when applying orthogonal arrays. Linear graphs are the graphs that represent the effects considered in a specific design. They are capable of showing the main effects and the two factor interactions which are of interest. The main effects are shown by a node and two factor interaction effects are shown by an edge between two nodes. Taguchi proposed a catalog of linear graphs for basic orthogonal arrays such as L_8, L_{16}, L_{32}, L_9, L_{27}, etc. One of the linear graphs in the catalog is used to accommodate the effects considered in a design. The linear graphs for L_{16} prepared by Taguchi (1976) are shown in Fig. 1.

In Taguchi's approach, one has to decide which two factor interactions shall be estimated and which two factor interactions are negligible. The linear graph which takes all the effects into account is then drawn. This linear graph is called a 'necessary linear graph'. The necessary linear graph is compared to the catalog of the linear graphs which is readily prepared. The one which includes the necessary linear graph is selected for actually designing an experiment.

An example of a linear graph is shown in Fig. 2. Figure 2 shows that the effects considered in this design are: main effects **A**,**B**, and **C**, and two factor interactions **AB**, **AC**, and **BC**.

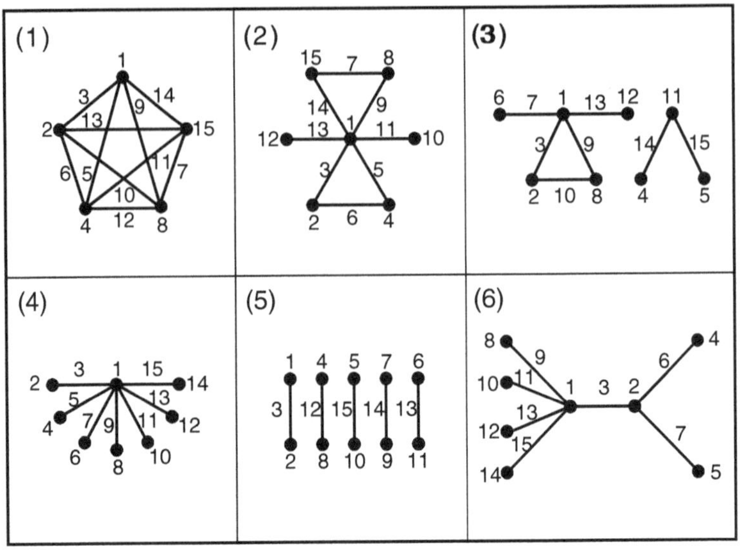

Fig. 1 Taguchi's Linear Graphs for L_{16}

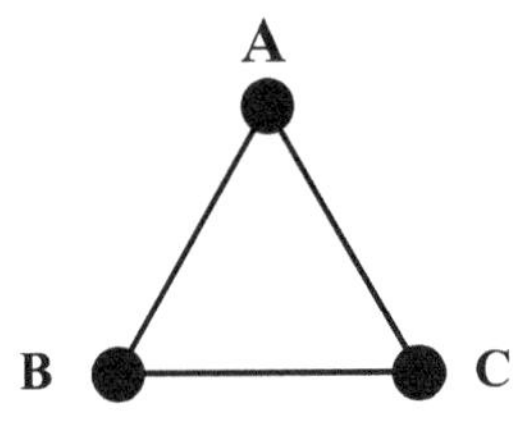

Fig. 2 A linear graph

Once the design is fixed, the experiment will be run and the analyses will be performed using cook-book style text books. The merit of Taguchi's approach is that no subtle knowledge on DOE is required in performing the analyses. This is why Taguchi's approach became popular, especially for practitioners who are usually not specialists in statistics. The most difficult part is the selection of the appropriate linear graphs after drawing a necessary linear graph, which is still generally easier than mastering DOE techniques. On the other hand, the drawback of Taguchi's approach is inflexibility of the designs. Prepared orthogonal arrays must be used so that the number of runs the designer can select is fixed. Also, the user might be performing the analyses without understanding the method which is not recommended when applying any statistical method.

3 Linear Graphs for Split-Plot Designs

3.1 Symbols Used in Linear Graphs

Taguchi proposed linear graphs not only for ordinary experiments but also for split-plot designs. In this paper, we consider the cases when L_{16} is used. Taguchi defines groups in columns of orthogonal arrays. The orthogonal array L_{16} is shown in Table 2. The levels of the factors are expressed as 1's and 2's instead of +'s and −'s.

The group corresponds to plots in split-plot designs. The whole plot can be allocated to group 1, subplots to group 2, sub-subplots to group 3 and sub-sub-subplots to group 4. In some cases, groups can be combined to form a new group. When group 1 and group 2 are combined, whole plots can be allocated to group 1 and/or group 2, subplots can be allocated to group 3, and sub-subplots can be allocated to group 4. Ojima et al. (2004) also give explanation on this matter.

Linear graphs for split-plot designs consider the level of groups. Instead of the usual circles, Taguchi used various kinds of circles to depict the level of groups. Watanabe et al. (2006) proposed to draw lines which depict the levels of groups for two factor interactions. Since Taguchi used only solid lines for two factor interactions even for split-plot designs, the proposal would be better fitted for depicting the split-plot designs. These symbols are shown below (Fig. 3).

Using these symbols, it became possible to precisely express the designed split-plot experiments. Regarding the levels of the groups of main effects and two factor interactions, the following important facts are known.

1. The two factor interaction between factors from different levels of groups will appear at the higher level of groups. For example, if **A** is allocated to group 2 and **B** is allocated to group 3, then the two factor interaction **AB** will appear in group 3.
2. The two factor interaction between factors from the same level of groups will appear at one of lower levels of groups. For example, if both **A** and **B** are allocated to group 3, then the two factor interaction **AB** will appear either in group 1 or in group 2.

It is important to know which two factor interactions appear as which level of group, because the standard errors of the factors might differ largely among the factors of different levels of groups. In other words, if a particular two factor interaction effect shall be more accurately estimated than other effects, the design can be selected so that the standard error for that effect becomes smaller. This is also the reason why distinguishing the edges of linear graphs is meaningful in split-plot designs. Furthermore we can select the source of variation for factors considered in the experiment. Of course, there are trade-offs among factors, so this idea is much more effective when priorities are different among the factors.

Table 2 Taguchi's L_{16} orthogonal array

No.	[1]	[2]	[3]	[4]	[5]	[6]	[7]	[8]	[9]	[10]	[11]	[12]	[13]	[14]	[15]
1	1	1	1	1	1	1	1	1	1	1	1	1	1	1	1
2	1	1	1	1	1	1	1	2	2	2	2	2	2	2	2
3	1	1	1	2	2	2	2	1	1	1	1	2	2	2	2
4	1	1	1	2	2	2	2	2	2	2	2	1	1	1	1
5	1	2	2	1	1	2	2	1	1	2	2	1	1	2	2
6	1	2	2	1	1	2	2	2	2	1	1	2	2	1	1
7	1	2	2	2	2	1	1	1	1	2	2	2	2	1	1
8	1	2	2	2	2	1	1	2	2	1	1	1	1	2	2
9	2	1	2	1	2	1	2	1	2	1	2	1	2	1	2
10	2	1	2	1	2	1	2	2	1	2	1	2	1	2	1
11	2	1	2	2	1	2	1	1	2	1	2	2	1	2	1
12	2	1	2	2	1	2	1	2	1	2	1	1	2	1	2
13	2	2	1	1	2	2	1	1	2	2	1	1	2	2	1
14	2	2	1	1	2	2	1	2	1	1	2	2	1	1	2
15	2	2	1	2	1	1	2	1	2	2	1	2	1	1	2
16	2	2	1	2	1	1	2	2	1	1	2	1	2	2	1
Component	a		a		a		a		a		a		a		a
		b	b			b	b			b	b			b	b
				c	c	c	c					c	c	c	c
								d	d	d	d	d	d	d	d
Group	1	2		3				4							

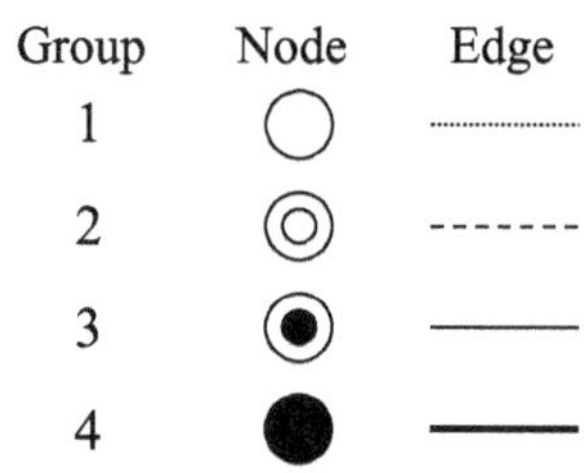

Fig. 3 Symbols for linear graphs for split-plot designs

3.2 *Linear Graphs for Split-Plot Designs*

Taguchi proposed altogether 18 linear graphs in his book. Topologically there are six different linear graphs. For each of the linear graph with the same topology, Taguchi presented three linear graphs. But this catalog covers only a small part of the possible split-plot designs using L_{16} orthogonal arrays. Washio (1988) gives 26 linear graphs which are capable of performing resolution IV designs, including three linear graphs proposed by Taguchi. Both Taguchi and Washio do not use different symbols for depicting the level of two factor interactions.

It is possible to improve the existing linear graph catalogs in two aspects. The first is to list linear graphs which are topologically different. The second is to list linear graphs which are topologically identical but have different structure when the levels of the group of the effects (main effects and two factor interaction effects) are considered. Regarding the former aspect, Sekino et al. (2008) derived all the linear graphs that can be allocated to L_{16} orthogonal arrays, which leaves the latter aspect to be realized.

In order to examine the linear graphs for split-plot designs for the same topology, we have introduced some measures for evaluating the designs.

- Interaction matrix and Interaction pattern

To express two-factor interactions in each design, we use the interaction matrix. It shows the number of interactions associated with each factor. The rows and the columns are arranged alphabetically. If we decide to consider a two-factor interaction, value 1 is given to the corresponding cell. If we decide not to consider a two-factor interaction, value 0 is given to the corresponding cell. The interaction matrix can express all the two factor interactions which are of interest. The interaction pattern expresses how many interactions each factor has, and these were arranged in descending order. It is introduced by Li et al. (1991). Even if the interaction matrix is different, the same interaction pattern might appear.

- Allocation pattern

Count the number of main effects within each level of group. When we have four groups, list the number according to the level of group, from group 1 to group 4. Those numbers can include zeros. These numbers are called allocation pattern.

Using these two measures, we examined the linear graphs. Of course, there are linear graphs where the main effects are allocated to different levels of group. But it might be possible to find linear graphs where the main effects are allocated to the same group but the two factor interactions appear in different groups. In other words, in such linear graphs, the main effects shown by nodes are the same, but the two-factor interactions shown by edges are different.

Some of the linear graphs are shown in Fig. 4. The topology of the linear graph is the same for all eight linear graphs. This topology is one of the six linear graphs proposed by Taguchi. This is the case when there are seven main factors. Thus, for each linear graph, there are eight two factor interaction effects. This is the case where the interaction pattern is 6222211 and the allocation pattern is [0115]. We can see there are designs where the main effects are allocated at the same places yet the level of group for some interactions are different. For all the possible allocation patterns, this topology gives total of 62 linear graphs. By using the obtained linear graphs, there are wide options available when designing this experiment.

3.3 Design Example

In this section a design example using the proposed linear graphs are shown. Let us look for a split-plot fractional factorial design of run size 16, which implies the use of the L_{16} orthogonal array. In this example seven main effects (A, B, C, D, E, F, G) are of interest, where the factor B is the whole plot, the factor G is the subplot, and

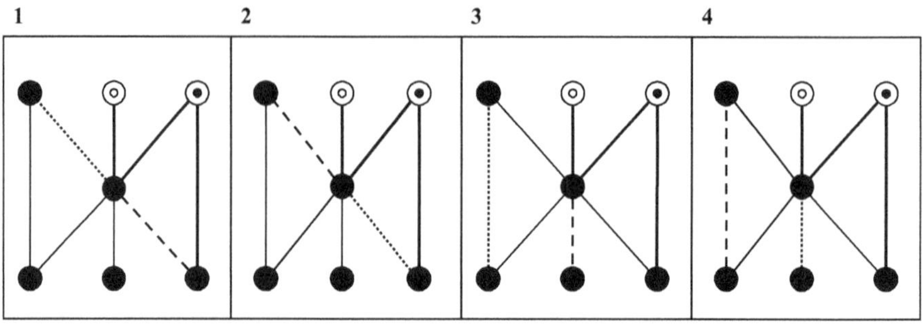

Fig. 4 Linear graphs for split-plot designs

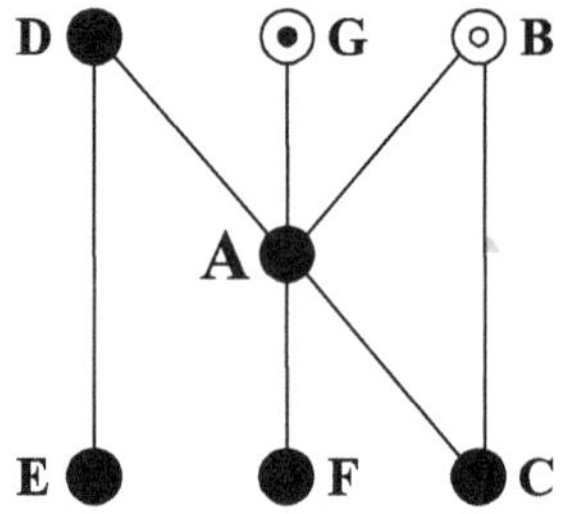

Fig. 5 The necessary linear graph for this experiment

the factors A, C, D, E, F are sub-subplots. Regarding interactions, seven two-factor interaction effects (AB, AC, AD, AF, AG, BC, DE) are of interest, where effects AB, AF, BC are considered very important, effects AD and BC are considered important, and effects AG and DE are not so important.

When the whole plot is allocated to Group 1 and 2, subplot to Group 3, and sub-subplot to Group 4 of the L_{16} orthogonal array, the necessary linear graph is shown as Fig. 5. In this case we can use the linear graphs 5–8 of Fig. 4.

In this particular situation, it is suggested to use the linear graph No. 7 of Fig. 4, because the effect of the very important interaction effect AF can be estimated better than when the linear graph No. 5 of Fig. 4 were used. In other words, the standard

Table 3 Allocation of effects for this experiment

Whole Plot			Subplot				Sub-subplot							
[1]	[2]	[3]	[4]	[5]	[6]	[7]	[8]	[9]	[10]	[11]	[12]	[13]	[14]	[15]

Allocation using linear graph No.7 of Fig.4

DE	B	AC	G		AF	AD	D	E	F	AG	C	AB	BC	A
[1]	[2]	[3]	[4]	[5]	[6]	[7]	[8]	[9]	[10]	[11]	[12]	[13]	[14]	[15]

Allocation using linear graph No.5 of Fig.4

AF	B	AD	G	DE		AC	F	A	D	AB	BC	AG	C	E
[1]	[2]	[3]	[4]	[5]	[6]	[7]	[8]	[9]	[10]	[11]	[12]	[13]	[14]	[15]

error of the effect of interaction effect AF is smaller using No. 7 than No. 5 because the interaction effect appears as subplots and not as whole plots. The allocation of effects to the columns of L_{16} is shown in Table 3.

Thus, by use of appropriate linear graphs, experimental designs considering the importance of the interaction effects are possible.

4 Conclusions

In this paper, we proposed an improved version of linear graphs, namely how we should distinguish the levels of group of the two factor interaction effects. We examined the possible linear graphs based on the measures to support expressing the split-plot designs. We found many linear graphs where the main effects are allocated to the same group but the two factor interactions appear in different groups. We were able to list up the linear graphs when the number of main effects is up to seven. We still need to improve our way of searching for linear graphs when the number of the main effects is larger.

Using the obtained linear graphs, the user can control the source of variation so that the standard errors of the estimated effects can be set to meet the preference of the designer.

References

Box, G. E. P., Hunter, J. S., & Hunter, W. G. (2005). *Statistics for experimenters – Design, innovation, and discovery* (2nd ed.). New York: Wiley.

Hedayat, A. S., Sloane, N. J. A., & Stufken, J. (1999). *Orthogonal arrays – Theory and applications*. New York: Springer.

Li, C. C., Washio, Y., Iida T., & Tanimoto S. (1991). *Linear graphs of resolution IV for orthogonal array L_{16}*. Technical Report of Keio University.

Ojima, Y., Suzuki, T., & Yasui, S. (2004). An alternative expression of the fractional factorial designs for two-Level and three-Level factors. *Frontiers in Statistical Quality Control, 7*, 309–316, Physica-Verlag.

Sekino, M., Suzuki, T., Yasui, S., & Ojima, Y. (2008). *Deriving all the linear graphs for L16*. Technical Report of Department of Industrial Administration Faculty of Science and Technology, Tokyo University of Science (In Japanese).

Taguchi, G. (1976). *Experimental design*. Tokyo: Maruzen (In Japanese).

Washio, Y. (1988). *Design of experiment and analysis*. Tokyo: Iwanami-Shoten (In Japanese).

Watanabe, G., Suzuki, T., Togashi, T., Yasui, & S., Ojima, Y. (2006). Allocating factors in split-plot designs for orthogonal arrays. In *Proceedings of ENBIS 2006 (European Network for Business and Industrial Statistics)*, Wroclaw (CD-ROM).

Wu, C. F. J., & Hamada, M. (2000). *Experiments – planning, analysis, and parameter design optimization*. New York: Wiley.

A Practical Variable Selection for Linear Models

Hidehisa Noguchi, Yoshikazu Ojima, and Seiichi Yasui

Abstract In the analysis of experiments, there are many variable selection algorithms for linear models. Most of these approaches select the best model based on some criteria such as AIC. These criteria do not allow for any relationship between predictors. However, in practice, the analysis is driven by following three principles: Effect Hierarchy, Effect Sparsity, and Effect Heredity Principle. The approach depending solely on those criteria ignore these principles, so it would often select a hard to interpretable models, for instance, which are consisted with only interaction terms. In this article, we extend the LASSO method to identify significant interaction terms mainly focusing on the heredity principle. And we compare the proposed method with ordinary LASSO and traditional variable selection approach. In the example, we analyze the data obtained from designed experiments such as Placket-Burman design and supersaturated design.

Keywords Variable selection • Lasso • Design of experiments • Screening designs • Effect heredity principle

1 Introduction

We consider the analysis of experiments where numerous predictors are examined. Variable selection is important when we would like to identify a subset of the predictors that exhibit the strongest effects. A screening experiment should be conducted to eliminate the unimportant predictors. Such an experiment is based on highly fractionated design like the Placket-Burman designs. Experimental data can usually be modeled by the following general linear model.

H. Noguchi (✉) · Y. Ojima · S. Yasui
Department of Industrial Administration, Tokyo University of Science, 2641 Yamazaki, Noda, Chiba, 278-8510, Japan
e-mail: j7410702@ed.tus.ac.jp; ojima@rs.tus.ac.jp; yasui@rs.tus.ac.jp

H.-J. Lenz et al. (eds.), *Frontiers in Statistical Quality Control 10*,
Frontiers in Statistical Quality Control 10, DOI 10.1007/978-3-7908-2846-7_23,

$$\mathbf{Y} = \mathbf{X}\beta + \boldsymbol{\varepsilon} \tag{1}$$

where $\mathbf{Y}$ is a vector of N responses, X is an $N \times (f+1)$ matrix of predictors, $\boldsymbol{\beta} = (\beta_0, \beta_1, \cdots, \beta_f)^T$ a vector containing an intercept β_0 and f regression coefficients, and $\boldsymbol{\varepsilon}$ a vector of N error variables $\varepsilon_1, \ldots, \varepsilon_N$ each of which are assumed to be independent and identically distributed as $N(0, \sigma^2)$. To model the response y in terms of the predictors $x_1, \cdots, x_f$, one may consider the following model

$$y = \beta_0 + \sum_{k=1}^{f} \beta_k x_k + \varepsilon \tag{2}$$

For the purpose of factor screening, it is usually sufficient to identify the main effects of the important factors and to obtain some insight about which factors may be involved in two-factor interactions. Therefore, we focus on a regression model with main effects and all possible two-factor interactions.

$$f(x) = \beta_0 + \sum_{i=1}^{p} \beta_i x_i + \sum_{i<j} \sum_{j=2}^{p} \beta_{ij} x_i x_j \tag{3}$$

where p is the number of main factors. A more simple and understandable model is preferred as long as it can well explain the data. Thus, in the regression fitting of the linear model Eq. 3, those covariates whose regression coefficients are not significant may be removed from the full model. The goal of variable selection in regression analysis is to identify the smallest subset of the covariates that explains the data well. The model hunter hopes to capture the true model or at least the covariates of the true model with significant regression coefficients. The rest of paper is organized as follows. In Sect. 2, we discuss the penalized least squares and principles in the variable selection. In Sect. 3, we introduce our model and an algorithm to fit the model. We demonstrate the proposed method through two examples in Sect. 4, and conclude with a discussion in Sect. 5.

2 Penalized Least Square and Variable Selection

There are two subjects in variable selection; prediction accuracy and interpretation. The least absolute shrinkage and selection operator (LASSO) which is member of the penalized least squares proposed by Tibshirani (1996). LASSO is a constrained version of ordinary least squares (OLS). The ordinary least squares estimators denoted by

$$\hat{\beta}^{OLS} = arg\min \sum_{h=1}^{N} (y_h - f(\mathbf{x}_h))^2$$

LASSO reduced the residual sum-of-squares by penalizing the L_1-norm of the regression coefficients.

$$\hat{\beta}^{LASSO} = arg\min\left\{\sum_{h=1}^{N}(y_h - f(\mathbf{x}_h))^2 + \lambda(\sum_{i=1}^{p}|\beta_i| + \sum_{i<j}\sum_{j=2}^{p}|\beta_{ij}|)\right\} \quad (4)$$

L_1-norm penalty can shrink some of the fitted coefficients to be exactly zero when making the tuning parameter sufficiently large. It may improve prediction accuracy and interpretation. Although LASSO can select a subset, it has a problem of not accounting for any relationships between predictors.

To enhance predictability and to select significant variables for "good" subset, in a screening context, there are following widely accepted principles (Wu and Hamada 2000):

1. Effect Sparsity Principle
   The number of active effects in a factorial experiment is small.
2. Effect Hierarchy Principle
   (a) Lower order effects are more likely to be active than higher-quadratic main effects or interaction effects.
   (b) Effects of the same order are equally likely to be important.
3. Effect Heredity Principle
   In order for an interaction to be significant, at least one of its parent factors should be significant. Subsets should obey heredity of active effects. For example, a subset with an active AB interaction but no A or B main effects may not be acceptable.

The third principle governs the relationships between predictors; interaction and its corresponding main factors. Although there are a lot of variable selection methods, most of them ignore this principle. Because of this, they often select models have an interaction term but not the corresponding main terms. Such models are usually difficult to interpret in practice. Chipman et al. (1997) proposed the variable selection method using Bayesian priors to focus the search on models obeying effect heredity principle. Yuan et al. (2007) apply the principle to Least Angle Regression (Efron et al. 2004). Zhao et al. (2009) and Nam et al. (2010) also address the variable selection problem using penalized least square with heredity constraint.

All these approaches divide heredity principle into two types; strong heredity and weak heredity principle. Strong heredity principle allows an interaction to be active only if both corresponding main effects are active. Whereas weak heredity principle allows an interaction to be active if one or more of its parents are active. For p factors, when we consider linear main effects and two-factor interaction,the number of effects is

$$f = p + \binom{p}{2} = \frac{(3p + p^2)}{2}$$

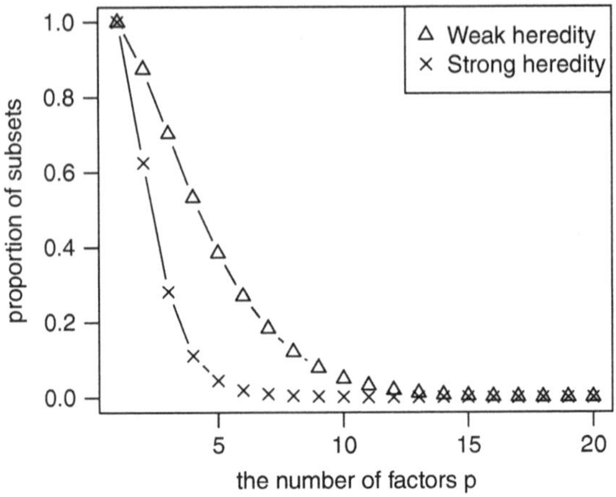

Fig. 1 The proportion of the number of subsets under strong heredity principle 2^{fs} and weak heredity principle 2^{fw} to all 2^{f} subsets

In a similar way, the number of effects under the strong heredity principle is as follows.

$$fs = \sum_{i=0}^{p} \binom{p}{i} 2^{\binom{i}{2}}$$

While, under the weak heredity principle, there are

$$fw = \sum_{i=0}^{p} \binom{p}{i} 2^{pi-i(i+1)/2}$$

effects. The number of possible subsets of active effects will be 2^{f}. Figure 1 shows the total number of subsets under different types of effect heredity is compared to number of all possible subsets. Strong heredity principle reduce all possible subsets that can occur to 2^{fs}, while weak heredity principle reduce it to 2^{fw}. Nelder (1998) was arguing for a rejection of the weak heredity principle as part of strategy for model selection. However, the analysis of blood glucose experiment in Yuan et al. (2007) indicates the weak heredity principle is more likely to be true. Chipman et al. (1997) suggest that in exploratory stages, it may be desirable to relax the restrictions of strong heredity. It leaves small probability to interactions which have active effect without corresponding main effects. In this paper, we extend the LASSO to select a subset obeying weak heredity constraint and compare it to the model obeying strong heredity by applying each model to data obtained from Placket-Burman design and super-saturated design. In this paper, we extend the LASSO to select a subset obeying weak heredity constraint and compare it to the model obeying strong heredity by applying each model to data obtained from Placket-Burman design and super-saturated design.

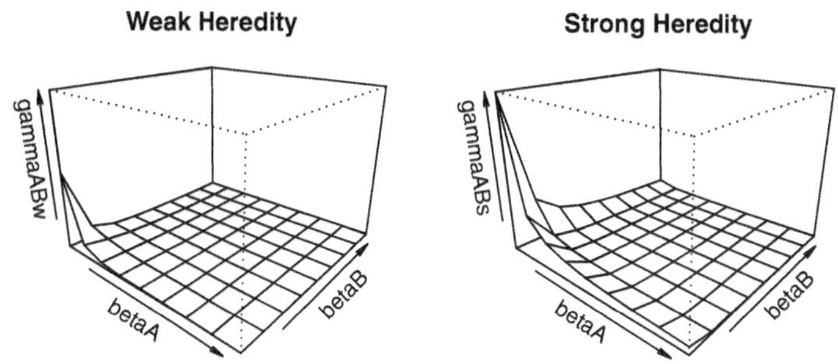

Fig. 2 Relationship of parameters between β_A and β_B ($\beta_A > 0, \beta_B > 0$) and γ_{AB} under Strong heredity (*right panel*) and Weak heredity (*left panel*)

3 Model Formulation

In this section we consider an expanded representation of the LASSO which obeys weak heredity principle.

3.1 Model

We consider the following model:

$$g_w(x) = \beta_0 + \sum_{i=1}^{p} \beta_i x_i + \sum_{i<j} \sum_{j=2}^{p} \gamma_{ij} (\beta_i^2 + \beta_j^2) x_i x_j \tag{5}$$

In Eq. 5, the coefficient for the interaction terms $x_i x_j$ is expressed by the parameter γ_{ij} and corresponding main effects parameter β_i and β_j. This expression can enforce weak heredity principle. Compared to this, in Nam et al. (2010), strong heredity was considered as following model:

$$g_s(x) = \beta_0 + \sum_{i=1}^{p} \beta_i x_i + \sum_{i<j} \sum_{j=2}^{p} \gamma_{ij} \beta_i \beta_j x_i x_j \tag{6}$$

We can assume without loss of generality that constant term β_0 can be omitted by centering all the response variables. Figure 2 shows the changes of γ_{AB} value against β_A and β_B ($\beta_A > 0, \beta_B > 0$) value under two types of heredity principle; Strong heredity and weak heredity. In Fig. 2, we assume the true coefficient value of the interaction term $x_A x_B$ is known and it has a positive value. Note that $\gamma_{AB} = 0$ when $\beta_A = \beta_B = 0$.

As given in Fig. 2, in strong heredity version, if the main effect β_A or β_B is going to be 0, it is equal to approaching the true value of the coefficient of the term

$x_A x_B$ to infinity. Therefore the parameter γ_{AB} must be large value. It may have an undesirable influence on finding other significant effects. Whereas, in weak heredity version, if the main effect β_A or β_B is going to be 0, it has no effects on the true value of the coefficient. Thus, γ_{AB} just tackle the active main effect.s

For the purpose of variable selection, we consider the following penalized least squares estimates:

$$(\hat{\beta}, \hat{\gamma}) = arg\min \left\{ \sum_{h=1}^{N} (y_h - g(\mathbf{x}_h))^2 + \lambda (\sum_{i=1}^{p} |\beta_i| + \sum_{i<j} \sum_{j=2}^{p} |\gamma_{ij}|) \right\} \tag{7}$$

where $g(\mathbf{x})$ is from Eq. 5, and the penalty is the L_1-norm of the parameters, as in LASSO Eq. 4. There are two tuning parameters, λ_β and λ_γ. The first tuning parameter λ_β controls the amount of shrinkage that is applied to the estimates for the main effect. If coefficients β_i and β_j are shrunken to zero, variable $x_i x_j$ is removed from the model. The second tuning parameter λ_γ controls the amount of shrinkage that is applied to the estimates for the interaction effect. If either β_i or β_j is equal to zero but the corresponding interaction effect is strong, γ_{ij} still has the possibility of being nonzero, so it has the flexibility of selecting interaction with one of corresponding main effect. Controlling the tuning parameter λ_γ means setting the value $s \geq \sum |\gamma_{ij}|$. In Eq. 7, it is equal to setting lower limits of $|\beta_i^2 + \beta_j^2|$. With increasing the value of s, lower limits of $|\beta_i^2 + \beta_j^2|$ get smaller.

3.2 Algorithm

To estimate the parameters β_i and γ_{ij}, we can extend an iterative approach of Nam et al. (2010). We first fix β_i and estimate γ_{ij}, then we fix γ_{ij} and estimate β_i, and we iterate between these steps until the solution gives convergence. Since at each step, the value of the objective function (7) decreases, the solution is guaranteed to converge to a local minimum. When $\beta_i = 1, \ldots, p$ are fixed, Eq. 7 becomes a LASSO problem, hence we can use either the LARS/LASSO algorithm (Efron et al. 2004) or a quadratic programming package to efficiently solve for γ_{ij}. When γ_{ij} are fixed, we can sequentially solve for β_i: for each $i = 1, \ldots, p$, we fix γ_{ij} and $\beta_{[-i]} = (\beta_1, \ldots, \beta_{i-1}, \ldots, \beta_p)$, then Eq. 7 becomes a simple LASSO problem with only one parameter β_i, and we can solve it with a closed form formula. In summary, the algorithm proceeds as follows:

1. Standardization:
   $\sum_h y_h/N = 0, \quad \sum_h x_{hk}/N = 0, \quad \sum_h x_{hk}^2/N = 1 \quad h = 1, \ldots, N, \ k = 1, \ldots, f$
2. Initialization:
   Initialize $\hat{\beta}_i^{(0)}$ and $\hat{\gamma}_{ij}^{(0)}$, $i < j$, $i = 1, \ldots, p$ with some plausible values. For example, we can use the least square estimates or the simple regression estimates by regressing the response $\mathbf{y}$ on each of the terms. Let $m = 1$.

3. Update $\hat{\gamma}_{ij} \quad i, j = 1, \ldots, p, i < j$
   First, to calculate $\tilde{y}_h$, we subtract main effects from the response.

$$\tilde{y}_k = y_k - \sum_{i=1}^{p} \hat{\beta}_i^{(m-1)} x_{hi}$$

Second, $\tilde{x}_{h,ij}$ will be obtained.

$$\tilde{x}_{h,ij} = \left((\hat{\beta}_i^{(m-1)})^2 + (\hat{\beta}_j^{(m-1)})^2 \right) (x_{hi} x_{hj})$$

Then γ_{ij} is established with $\hat{\gamma}_{ij}$ with using Lasso algorithm.

$$\hat{\gamma}_{ij}^{(m)} = arg \min_{\gamma_{ij}} \sum_{h=1}^{N} \left\{ (\tilde{y}_h - \sum_{i<j} \gamma_{ij} \tilde{x}_{h,ij})^2 + \lambda_\gamma \sum_{i<j} |\gamma_{ij}| \right\}$$

4. Update $\hat{\beta}_i$ Let $\hat{\beta}_i^{(m)} = \hat{\beta}_i^{(m-1)}$
   For each i in $1, \ldots, p$

$$\tilde{y}_k = y_h - \sum_{i=1}^{p} \hat{\beta}_i^{(m)} x_{hi} - \sum_{i<j} \left((\hat{\beta}_i^{(m)})^2 + (\hat{\beta}_j^{(m)})^2 \right) (x_{hi} x_{hj})$$

$$\tilde{x}_h = x_{hi} + \sum_{i<j} \hat{\gamma}_{ij}^{(m)} \hat{\beta}_j^{(m)} (x_{hi} x_{hj}) + \sum_{i<j} \hat{\gamma}_{ij}^{(m)} \hat{\beta}_i^{(m)} (x_{hi} x_{hj})$$

then

$$\hat{\beta}_i^{(m)} = arg \min_{\beta_i} \sum_{h=1}^{N} \{ (\tilde{y}_h - \beta_h \tilde{x}_h)^2 + \lambda_\beta |\beta_i| \}$$

5. Compute the relative difference

$$\Delta^{(m)} = \frac{|Q_n(\hat{\boldsymbol{\theta}}^{(m-1)}) - Q_n(\hat{\boldsymbol{\theta}}^{(m)})|}{Q_n(\hat{\boldsymbol{\theta}}^{(m-1)})}$$

where

$$Q_n(\hat{\boldsymbol{\theta}}) = \sum_{h=1}^{N} (y_h - g(\mathbf{x}_h))^2 + \lambda_\beta \sum_{i=1}^{p} |\beta_i| + \lambda_\gamma \sum_{i<j} \sum_{j=2}^{p} |\gamma_{ij}|$$

for $\boldsymbol{\theta} = (\beta_0, \beta_1, \ldots, \beta_p, \gamma_{12}, \ldots, \gamma_{p-1,p})$.

6. Stop the algorithm if $\Delta^{(m)}$ is within the range of 10^{-8}. Otherwise, let $m = m + 1$ and go back to step 2.

Table 1 Screening experiments with Plackett-Burman 12-run design and response data

No.	A	B	C	D	E	F	G	H	I	J	K	Y
1	+	+	−	+	+	+	−	−	−	+	−	1.0580
2	+	−	+	+	+	−	−	−	+	−	+	1.0041
3	−	+	+	+	−	−	−	+	−	+	+	−5.2001
4	+	+	+	−	−	−	+	−	+	+	−	5.3202
5	+	+	−	−	−	+	−	+	+	−	+	1.0216
6	+	−	−	−	+	−	+	+	−	+	+	−2.4711
7	−	−	−	+	−	+	+	−	+	+	+	2.8089
8	−	−	+	−	+	+	−	+	+	+	−	−1.2721
9	−	+	−	+	+	−	+	+	+	−	−	−0.9546
10	+	−	+	+	−	+	+	+	−	−	−	0.6441
11	−	+	+	−	+	+	+	−	−	−	+	−5.0251
12	−	−	−	−	−	−	−	−	−	−	−	3.0604

4 Example

In this section, examples of two experiments are given illustrating situations in which complex aliasing arises are given. These are a screening experiment using a Plackett-Burman 12-run design and supersaturated design.

4.1 Screening Experiments

Table 1 presents a 12-run PB design and illustrates its use in a screening context which can accommodate up to 11 factors labeled $A - K$. The data are constructed in Hamada and Wu (1992) based on the true model $Y = A + 2AB + 2AC + \varepsilon$, here $\varepsilon \sim N(0, \sigma = 0.25)$ that is, factor A has an active main effect and there are active interaction between A and B and between A and C, while the remaining factors $D - K$ are inactive.

Figure 3 compares the coefficient profiles obtained by the each model and shows main and interaction effects separately. The proposed weak heredity version is able to pick up the true effects A, AB, and AC, whereas the strong heredity version can not identify the correct model, because the interaction effects AB and AC can not be selected until the main effect A is selected. This should be expected because the correct model does not contain main effects B and C. The strong heredity constraints are so strict that the selection procedure cannot find the active main effect if it is small compared to interaction effect. In this example, selecting the interaction effect IK make it difficult to select the correct main effect A. With the strong heredity constraints, once the incorrect interaction effect is selected, it lead to pick up the corresponding main effects. Due to this, it is hard to identify the correct main effects in such a case. In contrast, the weak heredity constraint has flexibility on that point. We note that the ordinary forward selection methods (Hamada and Wu

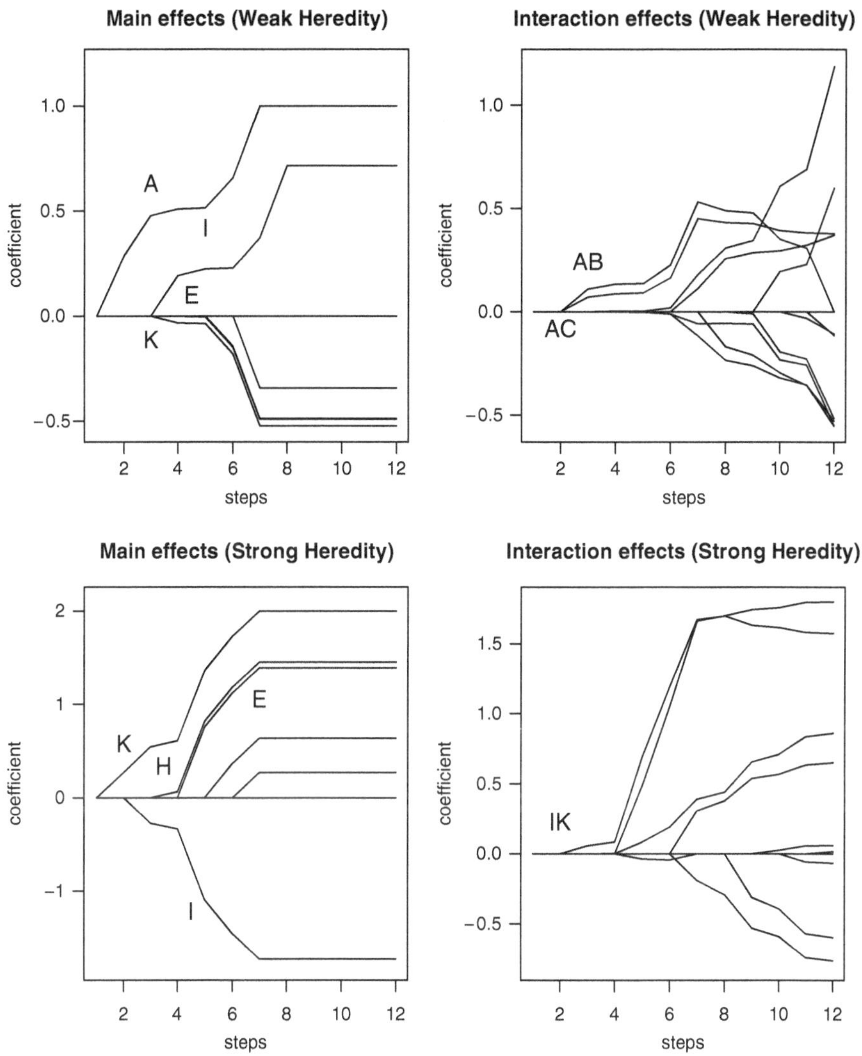

Fig. 3 Coefficient profiles for the PBD data

1992) could not identify any of the significant effects. This shows the advantages of the proposed model.

4.2 Supersaturated Design

A supersaturated design is a design in which the degrees of freedom for all its main effects and the intercept term exceed the run size. Here we consider the 14×23 design matrix in Table 2. The degrees of freedom for the 23 main effects

Table 2 Experiments with supersaturated design and response data

Run	a	b	c	d	e	f	g	h	i	j	k	l	m	n	o	p	q	r	s	t	u	v	w	y
1	+	+	+	−	−	−	+	+	+	+	+	−	+	−	−	+	−	−	+	−	−	−	+	133
2	+	−	−	−	−	−	+	+	+	−	−	−	+	+	+	−	+	−	−	+	+	−	−	62
3	+	+	−	+	+	−	−	−	−	+	+	+	+	+	+	+	−	−	−	−	+	+	−	45
4	+	+	−	+	−	+	−	−	−	+	+	−	+	−	+	−	+	+	+	−	−	−	−	52
5	−	−	+	+	+	+	−	+	+	−	−	−	+	−	+	+	−	−	+	−	+	+	+	56
6	−	−	+	+	+	+	+	−	+	+	+	−	−	+	+	+	+	+	+	+	+	−	−	47
7	−	−	−	−	+	−	−	+	−	+	+	+	+	+	−	+	+	+	+	+	−	−	+	88
8	−	+	+	−	−	+	−	+	−	+	+	−	−	−	−	−	−	+	−	+	+	+	−	193
9	−	−	−	−	−	+	+	−	−	−	−	+	−	−	+	+	+	−	−	−	−	+	+	32
10	+	+	+	+	−	+	+	+	−	−	−	+	−	+	+	+	−	+	−	+	−	−	+	53
11	−	+	−	+	+	−	−	+	+	−	−	−	−	+	−	−	+	+	−	−	−	+	+	276
12	+	−	−	−	+	+	+	−	+	+	+	+	+	−	−	−	−	+	−	+	+	+	+	145
13	+	+	+	+	+	−	+	−	+	−	−	+	−	−	−	−	+	−	+	+	−	+	−	130
14	−	−	+	−	−	−	−	−	−	−	−	+	−	+	−	−	−	−	+	−	+	−	−	127

(of two-level factors) and the intercept term exceed the run size 14. The main attraction for using supersaturated designs is their run size economy. An Experimenter should be used primarily for screening factor main effects. However, since we are interested in the variable selection with heredity principle, in this example, main effects and two-factor interactions are considered. Thus, there are a total of 276 candidate effects (23 main effects and 253 two-factor interactions). We use this example to illustrate how complicated heredity principle can be handled using the proposed methodology. And, in this complicated situation, we illustrate how the result will be different between weak heredity and strong heredity.

In this design, the effects are either orthogonal or completely aliased with others. Twelve pairs of aliasing relationships are arised. The results of the analysis are plotted in Fig. 4. In both cases, main effects (o, p) are selected in the first a few steps. In Chipman et al. (1997) and Lin (1993), they obtained the subset (d, l, o, u): note that their analysis focus on the main effects only. With LASSO focused on the only main effects without heredity principle, we obtained the subset (o, p, l, b). Here, we conform the number of effects making up the subset to one in Chipman et al. (1997) and Lin (1993). The subset become (o, ab, iu, rs) when main effects and interactions are considered without heredity principle. We can see this subset is hard to interpret. Applying both heredity principles, we obtain the subsets (o, p, m, mo) and (o, os, p, lo). Note that, these terms are not confounding with the others. Table 3 shows the value of R^2, adjusted R^2, and AIC for each subsets. We see that the weak heredity version could select the best subset in this example. Although the subset constructed by the strong heredity version selects main effects and appropriate interaction effect, it is the worst of all.

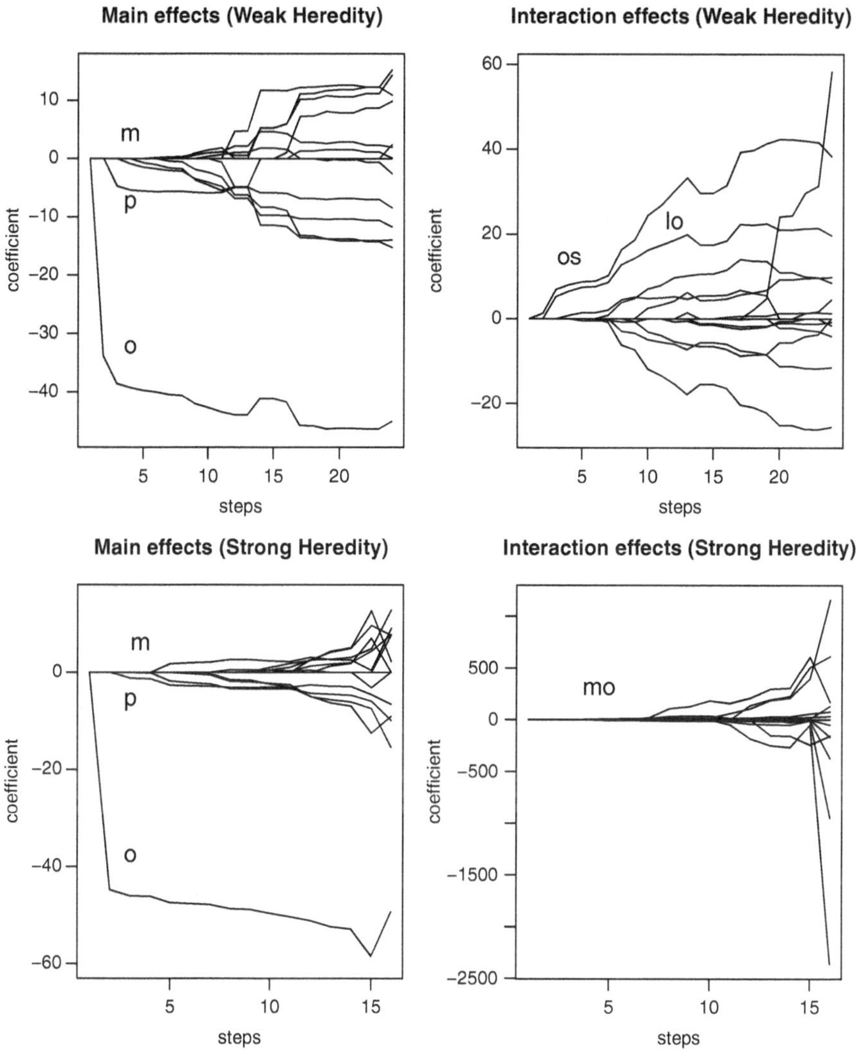

Fig. 4 Coefficient profiles for the supersaturated design

Table 3 Selected subsets and their criterion

	Active effects	R^2	adjR^2	AIC
No heredity (all effects)	(o, ab, iu, rs)	0.866	0.806	141.3
No heredity (main effects)	(o, p, l, b)	0.803	0.715	146.7
Strong heredity (all effects)	(o, p, m, mo)	0.743	0.628	150.4
Weak heredity (all effects)	(o, os, p, lo)	0.897	0.852	137.6
Chipman and Lin (main effects)	(d, l, o, u)	0.786	0.691	147.9

5 Discussion

We proposed a variable selection method based on the weak heredity principle. The results are compared with the strong heredity. In the last example, our procedure selects a more interpretable and better subset than any other approaches. Coefficients profiles can be different according to the type of heredity principle. Weak heredity is useful because of its flexibility. But, in practice, we will not know which version of the heredity principle to use, thus, we must try both of them. This make us consider that we should have to review on the relationships between the interactions and their parent terms. Modifying existing factors may lead to the best model. To further improved modelling, it will be possible to introduce new parameters into the models, which has been used by Breiman (1995), Zou (2006), and Zhang and Lu (2007).

References

Breiman, L. (1995). Better subset regression using the non-negative garrote. *Technometrics, 37*, 373–384.

Chipman, H., Hamada, M., & Wu, C. F. J. (1997). A Bayesian variable-selection approach for analyzing designed experiments with complex aliasing. *Technometrics, 39*, 372–381.

Efron, B., Hastie, T., Johnstone, I., & Tibshirani, R. (2004). Least angle regressionh (with discussion). *The Annals of Statistics, 32*(2), 407–499.

Hamada, M., & Wu, C. F. J. (1992). Analysis of designed experiments with complex aliasing. *Journal of Quality Technology, 24*, 130–137.

Lin, D. K. J. (1993). A new class of supersaturated designs. *Technometrics, 35*, 28–31.

Nam, H. C., William, L., & Ji. Z. (2010). Variable selection with the strong heredity constraint and its oracle property. *Journal of the American Statistical Association, 105*, 354–364.

Nelder, J. A. (1998). The selection of terms in response-surface models: How strong is the weak-heredity principle? *The American Statistician, 52*, 315–318.

Tibshirani, R. (1996). Regression shrinkage and selection via the Lasso. *Journal of the Royal Statistical Society Series B, 58*, 267–288.

Wu, C. F. J., & Hamada, M. (2000). *Experiments: Planning, analysis, and parameter design optimization*. New York: Wiley.

Yuan, M., Joseph, V., & Lin, Y. (2007). An efficient variable selection approach for analyzing designed experiments. *Technometrics, 49*(4), 430–439.

Zhang, H., & Lu, W. (2007). Adaptive LASSO for Cox's proportional hazard model. *Biometrika, 94*, 691–703.

Zou, H. (2006). The adaptive Lasso and its oracle properties. *Journal of the American Statistical Association, 101*(476), 1418–1429.

Capability of Detection for Poisson Distributed Measurements by Normal Approximations

Yusuke Tsutsumi, Hironobu Kawamura, and Tomomichi Suzuki

Abstract In the analysis of very small components, it is very important to know what concentration or amount of the analyte can be detected by the measurement method. When we determine whether the analysis sample is the same as the basic state, the capability of detection for the measurement method is defined as the amount that can be detected. The ISO 11843 series standardizes the capability of detection. ISO 11843 series has not described the capability of detection for Poisson distributed measures. The measurement results occasionally are count data. When the capability of detection can be exactly measured, it can be derived numerically, but it is very difficult to be derived analytically. There are various approximation methods having their own distinct features for a Poisson distribution. However, so far it is not known which approximation is best from the viewpoint of estimating the capability of detection. In this paper, the evaluation of the capability of detection for Poisson distributed measurements is described. A number of approximation methods are proposed from the viewpoint of the capability of detection for the measurement method. The best approximation method is then compared with the exact method.

Keywords Capability of detection • Poisson distribution • Normal approximation

Y. Tsutsumi (✉)
Mitsubishi Tanabe Pharma corporation, 2-2-6 Nihonbashi-Honcho Chuoku, Tokyo, Japan
e-mail: tsutsumi.yusuke@ms.mt-pharma.co.jp

H. Kawamura · T. Suzuki
Tokyo University of Science, 2641 Yamazaki Noda-city, Chiba, Japan
e-mail: kawamura@ia.noda.tus.ac.jp; suzuki@ia.noda.tus.ac.jp

H.-J. Lenz et al. (eds.), *Frontiers in Statistical Quality Control 10*,
Frontiers in Statistical Quality Control 10, DOI 10.1007/978-3-7908-2846-7_24,

1 Introduction

1.1 Backgrounds

In the analysis of very small components, it is very important to know what concentration or amount of the analyte can be detected by the measurement method. Recently, there have been reports of the detection of toxic substances such as pesticide residues and environmental endocrine disrupters. In such cases, it is necessary to judge whether the analysis sample contains a toxic substance that influences the human body. To determine if the toxic substance is present in the analysis sample, a statistical hypothesis testing procedure is applied to determine if the analysis sample can be regarded as blank (not containing the toxic substance). In an actual experiment, it is difficult to directly measure the amount of the analysis component, so the materials to be analyzed are often indirectly tested using a voltmeter or other similar apparatus. Very small amounts of the analyte are not detected directly. The measured value includes the error, and the basic state often includes the material to be analyzed. A spring balance is used as an example. The manual weight is categorized as the net state variable. The total weight, which includes the material, the pan, the spring, and any other parts, constitutes the state variable. The length of the spring would be the response variable. That is why the amount of the analyte is required to determine if the material to be analyzed contains toxic substance. In other words, we need to determine whether the amount of the analyte is in the basic state or not. The critical value is derived from the concepts of statistical test procedures. Basic state and critical values are different from analysis methods. When we determine whether the analysis sample is the same as the basic state, the capability of detection for the measurement method is defined as the amount that can be detected.

The ISO 11843 series standardizes the capability of detection. The series consists of five parts (Parts 1–5), and defines the capability of detection of the measurement method for various situations and assumptions. The ISO 11843 series describes the method of deriving the capability of detection in case the measurement error follows a normal distribution.

The measurement results occasionally are count data such as the measurement of the number of cells, pulse data, and the degree of degradation for electronic media. These count data follow a Poisson distribution instead of a normal distribution. Till date, the ISO 11843 series has not described the capability of detection for Poisson distributed measures. Furukawa et al. (2009) derived the capability of detection using square root approximation, which forms the basis for ISO/DIS (DIS: Draft International Standard) 11843 part 6.

1.2 Purpose

When the measured value follows a Poisson distribution, the capability of detection can be exactly measured. When the cumulative probability of a Poisson distribution is analytically derived, it is necessary to solve a nonlinear equation. If a nonlinear equation is exactly derived, it can be solved numerically, but such equations are very difficult, if not impossible, to be derived analytically. In a practical application, it is a necessary requirement that the derivation method is comprehensive. Hence, it is ideal to satisfy the following two conditions. First, all workers can understand the process. Second, the parameter can be derived manually or using an electronic spreadsheet (step by step). Therefore, when the method of capability of detection for Poisson distributed measurements is applied to practice, using an approximation method becomes a reasonable approach. In this paper, the capability of detection by normal approximation methods is described.

There are various approximation methods having their own distinct features for a Poisson distribution. However, so far it is not known which approximation is best from the viewpoint of estimating the capability of detection. Furukawa et al. (2009) derived the capability of detection using square root approximation of a pulse data that follows Poisson distribution and derived the applicability to a high measured value, but did not use various approximations or derive the minimum detected value of the net state value. It is not enough to reach a conclusion when the normal approximations are optimal in the case of a very small critical value.

In this paper, the evaluation of the capability of detection for Poisson distributed measurements is described. The capability of detection is evaluated when the measured value is small and follows a Poisson distribution. In addition, various normal approximation methods are compared. A number of approximation methods are proposed from the viewpoint of the capability of detection for the measurement method. The best approximation method is then compared with the exact method, which is discussed in Sect. 3.1.

2 Capability of Detection for the Measurement Method

The ISO 11843 series defines the critical value of the net state variable and the minimum detectable value of the net state variable, as shown in Fig. 1. This paper uses the definitions of the terminology defined in the ISO 11843 as follows:
Basic state:
Specific state of a system for use as a base for the evaluation of actual states of the system.
x_C: critical value of the net state variable:
Value of the net state variable, X, the exceeding, of which leads, for a given error probability α to the decision that the observed system is not in its basic state.

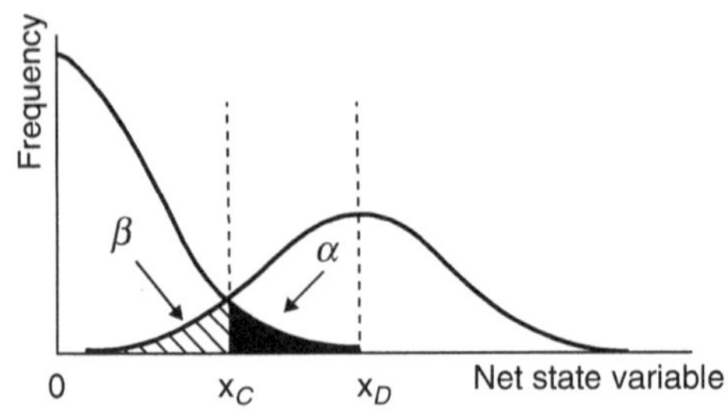

Fig. 1 Basic concept of capability of detection

x_D: minimum detectable value of the net state variable
True value of the net state variable, X, in the actual state will lead, with probability $(1 - \beta)$, to the conclusion that the system is not in its basic state.
The left curve in Fig. 1 shows the distribution of measurements in the basic state; the right curve shows the distribution of measurements in the actual state. Even though the system is in the basic state, the probability of erroneously determining it as not in the basic state is α In addition, even when the system is not in the basic state, the probability of erroneously determining it to be in the basic state is β. In this way, statistical hypothesis testing is applied for the definition of variables using the net state variable.

3 Capability of Detection for Poisson Distributed Measurements

3.1 *Exact Methods*

Figure 2, which is based on the concepts found in ISO 11843 part 1, shows the capability of detection for a discrete distribution. If measurements follow continuous distribution, the values x_C and x_D are determined so that the probability of making errors of the first kind and the second kind are exactly α and β respectively. In cases where measurements follow a discrete distribution, it is impossible to determine the values x_C and x_D to give exact values of α and β. Hence, the values x_C and x_D are determined so that the probability of making errors of the first kind and the second kind become smaller than α and β respectively. The α' and β' are defined as upper bounds for these probabilities.

It is necessary to be prudent when performing statistical tests. If the value x exceeds x_C, then x is detected otherwise x is not detected. When the basic state is λ_0, x_C is in the upper $100\alpha'$ percentile under the null hypothesis while x_D is in the lower $100\beta'$ percentile under the alternate hypothesis. Taking into consideration that x_D is analytically derived, it is necessary to derive a nonlinear equation.

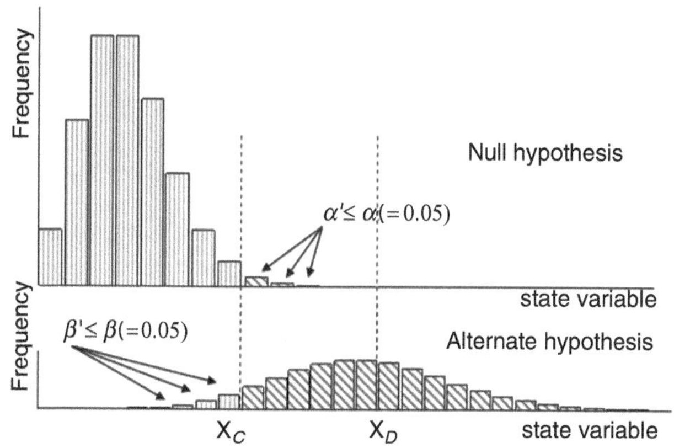

Fig. 2 The basic concept of capability of detection for discrete distribution

Table 1 Mean, variance and inverse transform of each approximation method

No	Methods	Mean	Variance	Inverse transform
1a	Direct	λ	$\lambda/n,$	y
1b	Direct[a]	λ^a	$(\lambda^a, \lambda^a/n, y)$	$(\lambda^a, \lambda^a/n, y)$
2a	Log	$\ln\lambda$	$1/n\lambda$	e^y
2b	Log[a]	$\ln\lambda^a$	$1/n\lambda^a$	e^y
3	Square root	$\sqrt{\lambda}$	$1/4n$	y^2
4	$\sqrt{x}+\sqrt{x+1}$	$\sqrt{\lambda}+\sqrt{\lambda+1}$	$\frac{\lambda}{4n}\cdot\left(\frac{1}{\sqrt{\lambda}}+\frac{1}{\sqrt{\lambda+1}}\right)^2$	$\frac{(y^2-1)}{4y^2}$
5	$2\sqrt{x+3/8}$	$2\sqrt{\lambda+3/8}$	$\frac{\lambda}{n(\lambda+3/8)}$	$\frac{1}{4}y^2-\frac{3}{8}$
6a	$(x+1/4)^{2/3}$	$(\lambda+1/12)^{2/3}$	$(2/3n)(\lambda)^{1/6}$	$y^{3/2}-\frac{1}{4}$
6b	$(x+1/4)^{2/3}$	$(\lambda+1/4)^{2/3}$	$\frac{4\lambda}{9n(\lambda+1/4)^{2/3}}$	$y^{3/2}-\frac{1}{4}$
6c	$(x+1/4)^{2/3}$	$(\lambda+1/12)^{2/3}$	$\frac{4\lambda}{9n(\lambda+1/12)^{2/3}}$	$y^{3/2}-\frac{1}{4}$
6d	$(x+1/12)^{2/3}$	$(\lambda+1/12)^{2/3}$	$\frac{4\lambda}{9n(\lambda+1/12)^{2/3}}$	$y^{3/2}-\frac{1}{4}$

[a]Continuity correction

3.2 *Normal Approximation Methods*

In this study, when the measurements follow a Poisson distribution, the critical value of the net state variable x_C and the minimum detectable value of the net state variable x_D are calculated using a normal approximation method of the Poisson distribution (Johnson and Kotz 1993).

The mean, the variance, and the values of the inverse transforms for each approximation method are given in Table 1. Apart from (6)a, the variance in all the methods is calculated using the delta method. Methods (1) and (2) are cases of continuity correction. Kittlitz (2006) proposed method (6) in which each option has a way of calculating the asymptotic variance using the delta method. Consequently, three types of approximation methods, (6)b, (6)c, and (6)d, are considered as the proposed methods.

3.3 Derivation of Net State Variables of Each Approximation Method and Minimum Detectable Net State Value

The critical value of the net state variable and the minimum detectable value of the net state value of each approximation method can be derived in the following steps.

Step 1: λ_0, which follows a Poisson distribution of the basic state, is transformed to each normal approximation. After transformation, two parameters corresponding to the mean and variance are generated. The parameter U_0 is defined as the mean of the basic state and the parameter σ_0^2 is defined as the variance of the basic state.

Step 2: U_C is derived from Eq. 1. K_0 is defined as the upper 100α percentile for the standard normal distribution and x_C can be derived using the inverse transform of U_C.

Step 3: The parameter U_D is defined as the mean of actual state, and the parameter σ_1^2 is defined as the variance of the actual state. K_1 is defined as the lower 100β percentile for the standard normal distribution. U_D can be derived from Eq. 2, and x_D can be derived by the inverse transform of U_D.

$$U_\mathrm{C} = U_0 + K_0\sqrt{\sigma_0^2} \tag{1}$$

$$U_\mathrm{C} = U_\mathrm{D} - K_1\sqrt{\sigma_1^2} \tag{2}$$

4 Comparison with Each Approximation Method

4.1 Evaluation Method of Each Approximation Methods

x_C is calculated by inverse transformation using each approximation method from the distribution of the basic state. The probability, that the approximation will be larger than x_C is α'. α' can be derived from the equation below and is judged by how close it approaches α without exceeding α.

$$\alpha' = P_r\left(x > [x_C]\right) = 1 - \sum_{i=1}^{[x_C]} \frac{x_0^i}{i} e^{-x_0} \tag{3}$$

The approximation methods are evaluated by Eq. 3, and by considering $\alpha = \beta = 0.05$. The number of measurements of the basic state and of the actual state is $J = K = 1$; the accuracy of each approximation method is evaluated while the value of the parameter λ_0 changes from 1 to 100 in steps of 1 for the Poisson distribution of measurements of the basic states. The criterion for evaluation of these approximation methods is defined from the viewpoint of the statistical test, namely that $\alpha < \alpha'$ is satisfied.

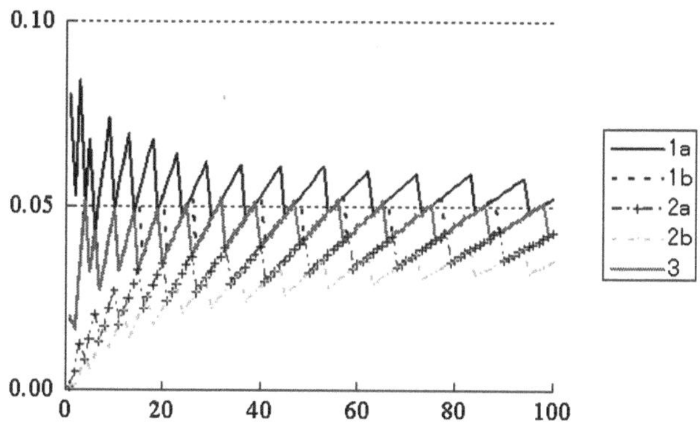

Fig. 3 The value of α' for the approximation methods

4.2 Results

Figure 3 shows α' calculated by each approximation method when the value of the parameter λ_0 changes from 1 to 100. The measurement is according to the Poisson distribution, and the cumulative probability can be calculated for α'. As a result, when λ_0 varies from 1 to 100, the value of α' is discrete, so the curve is not smooth. The overall result is that the value of α' broadly varies and is far from 0.05 when λ_0 is less than 20. Therefore, the approximation is less accurate. On the other hand, α' varies narrowly around 0.05, and the width of the variation is narrow when λ_0 is more than 20.

Regarding the value of α', it can be said that the direct approximation method (1) exceeds 0.05 significantly and the log approximation method (2) falls short. In the case where continuous correction was given for (1)b and (2)b, the $\lambda 0$ values for (1)a and (2)a are less than the value of α'. This result shows the concept of consistency of the statistical test. Meanwhile, the approximation methods (3) through (5) show almost the same shape. Regarding the approximation method (6), it can be stated that (6)a is located far from 0.05, but the others (6)b, (6)c, and (6)d are located around 0.05. Thus, the log approximation method (2) satisfies the necessary conditions. For (3) ~ (5), there are 9 points, 11 points, and 11 points, respectively, for which α' is more than α.

4.3 Discussion

The results of the preceding section prove that the log approximation method (2) is the best approximation method for calculating α' using Eq. 3. In Sect. 4.1, the value of x_C rounded down to the whole number was considered. However, the value of x_C rounded off to the whole number and rounded up to the whole number is also considered. Thus, x_C can be calculated by rounding the value off to the whole number or by rounding up to the whole number. The result will therefore change, and there may be a more accurate approximation method. Therefore, reevaluation of the approximation methods using α' derived by rounding the value off to the whole number or by rounding up to the whole number for each approximation method should be considered for the treatment of x_C.

5 Evaluation of Capability of Detection Considering the Treatment of Variable

Figure 4 shows the value of each α' for the critical value of the net state variable x_C derived by rounding up, rounding down, or rounding off.

In Fig. 4, for approximation method (1), the numbers never exceed $\alpha = 0.05$ for rounding off, suggesting that method (1) is the best way to obtain the critical value of the net state variable x_C. Similarly, the treatment of the variable x_C should be considered in each approximation method. The best treatment method for each approximation method needs to be selected. Therefore, since x_C has three different treatment methods and there are 11 approximation methods, there are a total of $33(3 \times 11)$ total treatment methods to be considered.

From the above methods, two approximation methods are selected such that $\alpha < \alpha'$. We select approximation method (1) and approximation method (6)c where x_C is rounded off in both cases. Figure 5 shows α' calculated using these two methods. From Fig. 5, for the approximation methods (1) and (6)c, the value of λ_0 changes from 1 to 100, and there is no case where α' is higher than 0.05. In

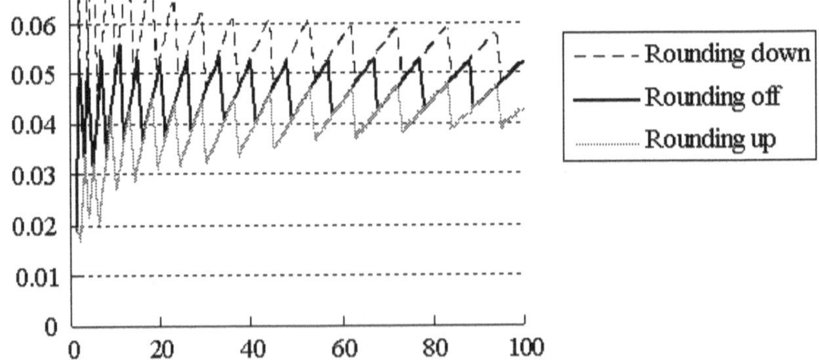

Fig. 4 The value of α' for the approximation methods

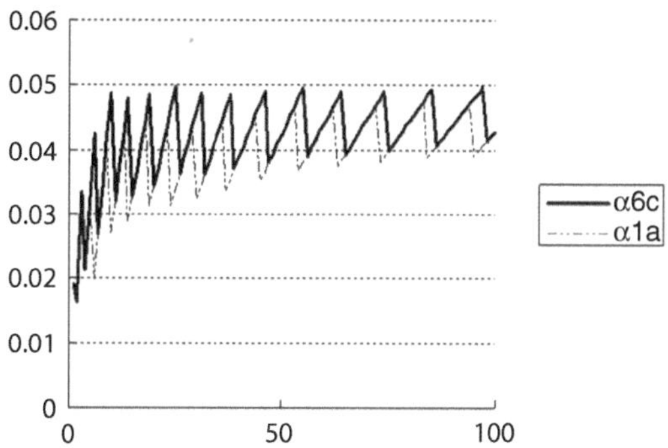

Fig. 5 α' of the two approximation methods

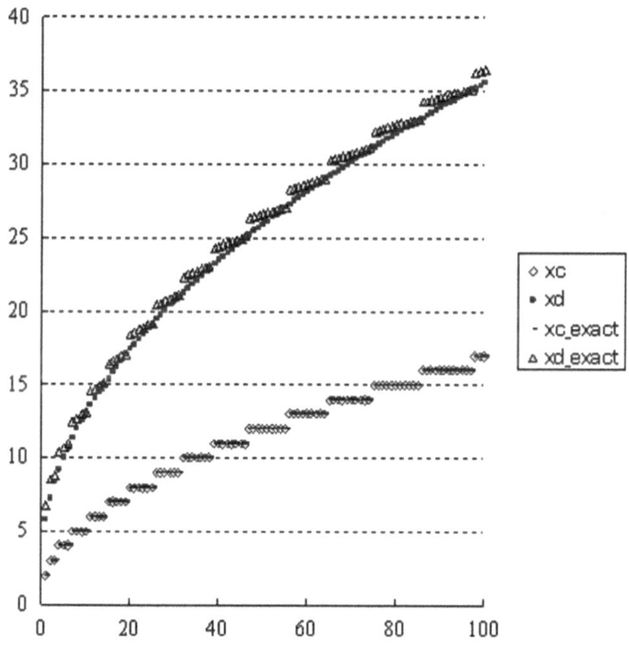

Fig. 6 Capability of detection derived from (6)c approximation method

addition, using Fig. 6 to compare both methods, we deduce that the approximation method (6)c is the most accurate one. Indeed, approximation method (6)c yields exactly the same value as when the cumulative probability was lower than 0.05 and is calculated directly using the Poisson distribution. This result proves that the approximation method (6)c is the best method, and that no approximation method is more accurate than (6)c.

Figure 6 shows the capability of detection where the parameters x_C and x_D are derived from method (6)c and the exact method. The x-axis represents λ_0 and the

y-axis represents the net state value. From Fig. 6, x_C is the same value as the exact x_C, and x_D is also nearly the same value as the exact x_D. Considering a practical application, these results do not impact the evaluation of the capability of detection. Therefore approximation can be applied satisfactorily in practice.

6 Conclusion

In this study, approximation methods are evaluated for the capability of detection of the measurement methods that follow a Poisson distribution. The critical value of the net state variable and the value of the minimum detectable net state variable are derived. When the measurements are small, they are compared with each approximation method, especially for $\lambda_0 = 1$ to 100. As a result, the approximation method (6)c, which rounded x_C off, was determined to be the best. Therefore, when the measurements follow a Poisson distribution, the approximation method (6)c can be used to derive the capability of detection. When the expectation value of the measurements is small, normal approximation of the Poisson distribution is not precise enough because the normal-approximated distribution diverges significantly from the actual distribution of measurements. Thus, we can conclude that when the expectation value increases, the Poisson distribution becomes precise because it would be close to the normal distribution. These examinations can provide feedback that can be used in the final version of ISO 11843 part 6.

References

International Organization for Standardization. (2000). *Capability of detection – part 2: Methodology in the linear calibration Case*. Geneva, Switzerland: ISO.

Furukawa, Y., Iwasaki, M., & Tanaka, A. (2009). A proposal for methodology for the determination of the critical value and the minimum detectable values in Poisson distribution measurements. ANQ Congress Tokyo 2009D.

Johnson, N. L., Kotz, S., & Kemp, A. W. (1993). *Univariate discrete distributions* (2nd ed., Vol. 4, pp. 151–198). New York: Wiley.

Kittlitz, R. G, Jr. (2006). Caluculating the almost exact control limits for a C-chart. *Quality Engineering, 18*, 359–366.

Business Data Quality Control: A Step by Step Procedure

Hans-J. Lenz and Esther Borowski

Abstract Modern information systems supply operative and analytic/statistical data for users. The system design and the usage must be done in such a way that high quality of the stored data is assured. This implies the necessity of fixing quality objectives, defining its characteristics, choosing appropriate measures and measurement techniques and, finally, of embedding this into a step by step procedure for data quality assurance. We start by examples of bad business data, discuss a data quality control methodology and its workflow, offer a first insight into the corresponding metadata model, and demonstrate DaRT – a data quality reporting tool on top of Oracle's Warehouse Builder (OWB).

Keywords Data quality control • DQC workflow • Quality indicators

1 Bad Business Data

According to ISO 9000 data quality can be defined as the degree of fulfilling requirements of quality criteria applied to a database given corresponding quality targets. A comprehensive version is "Fitness for use given intended purposes", cf. Tayi (1998). For example, in a marketing campaign addressing a client by "Mrs. Peter Brown" would evidently lead to a loss of goodwill. Of course, the final objective of data quality control must be "minimizing the long run total cost of data ownership".

H.-J. Lenz (✉)
Institute for Statistics and Econometrics, Freie Universität Berlin, Garystr.21, D-14195, Berlin, Germany
e-mail: hans-j.lenz@fu-berlin.de

E. Borowski
Institute of Information Science, Freie Universität Berlin, Garystr.21, D-14195, Berlin, Germany
e-mail: esther.borowski@gmx.de

H.-J. Lenz et al. (eds.), *Frontiers in Statistical Quality Control 10*,
Frontiers in Statistical Quality Control 10, DOI 10.1007/978-3-7908-2846-7_25,

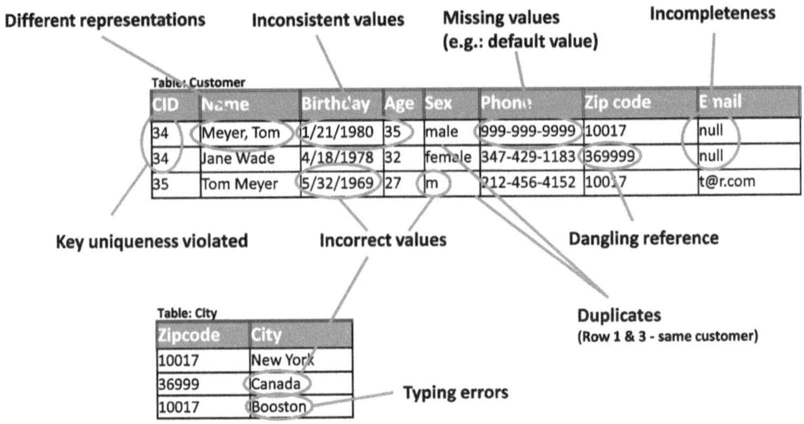

Fig. 1 Types of bad data quality in two tables of a database (cf. Borowski 2008)

Quality targets and criteria are heavily case dependent. The importance can be recognized in operating figures of budgeting, wages and salary as well as marketing campaigns. While even small errors of salary and wage computations are not allowed, marketing campaigns may be effective even if some estimates of the sales figures are erroneous. As will become clearer later, it is not only the data set alone with its syntactic rules which matters but also the inherent information with its semantic and pragmatic features which can be decrypted by interpretation.

In most business cases decision making is involved and typically some pragmatics come in. For example, the string $S = (2, 2, y, e, a, r)$ may be syntactically correct, but it is insignificant for business without interpretation, i.e. additional meta information. In terms of recruiting the first two characters of the string beforehand may represent the *age* of a candidate measured in *year*, and, as a result it increases the signification of the string example. Pragmatics is needed for decision making when a decision rule is applied like "Reject a candidate for a project manager position if he is younger than 22". Note, that real data like S need metadata as "data about data and methods" to fully utilize and understand a database by different user groups. Evidently, data quality control requires an understanding of all intentions of the data customers, cf. Tayi (1998).

Next, we will identify well-known incidents in terms of bad customer data. Quite obvious the location of bad data needs a lot of additional information to lead into success. The data consist of the two tables *City* and *Customer*. Many real life conflicts are exemplarily shown in Fig. 1. Evidently, various quality criteria are needed to detect, meter or clean poor data.

What are the big trouble makers in quality control of business data? Evidently, we have errors of various kinds like missing values, non-unique values of key attributes, duplicates, and inconsistencies related to business rules and balance equations. An example for a business rule is *discount* $=$ *3% if sales* $>$*1,000€ and payment period* $<$ *7 days, else discount* $=$ *2% if payment period* $<$ *14 days else discount* $=$ *0%*. A simple balance equation is *sales* $=$ *quantity* $\times$ *price per unit*. In the following we

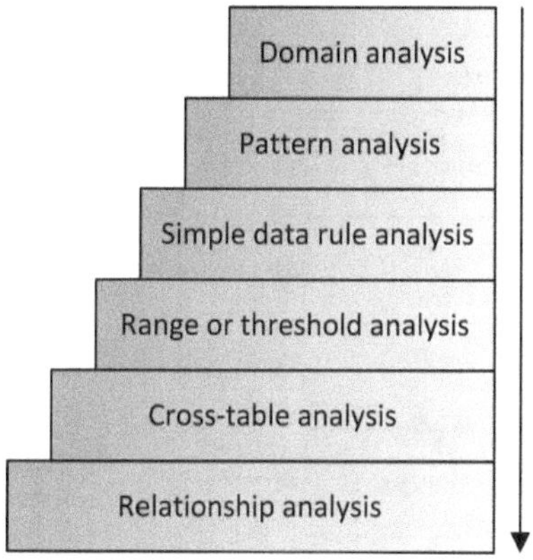

Fig. 2 Data Profiling due to Norris-Montanari (2007)

shall concentrate on these points in more detail. We define the intended quality criteria, give a hint to how to measure them, and integrate the indicators in a stepwise procedure as a workflow for data quality assurance.

The common idea is not new at all. Shepherd (1999) proposes a three step procedure of data profiling starting with checking columns (attributes), detecting dependencies of attributes and, finally, identifying redundancies. His approach includes the detection of homonyms and synonyms, too – a classical problem of database theory. Norris-Montanari (2007) integrates the analysis of thresholds and rules consistency, cf. Fig. 2. Note that the diagram of his approach is to be read "top-down".

Olson (2003) recommend a stepwise bottom-up procedure. His profiling workflow starts with detecting invalid values of single attributes and moves upwards by applying finally complex data rules to extract invalid combinations of valid data. In each step data are cleaned for achieving better data quality before stepping forward.

Schlaucher (2007) picks up the idea of a bottom-up procedure and embeds it into the services of the Oracle Warehouse Builder 10gR2. The advantage of his approach is that a variety of tools like Oracle profiler, Oracle cleanser etc. is offered. However, a streamlined workflow together with a user-friendly interface is missing. Proposals and a kind of guide for an inexperienced user are especially mandatory.

A joint venture between Oracle and the authors resulted in the prototype ***DaRT*** (Fig. 3).

2 Data Quality Criteria

The term "*data quality*" has many facets or dimensions. The situation is similar to the industrial domain. Even if several data sources are excluded, and only one source considered instead, the large number of single quality criteria is overwhelming.

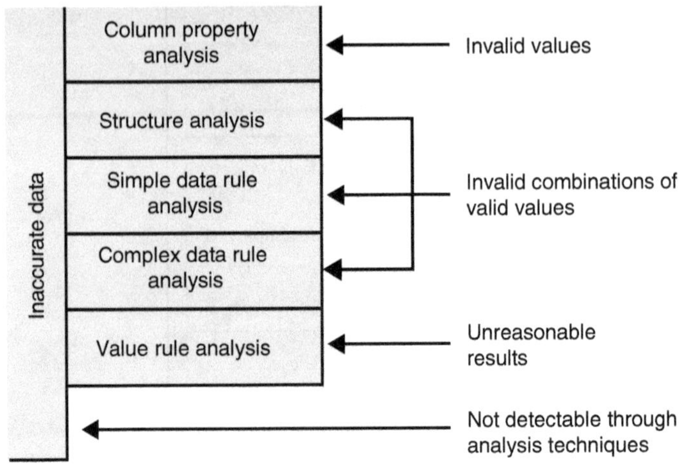

Fig. 3 Data profiling due to Olson (2003)

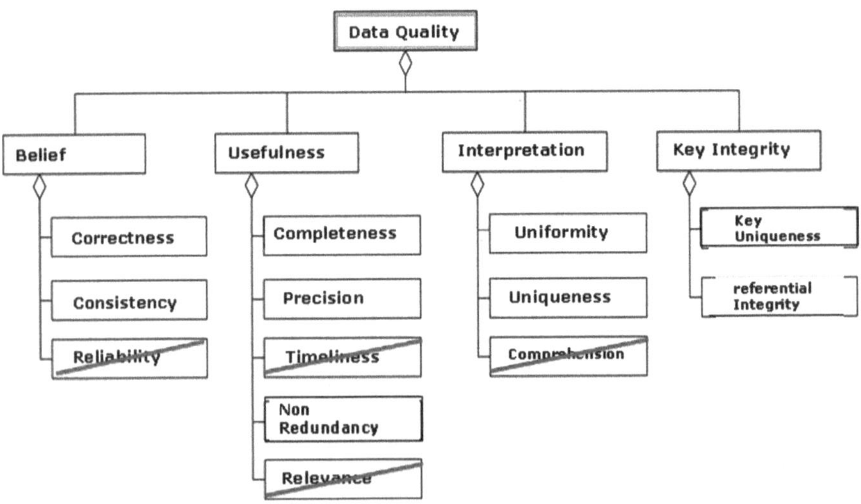

Fig. 4 Taxonomy of data quality criteria due to Hinrichs (2002)

To gain overview, there exist a lot of taxonomies for typing quality criteria. The proposal of Hinrichs (2002) seems advantageous for our purpose.

The main groups are Belief, Usefulness, Interpretation and Key Integrity. A description of these terms is extensionally given by the subordinated quality criteria. We skip four criteria demanding an intricate pattern for measurement, cf. Fig. 4. Data quality actions for measuring quality criteria can be classified into three groups:

- IT checks: Inspecting real data of given entities (objects) with respect to their corresponding metadata like definitions, types, domains, relations etc.
- Statistical checks: Explorative data analysis for a data profile reporting
- Rule checks: Data validation related to active business rules

Let us first turn to *Belief.* A simple definition of *correctness* leads to metering any deviation between an observation x_i and its true but possible unknown value μ_i given an object i. For instance, the first name "Peter" related to "Mrs." does not match, i.e. *gender (Peter)* $\neq$ *'male'* $\Rightarrow$ *truth-value(Peter, Mrs.)* = *'false'*. *Correctness* would be achieved, if the values "*Peter*" and "*Mr.*" were jointly observed. In this example we have a symbolic data space. Frequencies are used to handle cases with non-metric quality criteria use.

Definition 1. Let quality attribute A has *range* $(A) = \{a_1, a_2, \ldots, a_{|A|}\}$. Let χ: *range(A)*$\rightarrow \{0, 1\}$ be a characteristic function with $\chi(a) = 1$ if $cond_A(a)$ is true, and 0 else. Then the corresponding quality frequency indicator of table T with sample size $n = |T|$ is given by

$$q(A) = 100 \sum_{i=1}^{n} \chi(a_i)/n \tag{1}$$

In metric data spaces appropriate distance functions $||x - \mu||$ like Euclidean distance, Mahalanobis distance, Kullback-Leibler divergence etc. are available, cf. Lenz (2008).

The criterion *consistency* has itself many facets. First of all, it measures the deviation between the observed metadata of variables and its technically intended one, cf. Hinrichs (2002). For example the data type "float" used for commercial application would be misleading, but the type "decimal" with only two decimals right to decimal point would be admissible. Data type, length and domain are some of the metadata of main software concern.

We pick up *range consistency* of an attribute A. For illustrative purposes think of the set range$(A) = \{$male, female$\}$ labelled as "*Gender*". If a record exists where the gender shows a value equal to "f", "*Mann*" or "*weiblich*" the characteristic function $\chi(a)$ would flag a zero. Therefore we define the quality characteristic ("measure"):

$$q_{range}(A) = 100 \sum_{i \in A} \chi_{range(A)}(a)/|T| \tag{2}$$

where

$$\chi_{range(A)}(a) = \begin{cases} 0 & \text{if} \quad a \notin \mathrm{A} \\ 1 & \text{if} \quad a \in \mathrm{A} \end{cases} \tag{3}$$

Secondly, the collected data must be (semantically) coherent with respect to existing business rules or balance equations. An example of this kind of multi-attribute relationships is the inventory equation $I_t = I_{t-1} + p_t - d_t$ where the variables represent "inventory" I_t, rate of production p_t and demand d_t in period t.

The subgroup *Usefulness* contains the criteria *Completeness*, *Precision* and *Redundancy*.

Completeness counts the "not missing" or "not null" values and uses a formula of type (1). The corresponding indicator function is

$$\chi_A(a) = \begin{cases} 0 & \text{if} \quad a \quad \text{is missing (null)} \\ 1 & \text{if} \quad a \quad \text{is not missing} \end{cases} \tag{4}$$

Precision measures the length of a numerical attribute and its decimals. Deviations from given requirements are detected if appropriate frequency statistics are "too high".

In our context *Redundancy* is caused by duplicates, i.e. by more than one record related to the same real entity. For example, think of the first and third record in Fig. 1. They identically refer to the customer 'Tom Meyer'. There are mainly two approaches for duplicate detection or "decoupling". If a primary key (unique identifier) *Id* like SSN, tax identifier, zip code etc. is available and error-free, it should be used to avoid duplicates. If such an identifier is not at hand, one must take non-key attributes with high discrimination power, and can use statistical classification methods for deciding whether or not two records match or not, cf. Neiling (2004). A useful indicator function signals a zero if at least one duplicate exists in a data set and if not it should signal a one instead, i.e.

$$\chi_{dup}(t) = \begin{cases} 0 & \text{if a duplicate exists for record } t \\ 1 & \text{else} \end{cases} \tag{5}$$

Let N^* be the number of unique records (duplicates counted only once) in a given table T. Then we define as a *non redundancy indicator*:

$$q_{dup}(T) = 100 \sum_{t \in T} \chi_{dup}(t)/|T| = 100N^*/|T| \tag{6}$$

Assume $T_0 = \{$*Hans, Dave, Bernd, Claus, Hans, Hans, Bernd, Mike*$\}$ is given. It follows $N* = 5$ and $q_{dup}(T_0) \approx 60\%$.

Uniformity and *Uniqueness* can be defined in a straightforward way and will be disregarded. Finally we turn to *Integrity Constraints*, cf. Hinrichs (2002). Entity integrity refers to the uniqueness of the primary key *Id* or a key candidate P. Contrarily to the duplicate detection problem key uniqueness is reduced to one (possible concatenated) attribute, i.e. to the primary key which uniquely identifies each record. The appropriate indicator function χ_{dist} signals 0 if more than one specific value $p \in$ range (Id) ensues in a data set (table) T. The percentage of key distinct records is

$$q_{dist}(T) = 100 \sum_{t \in T} \chi_{dist}(t)/|T| \tag{7}$$

The referential integrity checks whether or not for each value f of a foreign key F there exists a related value p of the primary key P. For example, for each customer alive a city is required where he belongs to. Consequently, we use the dangling function as an indicator function
$\chi_{dangling}$: range(Id) $\cup\{null\} \rightarrow \{0, 1\}$.

$$\chi_{dangling}(f) = \begin{cases} 0 & \text{if} \quad f \notin \{null, range(Id)\} \\ 1 & \text{else} \end{cases} \tag{8}$$

Then we get as quality indicator of referential integrity

$$q_{dangling}(F) = 100 \sum_{f \in range(F)} \chi_{dang}(f)/|F| \tag{9}$$

3 Data Quality Workflow

In the preceding chapters we introduced a taxonomy for quality criteria, their definitions and measurements using a frequency based concept. From this it is evident that metadata is essential for any data quality control. Moreover, the framework which will be presented next is a step by step procedure and highly interactive, i.e. the software scans the database and delivers the (real) values of the quality criteria together with hints to nonconforming records. Such a messaging or signalizing is possible if clear target values of the quality characteristics are predefined. Of course, the data quality analyst must autonomously decide about acceptance or rejection including "repair" of records and tables. Such a workflow system can only be operated semi-automatically since a large fraction of metadata is strongly case-dependent (Fig. 5).

The numbering of boxes represents the single steps of the DQ workflow. Single attribute analysis is performed in steps 1–4 as shown on the bottom line of the pyramid. The next level is devoted to the dependency analysis with steps 5–7. Steps 8 and 9 include checks of referential integrity and simple statistics like extreme values, means and standard deviation. The top level is step 10 applying business rules to the data. The sequencing within any level is optional, and is used as a default value for the naive user. Before presenting some screenshots of data quality reports produced by DaRT we show in Fig. 6 how DaRT is integrated into the "quality cycle" as favoured by Oracle (2007).

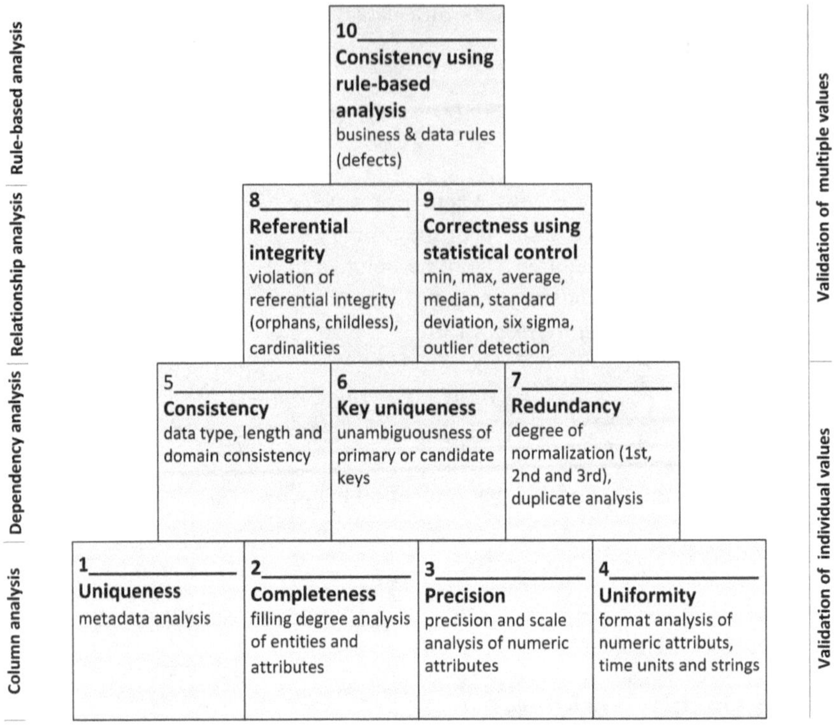

Fig. 5 Data quality pyramid and the ten embedded steps of the DQ procedure

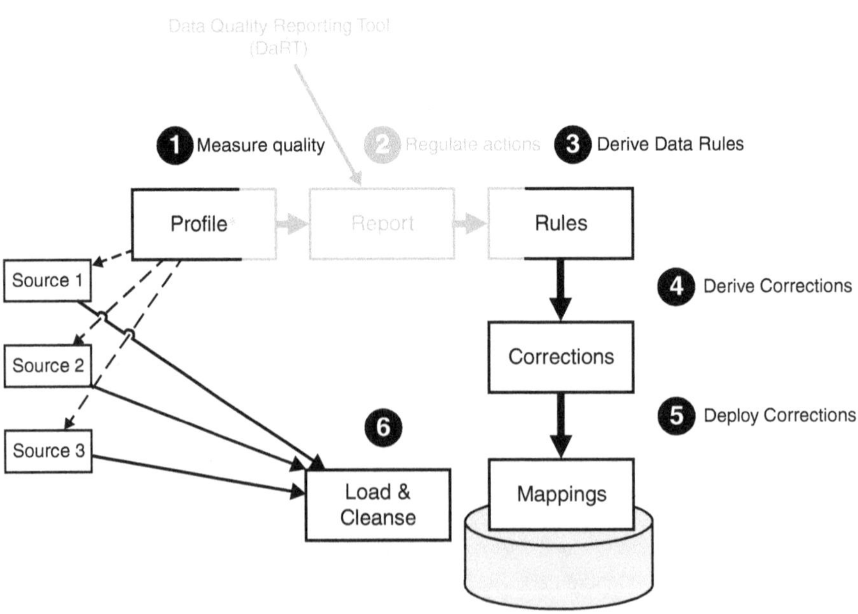

Fig. 6 Data quality workflow of Oracle with an embedded reporting step

4 Data Quality Reporting

The data quality report provides information about the real (status-quo ante) data quality of a given database. Using the "traffic light" principle the report builds a judgement about the degree of usefulness of the data related to each quality characteristic. Furthermore a quality action is displayed as an optional for the analyst. We present the analysis of *completeness* as an example (Table 1).

The recommended actions can be used for data cleansing, i.e. repair or improvements. We use an (extended) record from *Customer* to illustrate the "orchestration" of data profiling and cleansing, cf. Oracle (2008). In this example data cleansing applies parsing, standardization for getting uniformity, postal address validation and augmentation. The last two data transformations augment a fragment of the zip code and infer the gender from the first name as this mapping leads to a unique value in this case. Note that this is true for "Josephine" but not for "Chris", because of a lack of functional dependency *first_name→gender* (Table 2).

Next we show a report of level one analysis focussing on *Precision*.[1] For all attributes of numeric type the *length* and the number of decimals (called *scale*) is measured and compared with the requirement as given in the repository (meta database) for the table labelled *KUNDEN_STAMM* which stores customer records. This table has 1,032 entries. The quality feature *Precision* has values *accurate*, *moderate* or *imprecise*. From Fig. 7 it becomes evident that the first two attributes and the last five have imprecise values relative to their allowed maximum. With respect to the number of decimals all attributes are set to the default value *accurate* due to missing decimals.

In all cases of insufficient precision the software DaRT signals an updating of precision displayed in the most right column.

Next we show an example of a level two analysis where the consistency of data type, field length and value set are checked. We select the attribute *GESCHLECHT* (gender) as one out of 14 variables. The report shows that this attribute does not need type "VARCHAR2" with length 10, but simply NUMBER of length 1. About *3%* of all 1,032 records are contaminated. The range analysis (last block) shows that less than *3%* of all records have nonconforming characters instead of 0 and 1. Because the tolerance level is set equal to *95%* this nonconformity is not signalled by a red figure. DaRT is a "liberal" system in so far as its only locates an inconsistency but does not automatically repair it. Actions are exclusively started by the data quality analyst only (Fig. 8).

On level 3 *Correctness* and *Referential Integrity* is checked. We limit ourselves to the analysis of the customer data using simple descriptive statistics as displayed in Fig. 9. Two points are worthwhile mentioning. A *3*σ-check of the attributes ANZ_KINDER (no of kids) and FIRMENRABATT (discount rate) signals "too

[1]The (synthetic) database of (fictive) Service GmbH used was kindly made accessible by A. Schlaucher, Oracle Deutschland GmbH.

Table 1 Sample from a data quality report on *Completeness*

Attribute	Null values	Completeness (%)	DQ statement	DQ action
Family name	Allowed	100	Moderate	Null not allowed
Customer type	Yes	1	Bad (<75%)	Individual
State	No	100	Good	No action

Table 2 Cleansing of an incomplete, non-standardized and postal incorrect record (Source: Oracle (2008) -modified-)

Input		Output	
Attribute	Value	Attribute	Value
Name	Josephine Random	Family Name	Random
		First Name	Josephine
Title	Senior Manager	Title	Sr. Mgr.
Company	Oracle	Company	Oracle Corp.
Address	500 oracle parkay	Address	500 Oracle Pkwy
Address 1	Redwood, az 94065	City	Redwood City
		State	CA
		Zip code	94065-1675
Address2	USA	Country	USA
Phone	5067000	Phone	605 506 7000
Email	Joe.random@oracle.com	Email	Joe.random@oracle.com
		Gender	f

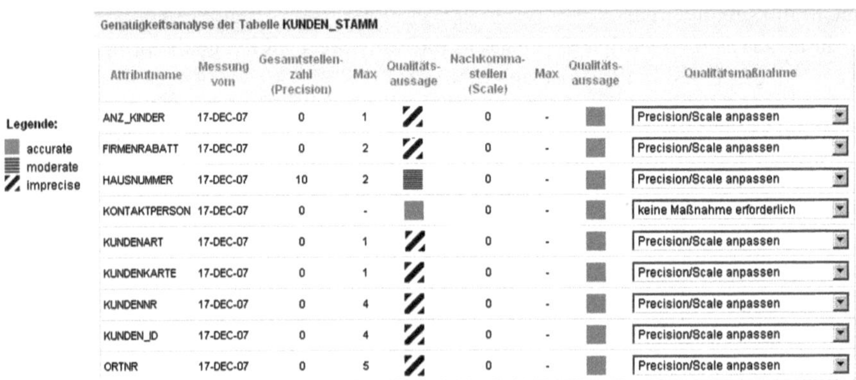
Genauigkeitsanalyse der Tabelle **KUNDEN_STAMM**

Attributname	Messung vom	Gesamtstellenzahl (Precision)	Max	Qualitätsaussage	Nachkommastellen (Scale)	Max	Qualitätsaussage	Qualitätsmaßnahme
ANZ_KINDER	17-DEC-07	0	1		0	-		Precision/Scale anpassen
FIRMENRABATT	17-DEC-07	0	2		0	-		Precision/Scale anpassen
HAUSNUMMER	17-DEC-07	10	2		0	-		Precision/Scale anpassen
KONTAKTPERSON	17-DEC-07	0	-		0	-		keine Maßnahme erforderlich
KUNDENART	17-DEC-07	0	1		0	-		Precision/Scale anpassen
KUNDENKARTE	17-DEC-07	0	1		0	-		Precision/Scale anpassen
KUNDENNR	17-DEC-07	0	4		0	-		Precision/Scale anpassen
KUNDEN_ID	17-DEC-07	0	4		0	-		Precision/Scale anpassen
ORTNR	17-DEC-07	0	5		0	-		Precision/Scale anpassen

Fig. 7 Precision/Scale Analysis on customer data

large deviations" and gives a red light. The observed maxima support this statement of incorrectness. The application of the statistics *Mittelwert* (arithmetic mean) or even the standard deviation σ to attributes with a nominal scale like HAUSNUMMER (house no) or KUNDENART (customer type) is statistically nonsense – even when used for data profiling only. This problem can simply be avoided by skipping such computations if the scale of an attribute is non metric.

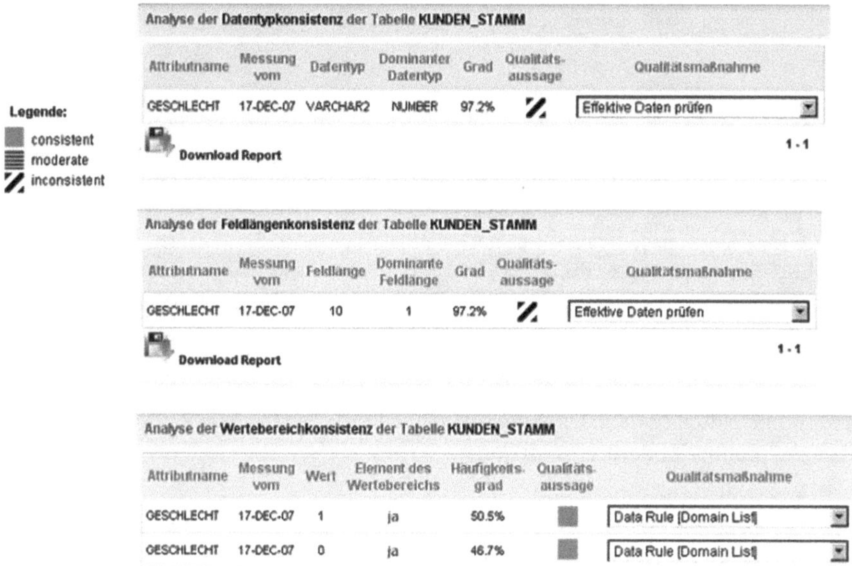

Legende:
consistent
moderate
inconsistent

Analyse der **Datentypkonsistenz** der Tabelle **KUNDEN_STAMM**

Attributname	Messung vom	Datentyp	Dominanter Datentyp	Grad	Qualitäts-aussage	Qualitätsmaßnahme
GESCHLECHT	17-DEC-07	VARCHAR2	NUMBER	97.2%		Effektive Daten prüfen

Download Report 1 - 1

Analyse der **Feldlängenkonsistenz** der Tabelle **KUNDEN_STAMM**

Attributname	Messung vom	Feldlänge	Dominante Feldlänge	Grad	Qualitäts-aussage	Qualitätsmaßnahme
GESCHLECHT	17-DEC-07	10	1	97.2%		Effektive Daten prüfen

Download Report 1 - 1

Analyse der **Wertebereichkonsistenz** der Tabelle **KUNDEN_STAMM**

Attributname	Messung vom	Wert	Element des Wertebereichs	Häufigkeits-grad	Qualitäts-aussage	Qualitätsmaßnahme
GESCHLECHT	17-DEC-07	1	ja	50.5%		Data Rule [Domain List]
GESCHLECHT	17-DEC-07	0	ja	46.7%		Data Rule [Domain List]

Fig. 8 Consistency Analysis of customer data focussing on data type, length and range

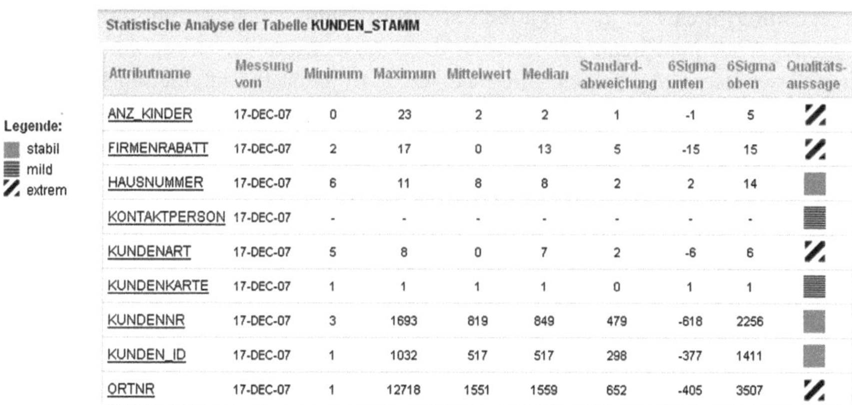

Legende:
stabil
mild
extrem

Statistische Analyse der Tabelle **KUNDEN_STAMM**

Attributname	Messung vom	Minimum	Maximum	Mittelwert	Median	Standard-abweichung	6Sigma unten	6Sigma oben	Qualitäts-aussage
ANZ_KINDER	17-DEC-07	0	23	2	2	1	-1	5	
FIRMENRABATT	17-DEC-07	2	17	0	13	5	-15	15	
HAUSNUMMER	17-DEC-07	6	11	8	8	2	2	14	
KONTAKTPERSON	17-DEC-07	-	-	-	-	-	-	-	
KUNDENART	17-DEC-07	5	8	0	7	2	-6	6	
KUNDENKARTE	17-DEC-07	1	1	1	1	0	1	1	
KUNDENNR	17-DEC-07	3	1693	819	849	479	-618	2256	
KUNDEN_ID	17-DEC-07	1	1032	517	517	298	-377	1411	
ORTNR	17-DEC-07	1	12718	1551	1559	652	-405	3507	

Fig. 9 Analysis of correctness applied to customer data

The final demonstration of data profiling as part of data quality control is devoted to the fourth level where business rules are applied to a project, i.e. to the database owned by company *Service GmbH*. Three business or data rules (*DR*) are activated. We pick up the data rule ANREDE_DR which refers to the salutation address of customers. Like the data type "enumeration" it checks whether or not the application of the salutation address uses a value from the defined range {*Herr, Frau, Firma*}.

As can be seen in Fig. 10 this rule is valid in *98%* of the customer records. Due to the preselected threshold 23 defects or a percentage of *2%* is a moderate error

Regelbasierte Analyse

Datenregel	Typ	Grad	Defekte	Qualitäts-aussage	Qualitätsmaßnahme
ANREDE_DR	DOMAIN_LIST_RULE	98%	23	■	Effektive Daten prüfen
LAGER_DR	DOMAIN_NO_NULL_RULE	100%	0	■	keine Maßnahme erforderlich
BILDUNG_DR	DOMAIN_PATTERN_LIST_RULE	23%	798	⁄	benutzerdef. Maßnahme

Legende:
consistent
moderate
inconsistent

Fig. 10 Rule-based analysis with data rules for salutation address (ANREDE_DR), inventory (LAGER_DR), and education (BILDUNG_DR) selected from tables *Customer* and *Inventory*

rate and signalled by a yellow symbol. The recommendation of the system is to inspect the contaminated records. This can simply be achieved by just clicking on an appropriate button of the GUI. Note that in the third case the red symbol is displayed since the number of defects is too large.

5 Conclusions

The methodology presented is tailor-made for the business not primarily for science or technology area. Our investigation starts from defining quality criteria and targets. It proposes adequate quality indicators which are measurable. The approach is built into a user friendly graphical interface (GUI) to assist the interaction between a data quality analyst and the software system ***DaRT***. It makes full use of the power of Oracle's profiler. The role of quality targets, thresholds, computing of quality indicator values and the massive utilization of metadata stored in a repository becomes evident.

References

Borowski, E. (2008). *Entwicklung eines Vorgehensmodells zur Qualitätsanalyse mit dem Oracle Warehouse Builder*. MSc thesis, Freie Universität Berlin.

Borowski, E., & Lenz, H.-J. (2008). Design of a workflow system to improve data quality using Oracle Warehouse Builder, *Journal of Applied Quantitative Methods, 3*, 198–206.

Hinrichs, H. (2002). *Datenqualitätsmanagement in: Data-warehouse-systeme*. Doctoral dissertation, Universität Oldenburg.

Lenz, H.-J. (2008). Proximities in statistics: Similarity and distance. In G. Della Riccia et al. (Eds.), *CISM courses and lectures: Vol. 504. Preferences and similarities* (pp. S. 161–177). Berlin: Springer.

Neiling, M. (2004). *Identifizierung von Realwelt-Objekten in multiplen Datenbanken*. Doctoral dissertation, TU Cottbus.

Norris-Montanari, J. (2003). Where to start – Data profiling. http://www.twdi.org/Publications/display.aspx?id=6807&t=y#a2.

Olson, J. E. (2003). *Data quality. The accuracy dimension*. San Francisco: Morgan Kaufmann.

Oracle (2007). Oracle warehouse builder 11g – An overview. http://www.oracle.com/technology/products/warehouse/11gri/presentations/owb11gr1-overview.ppt.
Schlaucher, A. (2007). Der Datenqualität auf der Spur. *DOAG News, Q1*, 24–28.
Shepherd, J. B. (1999). Data migration strategies. DM review magazine.
Tayi, G. K., & Ballou, D. P. (1998). Examining data quality. *Communications of the ACM, 41*(2), 54–57.

Data Quality: Algorithms for Automatic Detection of Unusual Measurements

Ross Sparks and Chris OkuGami

Abstract The paper offers simple robust algorithms for checking consistency of large volumes of measured data. The checks differentiate between data collected on a spatial grid at one time point; and data collected on a spatial grid over many time points, as well as several related measurements collected on a spatial grid over time. The checking process involves computationally efficient methods of estimating expected values and variances used to judge measurement consistency. Three-sigma control limits are applied to flag inconsistent measurements. CUSUM and EWMA plans are advocated for flagging consistently small biased measures.

Keywords Consistent measurement • Multivariate • Process monitoring • Spatial

1 Introduction

Data quality is defined as data being fit for purpose (Borowski and Lenz 2008). This paper focuses on the narrower aspect of measurement consistency checking. This checking process helps isolate poor measurements in large datasets. This is meant to make the task of ensuring quality data easier by focussing the manual data checking effort on inconsistent measurements only. Measurement inconsistency is assessed in terms of:

- **Spatial consistency:** The measurement is checked against an interpolated value using measures made at several geographical close locations. An example is thickness measures at several locations on a single sheet metal.
- **Temporal consistency:** Measurements made at one geographic location over time, and checked against their one-step ahead forecasts in real-time, or checked

R. Sparks (✉) · C. OkuGami
CSIRO Mathematics, Informatics and Statistics, North Ryde, NSW, Australia
e-mail: Ross.Sparks@csiro.au; Chris.OkuGami@csiro.au

H.-J. Lenz et al. (eds.), *Frontiers in Statistical Quality Control 10*,
Frontiers in Statistical Quality Control 10, DOI 10.1007/978-3-7908-2846-7_26,

against its predicted (interpolated) value made using measurements directly before and after for batch checking. An example is the daily milk volume produced by single dairy farm.
- **Multivariate consistency:** Here consistency is checked against different but related measurements either made at the same geographical location or spatially elsewhere. An example is stream flow at a single site in a river in relation to rainfall, humidity and temperature in the catchment. Multivariate consistency checks whether flows are consistent with the catchment: rainfall, humidity and temperature values.

Often we are interested in the joint spatial, temporal and multivariate consistency checking rather than marginal consistency checks. An example of this is the checking of river flows at a specific location given several upstream flows and rainfall measurements. If all stream flow measurements are unusually large in the catchment at a specified time, but all are spatially consistent with each other in the whole catchment and consistent with the local rainfall values, then these measurements will not be classified as unusual even when they are unusual in the marginal sense. Thus, avoiding a paradox very similar to Simpson's paradox (Blyth 1972). However, if one of the stream flows is unusual relative to local flow measurements made further upstream or downstream, then this will be flagged as inconsistent. This joint multivariate, spatio-temporal consistency checking is where we depart from other recommended univariate data quality checking.

The process of checking for unusualness is based on measurement departures from their conditional expected values given all the other local related measurements. The measures we condition on are called explanatory variables (e.g. for river flows, rainfall is an explanatory variable). These explanatory variables are usually measured at the same time or before the measured value being considered. Previous measurements made at the same site are also conditioned on (i.e. the measure made at the time before) to check for temporal consistency in a real-time consistency checking process.

The explanatory variables are selected to provide the interpolation of y with the least absolute error. Find the conditional expected value for the measured value (y) given the set of explanatory variables (x) (denoted $\mathrm{E}(y|x)$). Unusualness is measured in terms of how much y departs from $\mathrm{E}(y|x)$. Let the variance for this departure be denoted by σ^2, then assign unusualness codes on the scale of extremely unusual, very unusual, unusual, within the expected range, and close to expected using the following rules:

1. $||y - \mathrm{E}(y|x)|| > 5\sigma$ **extremely unusual**
2. $4\sigma < ||y - \mathrm{E}(y|x)|| < 5\sigma$ **very unusual**
3. $3\sigma < ||y - \mathrm{E}(y|x)|| < 4\sigma$ **unusual**
4. $2\sigma < ||y - \mathrm{E}(y|x)|| < 3\sigma$ **somewhat unusual**
5. $\sigma < ||y - \mathrm{E}(y|x)|| < 2\sigma$ **within expected range**
6. $||y - \mathrm{E}(y|x)|| < \sigma$ **close to expected**

We consider the most common more complicated situation where no duplicate measurements are available. In cases where duplicate pairs of measurements are made, one measurement can be considered as y and the other $\mathrm{E}(y|x)$. Let σ be the variance of the differences between the two measurements, then the above classification process for flagging potential measurement errors can be applied. In other words, the same framework used in this paper can be applied.

Section 2 discusses the consistency checking process. Section 3 looks at detecting persistent errors commonly caused by measurement device failure. Section 4 discusses how our quality consistency check fit into a larger QAQC framework.

2 Algorithms for Measurement Consistency Checking

As mentioned earlier, consistency checks can be broken down into temporal, spatial, and multivariate checks. Multivariate spatio-temporal consistency checks are typical in environmental or hydrological applications. Algorithms appropriate for each of these situations are described in the subsections to follow. Note that the emphasis here is on automatic consistency checking in large scale data collection applications such as sensor networks.

2.1 *Checking Univariate Temporal Consistency*

Checking whether a site's measurement process is consistent with past history is typical in data quality assessments. Exponentially weighted moving averages (EWMA) is simplest way to get an expect value of a measurement from historical data. Let $ewma_0$ and $ewmv_0$ be the sample mean and variance, respectively, for training data. Then the EWMA statistic is defined by

$$ewma_t = \lambda_1 y_t + (1 - \lambda_1) ewma_{t-1} \tag{1}$$

and the exponentially weighted moving variance by

$$ewmv_t = \lambda_2 (y_t - ewma_{t-1})^2 + (1 - \lambda_2) ewmv_t \tag{2}$$

where $0 < \lambda_i < 1$, $i = 1, 2$. We select $\lambda_1 = \lambda_2$, however there may be occasions where smoothing $(y_t - ewma_{t-1})^2$ more than y_t is appropriate. Here we use $\mathrm{E}(y_t|x_t) = ewma_{t-1}$ and $\sigma_t^2 = ewmv_{t-1}$ for assessing the unusualness of measurement y_t. More complicated time series models, such as GARCH models (see Silvennoinen and Terasvirta 2009), can be used to establish $\mathrm{E}(y_t|x_t)$ and variances σ_t^2. However, these models need more maintenance. The robustness of the GARCH models in applications needs testing prior to its use.

2.2 Checking Spatial Consistency

Let identical measurements made at the same time in other locations in the catchment be denoted explanatory variables $x_1, \ldots, x_k$. Assume $\mathrm{E}(x_i) = \mu = \mathrm{E}(y)$ for all i. Let d_i be the distance of measurement site x_i from the target site measurement y. Let $\mu_{y|x} = \mathrm{E}(y|x) = \mathrm{E}(y|x_1, \ldots, x_k)$ and $\sigma^2 = \mathrm{E}(y^2|x_1, \ldots, x_k) - \mathrm{E}(y|x_1, \ldots, x_k)^2$. The simplest estimate of $\mathrm{E}(\mu_{y|x})$ assumes that y is proportional to the inverse to the distance (d_i) x_i's and y are apart, and that $\mathrm{E}(x_i) = \mathrm{E}(y)$, then

$$\widehat{\mu}_{y|x} = \left(\sum_{i=1}^{k} x_i/d_i\right) \Bigg/ \left(\sum_{i=1}^{k} 1/d_i\right) \tag{3}$$

Let the estimate of $\mathrm{E}(y^2|x_1, \ldots, x_k)$ be given by $(\sum_{i=1}^{k} x_{it}^2/d_i)/(\sum_{i=1}^{k} 1/d_i)$ and then an estimate of the variance for this interpolated value is

$$\widehat{\sigma}^2 = \left(\sum_{i=1}^{k} x_i^2/d_i\right) \Bigg/ \left(\sum_{i=1}^{k} 1/d_i\right) - \left[\left(\sum_{i=1}^{k} x_i/d_i\right) \Bigg/ \left(\sum_{i=1}^{k} 1/d_i\right)\right]^2 \tag{4}$$

Use $\sigma^2 = \widehat{\sigma}^2$ to assess the unusualness of measurement y. Universal Kriging models (Cressie 1993) may offer more sophisticated spatial consistency checks for spatial measurements $x_1, \ldots, x_k$ to estimates of $\mathrm{E}(y|x)$ and σ^2. These models will need more maintenance than other options of kernel smoothing (Zheng et al. 2004), semi-parametric models (Ruppert et al. 2003), or two-dimensional splines (Whitten and Koelling 1973).

2.3 Simple Way of Checking Spatio-Temporal Consistency

A spatio-temporal model is required to establish a reasonable estimate of $\mathrm{E}(y_t|x_{1t}, \ldots, x_{kj})$. Assume $\mathrm{E}(y_t) = \mathrm{E}(x_{it})$ for all i. $\mathrm{E}(y_t|x) = \mathrm{E}(y_t|x_{1t}, \ldots, x_{kt}, y_{t-1}, \ldots, y_1)$ A very simple estimates of $\mathrm{E}(y_t|x)$ is, for a given weight $0 < \eta < 1$, taken from Eqs. 1 and 3 and given by

$$\eta\left(\sum_{i=1}^{k} x_{it}/d_i\right) \Bigg/ \left(\sum_{i=1}^{k} 1/d_i\right) + (1-\eta) ewma_{t-1}.$$

Under the assumption of independence, the respective consistency estimator from Eqs. 2 and 4 is

$$\widehat{\sigma}^2_{e,t} = \eta^2\, \widehat{\sigma}^2_t + (1-\eta)^2\, ewmv_{t-1}.$$

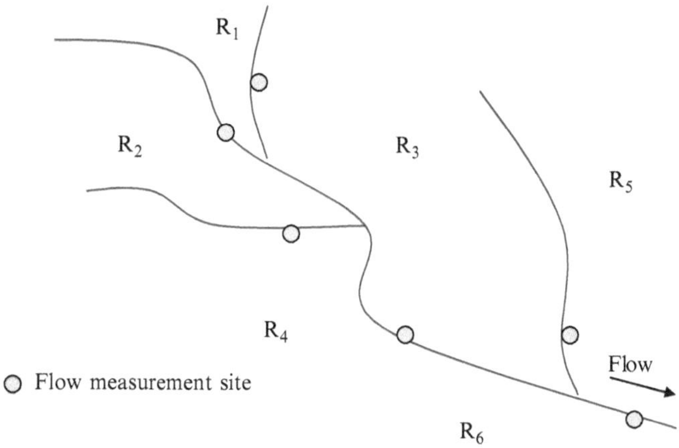

Fig. 1 Catchment with six rainfall sites and six steam flow sites – R_i = rainfall measurement site

The value for η is selected to give the smallest value for $\sigma^2_{e,t}$ for the training sample. This optimal value is estimated as $\eta = ewmv_{t-1}/(ewmv_{t-1} + \widehat{\sigma}^2)$. Under the assumption of independence, the estimate of the minimum value for $\sigma^2_{e,t}$ is

$$\eta^2\,\widehat{\sigma}^2 + (1-\eta)^2\,ewmv_{t-1} = \widehat{\sigma}^2 ewmv_{t-1}/(\widehat{\sigma}^2 + ewmv_{t-1}).$$

The more difficult option is to fit a spatio-temporal model (Finkenstadt et al. 2007) that can be sequentially updated and used.

2.4 *Checking Multivariate Spatio-Temporal Consistency*

Assume an example of consistency checking stream flow measures in a catchment (Fig. 1). We develop algorithms that will check consistency of these jointly with:

- Identical measurements made earlier (or later) in time at the same site, i.e. temporal consistency;
- Identical measurements made at neighbouring sites (at the same time or earlier), i.e. temporal consistency;
- Related measurements made at the same site or neighbouring sites, i.e. multivariate consistency (e.g. checking consistency of flow considering local rainfall as a related measure).

It is important to emphasis that we build interpolating time series models as opposed to forecasting models. In other words, we allow explanatory variables collected at the same time as the response. Therefore, the models used are generally more accurate than the usual time series models.

Simple regression (black box) models for predicting the measurement is used to check for consistency. The model predicted value is taken as the conditional expected value. The spatial variables are considered as explanatory variables. Upstream flow measurements in a river used to explain downstream flows are time adjusted for the duration it takes for water to flow between the two sites, i.e. if y_t is the downstream flow measurement made at time t, x_t is the upstream flow measurement made at time t, and it takes q units of time for water to flow between the upstream site and the downstream site in the river, then the correlation between y_t and x_{t-q} should maximize the $\text{cor}(y_t, x_{t-\tau})$ over all selections of τ.

Two data sources are used: training data (assumed clean) and test data. The training data should be representative of the time period to be assessed for consistency. The test data are to be checked for consistency. Training data provides starting values for estimated regression parameters and are used to estimate the best time lags between explanatory variables and response values. These parameters are used to find expected values and variances for the first measurements in the test data.

The algorithms supplied below have two phases:

1. The start up phase used to provide initial estimates of model parameters.
2. The recursive estimation phase used to update parameter estimates with each new observation in the test data.

The recursive estimation phase uses discounted least squares to fit the regression model. These fitted models are used for predicting the test data values. The approach emphasises simplicity and assumes that variables are locally (in time and space) linear. Estimating the time flows take from one site in the river to the next allows the user to appropriately condition on the upstream flows for predicting downstream flow measurement. This means that the spatial structure is included in the mean rather than the covariances as in Universal Kriging. Below it is assumed that upstream sites used as explanatory variances will have the appropriate temporal adjustment (denoted q_i for the ith upstream flow measurement) to improve their explanatory information.

2.4.1 The Start Up Phase

The training data are used to establish the starting estimates for parameters. Cleaned training data directly prior to the time period being checked is the default. The training data are also used to establish the explanatory variable temporal lags which best explained the response variable. Data checks are carried out for each catchment separately as described below.

The start up process has the following steps (stream flows used as an example):

Step 1: Specify the sequence measures will be checked in the catchments.

Recommendation: Start from the rainfall/stream flow sites that are furthest up stream from the mouth of the river.

Step 2: Specify the set of explanatory variables to be used. Our recommendation for a stream flow site is to use:

- **Rainfall:** Neighbouring rainfall sites to the stream flow site.
- **Stream flow:** Upstream sites for flows.

Include any other potentially causal explanatory variables, e.g. wind directions for rainfall. The general idea is to check the 'mass-balance' of flows. That is, for stream flow, we assume that no water is lost into aqueducts, there are no measurement errors, no water in rivers is lost due to evaporation, and all lag flows are known. Under these assumptions we should be able to check measures precisely when it does not rain. Selecting the explanatory variables for river flows in Fig. 1 involves upstream flow measurements, e.g. for the site furthest downstream in Fig. 1 select the two stream flow measurement sites upstream (one on the tributary and one on the main river). Select neighbouring rainfall stations, e.g. R4, R5 and R6. Thereafter, find the lag time for water to flow from the upstream sites to the target site. In this paper, we denote these time lags by q_1 and q_2. The assumption made in selecting q_i the lag flows is that flow-rate is constant. Whenever this assumption is unrealistic, use velocities measurements to dynamically adjust these lags. For simplicity we assume that velocity is homogeneous over time in the remainder of the paper because velocity data were unavailable to the authors.

Step 3: Specify the checking start and finish date.

Step 4: Specify the training dataset.

Recommendation: If the processes are strongly seasonal then use last seasons data to initialise estimates for checking this year's measurement accuracy (considering setting this as the default training dataset).

Step 5: Estimate the time delay (q_i) for each variable ($x_{i\,t-q_i}$) that best predict the response. Measures of explanatory usefulness consider here are correlation measures. That is, we examine correlation measures as a function of q_i; selecting the q_i that maximises the correlation. The following measures of correlation are examined:

Rank correlation: For the training sample, denote the rank correlation between measures x_{it} for all $t = 1, 2, \ldots, n - q_i$ and y_t for all $t = 1 + q_i, \ldots, n$ by rr_{q_i}.

Pearson correlation: For the training sample, denote the Pearson correlation coefficient between measures x_{it} for all $t = 1, 2, \ldots, n - q_i$ and y_t for all $t = 1 + q_i, \ldots, n$ by r_{q_i}.

Pseudo correlation: We make the variables invariant of scale by standardising them as follows

$$z_{i,t} = (x_{i\,t} - \bar{x}_{i\,q_i}) \Big/ \sqrt{\sum_{t=1+q_i}^{n} (x_{i\,t-q_i} - \bar{x}_{i\,q_i})^2/(n-q_i)}$$

$$z_{y,t} = (y_t - \bar{y}_{q_i}) \Big/ \sqrt{\sum_{t=1+q_i}^{n} (y_t - \bar{y}_{q_i})^2/(n-q_i)}$$

Let s_{q_i} is robust measure of similarity defined by

$$s_{q_i} = 1 - \sum_{t=1+q_i}^{n} |z_{i,t-q_i} - z_{y,t}|/(n - q_i)$$

Select the delay (q_i) that either maximizes: rr_{q_i}, r_{q_i} or s_{q_i}. The pseudo correlation measure s_{q_i} proved more robust that the other correlation measures at detecting known, but hidden delays in simulated data. In addition, a simulation study demonstrated improved accuracy in determining q_i if both times series are smoothed prior to estimating q_i. If two of these delay estimates coincide, then use this delay in the model for Step 6, otherwise use the delay in Step 6 that is half way between the q_i value which maximizes s_{q_i} and maximizes r_{q_i}. When the training sample gets very large then calculating the rank correlation is slow.

Step 6: Fit the following regression models:

Flows:

$$y_t = \beta_0 + \beta_1\, x_{1\,t-q_1} + \cdots + \beta_k\, x_{kt-q_k} + \alpha_1 y_{t-1} + \alpha_2\, y_{t-2} + e_t$$

where:

- y_t is the flow being checked for consistency measured at time t;
- $x_{i\,t-q_i}$ is either an upstream flow or an upstream rainfall measurement measured at time $t - q_i$ to account for delays;
- β_i are the regression parameters associated with the influence the ith explanatory variable;
- α_i are the regression parameters associated with carryover influences of measurements made at the same site at earlier times.

Note the time series model has no moving average component only autoregressive components; this is to avoid the need for iterative estimation.

Rainfall: For consistency checking of rainfall measurements the, zero inflated model is fitted as follows:

$$logit(p_t) == \beta_{10}\, x_{1\,t-q_1} + \cdots + \beta_{k0}\, x_{kt-q_k} + \alpha_{10} y_{t-1} + \alpha_{20}\, y_{t-2}$$

where $p_t = \Pr(Y_t = 0)$, and the model for $Y_t > 0$ is given by

$$y_t = \beta_1\, x_{1\,t-q_1} + \cdots + \beta_k\, x_{kt-q_k} + \alpha_1 y_{t-1} + \alpha_2\, y_{t-2} + e_t$$

for $y_t > 0$. Note that the model for positive rainfall is unaltered from the linear model used earlier in this paper, that is, the random error term of the model is assumed to be normally distributed with mean zero. An alternative approach worth trying is fitting a zero adjusted inverse Gaussian distribution using the *gamlss* package (Rigby and Stasinopoulos 2001, 2005; Akantziliotou et al. 2002). The zero adjusted inverse Gaussian model has the advantage of always produces positive value predictions. For zero adjusted measures, the measurements that are zero in value are classified as extremely unusual, very unusual and unusual, when the estimated p_t value is less 0.0001, 0.001 and 0.005 respectively. However if $y_t > 0$ then its unusualness is evaluated in the normal way described before. The preferred model is to use a zero adjusted model.

Let the training sample have n observations finishing at time t. Then the design matrix of explanatory variables is given by

$$X_{n\,t} = \begin{pmatrix} 1 & x_{1\,t-n+1} & \cdots & x_{k\,t-n+1} & y_{t-n} & y_{t-n-1} \\ 1 & x_{1\,t-n+2} & \cdots & x_{k\,t-n+2} & y_{t-n+1} & y_{t-n} \\ \vdots & \vdots & \ddots & \vdots & \vdots & \vdots \\ 1 & x_{1\,t} & \cdots & x_{k\,t} & y_{t-1} & y_{t-2} \end{pmatrix} = \begin{pmatrix} x^t_{t-n+1} \\ x^t_{t-n+2} \\ \vdots \\ x^t_t \end{pmatrix};$$

$$y_{n\,t} = \left(y_{t-n+1}\; y_{t-n+2} \cdots y_t \right)^t.$$

Check whether any column of $X_{n\,t}$ is a vector of a constant, e.g. rainfall is always zero during the time period the training data are collected. If it is nearly always true, then exclude this column from the matrix and exclude this variable from the model checking for consistency. If any of the Eigen-values of $X'_{n\,t}X_{n\,t}$ are zero then the inverse does not exist and we need to refer to Section 1.2.4.3 for one option on how to deal with this problem. Let

$$\theta = \begin{pmatrix} \beta_0 \\ \beta_1 \\ \vdots \\ \beta_k \\ \alpha_1 \\ \alpha_2 \end{pmatrix}; \quad P_{nt} = (X'_{nt}X_{nt})^{-1}; \quad b_{nt} = X'_{nt}y_{nt}.$$

then the ordinary least squares estimate of the regression coefficients are

$$\widehat{\theta}_{nt} = P_{nt}b_{nt}$$

and

$$\widehat{\sigma}^2_{nt} = x'_{t+1}P_n t x_{t+1}(y_{nt} - X_{nt}\widehat{\theta}_{nt})'(y_{nt} - X_{nt}\widehat{\theta}_{nt})/(n-k-3).$$

Step 7: Check the $t+1$ observation for unusualness by first calculating the model residuals (i.e. estimated errors) as follows

$$\widehat{e}_{t+1|t} = y_{t+1} - x'_{t+1}\widehat{\theta}_{nt}.$$

Assign consistency codes using rules: $||\widehat{e}_{t+1|t}|| > 5\widehat{\sigma}_{nt}$ as **extremely unusual** through to $||\widehat{e}_{t+1|t}|| < \widehat{\sigma}_{nt}$ as **close to expected**.

Step 8: Repeat steps 6 and 7 for all rainfall and flow measurements in a catchment for time $t + 1$.

The Recursive Estimation Phase

The start up phase is designed for checking the first observation in test sample (i.e. observation $t + 1$). Note that the current formulation of the training data are that it is the period just prior to the test data, however we think we want to have greater flexibility than this. For example, we may want the option of specifying last year‘s identical seasonal data.

The recursive estimation phase now updates the parameter estimates to provide updated predictions for the test sample observations. These predictions are used as expected values to check for consistency. The next sequence of steps is those necessary to achieve this aim.

Step 9: Update estimates using observation $t + 1$. We use a forgetting factor very similar to that applied in Guo et al. (1993) and Campi (1994). The updating equations are as follows:

$$\begin{aligned} P_{n\,t+1|t} &= (\lambda X'_{nt} X_{nt} + x_{t+1} x'_{t+1})^{-1} \\ &= P_{nt}/\lambda - P_{nt} x_{t+1} x'_{t+1} P_{nt} / [\lambda(\lambda + h_{n\,t+1|t})] \\ h_{n\,t+1|t} &= x'_{t+1} P_{nt}\, x_{t+1}, \end{aligned}$$

with forgetting weights given by $0.95 < \lambda < 1$. Then

$$b_{n\,t+1|t} = \lambda b_{nt} + x_{n+1} y_{t+1}$$

and

$$\widehat{\theta}_{n\,t+1|t} = \widehat{\theta}_{nt} + P_{nt} x_{t+1}(y_{t+1} - x'_{t+1}\widehat{\theta}_{nt})/(\lambda + h_{n\,t+1|t}) = P_{n\,t+1|t} b_{n\,t+1|t}$$

$$\widehat{\sigma}^2_{n\,t+1} = (1-\lambda)(y_{t+1} - x'_{t+1}\widehat{\theta}_{n\,t+1|t})^2 + \lambda \widehat{\sigma}^2_{nt}$$

An alternative approach is to use the *gamlss* models (Rigby and Stasinopoulos 2001, 2005; Akantziliotou et al. 2002) to fit both the mean and variance over time using the appropriate location-scale distribution. However this model can not easily be updated using a recursive approach, but would need to be fitted using a moving window of data.

Step 10: Check the $t + 2$th observation for unusualness by calculating the forecast error

$$\widehat{e}_{t+2|t+1} = y_{t+2} - x'_{t+2}\widehat{\theta}_{n\,t+1|t}$$

and assign consistency codes as before with $\widehat{e}_{t+2|t+1}$ replacing $y - \mathrm{E}(y|x)$ and $\widehat{\sigma}_{n\,t+1}$ replacing σ.

Step 11: Repeat steps 9 and 10 until all the observations in the test set have been checked. At each cycle update the parameter estimates, and then check the next observed set of measurements for unusualness and code them accordingly.

Step 12: Repeat these steps for each rainfall site first followed by stream flow site measurements. The updating equations now become as follows:

$P_{n\,j+1|j} = P_{n\,j|j-1}/\lambda - P_{n\,j|j-1}x_{j+1}x'_{j+1}P_{n\,j|j-1}/[\lambda(\lambda + h_{n\,j+1|j})]$ where

$$h_{n\,j+1|j} = x'_{j+1}P_{n\,j}\,x_{j+1}$$

$$b_{n\,j+1|j} = \lambda b_{n\,j|j-1} + x_{j+1}y_{j+1}$$

and

$$\begin{aligned}\widehat{\theta}_{n\,j+1|j} &= \widehat{\theta}_{n\,j|j-1} + P_{nj|j-1}x_{j+1}(y_{j+1} - x'_{j+1}\widehat{\theta}_{n\,j|j-1})/(\lambda + h_{n\,j+1|j}) \\ &= P_{n\,j+1|j}b_{n\,j+1|j} \\ \widehat{\sigma}^2_{n\,j+1} &= (1-\lambda)(y_{j+1} - x'_{j+1}\widehat{\theta}_{n\,j+1|j})^2 + \lambda\widehat{\sigma}^2_{n\,j} \text{ for } j > t\end{aligned}$$

At each stage, record the 'expected value' $(x'_{j+1}\widehat{\theta}_{n\,j+1|j})$ for observations that are classified as either, unusual, very unusual or extremely unusual. In addition, in the sequential checking process, we use the expected value in the place of identified measurement errors when using the measure as an explanatory variable for measurements checked later in the sequence.

Alternatives to Ordinary Least Squares (ols) Regression When $X'_{nt}X_{nt}$ Is Singular or Near Singular

Two options were explored when the explanatory variables produce a near singular design matrix. This problem should seldom arise if users are careful about selecting their explanatory variables and training dataset. However, those users that are unaware of the potential dangers of collinearities are likely to fall into this trap, and alternatives that avoid the collinearity problem are needed. The first approach to avoiding this problem is now offered based on principal component regression: this is the more complicated option, but it is the preferred option in terms of ensuring multivariate spatio-temporal consistency. The second approach uses weighted averages, with weights inversely proportional to distance, to establish the expected values. This approach has value in terms of its simplicity but it only preserves spatial consistency. It failed to ensure multivariate consistency or temporal consistency. It proved sub-optimal in terms of efficiency and therefore is not reported in this paper.

Principal Component Regression (PCR)

Use principal component regression when the design matrix is nearly singular. Introducing the mathematics necessary for implementing principal component regression; let an eigenvector and corresponding eigenvalue of design matrix $X'_{nt}X_{nt}$ be v and κ. Note that

$$X'_{nt}X_{nt}\,v = \kappa^2 v \quad \text{such that} \quad v'v = 1.$$

Let eigenvectors $v_i, i = 1, 2, \ldots, k+3$ correspond to eigenvalues $\kappa_1^2 > \kappa_2^2 > .. \geq \kappa_{k+3}^2 \geq 0$ with $v_i'v_j = 0,\ v_j'v_j = 1$ and $X'_{nt}X_{nt}\,v_j = \kappa_j^2 v_j$ for $j = 1, \ldots, k+3$. Let diagonal matrix of eigenvalues be defined by

$$D_\kappa = \begin{pmatrix} \kappa_1 & \ldots & 0 \\ \vdots & \ddots & \vdots \\ 0 & \ldots & \kappa_{m-1} \end{pmatrix},$$

and the matrix of corresponding eigenvectors be given by $V_\kappa = \begin{pmatrix} v_1 & v_2 & \ldots & v_{m-1} \end{pmatrix}$. Defined $U_\kappa = X_{nt}V_\kappa D_\kappa^{-1}$, and then the PCR coefficient estimate is

$$\widehat{\theta}_{pc} = V_\kappa D_\kappa^{-1} U_\kappa' y_{nt}.$$

The PCR estimate $\widehat{\theta}_{pc}$ will be used in the place of $\widehat{\theta}_{nt}$ whenever near zero eigenvalues are encountered. Also replace P_{nt} by starting the recursive estimation process with

$$P_{nt} = V_\kappa D_\kappa^{-2} V_\kappa'.$$

Other approaches such as ridge regression were not investigated.

2.5 *Batch Consistency Checking*

Real-time checking can only condition on past identical measurements at the same site. However, batch checking allows the option of conditioning on all identical measurements made at the same site besides the one that is being checked. For example, adding explanatory variables y_{t+1} and y_{t+2} to the model in Step 6 of Sect. 2.4.1 would be a reasonable choice for batch checking of the measurement y_t.

3 Detection of Errors That Persist

There is a need for detecting measurement device problems early. Two device problems will be investigated. The first is where the measurements depart on the low side or high side of the expected values, which is typical when say the battery fails in the device. The other is when the measurement device becomes more uncertain by increasing its variance significantly from a target acceptable variance.

3.1 Early Detection of Persistent Biases

The CUSUM plan of Page (1954) is useful for accumulating enough memory of small persistent one-sided departures from expected for detecting them early. The CUSUM for high sided departures is given by ($S_t^U = 0 = S_t^L$)

$$S_\tau^U = \max(0, S_{\tau-1}^U + \widehat{e}_{\tau|\tau-1}/\widehat{\sigma}_{n\,\tau} - k)$$

and CUSUM for low sided departures is given by

$$S_\tau^L = \min(0, S_{\tau-1}^L + \widehat{e}_{\tau|\tau-1}/\widehat{\sigma}_{n\,\tau} + k)$$

for $\tau = t + 1, t + 2, \ldots$, where $k (> 0)$ is some suitable offset. Measurements are classified as persistently biased at time τ and before if either

$$S_\tau^L < -h_c \quad \text{or} \quad S_\tau^U > h_c$$

where h_c is a suitable threshold designed to deliver an acceptable false alarm rate. After each bias signal, the CUSUM statistic is restarted at zero for the calculation of the next CUSUM value. This allows users to assess whether the bias persists after a signal. Other work on CUSUM of recursive residuals (Ploberger and Kramer 1992) differs slightly from our approach outlined above. We use standardised interpolation errors whereas they use recursive residuals, the standardisation we use involves only past data for real-time consistency checking, whereas their approach used all the data, and unlike our CUSUM statistic, their CUSUM is not based on Page (1954) with an offset k.

An alternative to the CUSUM is to use the EWMA plan (Wortham and Ringer 1971). Here we advocate a two sided EWMA plan as follows ($ewma_t^U = 0 = ewma_t^L$):

$$ewma_\tau^U = \max(0, (1 - \lambda_3) ewma_{\tau-1}^U + \lambda_3 \widehat{e}_{\tau|\tau-1}/\widehat{\sigma}_{n\,\tau})$$

and low sided departures is given by

$$ewma_\tau^L = \min(0, (1 - \lambda_3) ewma_{\tau-1}^L + \lambda_3 \widehat{e}_{\tau|\tau-1}/\widehat{\sigma}_{n\,\tau})$$

for $\tau = t + 1, t + 2, \ldots$, where $0 < \lambda_3 < 1$. Measurements are classified as persistently biased if either

$$ewma_\tau^L < -h_e \quad \text{or} \quad ewma_\tau^U > h_e$$

where h_e is a suitable threshold designed to deliver an acceptable false alarm rate. After a signal, the next ewma value is calculated by resetting the preceding ewma value to zero.

3.2 Early Detection of Unacceptable Uncertainty in Measurements

Although the CUSUM can be applied to signal increases in measurement uncertainty, we advocate the EWMV statistic because it is easy to interpret. Assume that the expected variance is given by σ_T^2, and that the unacceptable high variance is given by $\sigma_B^2 (> \sigma_T^2)$. The EWMV (Harris and Ross 1991) statistic is given by $(ewmv_t = \sigma_T^2)$

$$ewmv_\tau = (1 - \lambda_4) ewmv_{\tau-1} + \lambda_4 \widehat{e}^2_{\tau|\tau-1}$$

Unacceptable uncertainty is flagged whenever $ewmv_\tau \geq \sigma_B^2$. The alternative is to flag any significant increased in variance (uncertainty) of the measurements. In this case, we recommend the following statistic

$$V_\tau = max(\sigma_T^2, (1 - \lambda_4) ewmv_{\tau-1} + \lambda_4 \widehat{e}^2_{\tau|\tau-1})$$

Flag a significant increase in uncertainty from the target whenever $V_\tau > \sigma_T^2 h_v$ where $h_v (> 1)$ is a suitable threshold designed to deliver an acceptable low false alarm rate.

4 Concluding Remarks

The paper discusses the development of simple Quality Assurance/Quality Control algorithms for automatically detecting and flagging unusual measurements. When data are missing, a first pass interpolated value is used to patch the missing data. However, it is expected that this patched value would be revised at a later stage using more sophisticated modelling techniques with cleaner data (preferably using multiple imputation).

Future date quality challenges include:

1. Optimal design for multivariate spatio-temporal consistency checking.
2. Scalable real-time systems and algorithms for increasing volumes and complexity of data.
3. Integration of the automatic consistency checking process into a data quality assurance plan.
4. Expand consistency checking to real-time checking for model biases, checking for temporal biases in the model, thus judging whether deterministic or empirical models provide information that are fit for purpose.
5. Periodic reviews to improve the measurement process: check whether the measurement process is optimal, i.e. the right measures, taken at the right geographical locations, at the appropriate temporal resolution and with the right level of accuracy.
6. Monitoring the number of errors and the percentage of missing data over time.

Acknowledgements The authors would like to acknowledge Hydro Tasmania for applied hydrological knowledge that helped facilitate the technology emphasis of this paper and supplying the data used for development and validation, Bureau of Meteorology for their funding of this work under their Modernisation and Extension of Hydrologic Monitoring Systems Program, and Dr Mark Westcott for his helpful comments on an earlier draft of the paper.

Appendix

Finding an ensemble estimate of a measurement derived by combining estimates of this measurement using Eqs. 2 and 4 is

$$\eta\left(\sum_{i=1}^{k} x_{it}/d_i\right) \Big/ \left(\sum_{i=1}^{k} 1/d_i\right) + (1-\eta)ewma_{t-1}.$$

The estimates Eqs. 2 and 4 are assumed to be independent unbiased estimates of the true measured value y_t with variances σ_t^2 and σ^2, respectively. The variance of this ensemble estimate is given by

$$\sigma_e^2 = \mathrm{Var}\left(\eta\left(\sum_{i=1}^{k} x_{it}/d_i\right) \Big/ \left(\sum_{i=1}^{k} 1/d_i\right)\right) + \mathrm{Var}((1-\eta)ewma_{t-1})$$

since the estimates are independent. Then

$$\begin{aligned}\sigma_{e,t}^2 &= \eta^2 \mathrm{Var}\left(\left(\sum_{i=1}^{k} x_{it}/d_i\right) / \left(\sum_{i=1}^{k} 1/d_i\right)\right) + (1-\eta)^2 \mathrm{Var}(ewma_{t-1}) \\ &= \eta^2\sigma^2 + (1-\eta)^2\sigma_t^2.\end{aligned}$$

The best linear unbiased (ensemble) estimate is the estimate which minimises the variance $\sigma_{e,t}^2$. This estimate is found by differentiating $\sigma_{e,t}^2$ with respect to η, equating this to zero, and solving for η. That is,

$$\partial\sigma_{e,t}^2/\partial\eta = 2\eta\sigma^2 - 2(1-\eta)\sigma_t^2 = 0,$$

then solving for η gives

$$\eta = \sigma_t^2/(\sigma^2 + \sigma_t^2).$$

Since σ^2 and σ_t^2 are unknown, we use plug-in estimates for these unknown parameters to estimate η.

References

Akantziliotou, K., Rigby, R. A., & Stasinopoulos, D. M. (2002). The R implementation of generalized additive models for location. In M. Stasinopoulos, & G. Touloumi (Eds.), *Scale and shape in statistical modelling in society: Proceedings of the 17th international workshop on statistical modelling* (pp. 75–83), Chania.

Blyth, C. B. (1972). On Simpson's Paradox and the Sure-Thing principle. *Journal of the American Statistical Association, 67*, 364–366.

Borowski, E., & Lenz, H.-J. (2008). Design of a workflow system to improve data quality using oracle warehouse builder. *Journal of Applied Quantitative Methods, 3*, 198–206.

Campi, M. C. (1994). Exponentially weighted least squares identification of time-varying systems with white disturbances. *IEEE Transactions on Signal Processing, 42*, 2906–2914.

Cressie, N. A. (1993). *Statistics for spatial data*. New York: Wliey.

Finkenstadt, B., Held, L., & Isham, V. (2007). *Statistical methods for spatio-temporal systems*. Boca Raton: Chapman/CRC.

Guo, L., Ljung, L., & Priouret, P. (1993). Performance analysis of forgetting factor RLS algorithm. *International Journal of Adaptive Control and Signal Processing, 7*, 525–537.

Harris, T. J. & Ross, W. H. (1991). Statistical process control procedures for correlated observations. *Canadian Journal of Chemical Engineering, 69*, 139–148.

Page, E. S. (1954). Continuous inspection scheme. *Biometrika, 41*, 100–115.

Polberger, W., & Kramer, W. (1992). The CUSUM test for OLS residuals. *Econometrica, 60*(2), 271–285.

Rigby, R. A., & Stasinopoulos, D. M. (2001). The GAMLSS project: A flexible approach to statistical modelling. In B. Klein, & L. Korsholm (Eds.), *New trends in statistical modelling: Proceedings of the 16th international workshop on statistical modelling* (pp. 249–256), Odense.

Rigby, R. A., & Stasinopoulos D. M. (2005). Generalized additive models for location, scale and shape, (with discussion). *Applied Statistics, 54*, 507–554.

Ruppert, D., Wand, M. P., & Carroll, R. J. (2003). *Cambridge series in statistical and probabilistic mathematics: Vol. 12. Semiparametric regression*. Cambridge, UK: Cambridge University Press.

Silvennoinen, A., & Terasvirta, T. (2009). In T. G. Andersen, R/A. Davis, J.-P. KreiB, & T. Mikosch (Eds.), *Handbook of financial time series. Multivariate GARCH models*. Berlin/Heidelberg: Springer.

Whitten, E. H. T., & Koelling, M. E. V. (1973). Trend surfaces for geological mapped variables. *Mathematical Geology, 5*, 111–126.

Wortham, A., & Ringer, L. (1971). Control via exponential smoothing. *The Transportation and Logistic Review, 7*, 33–39.

Zheng, P., Durr, P. A., & Diggle, P. J. (2004). Edge-correction for spatial kernel smoothing methods; when is it necessary? *Proceedings of the GisVet conference 2004*. Ontario, Canada: University of Guelph.

Uncertainty and Quality Control

Elart von Collani

Abstract Uncertainty about future developments constitutes the most important and most difficult challenge for mankind. Despite this fact, uncertainty is not a part of general science. General science assumes that the future development follows cause-effect relations which can be described by mathematical functions, where the argument represents the cause and the image represents the effect. Scientific theories have exactly this form and it is widely believed that these functions represent "truth". Of course, this is nonsense as all the scientific theories are with certainty wrong and cannot describe the real evolution correctly. The inappropriate handling of uncertainty in science has produced a strange variety of "uncertainty theories" that causes confusion and helplessness. A friend of mine has expressed his confusion by the following words:

> *'Crisp sets', 'fuzzy sets', 'rough sets', 'grey sets', 'fuzzy rough sets', 'rough fuzzy sets', 'fuzzy grey sets', 'grey fuzzy sets', 'rough grey sets', 'grey rough sets', and now 'affinity sets'. My goodness! Is there anybody around who can enlighten me, i.e., help me to see a clear pattern in this set of sets, allegedly providing powerful tools to model various kinds of uncertainty?*

This paper examines the role and the handling of uncertainty in quality control. How is uncertainty quantified in quality control for making decisions aiming at maintaining or improving quality of processes and products.

Keywords Randomness • Ignorance • Knowledge • Fuzzyness • Quantification • Probability • Credibility • Probability space • Uncertainty space • Bernoulli space

---

E. von Collani (✉)
University Würzburg, Sanderring 2, D-97070 Würzburg, Germany
e-mail: collani@mathematik.uni-wuerzburg.de

H.-J. Lenz et al. (eds.), *Frontiers in Statistical Quality Control 10*,
Frontiers in Statistical Quality Control 10, DOI 10.1007/978-3-7908-2846-7_27,

1 What Is Uncertainty?

Before uncertainty can be analyzed it must be clearly explained and defined. This is necessary since in everyday speech, the word "uncertainty" may have different meanings according to the object of uncertainty. Therefore, it makes sense to start with a list of objects about which there might be uncertainty.

- **Facts**
  The word fact derives from the Latin Factum, and was first used in English with the meaning: "something that has really occurred or is the case". Fact is sometimes used synonymously with "truth" or "reality"[1] Thus, a fact is the true outcome of a completed process. It exists and represents therefore "truth". If a fact is unknown, then the ignorance makes one feel uncertain.

  As to quality control, almost all activities aim at reducing the ignorance about quality-related facts. The facts are given as values of certain quality characteristics and they are needed to make decisions concerning products and processes. For example:

  - Ignorance about the fraction nonconformance in a given lot may be reduced by acceptance sampling.
  - Ignorance about the nonconformance probability of a given production process may be reduced by means of a p-chart.
  - Ignorance about the value of the first moment (expected value) of the lifetime of a product may be reduced again by sampling.

  For being linguistically more precise uncertainty about facts shall be called "ignorance" in the remainder.
- **Statements**
  Any statement is made by using a language and therefore may be ambiguous and therefore uncertain. There are two types of languages, natural languages and artificial (formal) languages such as mathematics, and there might be uncertainty about the meaning of statements in both languages.

  - Uncertainty about the meaning of a statement made in a natural language is generally due to the vagueness and ambiguity that are characteristic for all natural languages.
  - Uncertainty about the meaning of a statement made by using mathematics is generally due to ambiguity of the involved variables and functions. Quantum mechanics may serve as a prominent example. The interpretation is uncertain and according to Richard Feynman,[2] there is no one who understands quantum mechanics. One example from quality control are statements about

[1] See http://en.wikipedia.org/wiki/Fact.

[2] See: http://www.spaceandmotion.com/quantum-mechanics-richard-feynman-quotes.htm.

probabilities since the concept probability has many different and inconsistent meanings.

Again, for being linguistically more precise uncertainty about the meaning of statements will be called "ambiguity" in the remainder.

- **Observations**
  Observations serve to get knowledge about something or to determine some unknown facts. However, whether by means of human senses or by technical devices, observations are always made with a finite resolution implying that the underlying fact which is observed cannot be determined exactly, i.e. leaving a certain degree of ignorance. Uncertainty about an observation is therefore subsumed under ignorance.
- **Future**
  The future outcome of a development does not exist so far and is therefore not a fact since a fact must already exist. The future is indeterminate and from experience it is well-known that any development will end in one of many possible outcomes. Repeating a process will yield different outcomes and some of them occur more frequently than others. Uncertainty about future is different from ignorance and ambiguity since man is the source of ignorance and ambiguity while indeterminacy of the future is independent of mankind. To make this difference clear, only uncertainty about the future development shall be called "uncertainty" here.

  In quality control the activities aim at determining facts. The gained knowledge about facts is utilized to reduce uncertainty about the future development to allow better decisions.

Ignorance, ambiguity and uncertainty may lead to problems in any decision-making process not only in quality control. Therefore, reducing ignorance, ambiguity and uncertainty are the central tasks of quality control.

2 Handling Ambiguity

Ambiguity refers to statements. There are at least two ways to handle ambiguity of natural languages, a more traditional one and a seemingly more advanced one. Both shall be briefly outlined, starting with the latter approach.

2.1 Describing Ambiguity by Fuzzy Sets

In 1965 Lotfi Zadeh introduced the theory of fuzzy sets to quantify the "fuzzyness" of natural languages. Consider the statement "A is nearly six feet tall". The ambiguity is contained in the word "nearly" and the problem is how to describe the ambiguity about the height of A quantitatively. To this end it is assumed that the set of "tall" people and the set of "not-tall" people are overlapping. The members of both (fuzzy) sets satisfy "imprecise" properties and this is expressed by certain "degrees of membership" in a fuzzy set. The degree of membership is specified by a corresponding "membership function", where the degrees of membership 0 and 1 mean no membership and full membership, respectively.

Clearly, neither fuzzy sets nor membership functions are unique, but depend on the subjective understanding of the underlying linguistic imprecision. In other words, linguistic ambiguity is a relative property. Thus, the selection of fuzzy sets and membership functions are subject to individual preferences. Furthermore, neither a specified fuzzy set nor the corresponding membership function reduces the linguistic ambiguity.

2.2 Avoiding Ambiguity by Quantification

The traditional way to avoid ambiguity of a statement is to quantify all involved characteristics and thereby replace the natural language by the language of mathematics. Mathematics is build on numbers which have a unique meaning. Moreover, mathematics is developed according to logic, and mathematical statements have therefore not only unique meanings but are also consistent.

The interpretation difficulties which may occur when using mathematics as language are caused by an inappropriate quantification, i.e., translation into mathematics. Unfortunately, science has so far not understood the significance of quantification, and as a consequence scientific results are in general ambiguous.

To sum up, describing ambiguity by fuzzy sets and membership functions does not reduce ambiguity, but may lead to further difficulties and confusion. In contrast, an appropriate quantification avoids completely any linguistic ambiguity.

3 Handling Ignorance

Ambiguity is a characteristic property of languages and can be overcome by appropriately using mathematics which means that a given situation is described by variables, functions, numbers and sets of numbers.

Ignorance on the other hand is a characteristic property of human beings. It refers to facts and means that the fact is not known. Any fact relates to a characteristic and if the corresponding characteristic has been quantified, then ignorance means that the actual value of the characteristic is unknown. But ignorance does not mean that there is nothing known about the considered characteristic and its actual value since it is always possible to specify values which cannot be the actual one. The more values can be excluded as being not the actual one, the smaller is the level or degree of ignorance.

Thus, the degree of ignorance can be described by the set of values which cannot be excluded. It immediately follows that ignorance can be reduced by learning or measurement processes that are used to exclude further values. The larger the set of values that cannot be excluded the larger is the degree of ignorance. No ignorance, i.e. complete knowledge, is achieved, if the set of possible values is given by a singleton.

Almost all activities in quality control represent efforts to reduce the degree of ignorance with respect to various quality characteristics of the considered systems, processes or products.

4 Handling Uncertainty

As specified above, uncertainty refers to future developments and represents the main problem not only for quality control. The future is related on the past and this dependence is the key for understanding and handling uncertainty. Uncertainty about the future has two distinct sources:

- Each week the drawing of lotto numbers is shown in the German TV. The initial conditions are always exactly the same, nevertheless, the outcomes are practically always different. The same can be observed for manufactured products. The values of the characteristics differ even for exactly the same production conditions. The observed variability of future developments is a characteristic feature of the evolution of universe, it is observed in the micro as well as in the macro world. This universal feature is generally called randomness and for handling uncertainty it is therefore necessary to analyze randomness.
- Any future development stochastically depends on the past, i.e., on the conditions at the start of the development. If these starting conditions are known, then the dependence can be exploited to describe the existing uncertainty. For instance, the starting conditions of drawing the lotto numbers are completely known and thus it is possible to describe the existing uncertainty by stochastic laws concerning the future variability. If there is ignorance about the starting conditions, then there is also ignorance about the laws of the future variability and the uncertainty about the future development increases.

Accordingly, human uncertainty is caused by randomness which is a characteristic feature of universe and by ignorance which is a characteristics feature of mankind.

There are many different ways to quantify uncertainty in order to analyze it. The traditional quantified model was introduced by Kolmogorov (1956) and is known as "probability space". Besides the probability space there are numerous other "uncertainty spaces" which are applied to analyze uncertainty. Some of these models are described and evaluated below. If uncertainty shall not only be analyzed but reduced then according to the two sources of uncertainty there are two ways: The first way consists of reducing the ignorance about relevant facts, and the second way is to change randomness by changing the relevant facts. Both ways are exercised in quality control. All monitoring activities aim at reducing ignorance, all repair and maintenance actions aim at changing the relevant facts.

5 Mathematical Models

Traditional science is based on the assumptions of a deterministic evolution of universe and of being able to determine "truths". Therefore there are no appropriate models available in science for describing and analyzing uncertainty. However, in mathematics there are several models developed which are applied to quantify and subsequently analyze ignorance, ambiguity and uncertainty.

5.1 Probability Space

The probability space was introduced by Kolmogorov as a purely mathematical construct which is not related to anything real. This fact was very clearly expressed by Kolmogorov himself who states in Kolmogorov (1956): [...] *the concept of a field of probabilities is defined as a system of sets which satisfies certain conditions. What the elements of this set represent is of no importance in the purely mathematical development of the theory of probability.*

The probability space consists of three components $(\Omega, \mathfrak{A}(\Omega), P)$ with are defined as follows:

- Ω is a non-empty (abstract[3]) set,
- $\mathfrak{A}(\Omega)$ a σ-algebra over the set Ω, and
- P is a set function defined on $\mathfrak{A}(\Omega)$ with the following properties:
  - Nonnegativity:
$$P(A) \geq 0 \quad \text{for } A \in \mathfrak{A}(\Omega) \tag{1}$$
  - Normality:
$$P(\Omega) = 1 \tag{2}$$
  - σ-Additivity:
$$P\left(\bigcup_{n=1}^{\infty} A_n\right) = \sum_{n=1}^{\infty} P(A_n) \quad \text{for } A_i \in \mathfrak{A}(\Omega) \text{ and } A_i \cap A_j = \emptyset \text{ if } i \neq j \tag{3}$$

As to possible applications of the probability space, Kolmogorov states in Collani (2004): *Every axiomatic (abstract) theory admits, as is well known, of an unlimited number of concrete interpretations* [...]. Therefore, for using the probability space, one must first give a concrete interpretation of the involved mathematical quantities.

[3]"Abstract" means that the elements have no real interpretation.

5.1.1 Probability Space as a Model of Randomness

Kolmogorov's probability space is the classical model for randomness with respect to a process, i.e. the change from the past to the future. It is taught all over the world and there is hardly any textbook dealing with randomness which does not contain Kolmogorov's probability space. The generally made interpretations are the following:

- Ω is the set of possible outcomes of the considered process.
- $\mathfrak{A}(\Omega)$ is the set of future events with respect to the considered process.
- P is called the probability measure. It assigns to each event (element of $\mathfrak{A}(\Omega)$) the corresponding probability.

The adequacy of the probability space for modelling randomness or uncertainty shall be checked by means of the simple process of throwing a coin. Each throw may be described by countless many features, for instance throwing range, throwing height, angle of impact, impact force, etc. In fact it is impossible to completely describe all possible features of an outcome of the process throwing a coin. It follows that the set of outcomes as well as the set of future events cannot be specified implying that the same holds for the probability measure P.

Besides the fact that it is not possible to specify a probability space even for simplest real processes, there are other deficiencies which make the probability space appear as being not suitable for serving as a mathematical model for randomness. As mentioned above applying a mathematical construct assumes that a clear interpretation of all the involved mathematical quantities has been given. Looking at the above stated interpretations reveals that there is no clear interpretation of the most important mathematical quantity "probability measure". Actually, in relevant textbooks there is either no interpretation or several inconsistent interpretations are given.

One of the interpretations is the so-called frequency interpretation which is looked upon as especially appropriate for explaining randomness. However, the frequency interpretation assumes a sequence of repeatable experiments and explains the 'probability of an event' as its relative frequency 'in the long run' (infinite frequency interpretation). Thus, according to the frequency interpretation, probability is explained by an infinite sequence of experiments, which obviously does not exist. In other words, the frequency interpretation does not explain the concept of probability, but leads itself to irresolvable interpretation difficulties.

If the probability space shall be used not only as a model for randomness but as a quantitative model for uncertainty, then the following additional weaknesses should be noted:

- The starting conditions do not explicitly appear in the model.
- The always existing ignorance about the starting conditions which mainly determine uncertainty is not included into the model.
- The probability space constitutes not a completely quantified model, since the the set Ω is not defined as a set of numbers but as a set of outcomes.

5.1.2 Probability Space as a Model of Ignorance

Besides being used as a model for randomness, the probability space is also used for a model of ignorance with respect to facts. In this case Ω and $\mathfrak{A}(\Omega)$ refer to the outcomes and events of a already completed process, and the probability of an event is explained as the degree of belief in the event.

Belief is one of the most dangerous forms of ignorance and many wrong decisions are made based on belief. Therefore, belief should be kept out of science! If two values cannot be excluded from being the actual value of a variable of interest, then both values have to be considered likewise. It is too dangerous to prefer the one over the other according to a "degree of belief". Such a degree is purely subjective and has no relation to the actual reality since nothing can be for example "half true".

It follows that the so-called subjective interpretation of probability which is the base of the so-called Bayesian statistics should not be used for making any decisions.

5.2 Credibility Space

The credibility space is another purely mathematical construct. Similar as the probability space, it is recommended to be applied for modelling uncertainty. Actually, it is the basic model for ambiguity as treated by fuzzy sets and membership functions. As Liu (2007) notes there are three aspects of vagueness or ambiguity considered in fuzzy theory namely possibility, necessity and credibility. In 1978 Zadeh proposed the possibility space as a basic model which was extended in 2002 to a credibility space by Liu and Liu (2002).

Liu and Liu's credibility space $(\Theta, \mathfrak{Q}(\Theta), Cr)$ follows Kolmogorov's approach. The three components are defined as follows:

- Θ is a non-empty (abstract) set,
- $\mathfrak{Q}(\Theta)$ is the power set over the set Θ, and
- Cr is a set function defined on $\mathfrak{Q}(\Theta)$ with the following properties:
  - Normality:
  $$Cr(\Theta) = 1 \tag{4}$$
  - Monotonicity:
  $$Cr(A) \leq Cr(B) \quad \text{for } A \subset B \text{ and } A, B \in \mathfrak{Q}(\Theta) \tag{5}$$
  - Self-Duality:
  $$Cr(A) + Cr(A^c) = 1 \quad \text{for } A \in \mathfrak{Q}(\Theta) \text{ and } A^c = \Theta \setminus A \tag{6}$$

– Sum Formula[4]:

$$Cr\left(\cup_{i=1}^{n} A_i\right) = \sup_{1\le i\le n} Cr(A_i) \quad \text{for } A_i \in \mathfrak{Q}(\Theta) \text{ and } Cr(A_i) \le \tfrac{1}{2}, i=1,\ldots,n \tag{7}$$

Often it is claimed that the probability space refers to future randomly occurring events, while the credibility space deals with the ambiguity or vagueness of the representation of an event. However, since the probability space is used not only in the case of future events, but also in the case of past events this differentiation does not really hold. The main reason for this ambiguity is the unclear and inconsistent definition of the concepts probability and credibility which are introduced based on (mathematical) axioms instead of requirements derived from reality. Actually, the differences between the probability space and the credibility space refer to the assumed set functions. Unfortunately, in neither case the assumed properties are backed by reality.

5.3 *Uncertainty Space*

In 2007 Liu introduced another mathematical construct $(\Xi, \mathfrak{A}(\Xi), \lambda)$ which he called 'uncertainty space'. The three components are defined as follows:

- Ξ is a non-empty (abstract) set,
- $\mathfrak{A}(\Xi)$ a σ-algebra over the set Ξ, and
- λ is a set function defined on $\mathfrak{A}(\Xi)$ with the following properties:
  - Normality:

$$P(\Omega) = 1 \tag{8}$$

  - Monotonicity:

$$\lambda(A) \le \lambda(B) \quad \text{for } A \subset B \text{ and } A, B \in \mathfrak{A}(\Xi) \tag{9}$$

  - Self-Duality:

$$\lambda(A) + \lambda(A^c) = 1 \quad \text{for } A \in \mathfrak{A}(\Xi) \text{ and } A^c = \Xi \setminus A \tag{10}$$

  - σ-Subadditivity:

$$\lambda\left(\cup_{n=1}^{\infty} A_n\right) \le \sum_{n=1}^{\infty} P(A_n) \quad \text{for } A_i \in \mathfrak{A}(\Xi) \text{ and } A_i \cap A_j = \emptyset \text{ if } i \ne j \tag{11}$$

[4] According to Hu et al. (2009)

The main weakness of all these theories and models of uncertainty/ambiguity /vagueness is the fact that none of them distinguishes clearly between future and past and that none of them starts from reality and develops the mathematical means appropriately. Each of them starts with formulating some mathematical axioms which leads to the necessity to adjust reality to the axioms.

This becomes clear in Liu's most recent book Liu (2010) on uncertainty theory. The question "What is Uncertainty?" is not posed an the beginning of the book, but only at the end in an appendix! His answer is at the same time surprisingly frank and extremely unsatisfactory. He states (Liu 2010, p. 251):

> In fact, I really have no idea how to use natural language to define the concept of uncertainty clearly, and I think all existing definitions by natural languages are specious just like a riddle. A very personal and ultra viewpoint is that the words like randomness, fuzziness, roughness, vagueness, greyness, and uncertainty are nothing but ambiguity of human language!

In other words uncertainty theories are developed and applied although there is at best a vague feeling about what uncertainty might be. Instead of a clear explanation, mathematics is used to solve the problem. Liu continues:

> However, fortunately, some "mathematical scales" have been invented to measure the truth degree of an event, for example, probability measure, capacity, fuzzy measure, possibility measure, credibility measure as well as uncertain measure. All of those measures may be defined clearly and precisely by axiomatic methods.

It is correct that all these measures are clearly defined as part of mathematics, however, this does not mean that there is a unique interpretation with respect to reality. In order to get an adequate (mathematical) description of something real, one has to start with reality and not with mathematics and axioms, i.e., postulates which need not be proved or demonstrated.

Although Liu realizes that applying the various uncertainty theories might lead to wrong or meaningless results, he does not give a solution. In order to illustrate the difficulties, we will take Liu's examples (Liu 2010, pp. 252–254) and use them later on to illustrate a solution.

Example "Fuzziness"

It is assumed that the distance between Beijing and Tianjin is "about 100 km". If "about 100 km" is regarded as a fuzzy concept, then we may assign it a membership function, say

$$\mu(x) = \begin{cases} \dfrac{x-80}{20} & \textit{if } 80 \le x \le 100 \\[2ex] \dfrac{120-x}{20} & \textit{if } 100 \le x \le 120 \end{cases} \qquad (12)$$

Based on this membership function, possibility theory (or credibility theory) will conclude the following proposition:

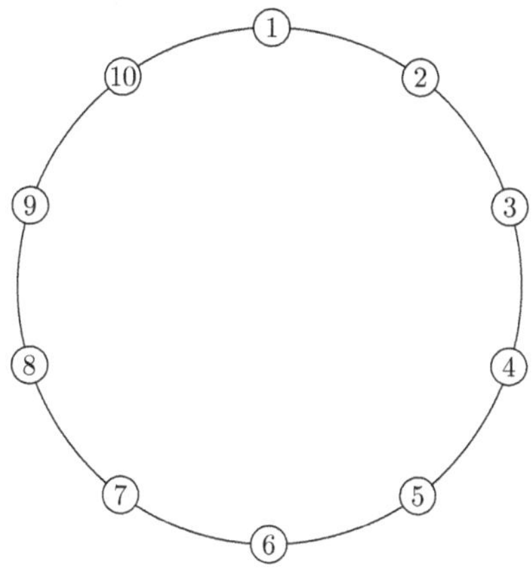

Fig. 1 Ten cities located on a ring in which the distance between any adjacent cities is 96 km, and the perimeter of the ring is 960 km

The distance between Beijing and Tianjin is "exactly 100 km" with belief degree 1 in possibility measure (or 0.5 in credibility measure).

However, it is doubtless that the belief degree of "exactly 100 km" is almost zero. Nobody is so naive to expect that "exactly 100 km" is the true distance between Beijing and Tianjin.

Example "Randomness"

Consider 10 cities located on a ring shown in Fig. 1 in which the distance between any adjacent cities is exactly 96 km, and the perimeter of the ring is just 960 km. Assume that the distance between adjacent cities is unknown and we have to acquire information from domain experts. A questionnaire survey shows that the distance between any adjacent cities is "100 ± 5 km".

Now let us treat "100 ± 5 km" as a random variable and assume it is uniformly distributed on [95, 105]. Then the perimeter of the ring is also a random variable whose 99% confidence interval is [977, 1023] and, unfortunately, the true perimeter 960 km is out of the 99% confidence interval. This is clearly an unreasonable conclusion. In other words, we cannot treat "100 ± 5 km" as a random variable.

Example "Ignorance"

If we treat it as an uncertain variable with uncertainty distribution

$$\Phi(x) = \begin{cases} 0 & \text{if } x < 95 \\ 0.5 & \text{if } 95 \leq x \leq 105 \\ 1 & \text{if } x > 105 \end{cases} \tag{13}$$

then the perimeter of the ring is also an uncertain variable whose uncertainty distribution is

$$\Psi(x) = \begin{cases} 0 & \text{if } x < 950 \\ 0.5 & \text{if } 950 \leq x \leq 1050 \\ 1 & \text{if } x > 1050 \end{cases} \tag{14}$$

In other words, the perimeter is "1,000 ± 50 km" and the true perimeter 960 km is within this range. This is a reasonable result.

These examples illustrate that without a clear understanding of ambiguity, randomness, ignorance and uncertainty and a blind reliance on mathematics wrong results and therefore wrong decisions are almost inevitable.

5.4 Bernoulli Space

Only recently an alternative to the probability/credibility/uncertainty spaces has been developed in Collani (2004). The alternative is named Bernoulli Space to commemorate Jakob Bernoulli who introduced the concept probability about 300 years ago.

The Bernoulli Space denoted $\mathbb{B}_{X.D}$ is the basic model of uncertainty as proposed in Bernoulli Stochastics. It is not an axiomatic mathematical construct such as, for example, the probability space, but starts from reality, and mathematics is only used to describe the identified real constituents.

As specified, uncertainty refers to the change from some determinate starting conditions to an indeterminate terminal condition (future outcome). The indeterminate future aspect of interest is quantified by a variable, say X, and the relevant starting conditions are quantified by another variable, say D. The variable X will randomly adopt one of several values and is called random variable, while D has a fixed but generally unknown value and is therefore called deterministic variable. For describing the change from the past given by D to the future represented by X, the ignorance with respect to D, the range of future variability with respect of X and the random structure of the future developments must be described quantitatively. The random structure means that if the process is repeated some developments will occur frequently and others infrequently. This phenomenon is due to randomness.

300 years ago, the Swiss theologian and mathematician Jakob Bernoulli explained randomness of a future event as the degree of certainty of its occurrence and named this degree *probability*. Note that the probability of a future event with respect to a given random variable X depends exclusively on the starting conditions. In particular, it is independent of any expert opinion and of the outcomes of any number of process repetitions.

The starting conditions are given by the actual value d of the deterministic variable D and the possible events refer to a random variable $X|\{d\}$. The probability of an event E is denoted $P_{X|\{d\}}(E)$. The probability represents a fraction of certainty, any certain result has therefore the probability 1 and any impossible event the probability 0:

$$P_{X|\{d\}}(E) = 1 \text{ if } E \text{ is a certain event} \tag{15}$$

$$P_{X|\{d\}}(E) = 0 \text{ if } E \text{ is a impossible event} \tag{16}$$

For an event consisting of two mutually exclusive events E_1 and E_2, the probability is obtained by adding the corresponding (probability mass) fractions:

$$P_{X|\{d\}}(E_1 \cap E_2) = P_{X|\{d\}}(E_1) + P_{X|\{d\}}(E_2) \tag{17}$$

Thus, the identified real properties are only expressed mathematically, without any axioms or principles. The above shall illustrate the appropriate proceeding in modelling. Modelling must start with the real situation and continue by expressing the recognized relations with mathematical terms. After the basic concept probability has been explained and defined, the stochastic model of uncertainty can be derived similarly. As noted above it must include ignorance (about the past) and randomness (of the future).

- Ignorance:
  Ignorance refers to the actual value of D and is adequately described by the set of values which cannot be excluded for D. This generally finite set denoted $\mathfrak{D}$ is called *ignorance space*. Each subset $\mathfrak{D}_0 \subset \mathfrak{D}$ specifies a certain degree of ignorance.
- Randomness:
  Randomness refers to the random variable X and is described by the variability and the random structure of X.
  - Variability: The range of variability of X depends on the starting conditions or, more precisely said, on the degree of ignorance about D. This dependence is described by a function $\mathfrak{X}$ named *variability function*, which is defined on an appropriate system of subsets $\mathfrak{T}(\mathfrak{D})$ over $\mathfrak{D}$ with images being the corresponding ranges of variability of the random variable X:

    $$\mathfrak{X} : \mathfrak{T}(\mathfrak{D}) \to \mathbb{R} \tag{18}$$

    with

    $$\mathfrak{X}(\mathfrak{D}_0) = \bigcup_{d \in \mathfrak{D}_0} \mathfrak{X}(\{d\}) \quad \text{for } \mathfrak{D}_0 \in \mathfrak{T}(\mathfrak{D}) \tag{19}$$

  - Random structure: When the considered process is repeated, then some of the outcomes occur often while others occur seldom. This phenomenon is described by a probability distribution which depends on the starting condition and defines a random structure on the corresponding range of variability of X. The random structure is given by the function $\mathfrak{P}$ which is named *random structure function* and which has the same domain as $\mathfrak{X}$ with images being probability measures:

    $$\mathfrak{P} : \mathfrak{T}(\mathfrak{D}) \to \mathbb{P} \tag{20}$$

    where $\mathbb{P}$ is the set of probability measures. $\mathfrak{P}$ has the following property:

    $$\mathfrak{P}(\mathfrak{D}_0) = P_{X|\mathfrak{D}_0} \tag{21}$$

with

$$P_{X|\mathfrak{D}_0}(A) = \frac{1}{|\mathfrak{D}_0|} \sum_{d \in \mathfrak{D}_0} P_{X|\{d\}}(A) \quad \text{for } A \in \mathfrak{X}(\mathfrak{D}_0) \tag{22}$$

where $|\mathfrak{D}_0|$ denotes the size (cardinality) of the set $\mathfrak{D}_0$, $X|\{d\}$ is the random variable X under the starting conditions d and $P_{X|\{d\}}$ the probability measure of $X|\{d\}$ from which follows that $P_{X|\mathfrak{D}_0}(A)$ is the average probability under the degree of ignorance given by $\mathfrak{D}_0$.

The Bernoulli Space is a mathematical model for the change from the relevant past represented by the deterministic variable D to the future of interest represented by the random variable X. Ignorance about the past is quantified by the ignorance space $\mathfrak{D}$; randomness of future is quantified by the variability function $\mathfrak{X}$ and the random structure function $\mathfrak{P}$:

$$\mathbb{B}_{X,D} = (\mathfrak{D}, \mathfrak{X}, \mathfrak{P}) \tag{23}$$

Each of the components of the Bernoulli Space is necessary, and together they are sufficient for a complete model of uncertainty.

For understanding the significance of the Bernoulli Space when compared with a "statistical model", one should consider the difference between an estimate, which is one point, and a confidence interval, which is given by a set of points. The only founded statement about an estimate is that it differs from the true value, while the confidence interval generally covers the true value. A statistical model consists of one probability distribution and the only founded statement that can be made is that it differs from the true probability distribution. In contrast the Bernoulli Space consists of a set of probability distributions, which generally cover the true one. If a Bernoulli Space is used instead of a statistical model, the risk of a wrong model is known just as in the case when using a confidence interval instead of an estimate.

6 Conclusions

Modern science as it has been developed during the last 500 years aims at discovering truth. Truth does not admit uncertainty as it is represented by a fact. Consequently, in science there is no need to develop models and methods for dealing appropriately with the real aspect of uncertainty.

Instead, uncertainty became a playing field of mathematicians. As listed by Liu[5] it started with probability theory by Kolmogorov in 1933, followed by Choquet's capacity theory in 1954 and Sugeno's fuzzy theory in 1974. In 1978, Zadeh introduced the possibility theory, which led to the credibility theory as proposed by Liu and Liu in 2002. Finally in 2007 Liu introduced his uncertainty theory. All of

[5] http://www.orsc.edu.cn/~liu/Lecture/Evolution/Evolution.pdf.

these theories do not differentiate clearly between future and past and more over do not clearly explain the used measures of uncertainty as real world aspects. Instead, Liu characterizes probability theory and uncertainty theory as follows:

- *Probability theory is a branch of mathematics for studying objective randomness.*
- *Uncertainty theory is a branch of mathematics for studying human uncertainty.*

Actually, these characterizations are contradictions. Mathematics and hence also its branches do not deal with anything real. Mathematics must be developed independent of all the real world restrictions only being subject to the constraints of logic. This requirement for mathematics was very clearly expressed by Kolmogorov (1956):

> The theory of probability, as a mathematical discipline, can and should be developed from axioms in exactly the same way as Geometry and Algebra. This means that after we have defined the elements to be studied and their basic relations, and have stated the axioms by which these relations are governed, all further exposition must be based exclusively on these axioms, independent of the usual concrete meaning of these elements and their relations.

The misconception to think that a mathematical discipline deals with something real has led to the unrealistic models and methods for uncertainty. But the various uncertainty theories represent also another probably even more questionable development, namely the attempt to establish belief as a scientific category. This development started by considering subjective probabilities and in Liu (2010) the following words can be found which factor out reality in favor of belief:

> An event has no uncertainty if its uncertain measure is 1 because we may believe that the event occurs. An event has no uncertainty too if its uncertain measure is 0 because we may believe that the event does not occur.

Quality control is mainly based on statistics, unfortunately employing both the frequency interpretation and the Bayesian interpretation. Furthermore, Zadeh's fuzzy approach is used although the enthusiasm for it during the 1980s and 1990s was followed by a certain disillusionment. However, as shown above the used models are not appropriate to describe uncertainty correctly and should therefore be given up.

References

Deng, J. L. (1985). *Grey systems (social economical)*. Beijing (in Chinese): The Publishing House of Defense Industry.

Guo, R., & Guo D. (2008). *Random fuzzy variable foundation for Grey differential equation modeling*. Focus. New York: Springer.

Hu, B. Q., Ip, W. C., & Wong, H. Fuzzy integral and credibility measure. In B. Cao, Z.-F. Li, & C.-Y. Zhang, (Eds.), *Fuzzy information and engineering* (Vol. 2, p. 254). Berlin: Springer.

Kolmogorov, A. N. (1956). *Foundations of the theory of probability* (N. Morrison, Trans., 2nd ed.). New York: Chelsea.

Liu, B. (2007). *Uncertainty theory* (2nd ed.). Berlin: Springer.

Liu, B. (2010). *Uncertainty theory.* Beijing: Uncertainty Theory Laboratory, Tsinghua University, http://orsc.edu.cn/liu/ut.pdf.
Liu, B., & Liu, Y.-K. (2002). Expected value of fuzzy variable and fuzzy expected value model. *IEEE transactions on Fuzzy Systems, 10*, 445–450.
von Collani, E. (2004). Theoretical stochastics. In E. von Collani (Ed.), *Defining the science of stochastics* (pp. 147–174). Lemgo: Heldermann Verlag.
Zadeh, L. (1978). Fuzzy sets as a basis for a theory of possibility. *Fuzzy Sets and Systems, 1*, 3–28.

FSC
www.fsc.org
MIX
Papier | Fördert
gute Waldnutzung
FSC® C083411

Kolibri 360 GmbH
Ferdinand-Jühlke-Straße 7
99095 Erfurt, Deutschland
produktsicherheit@kolibri360.de